TIME SERIES DATA ANALYSIS IN OCEANOGRAPHY

Chunyan Li is a course instructor with years of experience in teaching physical oceanography, estuarine dynamics, weather analysis, calculus, and time series analysis. This book is for students and researchers interested in oceanography and other subjects in the Earth sciences who are looking for complete coverage of the theory and practice of time series data analysis using MATLAB. This textbook provides an introduction to the topic's core theory with examples and exercises, many drawn directly from the author's own teaching and research experiences. The book discusses many concepts, including time; distance on Earth; wind, current, and wave data formats; finding a subset of moving platform-based data along planned or random transects; error propagation; Taylor series expansion for error estimates; the least squares method; base functions and linear independence of base functions; tidal harmonic analysis; Fourier series and the generalized Fourier transform; discrete Fourier transform and fast Fourier transform; power spectrum, cospectrum, and coherence; convolution and digital filtering techniques; sampling theorems; finite sampling effects, window functions, and reduction of side lobe effects; rotary spectrum analysis for velocity vector time series; short-term Fourier transform; wavelet analysis; and EOF analysis.

CHUNYAN LI is a professor in the Department of Oceanography and Coastal Sciences, College of the Coast and Environment at Louisiana State University. He is Director of the Wave–Current–Surge Information System (WAVCIS), part of the Gulf Coastal Ocean Observing System (GCOOS), which collects real-time meteorological and oceanographic data. Li's research focuses on coastal physical oceanography, estuarine dynamics, ocean observations, atmospheric–oceanic observing systems, storm surge, and severe weather-induced ocean response. He has published 120 peer-reviewed papers, many of which have involved ocean time series analysis and associated theories and dynamical processes in estuaries and the coastal ocean.

TIME SERIES DATA ANALYSIS IN OCEANOGRAPHY

Applications Using MATLAB

CHUNYAN LI
Louisiana State University

CAMBRIDGE
UNIVERSITY PRESS

University Printing House, Cambridge CB2 8BS, United Kingdom

One Liberty Plaza, 20th Floor, New York, NY 10006, USA

477 Williamstown Road, Port Melbourne, VIC 3207, Australia

314–321, 3rd Floor, Plot 3, Splendor Forum, Jasola District Centre, New Delhi – 110025, India

103 Penang Road, #05–06/07, Visioncrest Commercial, Singapore 238467

Cambridge University Press is part of the University of Cambridge.

It furthers the University's mission by disseminating knowledge in the pursuit of education, learning, and research at the highest international levels of excellence.

www.cambridge.org
Information on this title: www.cambridge.org/9781108474276
DOI: 10.1017/9781108697101

First published 2022

A catalogue record for this publication is available from the British Library.

Library of Congress Cataloging-in-Publication Data
Names: Li, Chunyan (Oceanographer), author.
Title: Time series data analysis in oceanography : applications using MATLAB / Chunyan Li.
Description: Cambridge, United Kingdom ; New York, NY : Cambridge University Press, 2022. | Includes bibliographical references and index.
Identifiers: LCCN 2021037860 (print) | LCCN 2021037861 (ebook) | ISBN 9781108474276 (hardback) | ISBN 9781108697101 (epub)
Subjects: LCSH: Oceanography–Statistical methods. | Time-series analysis. | MATLAB | BISAC: SCIENCE / Earth Sciences / Oceanography
Classification: LCC GC10.4.S7 L5 2022 (print) | LCC GC10.4.S7 (ebook) | DDC 551.46072/7–dc23
LC record available at https://lccn.loc.gov/2021037860
LC ebook record available at https://lccn.loc.gov/2021037861

ISBN 978-1-108-47427-6 Hardback

Contents

The plate section can be found between pp 241 and 242

Preface

While teaching an intro-level graduate course titled Oceanographic Data Analysis, I prepared teaching materials of various forms, organized according to topic, for students of diverse backgrounds. The course is designed with a focus on time series analysis, applicable to oceanographic and atmospheric time series data and model outputs. During the early years, most materials were presentation slides. Later, more writing was done, organized by topic to best suit the learning of a diverse group of students majoring mostly in oceanography, coastal civil engineering, coastal meteorology, climatology, fishery oceanography, geological oceanography, ocean remote sensing, and natural resources. This provided the materials for this book.

Throughout the years, as a physical oceanographer, my research has required field observations in the James River Estuary, Chesapeake Bay, South Atlantic Bight, Louisiana Shelf, estuaries and tidal inlets in South Carolina, Georgia, and Louisiana, and northern Alaskan Arctic lagoons and tidal inlets. Subsequent data analyses are important for interpretations of the dynamics, development and verification of theories, support of regression models, validation and application of numerical models, and implementation of numerical experiments. Some of the data files used in teaching the course Oceanographic Data Analysis, as well as in this book, are from these research activities.

The materials in this book can be taught in about one semester (about 15 weeks) for graduate students. I usually cover one topic in 1–2 lecture(s), each of 80 minutes in length, including some time for frequent in-class exercises. The in-class exercises are important for students to make the linkage among theories, examples, and applications: it is the best time to help the students directly and give them encouragement, because most mistakes they make are simple but can be frustrating if they are doing the work alone after class. The mistakes often can be fixed relatively quickly in class with my help, which gives them greater confidence when doing work independently. For the same reason, I included some exercises at the end of the chapters, many of which are used in my teaching as class examples, classwork, quizzes, or homework assignments.

With that history, this book is organized and ordered in a format following the course design. The implementation of the techniques for data analysis is done through the software package MATLAB by MathWorks®. I have included a chapter for MATLAB with practices, even though this book is not about MATLAB. I choose not to put a MATLAB overview in an appendix because I am trying to preserve the order of topics taught in the class for convenience of teaching as well as self-studying. I found it useful, as some students are not familiar with MATLAB, while other students know MATLAB but often need a quick review, especially of those commands useful for data analysis (such as those pertaining to matrix operations, functions, or commands for time and subsampling or selection of data, e.g. `datenum` and `find` etc.)

As a single semester course, most of the contents are limited to introductory level, for example, the content on wavelet analysis. The wavelet analysis can, however, have a wide range of applications. These applications may include, for instance, image processing, audio signal noise reduction, and data compression. Much of these applications are of less interest to geoscience, including oceanography, although some techniques may be modified for different purposes.

Throughout the book, efforts have been made to emphasize the basic ideas. Some content might appear easy for well-trained physical oceanography students. However, the subjects are selected to accommodate a broader spectrum of students of various backgrounds, including physical oceanography students.

This book was mainly written in the first year of the COVID-19 pandemic. During this time, my work in the Gulf of Mexico had to stop for a while. No travel was allowed for some time. All teaching moved online. The writing work seemed less important at times. The work nevertheless continued. I enjoyed the writing and making the figures. Although there is no sign of when the pandemic is ending and it is already Christmastime, let us hope that the following year will bring us back to normal, particularly with vaccines becoming available.

Dec. 2020.

Online Resources Available at: www.cambridge.org/chunyanli

Includes:
MATLAB code
Solutions manual for instructors
Figures from the book for use by instructors

Acknowledgments

Thanks to those who received me, taught me, and helped me. Thanks to my parents, siblings, and family.

1
Introduction

About Chapter 1

In this introductory chapter, we briefly go over the definitions of terms and tools we need for data analysis. Among the tools, MATLAB is the software package to use. The other tool is mathematics. Although much of the mathematics are not absolutely required before using this book, a person with a background in the relevant mathematics will always be better positioned with insight to learn the data analysis skills for real applications.

1.1 Some Definitions and Concepts

1.1.1 About Data

In most parts of this book, by *data* we specifically mean a *time series* or a set of time series – numbers or vectors in a sequence obtained through time from measurements, numerical model simulations, or predictions using certain methods or *algorithms* from computation by some theory or formula. In other words, data here are generally a series of numbers (usually real numbers) or a series of vectors that are lined up in time, i.e. each element in the series (either a number or a vector) is associated with an independent *time stamp*. Occasionally, we discuss data without an explicit time stamp (e.g. coastline data). Here, a vector is a group of independent numbers or variables such as the velocity components u, v, w defined in a three-dimensional space and time. A vector can also be a more abstract *collection of variables with different units*, e.g. the so-called *phase space* in statistical physics in which the instantaneous state of a system with n particles is described by the coordinates of all n particles and the velocities of the n particles $(x_1, y_1, z_1, \ldots, x_n, y_n, z_n, u_1, v_1, w_1, \ldots, u_n, v_n, w_n)$. This kind of vector is often seen in *empirical orthogonal function* (EOF) analysis. When a time series is discussed, we either imply that there is a series of time stamps associated with a series of numbers or vectors, or the data include explicitly the time stamps with a series of numbers or vectors.

Occasionally, data might be complex number time series with real and imaginary parts (the imaginary part is the product of a real number and $i = \sqrt{-1}$). In this book, we mostly work with real number time series. However, we will also work with complex number series in two subjects, especially the first subject: (1) *Fourier analysis* (Davis, 1963): when we do Fourier analysis and *fast Fourier Transform* for a time series, complex numbers are often introduced for operational computation purposes, although in theory that is not absolutely needed; (2) *rotary spectrum analysis*: in oceanography, for two-dimensional flow velocity vector time series, we sometimes use a technique called rotary spectrum analysis, which is a special kind of Fourier analysis for vector time series to examine the spectra of velocity in terms of the cyclonic and anticyclonic rotations. In this kind of analysis, the horizontal velocity components are expressed in a complex form for mathematical convenience ($w = u + iv, i = \sqrt{-1}$), taking advantage of the Fourier series in exponential form using the Euler formula $e^{ix} = \cos x + i \sin x$.

1.1.2 Function vs. Discrete Series of Numbers

Data are collected at discrete time instances and/or locations. For that reason, we sometimes use the phrases *digital data* and *discrete data*. Digital data is perhaps used more often to mean that data are saved in a "digital" form in a computer. In the context of discussion here, there is no difference between a data file saved in a computer and a data file with numbers written on a piece of paper. The point is that time series data are not "continuous" in the mathematical sense. However, we use a lot of theories and results from mathematical work on *continuous functions*. Of course, a continuous function is simply an idealized simplification and cannot be realized in real life, particularly in scientific data. No matter how fast one samples, there is always a finite sampling interval, and the data is discrete rather than continuous.

This intrinsic discontinuous nature of data does not really affect our application of mathematical results derived from studies on continuous functions. However, we sometimes do say something like "this time series has a discontinuity." That can be concluded by a quick eye-ball observation of the data file or a more sophisticated and objective computer-based scan of the data with certain criteria. This is, unfortunately, subjective. There may be an unusual outlier in the data that appears to be unreal due to instrument failure or some unexpected influence of other environmental factors. For example, say a water-level sensor is deployed at the bottom of an estuary, measuring the water level at an hourly interval. A storm arrives, and some strong waves unexpectedly push the instrument into a 5 m deeper water despite the fact that the instrument may be heavy and would not move under normal conditions. The data file would record at the next hourly data point an increase of 5 m in water level. The sudden jump is obviously "discontinuous." Of course, if the instrument is securely installed, this jump may not happen. But the reality is not always ideal. An instrument normally considered to be securely mounted may not be secure anymore if there is a record-breaking storm. The judgment of whether there is a discontinuity in a data file depends on the nature of the data under study and the experience of the person who does the analysis. Usually, a quick QA/QC with defined rules would be able to find "discontinuity" points in the given time series data.

1.1.3 Time Series vs. Spatial Series

Can numbers in space rather than in time be included in the analysis for either theoretical or application purposes? In general, yes. For instance, using the Fourier Transform as an example, instead of having variations expressed in time that can be converted into the *frequency domain*, a series of real numbers for different positions in space (i.e. *spatial series*) can be converted into the *wavenumber domain*. A spatial series can be in a one-dimensional, two-dimensional, or three-dimensional space, and thus the wavenumber can be either a scalar or a vector.

Our focus here in this book is time series, although at times we discuss the analysis of spatial data or space–time mixed data. In a sense, when one understands clearly time series analysis, it should become easier to understand spatial series analysis. Of course, these two types of data can be very different. One cannot simply use the time series analysis methods on spatial series without considering their differences. For one difference, a time series is ordered naturally from a start time (earlier) to an end time (later). Data in space, however, are not necessarily ordered "naturally." For example, for data defined on a rectangular grid, we could order the data according to the rows, or columns, or any other peculiar way. After all, spatial data may not be always given on a nice, rectangular grid. The data points could be totally random in distribution.

Full coverage of the analysis of spatial data is not within the scope of this book. In this book, we only include some aspects of the techniques dealing with data obtained in space, such as those obtained from a moving platform. Strictly speaking, they are not necessarily a pure spatial series – they are data mixed in time and space because a moving vessel's speed is limited.

It should be noted that we will discuss a method of analysis, the *empirical orthogonal function* or *EOF analysis*, which involves data mixed in time and space, or a collection of time series defined at different locations. As an example, a series of satellite images for the same region falls into this category – for each image, it presents a spatial series, but for each position in the region, there is a time series provided by the sequence of the images.

1.1.4 The Time Stamps and Time Intervals

When we are working with time series, the order of the sequence of numbers is important because that provides information about variations in time (e.g. an event or a trend) that determine the rate of change in time. We are interested in the variations, though the average values can be important as one parameter. How a measured or simulated variable would change with time (or space, for spatial series) is what we are often interested in.

To calculate the rate of change in time, we must have information about the unit of time and *time intervals* of the data. Therefore, a time series of temperature, for instance, is not just a series of temperature values but also a series of time values with a proper unit that goes side-by-side with the temperature values that are measured or defined at those times.

For time series data, the time intervals between data points can vary. It is more convenient in many applications that the time series data are obtained at constant time intervals. This is often the case with modern instruments that collect data automatically with a built-in *microprocessor* or mini-computer and *data logger*. In this case, and also in the case of model output of time series data, some data files might ignore explicit time stamps; however, information about time intervals and start time with proper units must be provided either inside the data file or separately, depending on the designer of the data collection device or the programmer of a computer model and nature of data (e.g. whether a uniformly spaced time series data or not).

Many of the data analysis techniques are directly applicable only to uniformly spaced data with constant time intervals. Otherwise, some treatment such as interpolations or resampling needs to be done first. There are some exceptions to this. For one example, the general *least squares method* does not require that data points are equally spaced in time. A specific example in oceanography is the *tidal harmonic analysis*, which does not require that the data have constant time intervals. Harmonic analysis is based on the least squares method. The following is an example of time series of velocity vector at 2-minute intervals.

Each line corresponds to one *record of data* from observations at a given time instance. Here the time stamps (the first six columns) are provided. The last two columns give the east and north components of velocity in cm/s. Alternatively, the data could have been provided without the first six columns, but the starting time

Year	Month	Day	Hour	Minute	Second	VE (cm/s)	VN (cm/s)
2016	04	05	21	32	56	−1.1	1.1
2016	04	05	21	34	56	−2.6	2.8
2016	04	05	21	36	56	−5.1	4.0
2016	04	05	21	38	56	−10.4	8.5
2016	04	05	21	40	56	−8.3	7.9
2016	04	05	21	42	56	−5.5	4.8
2016	04	05	21	44	56	−5.0	3.7
2016	04	05	21	46	56	−4.8	3.7
2016	04	05	21	48	56	−13.2	9.8
2016	04	05	21	50	56	−10.9	7.4
2016	04	05	21	52	56	−6.7	3.9
2016	04	05	21	54	56	−4.6	2.7
2016	04	05	21	56	56	−3.0	1.1
2016	04	05	21	58	56	−2.7	0.4
...							

and time intervals have to be provided for the time series data to be meaningful. For observational data, it is probably better to have the time stamps included explicitly for each record to avoid mistakes because real observations, especially the raw data files, often have gaps for various reasons. If the time stamp is not explicitly included in a data file for each record (or each sample), a single gap can mess up the entire dataset by introducing misalignment in time. With proper time stamps, after reliable quality control of the data, the gaps could be filled. For constant interval data, inclusion of the time stamps will make the data file larger. If data file size is to be minimized, the time stamps can be omitted, given that the start time and time interval are provided in the data file or with the data file. This is often practiced for saving numerical model output files, which can be extremely large.

1.1.5 Oceanographic and Other Data

By *oceanographic data*, we intend to limit our discussion to data of oceanic origin, i.e. those obtained from the ocean or coastal and estuarine waters. This, however, should not be taken too narrowly as far as the data analysis techniques are concerned. Generally speaking, the methods discussed in this book can be applied to data from other disciplines. For instance, the fundamental methods of analysis for atmospheric data should be very similar, if not the same. There would be no essential differences. The atmospheric and oceanic time series data sometimes need to be analyzed together, e.g. for storm surge problems, in which the atmosphere provides the forcing (air pressure, wind stress, precipitation), while ocean and coastal waters respond to the forcing. Oceanographic data do indeed have some unique aspects that are pertinent mainly, if not only, to the ocean dynamics. For example, tides occur in the ocean, and we will discuss the tidal analysis (i.e. the abovementioned harmonic analysis). Although there is a tidal signal in the atmosphere known as the *atmospheric tide*, harmonic analysis is usually meant for the ocean only. The technique, however, is applicable to the atmospheric tide as well.

1.1.6 The Tool We Use: MATLAB

MATLAB is a commercial software package that is suitable for calculations and related visualizations of *vectors* (one-dimensional), *matrices* (two-dimensional), and *arrays* (one-, two-, or multidimensional). Although the methods of analysis are independent of the computer programs that implement them, we use MATLAB exclusively in this book. It has many choices of mathematical tools. For example, it has specialized tools for Signal Processing and Wavelet Analysis Toolboxes, which we may need to use at some point in this book. MATLAB is easy to learn and use, and yet has lots of powerful capabilities. Alternative computer languages and or software packages also exist, e.g. IDL, R, C, C++, FORTRAN, Python, etc. Many resources on these computer languages are available online or in publications, but we only use MATLAB throughout this book for consistency and simplicity.

1.2 Background Knowledge

To learn the materials in this book well, some background knowledge in mathematics and physics and related basic skills are preferred, though a high-level skill of any of these subjects is not necessarily required. The subjects of study, particularly in terms of the techniques in this book, are quite broad. Background knowledge includes *linear algebra*, *calculus*, *Fourier theorem*, *numerical analysis*, and some basics of statistics. In addition, oceanographic data analysis is often aimed at the resolution of dynamical processes in the ocean. Therefore, some background knowledge in e.g. physics, fluid dynamics, and tidal theory would also help to understand some of the techniques (e.g. rotary spectrum analysis and tidal harmonic analysis). Assumptions are made here that the readers have some basic background knowledge of these subjects, but efforts have been made to provide enough information as standalone materials. Some selected basic information and review of background theories are provided in the book for convenience. Intuitive interpretations may be provided for a better understanding when some background information is discussed.

1.2.1 Linear Algebra

Linear algebra is a branch of mathematics dealing with arrays of numbers. A time series of a quantity is itself a special case of an array. Linear algebra includes theories and methods of solving linear sets of equations. This provides the basis for many data analysis techniques, such as linear regression, Fourier Transform, and harmonic analysis. Knowledge in linear algebra makes it much easier to work with matrix operations, which are the major backbone of MATLAB. MATLAB is best suited for working with vector, matrices, and arrays. Linear algebra with matrix operations greatly simplifies the mathematical expression of concepts. These simplifications are implemented by the design of MATLAB language, while inside MATLAB the computer is directed to do the heavy lifting for detailed calculations. With a combination of the two (simplified concepts and MATLAB), our brain power can be reserved for interpretation of the results instead of the details of the actual calculations.

1.2.2 Calculus

Calculus is an important branch of mathematics about the rate of change of functions (differentiation) and the inverse calculations of differentiation (integration). It is the mathematical backbone of almost all disciplines in modern physics. Calculus is fundamental to many theories and techniques of analysis and number crunching, in many ways. This can be seen mainly in two examples in this book: one is Fourier Transform, in which the basic knowledge of integration and *linear independence* of *base functions* will help greatly. Similarly, the least squares method is also based on the theory of calculus. The optimal solution of the *Fourier coefficients* or *best fit* in the least squares method is a great product of calculus. Another example of the application of calculus is *Taylor series expansion*, which is very useful in laboratory experiments for *error estimations*. The beauty of Taylor series expansion is that it can approximate an arbitrary differentiable function to any accuracy (at least in theory) by a polynomial, which is one of the simplest general functions one can find. In addition, the theories for the empirical orthogonal function and wavelet analysis are also based on calculus. Without calculus, none of these techniques would have been invented.

1.2.3 Fourier Theorem

Fourier theorem is the basis for the technique we discuss extensively in this book: the Fourier analysis. Although we will provide relatively complete, albeit brief, coverage of the theory, it will help if the readers have learned the subject before. The importance of Fourier theorem to time series analysis can be compared to that of Newton's Second Law to mechanics. This may be an overstatement, but the point is, Fourier theorem provides a great leap in understanding the characteristics of a function – in our case, a function of time. The theorem basically guarantees that, except for a peculiar function with infinite discontinuity points, a general (arbitrary) piece-wise continuous function that only has a finite number of finite-range jumps can be expressed (decomposed) in terms of a series of sine and cosine functions of different scales. Here, scales are either in terms of frequency for time series or wavenumber for spatial series.

In the case of a time function, if the length of time of the function is finite (which is usually the case, because no real observations can be infinitely long, except in idealized thought experiments), the series of sine and cosine functions generally are discrete in terms of the scales (discrete frequencies or wavenumbers), although there are infinite numbers of them. These (infinite number of) discrete sine and cosine functions are called the base functions. The series itself is called the *Fourier series*. The Fourier theorem tells us that such an infinite Fourier series converges to the original function in an averaged sense. Here, "average" means the average of the right and left limits of the function at one point, obtained by approaching from the right of the point and that the left of the point. In other words, at any point (or time, for time function) where the function has a finite discontinuity, the Fourier series converges to the average of the two points on both sides of the jump – i.e. the middle of the jump. If the function at the point is continuous, these two limits are the same. That is, if the function is continuous, the Fourier series will converge to the function at that point.

The Fourier theorem also helps us to understand the effect of finite sampling, such as the highest frequency that can be resolved (i.e. the *Nyquist frequency*), the frequency resolution, and the artificial oscillations or *Gibbs Effect*. The properties of the Fourier series allow a wide applicability and an extension into the *Fourier Transform*.

1.2.4 Numerical Analysis

Numerical analysis deals with some practical methods and techniques of calculations using computers, such as root finding from a nonlinear equation, *interpolation* and techniques using polynomials, numerical differentiation, numerical integration, calculations in linear algebra, and numerical solutions for ordinary differential equations. An example of relevance to time series analysis in this book is an understanding of interpolations, although we will not discuss the theories of interpolations; we will, rather, rely on just MATLAB's built-in functions for interpolations. Interpolations are oftentimes among the first steps of data processing, before conducting a Fourier Transform or spectrum analysis for a time series obtained from actual measurements from the ocean (or anywhere else). As mentioned earlier, observations usually contain gaps. Interpolations are useful for filling "small" *data gaps*, in which the gaps between adjacent data points are much smaller than the minimum useful *time scale*s. For large data gaps, interpolation may introduce significant errors. How do we determine if a gap is really "small"? This depends on the problem under discussion. It depends on the significant time scales contained in the signal, time intervals, number of gaps, and length of the gaps. Although actual situations can have an infinite number of possibilities, the decision is usually based on common sense. For instance, for a problem with tidal variations of water level, we have a time series with 10-minute time intervals. For this time series, if there are six consecutive *missing data* points in a row within a tidal cycle (12-hour period for a semi-diurnal, tide-dominated case), the gap would not pose a major problem; a simple interpolation would fill the data gap without altering the information in any significant way, meaning no major influence on the spectrum or temporal properties. However, if the time interval is 1 hour and there are six consecutive data points missing (e.g. a 6-hour data gap), the time series property and spectrum after an interpolation for that 6-hour gap would be highly questionable. That, of course, also depends on how we do the interpolation: Are we using a linear interpolation, or are we using some functions for the interpolation? The problem with a function is that for real world observations, if data are missing, there is no way we know exactly what happened during the time when valid data were not recorded. Any function used for interpolation is subjective and can never be verified, because what we do not know is exactly whether there was any "event" during the time with missing data. If we have additional information, e.g. observations from an adjacent position, then the story will be completely different, as then we might be able to assume a certain relationship for the interpolation based on the additional information.

1.2.5 Statistics

Statistics is needed here and there in the book. Statistics, however, can arguably be a double sword in this context. On the one hand, statistics is often needed for analysis on real data because of the intrinsic random nature of observations: real data are functions affected by *random processes*, at least to some extent. The random processes include intrinsic characteristics of the physical processes, such as chaos in the dynamics or random errors in measurements and systematic errors in instruments. On the other hand, many of the mathematical tools are developed first, without considering the random nature of numbers at all, i.e. only deterministic functions are considered. As we all know, calculus is presented originally and mostly with the differentiations and integrations of *deterministic functions* rather than random functions. Fourier Transform was also originally developed without any consideration of the randomness of functions. When these mathematical operations are applied to real data, however, the effects of randomness may have to be considered. This will make the problem a little more complicated, but we have to admit that there are some uncertainties, and any estimate will need to take that into account. Having said these, this book is not a statistics book. Some of the analysis will be easier to understand without considering randomness in the beginning. The statistical aspects can be better understood after one has grasped the basics of the techniques developed for deterministic functions.

1.3 About Wind, Current, and Wave Directions in 2-D

In oceanographic data analysis, we often work with two-dimensional vector time series data such as wind, current, and wave, when the vertical components of wind, current, and wave are not considered. Here we discuss and contrast the definitions of directions of two-dimensional wind, current, and ocean waves. It can be confusing as to what the definitions of directions of the two-dimensional wind, current, and wave are. Although two-dimensional wind and currents are vectors that can be expressed in two components (for the horizontal two-dimensional wind and current velocities), they are routinely reported with a magnitude and direction. The direction of wind and that of currents are, however, opposite in definition. It can be confusing and therefore is important that the definitions of these directions are well understood by the data users.

Wind velocity is a vector which tells us where an air parcel at a given location and time is moving to. In theory, wind velocity should have three components in a Cartesian coordinate: the *vertical wind velocity component* and two horizontal components. The vertical component, however, is not routinely reported unless for special studies of cloud physics, precipitation mechanism, and small-scale atmospheric processes in the troposphere. For this reason, when we talk about the *wind direction*, we are referring only to that of the *horizontal wind velocity*. This is the same for current velocity and waves. When the ocean current direction is mentioned, we usually mean that of the horizontal velocity. The direction of propagation of internal waves can be at an angle to the surface. Here we are only referring to the direction of surface waves.

1.3.1 Examples of Wind Data

The following shows some sample hourly (horizontal) wind data from an offshore station at (90°32′ W, 29°3.2′ N) obtained in April 2010.

```
%month,day,year,hour, Speed(m/s), Direction (degree)
  4,   1,  2010,  0,     4.82,       173
  4,   1,  2010,  1,     4.17,       175
  4,   1,  2010,  2,     4.33,       163
  4,   1,  2010,  3,     4.54,       153
  4,   1,  2010,  4,     5.97,       154
  4,   1,  2010,  5,     6.48,       156
  4,   1,  2010,  6,     7.09,       165
  4,   1,  2010,  7,     6.99,       162
  4,   1,  2010,  8,     6.51,       155... ...
```

The first line is a comment line in MATLAB and is self-explanatory: it explains the data in the following rows and columns, which include time (the first four columns), wind speed in meters per second (the fifth column), and wind direction in degrees (the last column). For wind speed, there should be no vagueness, as it is the wind speed relative to the Earth. For wind direction, there are two questions: what is *the reference direction of wind* with 0 degrees and how is the angle measured – clockwise or counterclockwise?

Before we answer these questions, let us look at the following wind data from the same location and same time:

```
%month,day,year,hour,      WindEast, WindNorth
   4,  1,  2010,  0,       -0.5874,     4.7841
   4,  1,  2010,  1,       -0.3634,     4.1541
   4,  1,  2010,  2,       -1.2660,     4.1408
   4,  1,  2010,  3,       -2.0611,     4.0452
   4,  1,  2010,  4,       -2.6171,     5.3658
   4,  1,  2010,  5,       -2.6357,     5.9198
   4,  1,  2010,  6,       -1.8350,     6.8484
   4,  1,  2010,  7,       -2.1600,     6.6479
   4,  1,  2010,  8,       -2.7512,     5.9001  ... ...
```

The only thing different here is that the wind data are provided as the east and north components, rather than the wind speed and wind direction. Both types of wind data are commonly used. Perhaps it is more common to use the speed and direction in observational data and east and north components from model output. Indeed, the standard weather data reported by NOAA and other agencies usually use the first format. The everyday weather forecast on TV and radio uses the wind speed and direction, rather than the east and north components. We say that "the wind tomorrow is predicted to be from the northeast with a maximum speed of 10 knots or 5 m/s," but not that "the wind tomorrow is predicted to have an east

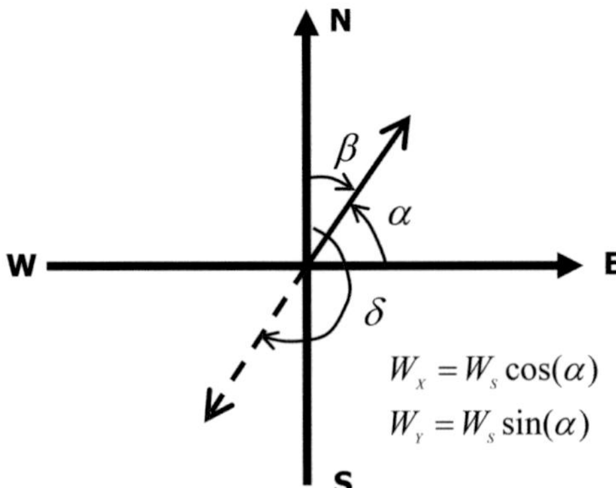

Figure 1.1 Definition of wind direction. The dashed line with an arrow indicates the wind direction, while the solid line with an arrow shows the wind velocity vector.

component of negative 7 knots and a north component of negative 7 knots," which would be awkward and inconvenient if not confusing for the general public.

1.3.2 Definition of Direction of Wind

As an international standard, the wind direction is defined as the direction from which the wind comes (therefore, wind direction is the opposite of the wind velocity vector). In some literature, including handbooks, the definition of direction of wind stops here; however, this is incomplete, because this definition does not tell us how the direction is counted – clockwise or counterclockwise – and from where, i.e. what is the 0 direction? To complete the definition, the wind direction is the direction in degrees from which the wind is coming, and the angle is counted from true north (0 degrees is defined from true north) and clockwise (Fig. 1.1). Wind direction is presented as a number in degrees from 0 to 359.9. With this definition, a 45-degree wind is from the northeast. If the wind direction is 90 degrees, it is coming from the east, while 180 degrees is from the south and going to the north.

1.3.3 Converting between Wind Speed and Direction to Wind Velocity Components

To derive the east and north wind components from the wind data with wind speed and direction, we notice that from Fig. 1.1, the wind direction δ and the wind velocity vector direction α have the relationship

$$\alpha = 270 - \delta. \tag{1.1}$$

This is the *angle mapping for wind* between wind direction (measured clockwise) and the angle of wind vector relative to the east direction (measured counterclockwise).

The wind vector's east and north components are then determined by

$$w_x = W \cos \alpha, \tag{1.2}$$

$$w_y = W \sin \alpha, \tag{1.3}$$

in which w_x, w_y, and W are the wind velocity components in the x and y directions and wind velocity magnitude, respectively. Here the positive x and y directions are the east and north directions, respectively.

1.3.4 Definition of Direction of Ocean Current

The *direction of ocean current* is defined as the direction to which the ocean water goes. Therefore, the ocean current direction is the same as the velocity vector. The ocean current direction is also counted from the north (0 degrees to the north) and measured clockwise. Again, here we are also just referring to the horizontal velocity.

1.3.5 Converting between Current Speed and Direction to Current Velocity Components

Ocean current data is very similar to wind data, such that the data can be given as either current speed and direction (where it goes to) or the east and north components of the horizontal flow vector. We can still use Fig. 1.1 to derive the equations. The only difference is that the ocean current direction is the direction to which it goes, and therefore β in Fig. 1.1 is the direction of the ocean current. The corresponding equation to map the angles is

$$\alpha = 90 - \beta. \tag{1.4}$$

This is the *angle mapping for current* between the current direction (measured clockwise) and the direction of current vector relative to the east direction (measured counterclockwise).

The east and north components of the current velocity are determined by

$$u = U \cos \alpha, \tag{1.5}$$

$$v = U \sin \alpha, \tag{1.6}$$

in which u, v, and U are the current velocity components in the x and y directions and flow speed, respectively. Here the x and y directions are the east and north directions, respectively.

The following shows some sample data for current obtained from Caminada Pass of Barataria Bay in speed-direction format:

```
%Date; Time; Pressure; Temperature; Speed(m/s); Dir(degree)
12 19  2017 16  35 28;  1.578; 20.10;    0.074;  55.45;
12 20  2017 08  50 28;  1.256; 19.41;    0.102;   9.55;
12 20  2017 09  05 28;  1.263; 19.46;    0.139;   2.47;
12 20  2017 09  20 28;  1.270; 19.52;    0.177; 358.38;
12 20  2017 09  35 28;  1.277; 19.51;    0.121;  14.84;
12 20  2017 09  50 28;  1.284; 19.46;    0.133; 359.14;
12 20  2017 10  05 28;  1.291; 19.36;    0.190;   6.94;
12 20  2017 10  20 28;  1.289; 19.04;    0.120;  21.45;
12 20  2017 10  35 28;  1.285; 18.91;    0.159;   6.86;
12 20  2017 10  50 28;  1.281; 18.78;    0.204; 358.60;
12 20 2017 11  05 28;  1.281; 18.71;    0.152;    5.30;
12 20 2017 11  20 28;  1.296; 18.59;    0.203; 357.74;
12 20 2017 11  35 28;  1.308; 18.61;    0.186; 351.65;
12 20 2017 11  50 28;  1.305; 18.64;    0.178;    1.61;
12 20 2017 12  05 28;  1.320; 18.59;    0.178; 347.68;
12 20 2017 12  20 28;  1.335; 18.54;    0.188; 334.52;
12 20 2017 12  35 28;  1.354; 18.55;    0.112; 354.34;
12 20 2017 12  50 28;  1.345; 18.54;    0.126;    4.09;
12 20 2017 13  05 28;  1.353; 18.56;    0.101;    5.65;
```

Like wind data, ocean current data can be presented in either speed–direction format or east–north format, for the horizontal velocity. It should be noted that it is more common to have the vertical velocity component measured in ocean current data than in the wind data. An acoustic Doppler current profiler (ADCP) usually measures the three-dimensional velocity components, except for the two-beam models, which only measure the horizontal components. Unlike horizontal velocity, the direction of the vertical velocity component is either up or down so there is no question about the definition of direction in the vertical.

1.3.6 Definition of Directions of Waves

For wave data, the definition is similar to that of wind: the *direction of waves* is defined as that from which the waves are coming. The units are degrees from true north, increasing clockwise, with north as the reference or 0 degrees. Not all wave data have information about the wave direction. Commonly reported wave data include the wave height, wave period, significant wave height or the average of the highest one-third of the wave height, etc. Only the *directional wave data* include the information about wave direction. Directional wave data are relatively rare. In addition to these wave data formats, *spectral wave data* are another format, which includes information for different frequencies. In the following, some sample directional wave data are shown. The data are obtained from the National Data Buoy Center (NDBC, www.ndbc.noaa.gov/ dwa.shtml).

MM	DD	TIME	WVHT	SwH	SwP	SwD	WWH	WWP	WWD	STEEPNESS	APD
month	date	(EDT)	ft	ft	sec	compass	ft	sec	degree		sec
06	30	7:40 am	2.6	2.3	10.8	SE	0.7	4.0	SSW	N/A	5.9
06	30	6:40 am	2.6	2.6	9.1	SE	0.7	3.8	SSW	N/A	5.8
06	30	5:40 am	2.6	2.6	10.0	ESE	0.7	2.9	SW	N/A	5.7
06	30	4:40 am	2.6	2.6	10.0	SE	0.7	4.0	SSW	SWELL	5.8
06	30	3:40 am	2.3	2.3	5.0	SSW	1.0	3.6	SSW	N/A	5.3
06	30	2:40 am	3.0	2.6	9.1	SE	1.0	3.3	SSW	SWELL	5.7
06	30	1:40 am	3.0	3.0	5.6	SSW	1.0	4.0	SSW	STEEP	5.6
06	30	12:40 am	3.0	3.0	5.9	SSW	1.0	3.7	SSW	STEEP	5.5
06	29	11:40 pm	3.0	3.0	5.6	SSW	1.0	4.0	SW	STEEP	5.4
06	29	10:40 pm	3.3	3.0	5.6	SW	1.0	3.2	SW	STEEP	5.3
06	29	9:40 pm	3.0	3.0	5.3	SSW	1.0	3.7	SW	STEEP	5.1
06	29	8:40 pm	3.0	2.6	5.3	SSW	1.3	4.0	SW	STEEP	5.1
06	29	7:40 pm	3.0	2.6	10.0	SE	1.3	3.8	SW	SWELL	4.9
06	29	6:40 pm	3.3	2.6	10.0	SE	1.6	3.6	WSW	SWELL	4.8
06	29	5:40 pm	3.3	2.3	5.3	SW	2.0	4.3	SW	STEEP	4.7
06	29	4:40 pm	3.3	2.3	5.3	SW	2.3	3.8	WSW	V_STEEP	4.5
06	29	3:40 pm	3.3	2.0	10.0	SSE	2.6	5.0	SW	V_STEEP	4.5
06	29	2:40 pm	3.3	2.6	10.0	SE	2.0	3.3	W	SWELL	4.7
06	29	1:40 pm	3.0	2.3	10.0	SE	2.0	4.2	SW	SWELL	4.9
06	29	12:40 pm	3.0	2.0	10.0	SE	2.3	5.3	SW	AVERAGE	4.9
06	29	11:40 am	2.6	2.6	10.0	SE	1.3	3.8	W	SWELL	5.1
06	29	10:40 am	2.6	2.0	10.0	SE	1.6	5.0	SW	SWELL	5.0
06	29	9:40 am	2.6	2.3	10.0	SE	1.3	3.7	W	SWELL	4.8

Here, the first two columns give the month and date. The third column is Eastern Daylight Time. Depending on the time of the year, it could be the Eastern Standard Time. The time zone may also be different, depending on the location of the buoy. A commonly accepted practice, however, is to use the world time (UTC, or previously GMT), but for some reason, this specific station used the Eastern time.

The next column is the significant wave height (WVHT): the average height in feet of the highest one-third of the waves during the one-hour sampling period. The fifth column is the swell height (SwH) in feet. It is the average of the highest one-third of the swells. It may be estimated as a function of wave periods or frequencies. Swell waves are waves with longer wavelength and period (compared to wind-waves) propagated into the region and not generated by local wind. The sixth column is the swell period (SwP) – the peak period in seconds. If more than one swell is present, this is the period of the swell containing the maximum energy. The seventh column is the swell direction (SwD), which is the direction from which the swells are coming. The direction is given (Fig. 1.2) on a *16-direction compass scale or 16-point compass scale*, i.e. the North (N), North-Northeast (NNE), Northeast (NE), East-Northeast (ENE), East (E), East-Southeast (ESE), Southeast (SE), South-Southeast (SSE), South (S), South-Southwest (SSW), South-West (SW), West-Southwest (WSW), West (W), West-Northwest (WNW), Northwest (NW), and North-Northwest (NNW). Sometimes, the direction of winds

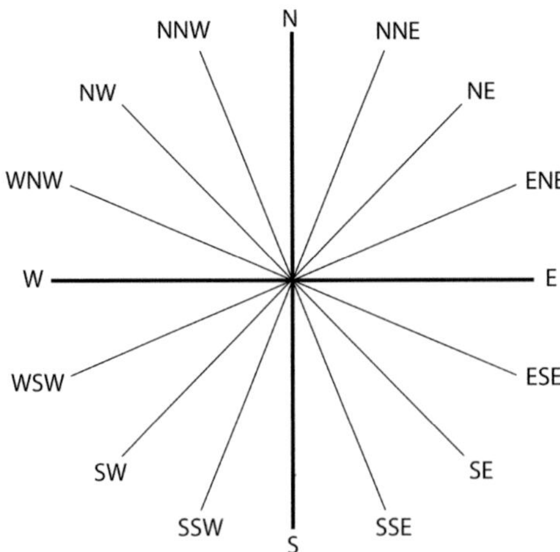

Figure 1.2 The 16-point compass scale for wind and wave.

or waves can be presented with an 8-point compass scale or 32-point compass scale. They can also be presented with digital values. The eighth column is the wind-wave height (WWH), which is the average height of the highest one-third of the wind-waves. Wind-waves are those waves generated by the local wind. The ninth column is the wind-wave period (WWP), which is the peak period in seconds of the wind-waves. The tenth column is the wind-wave direction (WWD), which is the direction from which the wind-waves are coming. Likewise, the direction is given on a 16-direction compass scale. The eleventh column is the steepness of the waves with just a few scales: very steep, steep, average, or swell. The last column is the dominant wave period (AVP) in seconds – the period with maximum energy, which is either the swell period or the wind-wave period.

1.3.7 Rotation of Coordinate System for Wind or Current Vectors

Sometimes we need to rotate a coordinate system, in which vectors will be expressed with different coordinates. For example, we may need to compute the along shelf and across shelf wind velocity or ocean current velocity components on a continental shelf, or along-channel and across channel flow velocity components at a tidal inlet. A simple transformation performed by a rotation of the coordinate system can accomplish this.

Assume that vector $\overrightarrow{OP}$ has coordinates (x, y) in the original coordinate system. The coordinate system is then rotated by an angle of θ. The coordinates of point P are (x', y') in the new system. There is a relationship between (x, y) and (x', y'), as shown below (Fig. 1.3):

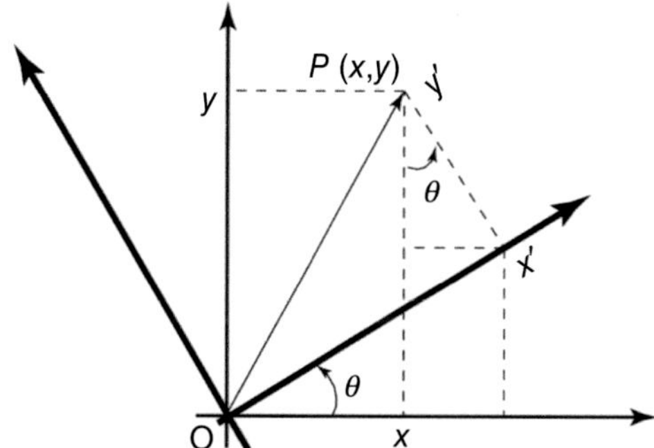

Figure 1.3 Coordinate system rotation and variable transform.

$$x' \cos\theta = x + y' \sin\theta, \tag{1.7}$$

$$y = y' \cos\theta + x' \sin\theta, \tag{1.8}$$

which lead to

$$x = x' \cos\theta - y' \sin\theta, \tag{1.9}$$

$$y = x' \sin\theta + y' \cos\theta. \tag{1.10}$$

This is a mathematical transform between the coordinates of the two systems. The inverse transform is

$$x' = x\cos\theta + y\sin\theta \tag{1.11}$$

$$y' = -x\sin\theta + y\cos\theta. \tag{1.12}$$

Review Questions for Chapter 1

(1) Can any real observational data be a continuous function of time?
(2) How can one determine if an observed time series contains a "jump"?
(3) Does it matter if a time series is ordered in time?
(4) What is calculus? What does it do? Why calculus is important?
(5) For time series data, do they have to be equally spaced in time (or have constant time intervals)?
(6) Contrast time series and data distributed in space ("space series"). Are there any differences or similarities?
(7) Assume that a collection of hourly time series data for wind has been obtained for a length of 1 year from Station 1. The time interval is generally 1 hour with

some gaps. If the data have a few randomly distributed 2-hour gaps, 5-hour gaps, 8-hour gaps, 1-day gaps, and 3-day gaps, would you be comfortable using the data? Further assume that there is a station (Station 2) 30 km away with data from the same period but also with similar gaps. As the gaps are randomly distributed, the gaps from Station 2 are generally not the same as those from Station 1. You may want to do interpolations to fill the (short) gaps and, alternatively, use the data from Station 2 to fill the (longer) gaps as a last resort. For obvious reasons, you want to minimize the use of Station 2 data. Make a decision about how you would like to fill the data gaps (which gaps to fill with interpolations and which with replacement data from Station 2) and provide your reasoning.

(8) Wind direction is defined as the direction from which the wind comes relative to true north, measured clockwise. When wind direction is measured, however, it is often first measured relative to magnetic north. What do you think one should do to convert the wind direction to that relative to true north?

(9) Current velocity and wind velocity measurements often rely on an electronic magnetic sensor (or electronic compass) which senses the Earth's magnetic field and determines the magnetic north. The installation of an electronic compass requires that it is away from powerlines, steel, or magnets. Why?

Exercises for Chapter 1

(1) If the wind direction were measured counterclockwise, with everything else being the same, how would you modify the equations for the east and north component of the wind velocity?

(2) If the 0 wind direction were defined from the east and measured clockwise, how would you modify the equations for the east and north component of the wind velocity?

(3) If the 0 wind direction were defined from the east and measured counter-clockwise, how would you modify the equations for the east and north components of the wind velocity?

2

Introduction to MATLAB

About Chapter 2

This chapter introduces MATLAB, aimed at the basic knowlcdgc and skills related to what may be needed in the following chapters for data analysis. This chapter, however, is far from a complete coverage of MATLAB; nor do we need everything provided by MATLAB. For those who are familiar with MATLAB already, this chapter may be either skipped or used as a quick review. The exercises at the end of the chapter may be useful for some data processing, e.g. the selection of a subset of dataset is often needed, and the MATLAB function `find` is particularly useful for that.

2.1 MATLAB: A Matrix Lab

MATLAB stands for matrix laboratory. It is also a general mathematical laboratory with many built-in functions and optional toolboxes. As some example toolboxes, the Signal Processing Toolbox and the Wavelet Toolbox are useful for time series data analysis. It has many easy-to-use functions to visualize different kinds of arrays. MATLAB can also do "direct" computation on complex numbers in very much the same way as for real numbers. With the built-in functions, and particularly with the many toolboxes, MATLAB is a sophisticated computation workshop for a variety of disciplines, such as image processing, optimization, symbolic math, mapping, test and measurements, controls, robotics system, navigation, electronics, differential equations, automated driving, wireless communications, and many more. As far as we are concerned, MATLAB is a software package that can be used to do mathematical calculations related to time series data analysis and to output publication-quality graphics.

MATLAB is an interactive system. It does not require a user to "compile" and "link" before running a program. It allows a command-line operation (issue one line of command at a time), multiple-line operation (cut and paste a sequence of script or code into the command window for multiline instructions), and the use of a standalone MATLAB script (the so-called m-file) in simple ASCII text format that can be run by MATLAB. It allows the user to easily build up a collection of functions and to share among colleagues.

The *MATLAB script* (or *m-file*) is a text file with a collection of sequential commands that the users design for mathematical calculations and other computational tasks. These can include actions of data reading, communication with scientific apparatus, mathematical calculations, visualizations, and data output, among many other functions. MATLAB can work with sensors, machines, and real-time controls. For our applications here, we will not discuss these capabilities. We will only use a small fraction of MATLAB's functions for our purposes of data analysis. For more information, the readers are referred to e.g. Attaway (2013), The MathWorks, Inc. (2020a; 2020b), or MATLAB's web site: www.mathworks.com/.

The good thing about MATLAB is that (1) it is very easy to learn and use; (2) it allows the user to continuously build up a collection of functions; (3) it is widely used so that users can share with colleagues and find information easily; and (4) it is very powerful, such that one can count on it when it comes to meeting requirements for quality products, such as results for theses, dissertations, meeting presentations, publication-quality graphics, and animations, etc. With respect to the application of MATLAB on time series data analysis – MATLAB is a natural choice. With MATLAB, we can do essentially any time series data analysis without looking for anything else.

It should be emphasized that although MATLAB is a fully loaded tool for data analysis, this book is not about MATLAB. We will use MATLAB as a tool for the objective of applying data analysis techniques. If the reader has previously learned MATLAB, it will be useful and convenient. If the reader has never had a chance to learn MATLAB, he or she should spend some time learning it to benefit the most from this book. For new users, it is perhaps not always a good idea to spend too much time trying to remember the commands and functions other than the commonly used ones without a clear purpose. Like many modern software packages, MATLAB has functions that can help users to find information when needed. With practice, users can usually remember needed commands before even realizing it. New users should pay attention to how to find information (a command or the syntax of a command). In the following, it is assumed that the reader knows the basics of how to start the MATLAB program and how to interact with it.

2.2 More on MATLAB

2.2.1 MATLAB Is Case Sensitive

MATLAB commands, as well as variables, parameters, and functions, are case sensitive. For example, a variable named `A1` is different from another variable named `a1`. For that reason, the user is cautioned to define and use variables with attention to the case-sensitive nature.

2.2.2 The Atoms of MATLAB

The smallest data units, or atoms, of MATLAB are vectors (one-dimensional), matrices (two-dimensional), or, generally, arrays (one-, two, or multidimensional) of real numbers or complex numbers. Array is the general name as it covers both vectors ($1 \times n$ or $n \times 1$) and matrices ($n \times m$), as well as multidimensional arrays (e.g. $n \times m \times l$).

Creating an array variable inside MATLAB does not require an explicit definition of the size or dimension of the array, as other programming languages do. It can read data files easily without an explicit definition as to the size of arrays – this can be better appreciated by those who have programmed in other traditional, high-level computer languages, such as C, C++, BASIC, and FORTRAN. The operations on these arrays in traditional computer languages require the user to define the arrays before any assignment of values or any computation. Definitions are usually rather strict and have a few drawbacks: it requires the use of an explicit index for mathematical operations; it is easy to make a mistake; it is relatively slower in computations; it may be difficult to debug (find an error); and it could be difficult to read by other users or even the author after some time.

MATLAB can read large data files in rectangular format by a *load* command, treat multiple variables, and do almost any type of calculation, especially matrix computations. This allows users to solve sophisticated technical computing problems with arrays more easily than traditional computer languages.

2.2.3 Operations on the MATLAB's "Atoms" (Arrays)

Here, operation means mathematical computations in general. Why do we need operations on *arrays*? Because data are simply series of numbers that are, exactly, "arrays." So, any operation on a set of data points is an operation on an array or arrays. The capability of MATLAB to work with an array as a whole, rather than only with individual *elements*, makes it much more convenient and efficient.

Column-Vector Vs. Row-Vector

Before we talk about the operation of MATLAB arrays, we need to understand how MATLAB defines the arrays. First, we look at the definition of a *vector* – a one-dimensional array. A vector in MATLAB can be either a *row-vector* or a *column-vector*. The difference between the two is trivial unless matrix operations are needed.

A row-vector is a series of numbers lined up in a horizontal direction from left to right when MATLAB prints its contents out on the computer screen. To define a row-vector we can e.g. write within the MATLAB working window like this:

```
a = [ 1, 2, 3, 12, 23, -9]
```

Each number inside the brackets is followed by a comma except for the final one.

After hitting the "enter" or "return" key of the keyboard, the following is shown in MATLAB:

```
>> a = [ 1, 2, 3, 12, 23, -9]
a =
      1     2     3    12    23    -9
```

The series of numbers is shown in a row.

This command is also equivalent to

```
a = [ 1  2  3  12  23  -9]
```

That is, by using at least one space to replace the comma, MATLAB defines the same row-vector. Adding more space between numbers is insignificant and does not affect the result. This feature can be used to add more space in the script to adjust the alignment for a better visual effect.

A column-vector can be defined like this:

```
b = [ 1;2;3;12;23;-9]
```

which yields

```
>> b = [ 1; 2; 3; 12; 23; -9]
B =
      1
      2
      3
     12
     23
     -9
```

This is shown along a vertical column. Again, adding more space between the numbers does not affect the outcome.

2.2.4 Suppressing Display on Screen

After each command, if there is a computation or assignment of an array, MATLAB prints out the result on the screen by default. The print can be suppressed by adding a semicolon ";" at the end of the command. For example, to suppress the display of an array named a, as defined below, a semicolon is included at the end of the assignment. For instance, the command

```
a = [1  2  3  12  23  -9];
```

will define the array and put it into memory, but MATLAB does not print it out on screen. This is useful if the user is producing a large array or many arrays that the user does not need to look at right away. Imagine an array with millions of records (rows) displayed on screen – it could be annoying and a waste of time and is not needed most of the time. If needed, one can always look at the contents on the screen at any time after the definition by typing the variable name without anything else and hit the enter key.

2.2.5 Display MATLAB Memory Contents

One of the most commonly used commands of MATLAB is whos. With this command, one can view all the variables and their attributes (name, dimension, and data type) in MATLAB's working memory. As an example, typing the command whos gives something like this:

```
>> whos
  Name      Size              Bytes  Class
  a         1 x 6              48  double array
  b         6 x 1              48  double array
```

The actual output depends on what variables are present in the MATLAB memory, which depends on the user's actions in MATLAB prior to the inquiry with the command whos. In the screen print out just given, 1×6 indicates that array a has one row and six columns; and 6×1 indicates that array b has six rows and one column.

2.2.6 The Difference between Row- and Column-Vectors

In the last example, the variables a and b are 1×6 and 6×1 arrays, respectively. These variables are different in format. They are, however, essentially the same if only some operations on a single one-dimensional array are used. For instance, command line instructions like these

```
>> figure; plot(a)
```

and

```
>> figure; plot(b)
```

will produce the same results, which are identical figures.

A row-vector or a column-vector can be changed to the other type with a simple operation of *matrix transpose* like this:

```
>> c = b'
```

which yields

```
>> c = b'
c =
        1      2      3     12     23     -9
```

The difference between a column-vector and a row-vector is insignificant unless a matrix operation is involved. For example, doing a fast Fourier Transform to a single array `y` using the MATLAB function `fft` does not restrict whether a column or row-vector is used. If `y` is a column-vector, then `fft(y)` yields a column-vector of complex numbers of the same length as `y`; on the other hand, if `y` is a row-vector, `fft (y)` yields a row-vector of complex numbers with the same length as `y`.

If the elements of an array have complex numbers, the transpose operator will actually do two things: the first is a transpose, and the second is changing all elements to their corresponding *complex conjugates*. For example, if `d` is a complex array defined as

```
>> d = [ 1 + i, 2 - i, 3 - 2i]
d =
  1     +    1i              2    -   1i         3    -   2i
```

and the transpose of the complex conjugate of `d` is

```
>> d'
ans =
        1         -     1i
        2         +     1i
        3         +     2i
```

Then, to do a *nonconjugate transpose* for the array `d`, one can use this command:

```
>> d.'
ans =
        1         +     1i
        2         -     1i
        3         -     2i
```

Here, the operator .' is the *nonconjugate transpose operator*. The difference between d' and `d.'` is that the latter does not have a change in the sign of the imaginary part of the complex numbers in the array. For real value arrays, these two operations are essentially the same.

In MATLAB, there is also an equivalent *nonconjugate transpose function* simply called `transpose`, which can be demonstrated by the following example:

```
>> transpose(a)
ans =
        1
        2
```

```
        3
       12
       23
       -9
>> transpose(d)
ans =
        1        +     1i
        2        -     1i
        3        -     2i
```

2.2.7 Higher Dimensional Arrays

For defining a two-dimensional array (or matrix), the following line of MATLAB command,

```
d = [ 1 2 3; 4 5 6] ;
```

is the same as

```
e = [ 1, 2, 3; 4, 5, 6] ;
```

and they produce

```
>> d = [ 1 2 3; 4 5 6]
d =
        1      2      3
        4      5      6
>> e = [ 1, 2, 3; 4, 5, 6]
e =
        1      2      3
        4      5      6
```

If one does not finish defining the matrix on the first line with a right square bracket] , then MATLAB expects you to input the second line, until it is finished with a] . This allows the user to write a MATLAB script in a neat way:

```
>> e = [ 1, 2, 3;
         4, 5, 6]
e =
        1      2      3
        4      5      6
```

One can use subscripts to define or access the arrays. Here we define a three-dimensional array as:

```
F(1:3,1:5,1:5) = 0
```

which yields

```
>> F(1:3,1:5,1:5) = 0
F(:,:,1) =
     0     0     0     0     0
```

```
       0     0     0     0     0
       0     0     0     0     0
... ...
F(:,:,5) =
       0     0     0     0     0
       0     0     0     0     0
       0     0     0     0     0
```

Here, each `F(:,:,i)` is a two-dimensional array. Note, when the colon ":" is used alone for index or subscript, it means all elements in the specific dimension.

2.2.8 Matrix Operations

Assuming that we have input the command as above, try the following to get a feeling about the matrix operations:

```
g = b' % (transpose of array),
a1 = a*3, a2 = b/3 %(general math operations)
a3 = a.*b' %(element by element multiplication)
h = a * b %(matrix multiplication)
A = b * a %(matrix multiplication - a different order gives
           %different results)
```

The texts after % are comments. Any text on the same line after this symbol is treated as a comment that has nothing to do with the MATLAB operations. These words will simply be ignored by MATLAB for any action. But they can be useful as notes for the programmer or whoever later uses the MATLAB code.

Note also that matrix multiplication is order-dependent, i.e. `a * b` and `b * a` are generally not the same. They may not even have the same dimensions if they both exist. Not every pair of matrices can have a meaningful multiplication. The condition that two matrices must satisfy for their multiplication to exist is that the number of columns of the first matrix must be equal to the number of rows of the second matrix. This is dictated by the definition of matrix multiplication: each row of the first matrix is doing an inner product (see Chapter 8) with each column of the second matrix (or the sum of element-by-element multiplication).

2.3 Programming Efficiency

Like many computer languages, MATLAB can do repeated computations using a `for` loop:

```
for i = 1:N
[... computations]   % ALL THE COMMANDS GO TO HERE
end
```

This does the computations defined by codes inside the brackets N times (`i` increases from 1 to N). However, when working with matrix computations, programming with the for-loops can increase the computational time (compared to that using a matrix format, if possible), although for-loops are widely used. If the for-loop includes extensive computations, the efficiency can be negatively impacted. The following shows an example comparing computational times using the for-loop and not using the for-loop for calculating the sum of squared of the natural numbers:

$$S = \sum_{i=1}^{n} i^2$$

```
clear
n = 1:10000;
% === Method #1: using For Loops ===
tic
for i = 1:length(n)
 s(i) = n(i) * n(i);
end
t = s(1);
for i = 2:lcngth(n)
 t = t + s(i);
end
t
toc
%=== end of using For Loops ===
%=== Method #2: not using For Loops ===
tic, t1 = sum(n. * n); toc
%=== end of using vector format ===
```

The output is

```
t =
333383335000
Elapsed time is 0.057416 seconds.
Elapsed time is 0.000039 seconds.
```

The first method using the for-loop elapsed 0.057416 second, while the second method without using the for-loop elapsed only 0.000039 second. The MATLAB function pairs `tic` and `toc` can be used together to calculate the time the computer takes. In the above example, the time ratio is 1400 times. This time ratio changes with different computers; it usually decreases as the number of loops `n` increases, but generally the for-loop computation is less efficient. The time ratio should be consistent for different runs on the same computer, although they are usually not exactly the same. If there is an extensive computation with matrices, the programmer should always try to minimize the use of for-loops.

2.4 MATLAB Built-In Functions and Constants

2.4.1 Built-In Functions

MATLAB has many built-in functions, which are either loaded in memory automatically at start-up that can be used as is or accessible in m-files that can be read and modified. If you type

```
help abs
```

it will produce the following explanation on the function abs, which calculates the absolute value of a given number or array. If the number is an imaginary or complex number, it will yield the magnitude or modulus of that number (a modulus of a complex number is the square root of the sum of the real part squared plus the imaginary part squared: $\|X\| = \sqrt{x_r^2 + x_i^2}$, in which x_r and x_i are the real and imaginary parts of the complex number $X = x_r + ix_i$, respectively.

```
>> help abs
 ABS     Absolute value.
    ABS(X) is the absolute value of the elements of X. When
    X is complex, ABS(X) is the complex modulus (magnitude) of
    the elements of X.
    See also sign, angle, unwrap, hypot.
```

Other commonly used MATLAB built-in functions include e.g.

```
clear - clear the memory
sin - sine function
cos - cosine function
sqrt - square root
exp - the natural exponential
log - logarithm with base 10 or the common logarithm
tan - tangent
atan - arctangent
```

Note that trigonometric functions such as sin and cos all use radian rather than degrees as the unit for their arguments, just like in FORTRAN and other languages. So, sine of 30 degrees should be written as

```
sin(30 * pi / 180)
```

which yields

```
>> sin(30 * pi / 180)
ans =
    0.5000
```

Alternatively, one can use the trigonometric functions with degrees: sin now is replaced by sind, cos by cosd, etc. As an example, the following shows the application:

```
>> sind(30)
ans =
    0.5000
```

2.4.2 Built-In Constants

MATLAB also has some built-in constants. For example, the built-in constant pi represents the constant π or the circumference ratio in mathematics: if you type pi and press enter, it gives the following line, which shows the values of π up to four digits:

```
>> pi
ans =
3.1416
```

Internally, the value of `pi` has an accuracy of a double precision real number. If you type `exp(1)` and press enter, it gives you the Euler Number `e`, defined by

$$e = \lim_{n\to\infty}\left(1+\frac{1}{n}\right)^n$$

```
>> exp(1)
ans =
2.7183
```

Like the circumference ratio π, `e` is also an irrational number. The reason that we put (1) on the right of `exp` is because `exp` in MATLAB is a built-in function, not just a constant. Here `exp` is the function e^x, in which x is a variable. It is the value within the parenthesis. If one needs to calculate e^3 and e^{-3}, the expression would be `exp(3)` and `exp(−3)`, respectively.

By default, `i` and `j` in MATLAB are defined as the unit imaginary number $\sqrt{-1}$, unless they are explicitly redefined by a user-provided assignment, e.g. `i = 3`, after which the variable `i` will have a value of 3 until it is changed again by another command or assignment or when MATLAB program is terminated and restarted, when `i` will be $\sqrt{-1}$ again. The following lists more built-in constants or parameters in MATLAB.

`eps` – this returns the spacing of double-precision floating point numbers, which is eps = 2^{-52}, or 2.2204×10^{-16}.

`d = eps(X)` is the positive distance from `abs(X)` to the next larger in magnitude floating point number of the same precision as `X`. `X` may be either double precision or single precision. For all `X`,

`eps(X)`, `eps(-X)`, and `eps(abs(X))` are the same.

`eps('double')` is the same as eps or `eps(1.0)`.

`eps('single')` is the same as `eps(single(1.0))` or `single(2^−23)`.

`realmin` – the smallest positive floating point number.

`realmax` – the largest positive floating point number.

`Inf` – the IEEE arithmetic representation for positive infinity. Infinity is also produced by such operations as a number divided by zero, e.g. `1.0 / 0.0`, or from an overflow, e.g. `exp(1000)`.

`NaN` – not-a-number; `NaN` is the IEEE arithmetic representation for not-a-number. An NaN is obtained as a result of mathematically undefined operations like 0.0 / 0.0 and `inf-inf`.

`NaN('double')` is the same as NaN with no inputs;

`NaN('single')` is the single precision representation of NaN;

NaN(N) is an N-by-N matrix of NaNs;
NaN(M,N) or NaN([M,N]) is an M-by-N matrix of NaNs.

The following lists some examples showing some of the above constants in MATLAB.

```
>> eps
ans =
  2.2204e-016
>> Inf
ans =
   Inf
>> realmax
ans =
  1.7977e+308
>> realmin
ans =
  2.2251e-308
>> NaN
ans =
   NaN
```

2.4.3 Examples of More Functions

Other built-in functions that are important in data analysis are included in the following, with examples:

A = zeros(2,20) defines an array named A with 2 rows and 20 columns (or 2 by 20) of 0s.
size(g) shows the dimension of g.
B = ones(3,3) defines an array named B with 3 rows and 3 columns (or 3 by 3) of 1s.
C = rand(1,100) defines a 1 by 100 array of pseudorandom numbers.
D = randn(1,50) defines a 1 by 50 array of normally distributed pseudorandom numbers.
fix(X) rounds the elements of X to the nearest integers toward zero.
floor(X) rounds the elements of X to the nearest integers toward minus infinity.
ceil(X) rounds the elements of X to the nearest integers toward infinity.
round(X) rounds the elements of X to the nearest integers.
num2str(X) converts numbers (here matrix X) to a string.
str2num is a reversed operation of num2str; it converts a string to a corresponding number.

2.4.4 More Commonly Used Functions and Commands

The best way to learn and use MATLAB built-in functions and commonly used commands is perhaps to start with the book *MATLAB Primer* and online tutorials provided by MathWorks®. In addition, the User Manuals for specific topics can also be helpful. In the following, some selected commonly used built-in functions are introduced. For those who have no experience or limited experience with MATLAB, this should allow a peek into the powerful program and offer an idea of its ease of use, while at the same time building up the basic knowledge, vocabulary, and skills of MATLAB.

Better Orient Yourself Working in MATLAB

`pwd`: When working with MATLAB, one should first know which directory she/he is working with. With MATLAB, it is easy to find out the user's default working directory. Simply issue a command `pwd`, which stands for *present working directory*; MATLAB will respond with a couple of lines, with the last line showing the default directory, e.g. *C:\Users\CL\Documents\MATLAB.*

To change the working directory, one can use the command `cd`, which is very much the same as the DOS command `cd`. For example,

```
>>cd ..
```

would change the working directory to an upper-level directory. To verify that, use `pwd`:

```
>> pwd
ans =
C:\Users\CL\Documents
```

To change the working directory to one named d:\data (if this fold has been created earlier by a user), do this:

```
>> cd d:\data
```

To verify, use `pwd`; this is what the user would see:

```
>> pwd
ans =
d:\data
```

Keep Track of What You Typed

`diary`: After the MATLAB program is started, the user can at any time issue the command `diary` on the command line, followed by a space and a file name. It will start a file with this name in the default directory. Whatever the user types within the MATLAB command window, and whatever MATLAB displays on the computer screen afterwards, will be saved in this file. To designate a different directory where this file is saved, one can specify a path for the directory to store the file before issuing the command `diary`, e.g. *cd c:\Users* will change the working directory to a folder named *Users* on the *c* drive. With this method, as the user is working within MATLAB, the entire record is saved. Later, the user can edit this file and make corrections for a MATLAB script that can be reused or expanded. It saves time and effort. This function can be turned off by typing a command with an argument `off`: i.e. `diary off`, after which no command is saved. When the command `diary on` is issued, it turns back on again. There is no need to specify the name again unless attempting to save the commands in a different file. The user can also just type the command `diary` to toggle between on and off for this function if it is repeatedly executed. This is a useful feature. Alternatively, the user can select the commands and displayed screen in the command window and copy and paste to a text file to edit or develop a script for later usage.

Length of Digits Shown in MATLAB

`format`: Another useful command is `format`. If one issues the command `format long`, the screen display will show a real number with eight digits after the decimal point, e.g. 1/3 will be displayed as 0.33333333. If one uses `format short`, 1/3 will be displayed as 0.3333. If one types `format bank`, only two digits will be shown (i.e. 0.33 for this example), which is normally used when dealing with banking. Note that this does not mean that in the computer memory there is only a two-digit accuracy. It still has the same accuracy, but it just shows the first two digits after the decimal point. Likewise, `format rat` will give a rational expression: in this example, when 1/3 is entered, it will show 1/3 instead of 0.3333. This is also a nice feature. With `format rat`, the mathematical constant π, shown by entering `pi` within thc MATLAB window, is expressed as 355/113.

Save Variables in MATLAB

Assume that the user has started the MATLAB program and worked in the MATLAB environment for a while and has accumulated a number of variables inside the MATLAB environment. The user can choose to save all or some of the variables for later use with the command `save`. Once the MATLAB program is terminated, all variables will be lost unless they are saved. The saved variables will not be automatically loaded into the memory the next time MATLAB is started. They have to be loaded manually by the user with the `load` command. Note that here we are talking about saving the variables and their associated values in MATLAB, but not commands typed by the user or the screen display. The commands can be saved in the diary file if it is turned on earlier. To save the variables, one uses the `save` command followed by a file name:

```
save 'myworktoday'  % save file to disk
```

This will save everything into a file named "*myworktoday.mat*" in the default working directory for MATLAB. If one only needs to save certain variables, a list of variables should be included with the `save` command:

```
save('myworktoday', variablelist)
```

The `variablelist` can be one or more variables separated by commas, e.g. `a, b, c`, to save the variables named `a, b,` and `c`. The users are reminded to use the `help` command to find the syntax for a MATLAB command or function. For example, typing `help save` will result in a few lines of descriptions being displayed regarding the usage of the `save` command.

Load Saved File Back to MATLAB Memory

Should the user intend to reload the variables saved earlier into the memory again, the `load` command can be used to accomplish this. Just type `load` followed by the file name, i.e.:

```
load 'myworktoday'  % load saved file back to computer memory
```

This will let the user reloading everything saved earlier in the file.

Now let us look at the following example with some commonly used MATLAB commands:

```
t = 1:100;          % define a series of integer numbers
t1 = 1:0.5:60;      % the increment can be a fraction instead
                    % of an integer
x = t / 10;         % doing algebra
y = sin(x);         % define function y using a built-in MATLAB
                    % trigonometric function sine of x
y1 = cos(x);        % define function y1 using a built-in MATLAB
                    % trigonometric function cosine of x
y2 = sqrt(x);       % define function y2 using a built-in MATLAB
                    % function to calculate square root of x
y3 = x.*y;          % element by element multiplication of two arrays
y4 = y.*y.*x;       % element by element multiplication of three
                    % arrays
y5 = sqrt(abs(y))+ y.^2. / t;  % more computations
figure              % start a figure frame
plot(t, y)          % plot a solid line/curve with the variable
                    % t being the x-axis and y the y-axis with a
                    % default blue color
hold on             % this command allows for keeping the line just
                    % plotted and overlaying the following plots on
                    % the existing plot (without this command,
                    % any new plot will cause the earlier plot
                    % to be cleared
plot(t, y1, '-','LineWidth',2) % plot a solid line/curve with
                    % the variable t being the x-axis and y1 the
                    % y-axis with a default blue color and a line
                    % width of 2 (the solid line indicated by '-'
                    % which is the default and therefore is
                    % optional)
plot(t, y2, 'r-', 'LineWidth',3)  % plot a dashed line/curve
                    % with the variable t being the x-axis and y2
```

```
                  % the y-axis with a red color and a line width
                  % of 3
plot(t, y3, 'g-.','LineWidth',4) % plot a dash-dotted line/
                  % curve with the variable t being the x-axis and
                  % y3 the y-axis with a green color and a line
                  % width of 4
plot(t, y4, 'k.','LineWidth',5) % plot a dotted line/curve with
                  % the variable t being the x-axis and y4 the
                  % y-axis with a black color and a line width of
                  % 5
plot(t, y5, 'c','LineWidth',6) % plot a solid line/curve with
                  % the variable t being the x-axis and y5 the
                  % y-axis with a cyan color and a line width of 6
xlabel('x-axis (unit for x)', 'FontSize',12) % print x-axis
                  % (unit for x) under the x-axis with a font size
                  % 12
ylabel('y-axis (unit for y)', 'FontSize',12) % print y-axis
                  %(unit for y) along the left hand side
                  % of the y-axis with a font size 12
title('Title of this Figure', 'FontSize',15) % print Title of
                  % this Figure on top of the figure
legend('L1', 'L2', 'L3', 'L4', 'L5', 'L6', 'Location', 'Southwest')
                  % Label each of the 6 lines with L1, ... , L6,
                  % respectively on the lower left corner
save variables_example
```

The results are shown in Fig. 2.1. The variables are saved in a file named "variables_example.mat," which can be reloaded into MATLAB at a later time when needed. There are many more options that we can use with the `plot` command and associated commands such as `title, legend, axis, text`, etc.

Save a Plot to a Graphics File

To save a plot as a digital file (so that the plot can be included in a Word document or edited manually with a graphics software, for instance), one can use the `print` command. The following is a list of examples for the print command for different file formats.

```
print -djpeg test.jpg  % print the figure to a jpeg file
print -dps test.ps     % print the figure to a postscript file
print -dpsc test.psc   % print the figure to a colored
                       % postscript file
print -deps test.eps   % print the figure to an encapsulated
                       % ps file
print -dtiff test.tif  % print the figure to a tiff file
```

If the user intends to modify the figure using another software package (e.g. Adobe Illustrator), one of the postscript (*.ps, *.psc, or *.eps) files maybe the best choice.

One can basically save MATLAB plots in one of the common graphics file formats. To see all possible graphics files that MATLAB can save to, type `help print` in MATLAB's command window.

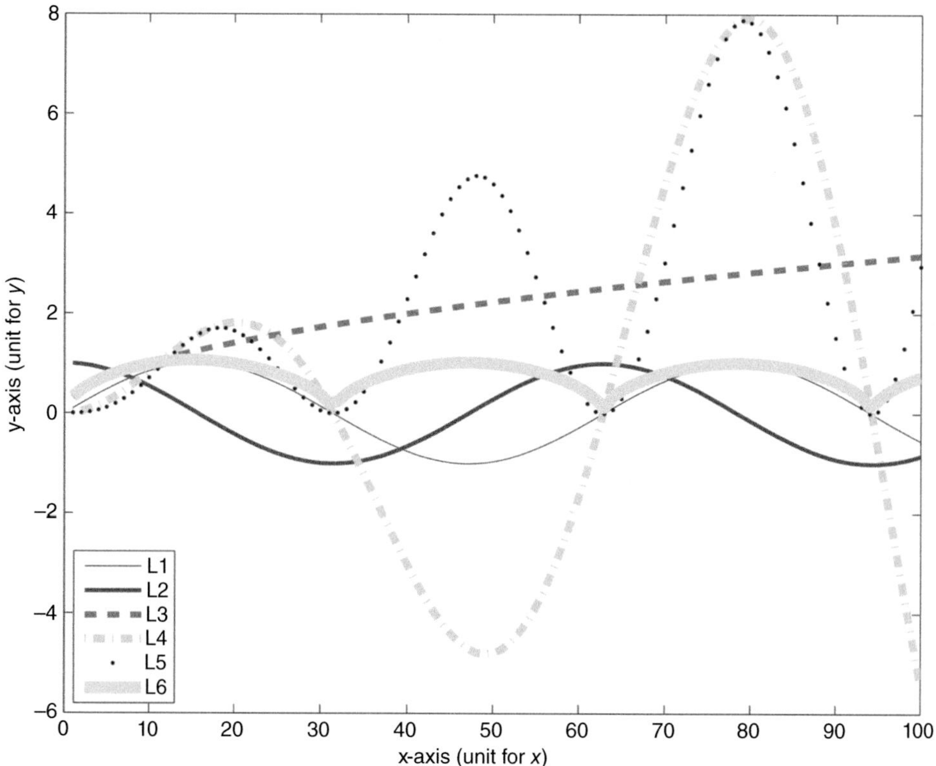

Figure 2.1 Example of figure produced by the MATLAB command `plot`. A black-and-white version of this figure will appear in some formats. For the color version, please refer to the plate section.

Basically, the syntax is `print` plus a space and `-d`, in which `d` stands for device followed by the graphics format indicator e.g. jpeg, followed by the file name with a proper file extension.

Find a Subset of Data

`find`: Another extremely useful command is `find`. This command is to quickly find a subarray from a given array under given conditions. This can be demonstrated clearly by a simple example. The data files used in the following and throughout the book are downloadable from the Cambridge University Press website.

```
%==============================
% Examples of data manipulation
%==============================
load CL1.txt; % load an ASCII MATRIX data file (the file has no
              % letters and only consists of a rectangular
              % matrix with the same number of real values in
              % all the columns and same number of real values
              % in all the rows
```

```
lon = CL1(:,1);   % define the first column of the matrix CL1 as
                  % lon (longitude)
lat = CL1(:,2);   % define the second column of the matrix CL1 as
                  % lat (latitude)
plot(lon, lat)  % plot longitude against latitude (coastline)
axis([ -81.7 -81.6 30.5 30.6] )   % zoom in the above figure with
                 % the minimum and maximum x-axis values to
                 % be -81.7 and -81.6, respectively, and the
                 % maximum and minimum y-axis values to be 30.5
                 % and 30.6, respectively.
grid on % add grids on the figure: dashed horizontal and
        % vertical lines

% "find" is a very useful command to limit the data range
IND = find(lon > -81.7 & lon < -81.6 & lat > 30.5 & lat < 30.6);
size(IND)
N = size(IND)
% use "for loop" to calculate the mean value
 meanLon = lon(IND(1));
for i = 2:N(1)
 meanLon = meanLon + lon(IND(i));
end
meanLon = meanLon / N(1)
% Alternatively, one can use the MATLAB's built-in "sum" function
meanLon = sum(lon(IND)) / N(1)
% or the MATLAB mean function
meanLon = mean(lon(IND))
```

In this example, we load an ASCII data file, in which we have a matrix of latitude and longitude. We then select a subset of the data from the variables to examine the data within certain ranges of longitude and latitude. Finding a subset of data is a common action in data processing and analysis. As another example, if one has a set of sea surface temperature (SST) data from the entire ocean but the interest is only the characteristics of the SST in a smaller region, then the `find` function can be used to select the subset of data. The `find` function can pick out a subset of data points by some conditions limiting the range of latitude and longitude. If one just wants to work on the SST along a straight line with a given width, a formula can be used to limit the data to be below one line and above another parallel line, i.e. between two parallel lines. The `find` function can accomplish this task.

2.5 Examples of Working with Spatial Data

In this section, we discuss some examples of working with spatial data and how to select a subset of data in space in theory and apply it using MATLAB. We say "theory"; it really is simple plane geometry and algebra. However, we should not underestimate the usefulness of simple theories. Readers may find these examples useful when working with data obtained from a moving vessel.

To illustrate, we start with an assumption that we have some data over a two-dimensional plane in an x-y Cartesian coordinate system, i.e. the data are obtained on a plane with coordinate (x, y). They may be multiple data points. The data could be any variable, for instance, water temperature or velocity profiles from an acoustic Doppler current profiler installed on a moving platform. The x-y coordinates could represent a position in the ocean, an estuary, or a bay.

2.5.1 Example 1: Data Above a Vertical Coordinate

If we are going to pick all data above a given vertical coordinate b or north of b, we are essentially requiring that the data position satisfies (Fig. 2.2)

$$y > b. \tag{2.1}$$

In MATLAB, we use the following command to find the index of array `y` satisfying the above condition:

```
>>idi = find(y > b);
```

Here the output from the above command is `idi`, the index for condition (2.1).

In Fig. 2.2, the small squares represent positions of data points, which are numbered for convenience of discussion. It can be seen that those points satisfying (2.1) are numbered from 1 to 10. The rest (numbered from 11 to 23) are all below $y = b$, or they satisfy the following instead:

$$y < b. \tag{2.2}$$

In MATLAB, we use the following command to find the index of array `y` satisfying the above condition:

```
>>idi = find(y < b);
```

The output from the above command `idi` gives the index for condition (2.2).

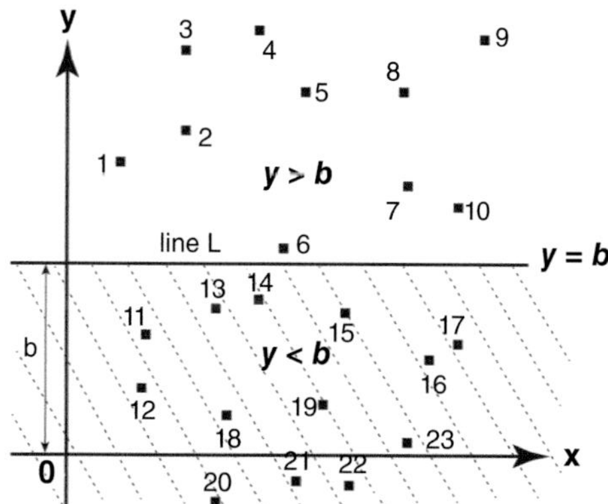

Figure 2.2 Subset of data defined by the positions. The shaded area are those positions below the line defined by $y = b$.

2.5.2 Example 2: Data on One Side of a Horizontal Coordinate

Likewise, if we only need to use data on the left side of a given horizontal coordinate x_0 on a 2-D plane, we are essentially requiring that the data position satisfies

$$x < x_0. \tag{2.3}$$

The index of subset data satisfying (2.3) is determined by the MATLAB command

```
>>idi = find(x < x0);
```

or if we need to select data on the right-hand side of a given horizontal coordinate x_0, the condition becomes

$$x > x_0. \tag{2.4}$$

The MATLAB command for this would be

```
>>idi = find(x > x0);
```

2.5.3 Example 3: Data on Either Side of a Sloped Line

Generally, if we need to use data above a line defined by

$$y = ax + b. \tag{2.5}$$

We are requiring that the data position satisfies (Fig. 2.3):

$$y > ax + b. \tag{2.6}$$

The MATLAB command for this would be

```
>>idi = find(y > a * x + b);
```

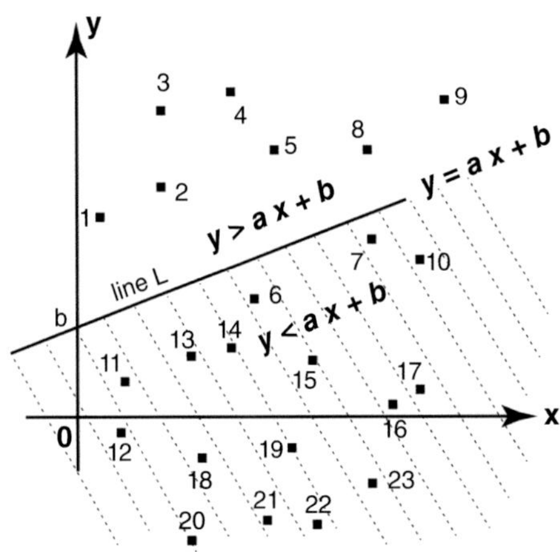

Figure 2.3 Subset of data defined by the positions. The shaded area is those positions below the line defined by $y = ax + b$.

Here a is the slope of the line, and b is the intercept of the line. In Fig. 2.3, we can see that the points that satisfy (2.6) are those numbered as 1, 2, 3, 4, 5, 8, and 9.

On the other hand, if we need to use data below the line defined by (2.5), we are requiring that the data positions satisfy

$$y < ax + b. \tag{2.7}$$

The MATLAB command for this would be

```
>>idi = find(y < a * x + b);
```

In Fig. 2.3, we can see that the points that satisfy (2.7) are those numbered as 6, 7, and 10 through 23.

2.5.4 Example 4: Data between Two Parallel Lines

If we have two parallel lines with the same slope a but different intercepts b_1 and b_2, and we need to select data points in between the two parallel lines, then we must have

$$ax + b_2 < y < ax + b_1. \tag{2.8}$$

The MATLAB command for this would be

```
>>idi = find(a * x + b2 < y & y < a * x + b1);
```

Any data points that are within the shaded region in Fig. 2.4 satisfy the above condition.

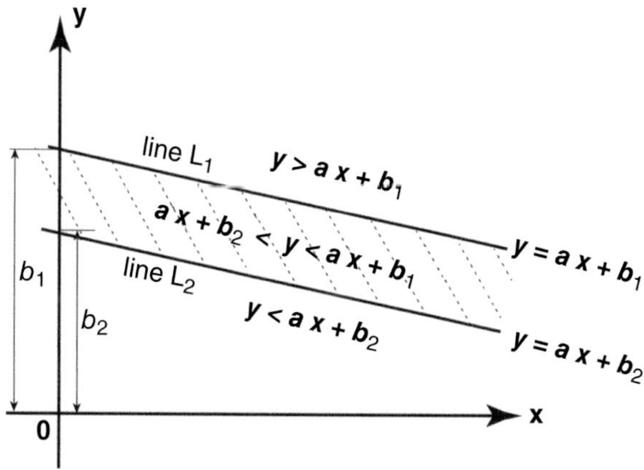

Figure 2.4 Subset of data defined by the positions. The shaded area is those positions between the lines defined by $y = ax + b_1$ and $y = ax + b_2$.

Now we examine the following example with the given MATLAB script

```
clear
x = [ 1 2 3 4 5 6 7 8 9 10 ...  % define the x-coordinates of
      1 2 3 4 5 6 7 8 9 10 ...  % data points
      1 2 3 4 5 6 7 8 9 10 ...
      1 2 3 4 5 6 7 8 9 10 ...
      1 2 3 4 5 6 7 8 9 10];
Y = [ 0 0 0 0 0 0 0 0 0 0 ...   % define the y-coordinates of
      1 1 1 1 1 1 1 1 1 1 ...   % data points
      2 2 2 2 2 2 2 2 2 2 ...
      3 3 3 3 3 3 3 3 3 3 ...
      4 4 4 4 4 4 4 4 4 4];
figure
plot(x,y, 'ko')
xlabel('x')
ylabel('y')
hold on
idi = find(y > = x) % SELECT DATA POINTS THAT SATISFY y > = x
plot(x(idi),y(idi),'ko','LineWidth',3)
x1 = [ 1:10];
y1 = x1;
plot(x1,y1, '-k')
print -dpng Chapter2.5.png
```

This example plots a matrix of dots on a plane, selects the points that satisfy

$$y \geq x$$

and highlights them with thicker edges (Fig. 2.5).

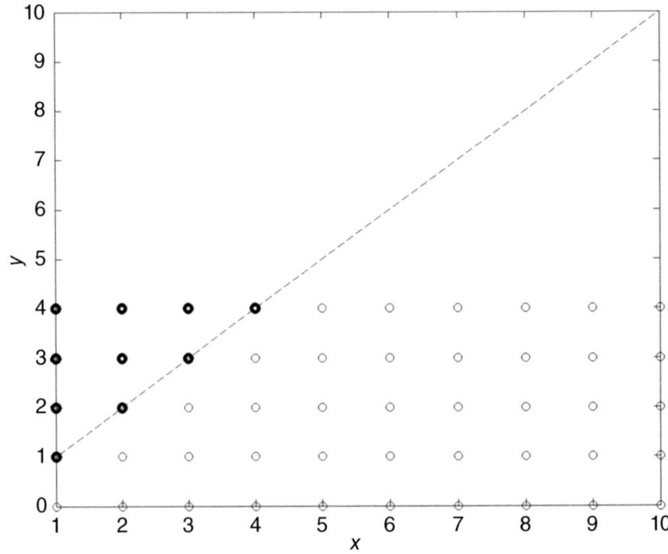

Figure 2.5 Subset of data points above the line $y = x$.

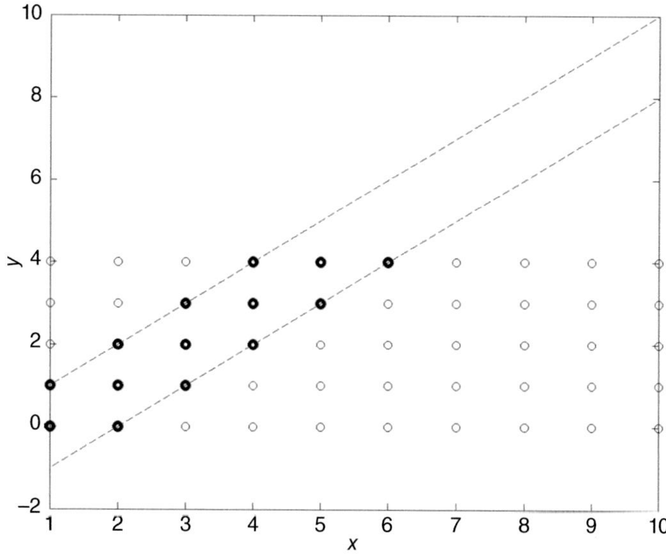

Figure 2.6 Subset of data points between the lines $y = x$ and $y = x - 2$ including data points on the lines.

If we modify the latter half of the script as below

```
figure
plot(x,y,'ko')
xlabel('x')
ylabel('y')
hold on
idi = find(x > = y & x - 2 < = y)
plot(x(idi),y(idi),'ko','LineWidth',3)
x1 = [ 1:10] ;
y1 = x1;
plot(x1,y1,'-k')
plot(x1,y1-2,'-k')
```

We select the data points that are within two parallel lines.

The highlighted data points (Fig. 2.6) are those between the two parallel lines $y - x - 2$ and $y = x$ or

$$x - 2 \leq y \leq x.$$

2.5.5 Data within Rectangles, Parallelograms, or Circles

Sometimes, we need to define a line that is perpendicular to a given line. The slope of a line that is perpendicular to a given line with a slope of a is the negative reciprocal of a, i.e. the slope of the perpendicular line is $b = -1/a$. This can be used to select data points that are within a rectangle. This can also be extended to

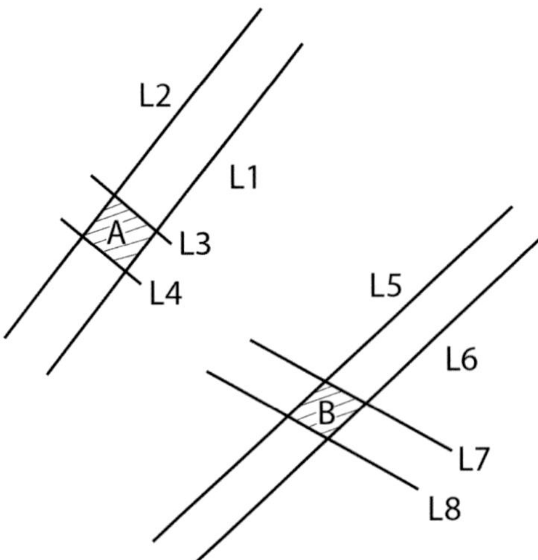

Figure 2.7 Using pairs of parallel lines to select data. A and B are areas defined by a rectangle and a parallelogram, respectively. L1 and L2 are parallel to each other; L3 and L4 are parallel to each other; so are L5 and L6, and L7 and L8. L1 is perpendicular to L3; while L5 is not necessarily perpendicular to L7.

select data within a parallelogram in which the sides are not necessarily perpendicular (Fig. 2.7). Usually, we anchor rectangles or parallelograms at fixed points or grid points. The data can be selected within a rectangle, a parallelogram, or even a circle centered at the grid point. For convenience of discussion, we can call these rectangles, parallelograms, or circles the *data stamps* or footprints.

For a rectangular or parallelogram data stamp, two pairs of lines parallel to each other can be used to select the data. In Fig. 2.7, if the lines L1, L2, L3, and L4 are expressed as

$$y_1 = ax + b_1, \tag{2.9}$$

$$y_2 = ax + b_2, \tag{2.10}$$

$$y_3 = bx + b_3, \tag{2.11}$$

$$y_4 = bx + b_4. \tag{2.12}$$

Then data points within the data stamps (the shaded areas of Fig. 2.7) can be selected by a MATLAB command using the `find` function:

```
idi=find(y> =a* x+b1&y< =a* x+b2&y> =b* x+b4&y<= b* x+b3);
```

Here `idi` is the index for the variables `x` and `y`. In oceanographic data, `x` and `y` are often longitude and latitude, respectively. Assume that there is a variable measured in the region, such as the water depth `h` and velocity components `u` and `v`, and that the time variable is t; we can get the time series within a selected data stamp by `t(idi)`, `h(idi)`, `u(idi)`, and `v(idi)`.

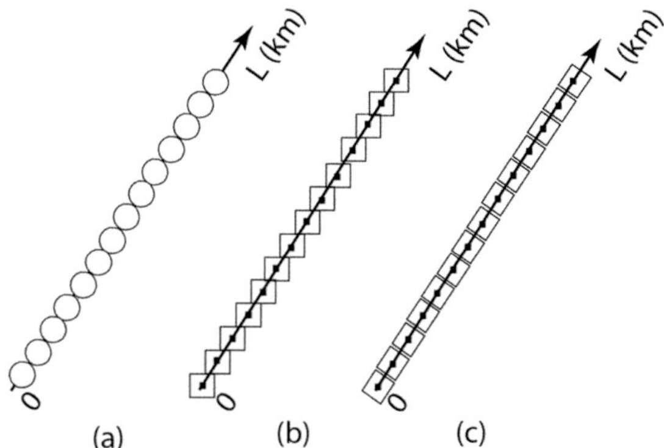

Figure 2.8 Data stamps. (a) circular data stamps; (b) rectangular data stamps; (c) rotated rectangular data stamps. The arrows show the direction of increasing distance with a hypothetical unit of km as an example.

For a circle, the condition is

$$(x - x_0)^2 + (y - y_0)^2 \leq r^2, \tag{2.13}$$

in which r is the radius of the data stamps and (x_0, y_0) is the anchor point or center of the circle. To select data points within a circle, a relevant MATLAB command would be

```
idi = find((x - x0).^2 + (y - y0).^2 < = r^2).
```

There can be many different types of data stamps. Fig. 2.8 shows a few examples of data stamps along a straight line transect: circles, rectangles, and rotated rectangles. Note that these data stamps do not have overlaps, but they may have some overlaps. In these examples, the data stamps are defined along a transect (a straight line), which has a starting point and a direction of increasing distance from the point. This is often used to make an along-transect plot (distance vs. variable values). Of course, the transect does not have to be a straight line.

Usually, selecting a subset of data points is not the final goal. The goal is doing some data analysis for the selected data points. For instance, if there is a variable for water temperature named T, and if we need to calculate the average temperature for the subset data points with index `idi` found using the MATLAB function `find`, we use the following MATLAB scripts after the above scripts:

```
meanT = mean(T(idi));
```

Here the function `mean` is a MATLAB built-in function for calculating simple arithmetic average of the array inside the parentheses.

2.6 Selection of Data from Moving Platforms

Many oceanographic surveys use research vessels or automated unmanned vehicles. These moving vessels or platforms usually follow a designed route to

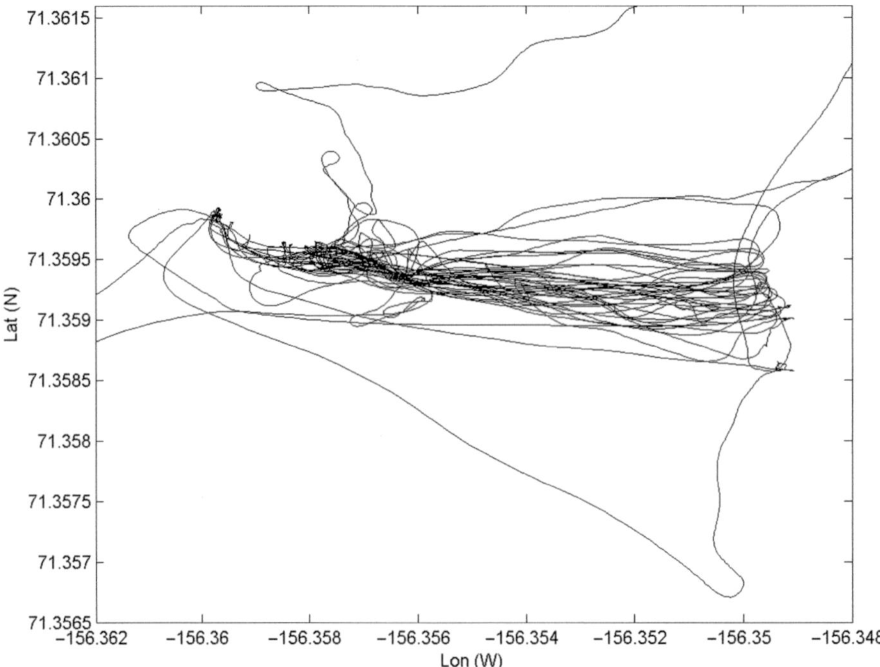

Figure 2.9 Track of survey made by a pontoon boat in the arctic Elson Lagoon in 2014.

collect data. Sometimes, repeated occupations of a designed route are made. The actual tracks are usually less regular than planned because of the unpredictable influence of varying sea state, wind, ocean current, difference in skills or fatigue of boat drivers, and inertia of platform when turns are made, etc. Fig. 2.9 is an example of such a complicated track of a survey done by the author using an inflatable pontoon boat across an inlet in an arctic lagoon (Elson Lagoon) in northern Alaska. Numerous repetitions were made and the planned track was a straight line across the Eluitkak Inlet. The objective of the repeated occupation of the transect was to measure the water depth and velocity profiles. The question is. how should we use the data? A commonly used method is to make some data stamps as discussed above, within each of which data points are considered as obtained from the same point (e.g. the center of the data stamp).

As another example, Fig. 2.10 is a map of the lower Chesapeake Bay mouth and a planned transect of survey (Li et al., 2000). The transect was repeatedly occupied during numerous surveys. Fig. 2.11 shows some actual ship tracks made continuously for 10 days in May 1999. The survey measured flow velocity profile along this transect with an acoustic Doppler current profiler. Fig. 2.11b shows the

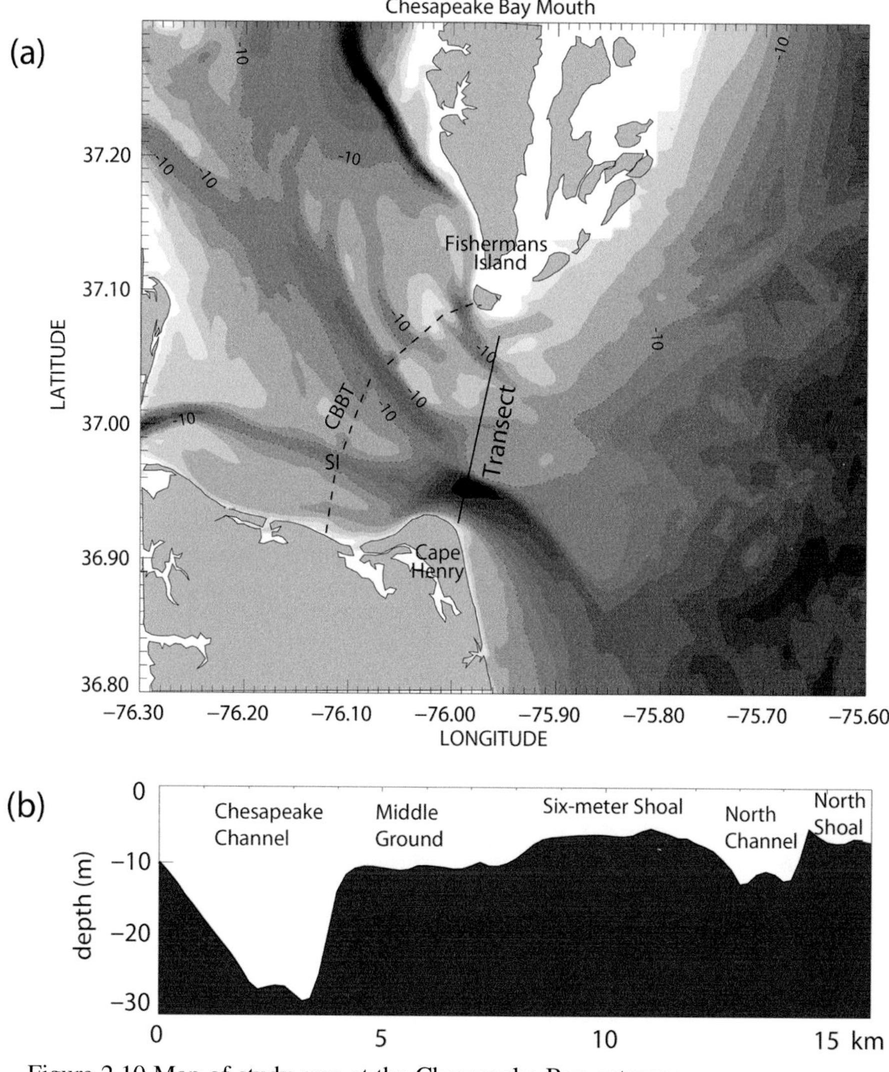

Figure 2.10 Map of study area at the Chesapeake Bay entrance.

data stamps as tilted rectangles 320 m × 80 m in size. The parameters in each of the data stamps are found using the MATLAB `find` command. These selected data are then grouped together and aligned in time as a single time series for each parameter. A time series analysis can then be performed for each data stamp.

Using this method, the water depth values measured at each of the data stamps are grouped into a time series. After the mean depth at each position (data stamp) is subtracted, the water depth variations show the tidal variation of the water surface elevation. The time series of water depth variation compared with the surface

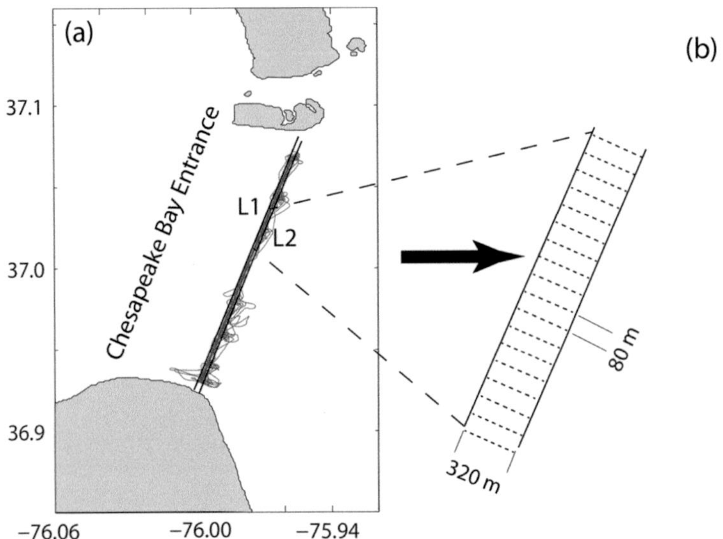

Figure 2.11 Ship transect line and conditions to select data for analysis. The arrow points to an idealized partitioning of the transect into rectangular bins for data selection and analysis.

elevation measurements at a nearby tide gauge do show a remarkable consistency (Fig. 2.12).

The measured depth at four selected locations along the transect (Fig. 2.10), i.e. the Chesapeake channel (CC), middle ground (MG), 6-m shoal (SMS), and north channel (NC), are separately plotted and fitted to tidal signals using harmonic analysis (see Chapter 11). The results are shown in Fig. 2.13.

Selected MATLAB Commands in This Chapter

`whos` – display memory contents.

`'` – complex conjugate transpose.

`.'` – nonconjugate transpose.

`transpose` – nonconjugate transpose function.

`diary` – record keystrokes and screen output into a text file for later usage such as edit.

`format` – define screen display format (number of digits etc.), which does not affect what is saved in the computer memory.

`save` – save memory contents in the MATLAB environment to a file

`load` – load a saved MATLAB file (.mat) back into memory, or load a text data file of rectangular form into memory.

`plot` – make a screen plot within MATLAB.

`print` – print a plot to a digital graphics file for later use.

`find` – find the index of a subarray from a given array with given conditions.

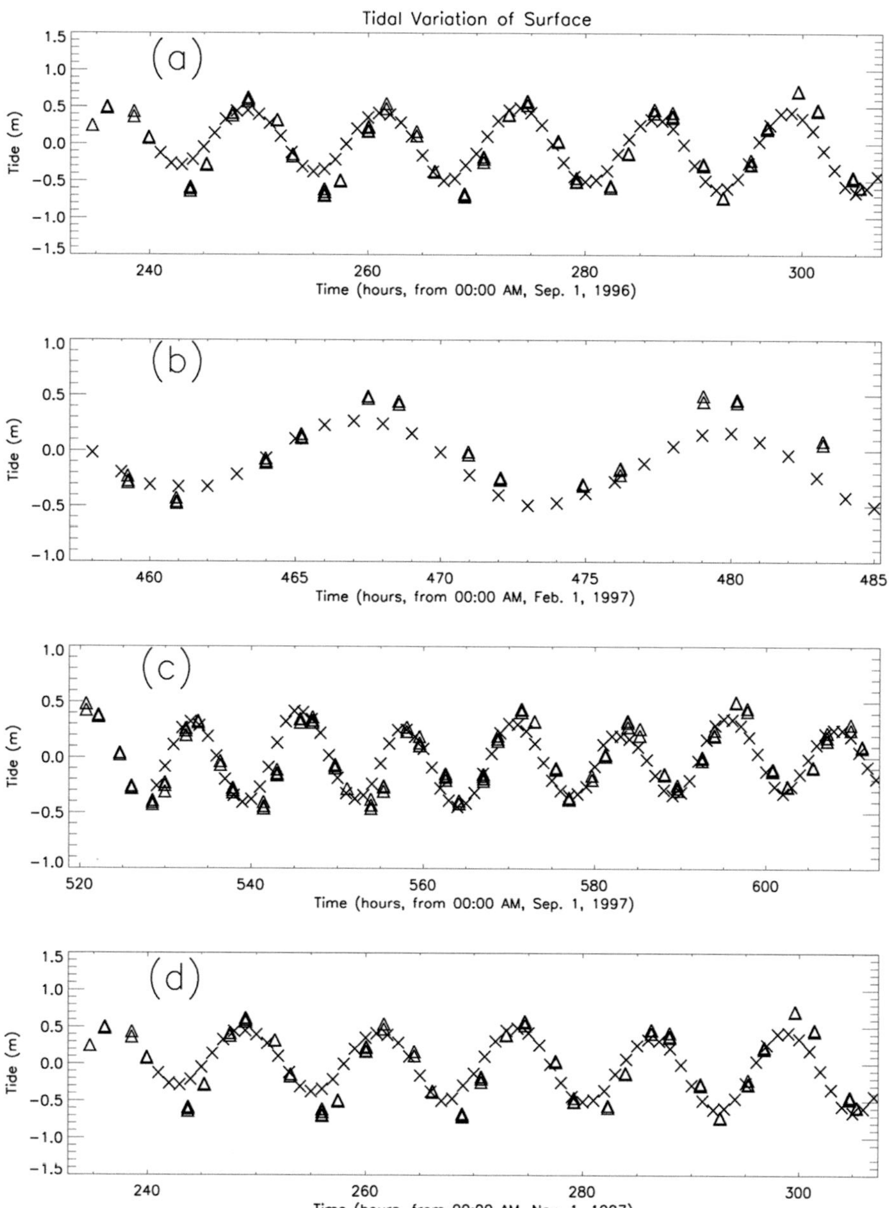

Figure 2.12 Comparison of water level time series between NOAA's tide gauge at the Chesapeake Bay Bridge Tunnel and the vessel-based water depth data measured by an ADCP. (a)–(d) are for four different surveys conducted in Sep. 1996, Feb. 1997, Sep. 1997, and Nov. 1997, respectively.

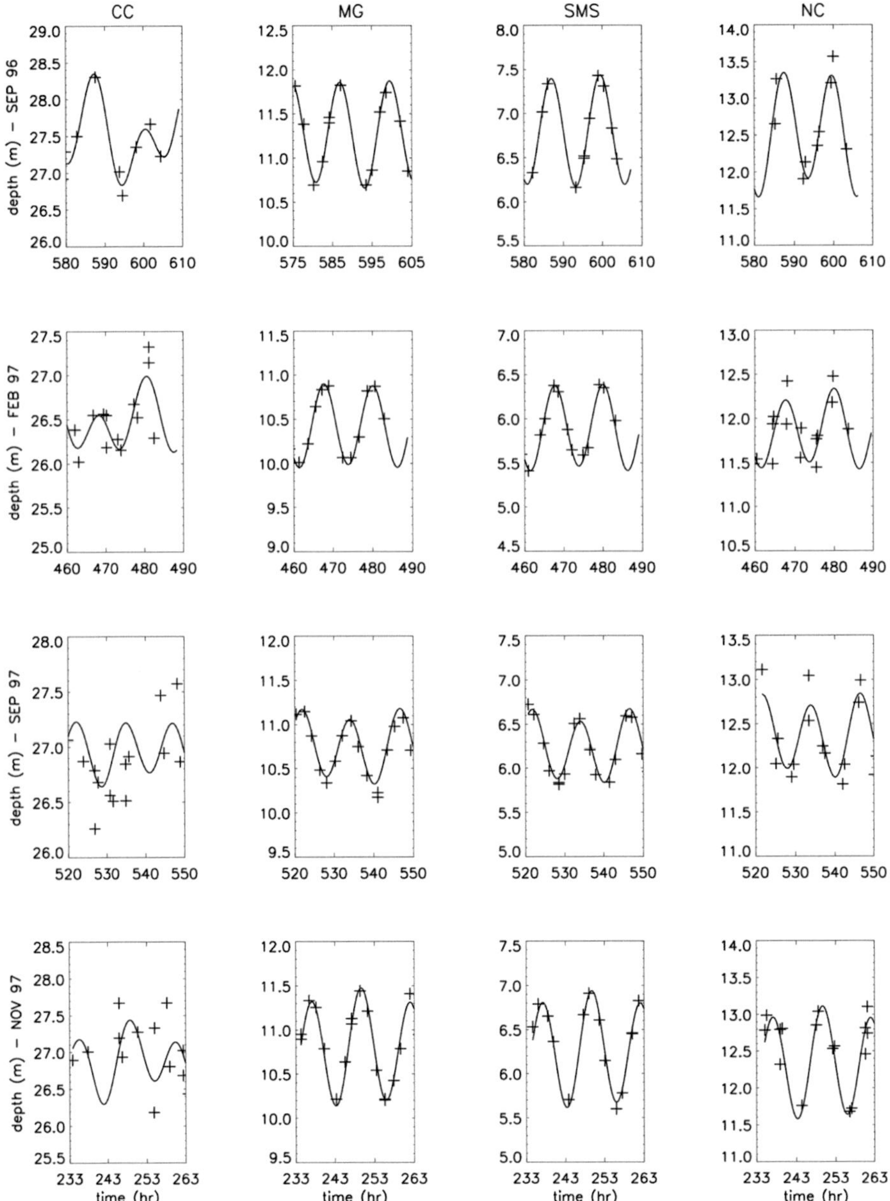

Figure 2.13 Time series of water depth data measured by an ADCP and harmonic fit for the four locations at CC, MG, SMS, and NC (the four columns) for the four different surveys (the four rows) conducted in Sep. 1996, Feb. 1997, Sep. 1997, and Nov. 1997, respectively.

Review Questions for Chapter 2

(1) For those inexperienced in MATLAB, check the MATLAB Primer or getting Started with MATLAB (downloadable from MathWorks web site), or go through the MATLAB tutorial from inside the MATLAB environment and learn MATLAB commands in the following categories: commands or functions for computation or matrix operation; commands or functions for 2-D graphics; and commands or functions for 3-D graphics.

(2) How do you define a column-vector and a row-vector in MATLAB (where do you use a space, a comma, or a semicolon)?

(3) How do you check what variables are in the MATLAB working environment?

Exercises for Chapter 2

(1) Working with coastline data and the use of MATLAB command, `find`.

- **(a)** Use the coastline file named "CL1.txt" – this is a data file for coastline along the Georgia coast;
- **(b)** Browse the file using a text editor or Word and then load the file into MATLAB by using the `load` command;
- **(c)** The first column of the data file is longitude, and the second column is latitude; name a variable of lon and assign longitude to lon, name a variable lat and assign latitude to lat;
- **(d)** Find out how many of the data points are NaN (not-a-number);
- **(e)** Exclude the NaN data points and calculate the mean longitude and mean latitude for the entire dataset;
- **(f)** For these latitude and longitude values, assume you only need to focus on an area between latitude 30.8 and 31.1, define lat1, lon1 and assign values of longitude and latitude for the condition that latitude is within [30.8, 31.1];
- **(g)** Calculate the mean values for lon1 and lat1;
- **(h)** Plot a clean graph with all the coastline in "CL1.txt" shown as blue color, while a strip of coastline within latitude = [30.8, 31.1] as red color (Fig. 2.E1);
- **(i)** Plot a graph with all the coastline in "CL1.txt" shown as blue color, except the upper left corner with latitude above 30.8 and longitude left of −81.4 as red color (Fig. 2.E2).
- **(j)** Can you make the following plot (Fig. 2.E3), with a circle centered at lon = −81.35 and latitude = 31.3 and radius of 0.05 degrees?

(2) Selecting a subset of data points.

The data file named "lat_lon_Fourchon_2010.mat" has latitude and longitude of tracks of a survey vessel. Load the data file into MATLAB and select a

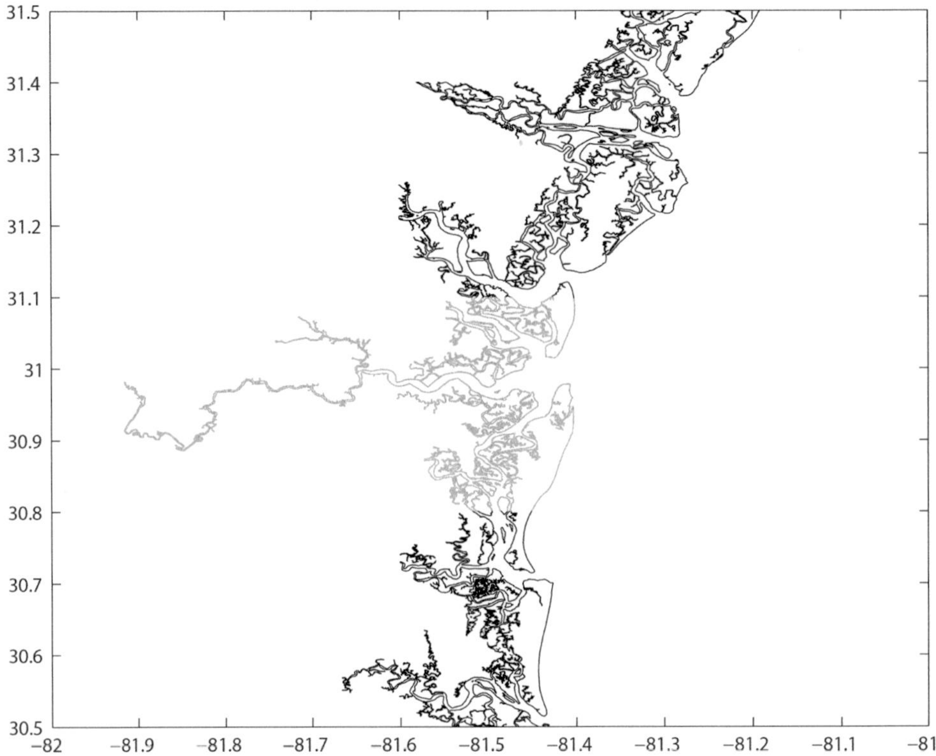

Figure 2.E1 Results after finishing step 8.

subset of data points, using the MATLAB function `find`, along a transect defined by those track lines within two parallel lines:

The first line passes these two points, P1: (−90.224, 29.099); P2: (−90.2205, 29.0975); while the second line passes these two points, P3: (−90.224, 29.0993); P4: (−90.2205, 29.09753);

Draw these two lines to show the boundary of the subset of the data points (Fig. 2.E4).

(3) MATLAB function.
Create a MATLAB function named mmm.m which returns the three-m's: the maximum, minimum, and mean of a given array. The first line of the function should read something like

```
function [ m1,m2,m3] = mmm(x)
```

Complete the function and save it as mmm.m in your MATLAB working directory. Test the function with an array

```
x = 1:99
```

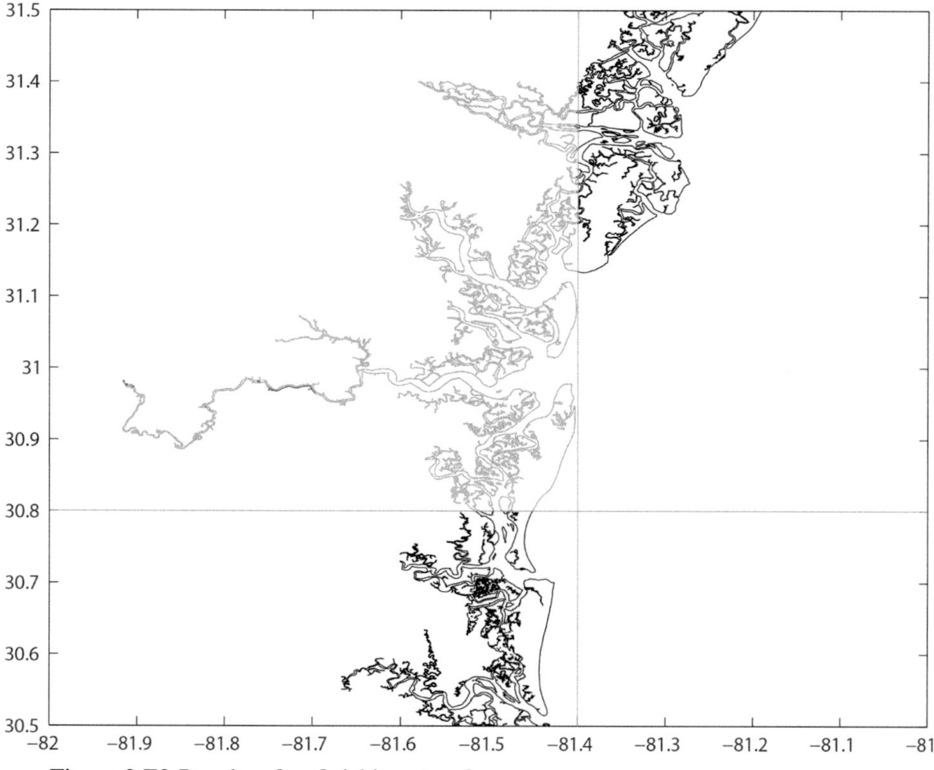

Figure 2.E2 Results after finishing step 9.

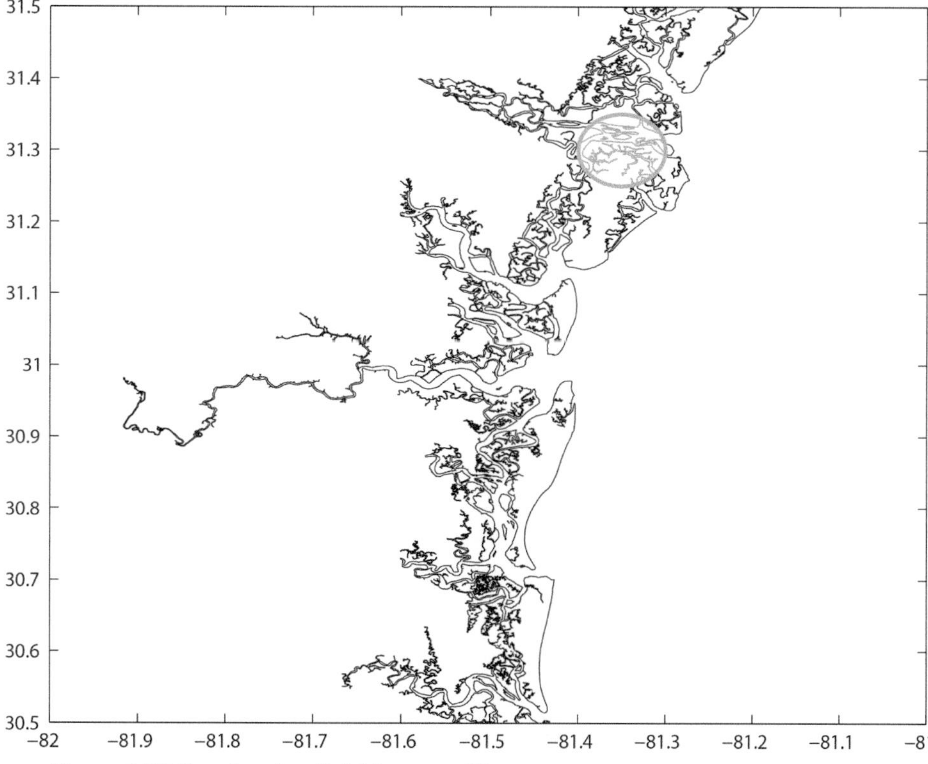

Figure 2.E3 Results after finishing step 10.

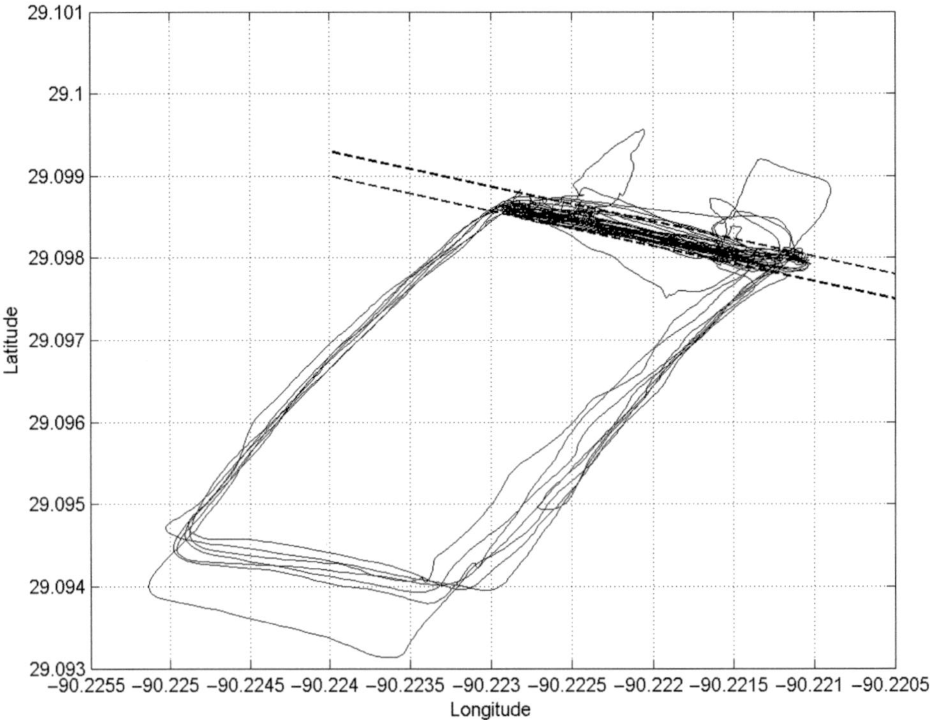

Figure 2.E4 Selecting a subset data points for exercise 2.

to verify that your function works as anticipated (the minimum, maximum, and mean values of x should be 1, 99, and 50, respectively).

(4) Work with wind data.

(a) Write a MATLAB function to convert wind speed and wind direction time series into the east and north components. Then apply the function to the wind data ("Wind_data.txt," in which the wind speed and direction are denoted as WSPD and WDIR, respectively). Make a plot showing your results for both the raw data (wind speed and direction time series) and the converted time series (wind velocity in east and north components in m/s).

(b) Modify the above MATLAB function for the ocean current data. Then apply the function to the current data ("Current_data-sample.txt"). Make a plot showing your results for both the raw data (current speed and direction time series) and the converted time series (current velocity in east and north components in m/s).

(c) Write a MATLAB function that does the coordinate system rotation with the input vector (u, v) and output vector (u', v') in the new system. Test the function with some examples.

3

Time and MATLAB Functions for Time

About Chapter 3

As mentioned earlier, time series data must include time stamps. It may seem trivial, but some attentions are needed to properly use time to avoid mistakes. The objective of this chapter is to review a few concepts of time so that when an analysis of time series data is performed, there is less chance to make mistakes with respect to data consistency, the result of analysis, and interpretation. We will discuss some basic astronomical concepts related to time; different definitions of day; and time measurements, GMT, and UTC. We will learn using MATLAB to construct a time sequence from civil time or time strings, i.e. the year, month, day, hour, minute, and second to a real number of time and vice versa. We will also briefly discuss the Positioning, Navigation, and Timing (PNT) data from the Global Positioning System (GPS).

3.1 Time Is of the Essence

Time is of the essence. Time stamps of a series of data allow the calculation of rate of change, frequency or period of oscillations, spectrum, and time scales, in order to capture the dynamics of the processes in question, such as those in the ocean.

As an example of the importance of physics data, the measurement of gravitational acceleration g (~9.8 m/s^2 at sea level) can be done with a highspeed camera taking a series of pictures of an oscillating pendulum in vacuum or a falling object through a transparent vacuum tube. In this specific example, the accurate measurement of the value of g depends on the accuracy of the time for each frame of pictures.

Imagine a set of water temperature values were obtained without time stamps – there are values of temperature but no record as to when they were measured. These data would have little value. Or, if we have a set of time series of water temperature but we do not know the unit of time, or we know the unit of time but not the *time zone* defining the time: we don't know whether it was measured with the Eastern time, Pacific time, or *Greenwich time*. Time can be recorded with a unit of seconds, or minutes, or hours, etc. But many applications for environmental data including oceanographic data use month/day/year/hour/minute/second or similar format and therefore are strings of text or a combination of numbers (i.e. using 1 to denote Jan., 2 to denote Feb., . . . , 12 to denote Dec.), rather than a single number for a time instance. MATLAB has some built-in functions to convert between the time strings and time values of certain origin (or the start time, which is 0).

In theory, a time zone is a zone of Earth surface of 15° longitude with an origin (or 0 longitude) passing Greenwich, England (Howse, 2003). The Earth rotates 15° longitude within 1 hour. However, the actual time zones are modified from the theoretical divisions by considering geography, political regions, and regional laws. The U.S. has six time zones: the Eastern, Central, Mountain, Pacific, Alaska, and Hawaii time zones. As an extreme example, Russia has 11 time zones. Geographically, China covers five time zones but has been unified into a single time zone since 1949. The complications do not stop there, considering the difference between *daylight saving time* and *standard time*. Some countries or regions use daylight saving time, but some do not, and the laws of some countries change over time.

Coming back to the time stamps for time series data, in a largescale survey across the entire ocean covering several time zones, not knowing the actual time zone used for time stamps could be a real problem for the usage of the data. This, of course, is less likely in modern times, as most scientific apparatus have an internal clock for data recording or automated recording, and often multiple sources of time may be available. For example, a vessel-based GPS may be recording meteorological data with time stamps derived from the GPS signal. In addition, rarely, a cross-ocean basin survey would use local time. In any case, the value of the data would be greatly compromised if there were any problems with the time stamp or we were not sure which time zone was used.

3.2 MATLAB Functions for Time Strings, Time Vectors, or Time Conversion

We now learn how to use MATLAB to convert time: computers only read digital numbers and do not deal directly with time written in Jan, Feb, . . . , Dec or 2020.1.31 (Jan. 31, 2020). For example, 10:15:16 AM, Jan. 31, 2020 is just a string of time, not a single number in the real axis. When we do the analysis, we can use MATLAB to convert these to a series of time values. MATLAB has several functions to convert time strings, such as `datenum`, `datestr`, `datevec`, `now`, and `clock`.

First, let us type the following command:

```
format rat
```

followed by

```
clock
```

which gives the following print in the MATLAB window

```
>> clock
ans =
  2020 5 16 11 36 6841/250
```

This is a *time vector* showing that the current year is 2020, the month is May, the date is the 16th, the hour is 11, the minute is 36, and the second is closest to the rational number 6841/250, which is roughly `27.3640` seconds.

If we do not use format rat, the output would look like this

```
>> clock
ans =
  1.0e+003 *
     2.0200     0.0050     0.0160     0.0110     0.0360     0.0274
```

which is essentially the same but a bit difficult to read or interpret.

One can also use `fix(clock)` to print out only the integer parts of the time vector:

```
>> fix(clock)
ans =
         2020            5           16           11           36           27
```

The `clock` command gives a time vector conformed to the traditional expression of *civil time*. It is not a single number. For data analysis, we often need to use a real number to express the time lapsed from a given origin or starting time. For that purpose, we can use another MATLAB command `now`:

```
now
```

would yield

```
ans =
  7.3793e+003
```

The unit of this output is days. To view the number with more digits, we can type `format long` followed by `now` and get the following:

```
ans =
    7.379274836500463e+005
```

The time defined this way has its origin at the 00:00 on "month 0 of the year 0." Of course, there is no month 0, nor year 0. This can be seen as a definition that is purely mathematical for convenience. The year after 1 BC (before Christ) or BCE (before common era) is AD 1 (here, AD is Anno Domini, or CE, which is common era) simply by definition (so there is no AD 0 between 1 BC and AD 1 in the Julian or Gregorian calendar). Therefore, year 0 is actually 1 BC. This definition is equivalent to saying that the MATLAB time origin is 0 o'clock on one day before the midnight (0 o'clock) of Jan. 1 of 1 BC.

The function `datestr` will convert the above number for time back to a time vector:

```
>>datestr(7.379274836500463e+005)
ans =
16-May-2020 11:36:27
```

To verify the time origin, one can type this:

```
>> datestr(0)
```

and get this

```
ans =
00-Jan-0000
```

Obviously, there is no date like this (there is no definition of Jan. 0).

The function `datevec` returns a time vector given a single time value or a time string. For example,

```
>>datevec(7.379274836500463e+005)
ans =
  1.0e+003 *
  Columns 1 through 4
   2.020000000000000  0.005000000000000  0.016000000000000  0.011000000000000
  Columns 5 through 6
   0.036000000000000   0.027363998413086
```

Or for a better print out:

```
>>fix(datevec(7.379274836500463e+005))
ans =
       2020           5          16          11          36          27
```

There is another commonly used function, `datenum`, which is quite flexible and can be used in different formats. For example, if we enter the following lines into the MATLAB command window:

```
datenum(2020,5,16,11,36,27)
datenum('16.05.2020','dd.mm.yyyy')
datenum('16-May-2020')
```

we will get

```
ans =
    7.379274836458333e+005
ans =
      737927
ans =
      737927
```

The first command includes the information for hour, minute, and second, but the last two commands do not (so the outputs are the same integers).

3.3 Caution on Using Local Time

If we go to the U.S. Geological Survey (USGS) webpage and download some time series data for at least one year, we will notice that there are two times each year when the time stamp for the time series data either changes from standard time (ST) to daylight time (DT) or from DT back to ST. This should be taken into account carefully when doing data analysis at least two times in a year. To illustrate this, it is better to use an example. The following is a segment of time series data from the USGS water data webpage: (https://maps.waterdata.usgs.gov/mapper/index.html):

```
USGS  293306092181800  2019-11-03 00:06  CDT  1.21  P
USGS  293306092181800  2019-11-03 00:12  CDT  1.28  P
USGS  293306092181800  2019-11-03 00:18  CDT  1.16  P
USGS  293306092181800  2019-11-03 00:30  CDT  0.83  P
USGS  293306092181800  2019-11-03 00:36  CDT  0.94  P
USGS  293306092181800  2019-11-03 00:42  CDT  1.04  P
USGS  293306092181800  2019-11-03 00:48  CDT  1.00  P
USGS  293306092181800  2019-11-03 00:54  CDT  0.96  P
USGS  293306092181800  2019-11-03 01:00  CDT  0.96  P
USGS  293306092181800  2019-11-03 01:00  CST  0.98  P
USGS  293306092181800  2019-11-03 01:06  CDT  0.88  P
USGS  293306092181800  2019-11-03 01:06  CST  0.91  P
USGS  293306092181800  2019-11-03 01:12  CST  0.81  P
USGS  293306092181800  2019-11-03 01:18  CDT  0.80  P
USGS  293306092181800  2019-11-03 01:18  CST  0.84  P
USGS  293306092181800  2019-11-03 01:24  CDT  0.85  P
USGS  293306092181800  2019-11-03 01:24  CST  0.90  P
USGS  293306092181800  2019-11-03 01:30  CST  0.97  P
USGS  293306092181800  2019-11-03 01:36  CDT  0.93  P
USGS  293306092181800  2019-11-03 01:36  CST  1.00  P
USGS  293306092181800  2019-11-03 01:42  CDT  0.90  P
USGS  293306092181800  2019-11-03 01:42  CST  1.00  P
```

```
USGS  293306092181800  2019-11-03 01:48  CDT  0.85  P
USGS  293306092181800  2019-11-03 01:48  CST  0.84  P
USGS  293306092181800  2019-11-03 01:54  CDT  0.95  P
USGS  293306092181800  2019-11-03 01:54  CST  0.71  P
USGS  293306092181800  2019-11-03 02:00  CST  0.66  P
USGS  293306092181800  2019-11-03 02:12  CST  0.88  P
```

What is shown here is from an actual data file downloaded from the above webpage. The first column (USGS) needs no explanation. The second column (293306092181800) is the station number. These two columns do not change for a fixed station. The next column is a variable (e.g. 2019-11-03 00:06) as the time stamp. This time stamp is not complete without the next column (CDT or CST). Here, CDT is Central Daylight Time and CST is Central Standard Time. The next column is also a variable (e.g. 1.21) and is the gage height or water level in feet. The last column (P) indicates it is provisional data subject to revision (insignificant to the discussion here). It can be seen that from the line with a time stamp of "2019-11-03 01:00 CDT" until the line with a time stamp of "2019-11-03 02:00 CST," the data are "messed up": the records alternate between CDT and CST a few times. This is probably because the computer program saving the data file lacks a little sophistication dealing with the time change: every year in the U.S., the civil time (for each time zone) is changed from the standard time (ST) to daylight time (DT) at 2:00 AM on the second Sunday in March and from DT back to ST at 2:00 AM on the first Sunday in November. This is equivalent to having a "leap hour" in March and a negative "leap hour" in November. Since 1:00 AM CST is really the same as 2:00 AM CDT, we need to rearrange the data as the following.

```
USGS  293306092181800  2019-11-03 00:06  CDT  1.21  P
USGS  293306092181800  2019-11-03 00:12  CDT  1.28  P
USGS  293306092181800  2019-11-03 00:18  CDT  1.16  P
USGS  293306092181800  2019-11-03 00:30  CDT  0.83  P
USGS  293306092181800  2019-11-03 00:36  CDT  0.94  P
USGS  293306092181800  2019-11-03 00:42  CDT  1.04  P
USGS  293306092181800  2019-11-03 00:48  CDT  1.00  P
USGS  293306092181800  2019-11-03 00:54  CDT  0.96  P
USGS  293306092181800  2019-11-03 01:00  CDT  0.96  P
USGS  293306092181800  2019-11-03 01:06  CDT  0.88  P
USGS  293306092181800  2019-11-03 01:18  CDT  0.80  P
USGS  293306092181800  2019-11-03 01:24  CDT  0.85  P
USGS  293306092181800  2019-11-03 01:36  CDT  0.93  P
USGS  293306092181800  2019-11-03 01:42  CDT  0.90  P
USGS  293306092181800  2019-11-03 01:48  CDT  0.85  P
USGS  293306092181800  2019-11-03 01:54  CDT  0.95  P
USGS  293306092181800  2019-11-03 01:00  CST  0.98  P
USGS  293306092181800  2019-11-03 01:06  CST  0.91  P
USGS  293306092181800  2019-11-03 01:12  CST  0.81  P
USGS  293306092181800  2019-11-03 01:18  CST  0.84  P
```

```
USGS  293306092181800  2019-11-03 01:24  CST  0.90  P
USGS  293306092181800  2019-11-03 01:30  CST  0.97  P
USGS  293306092181800  2019-11-03 01:36  CST  1.00  P
USGS  293306092181800  2019-11-03 01:42  CST  1.00  P
USGS  293306092181800  2019-11-03 01:48  CST  0.84  P
USGS  293306092181800  2019-11-03 01:54  CST  0.71  P
USGS  293306092181800  2019-11-03 02:00  CST  0.66  P
USGS  293306092181800  2019-11-03 02:12  CST  0.88  P
```

The above reordered data can then be used for further analysis. We still cannot directly use the hour values because after 1:54 it dropped back to 1:00. It is in fact 2:00 if CDT were used. The best thing to do, as a common practice, is to convert all the time, whether in CDT or CST, to the world standard time (the *Greenwich Mean Time* or *GMT* – more on this later).

In this example, the time interval is 6 minutes, but the data have some dropouts, so sometimes the time interval is 12 minutes or longer. Imagine if the time interval were 1 minute; there would be a lot of changes to make to reorder. Of course, one can write a computer script to do the job instead of doing it manually, which is subject to mistakes.

The following is another example for data with time change from ST to DT at 2:00 AM on the second Sunday of March.

```
USGS  07380249  2019-03-10 00:45  CST  2.57  A
USGS  07380249  2019-03-10 01:00  CST  2.61  A
USGS  07380249  2019-03-10 01:15  CST  2.62  A
USGS  07380249  2019-03-10 01:30  CST  2.62  A
USGS  07380249  2019-03-10 01:45  CST  2.64  A
USGS  07380249  2019-03-10 03:00  CDT  2.65  A
USGS  07380249  2019-03-10 03:15  CDT  2.64  A
USGS  07380249  2019-03-10 03:30  CDT  2.64  A
USGS  07380249  2019-03-10 03:45  CDT  2.64  A
```

Here the last column, “A,” stands for approved data (insignificant to the discussion here). The time interval of the time series is 15 minutes. We can see that the order of the data is fine – 3:00 CDT is 1 hour after 1:00 CST – but to make use of the data, we need to use the same time (e.g. GMT, Section 3.4).

The point is that when we deal with time series data, we have to make sure that the time is correct, whether we are using the GMT or local time. Otherwise, we could mess up with time, subsequent analysis, and interpretation of the data.

3.4 Solar Time, Mean Solar Time, and Greenwich Mean Time (GMT)

It is common sense that the civil time we use on a daily basis should be based on or consistent with the movement of the Sun. This is the same as it was in ancient times for agricultural activities and everyday life. The position of the Sun is used to define the local noon, which is when the Sun passes across the *local meridian*. A day is then divided into 24 hours, 1 hour into 60 minutes, and 1 minute into 60 seconds. Time defined in this way is called the *apparent* (or *true*) *solar time* or just solar time.

Solar time, however, is not uniform during a full year, and so it cannot be used for accurate time recording, nor can it be used for oceanographic observations. A day defined by just the visual motion of the Sun is called a *solar day*. In fact, during the ~365 days in a year, a solar day (or 24-hour solar time) would be 20 seconds shorter or 30 seconds longer than its annual mean. Because solar time is not uniform, it is not suitable for scientific research, though it may be convenient for everyday life, particularly in ancient times when there was no need for precision time measurements. The mean position of the Sun within a year, however, is quite robust. We can imagine a fictitious "dynamical mean sun" that is invisible, moving at a steady speed in the sky, slightly different from the real visual Sun. This gives the *mean solar time* (*MST*).

The difference between the apparent solar time and mean solar time is described by the Equation of Time (Milne, 1921). The accumulated error of the apparent solar time can cause a maximum difference with the mean solar time by 16 minutes within a year. The irregularity of the apparent solar time is because of several astronomical influences. These mainly include: (1) the Earth's orbit is not a perfect circle, but an ellipse, and (2) the Earth's axis of rotation is not perpendicular to the orbital plane. By the momentum conservation, the Earth's orbital speed varies with time in the year – when the Earth is closer to the Sun, it moves faster than when it is farther away from the Sun. That results in the change in speed with which the apparent Sun moves against the background stars, which leads to a change in the length of an apparent solar day. The tilt of the Earth's axis of rotation has random and small variations which can also result in a change in the length of apparent solar day in a year.

Greenwich Mean Time or GMT is the MST at the Royal Observatory in Greenwich, London. GMT is also called *Universal Time* or UT, or *Zulu Time*. The GPS data we receive nowadays often have a stamp Z at the end of each GPS sentence or record. GMT, Zulu Time, or UT is also called UT0. This is because it is still not quite regular (Wahr, 1988), and a corrected version is named UT1. This correction is done based on observations of the polar wandering, a phenomenon of a slight change (an irregular oscillation) with time of the Earth's axis of rotation because the Earth is not a truly rigid body. A further correction is made to UT1 to yield UT2 by considering the seasonal change of rotation speed of the Earth.

3.5 Stellar or Sidereal Time

Sidereal time is based on the movement of the stars, rather than that of the Sun. If we choose a far-away star and assume that it never moves, we can define how long a 24-hour period is in sidereal time. From the time the chosen star passes a local meridian to the time it passes the same meridian again, we could define the length of 24 hours in sidereal time (a sidereal day), which is also called a *stellar day*.

The assumption that the chosen distant star does not move can be problematic, as any star moves. In general, the further away a star is, the smaller its movement will appear to be within a given amount of time. At first sight, the sidereal time might be more regular than the solar time. However, sidereal time also can have irregularities because of the motion of the distant stars, albeit small. In addition, sidereal time is not quite consistent with civil time because of the Earth's orbital motion (see below).

3.6 The Difference between Stellar and Solar Times

Stellar/sidereal time is different from solar time. The Earth is moving around the Sun at a speed of ~365.2425 days for a full circle, and at the same time the Earth is rotating in space.

Fig. 3.1 is a view looking down from above the north pole. From this view, the Earth is rotating counterclockwise. It also happens to be orbiting in the same direction around the Sun. In other words, the Earth's rotation around its own axis and Earth's orbital movement around the Sun are all in the same direction.

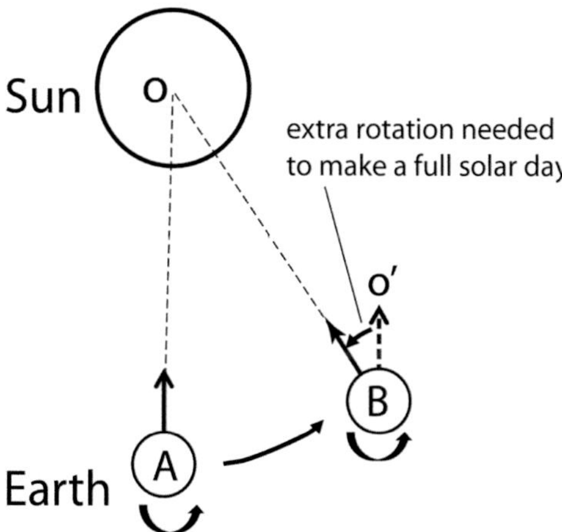

Figure 3.1 The difference between Solar and sidereal days.

Assume that an observer is standing at point *A* when the Sun is at the highest position from the observer's point of view (i.e. at the local noon time or when the Sun is crossing the local meridian). As the Earth is rotating around its own axis, it is also orbiting around the Sun. After the Earth finishes rotating in space for 360 degrees, it is one sidereal day later. But the Earth has already moved around the Sun for some distance, or roughly 1 degree. In order to get to the next noon or when the Sun crosses the local meridian again, the Earth has to rotate an extra angle of about 1 degree. From this, we can see that a solar day is longer than a corresponding stellar day. So, if we use distant stars as a reference, rotation in 360 degrees or a stellar day is a little bit shorter than a mean solar day: about 4 minutes shorter. As a result, one mean solar day or 24 hours in civil time is the time for the Earth to rotate in space for about 361 degrees.

3.7 Coordinated Universal Time (UTC)

The drawback of the conventional GMT is that it is irregular as well. The next update of the definition of time came in the 1960s with *Coordinated Universal Time* (*UTC*), using atomic oscillations to define precisely the length of a second. One might think that since UTC is using atomic time, it should be uniform and solve the irregularity problem once and for all. However, the irregularity of the length of a day will result in an accumulated "error" over time in terms of aligning the time with the visual solar movement when a fixed length of a second is used. In addition, because of the gradual slowing down of the Earth's rotation due to tidal friction, the length of a day is increasing over time.

It is found that the length of the mean solar day has been increasing at a rate of 1.70 ± 0.05 ms per century on average over the last 2700 years, between 700 BC and AD 1990 (Stephenson and Morrison, 1995). Tidal friction dissipates the kinetic energy of the rotating Earth, which causes the Earth to slow down its rate of rotation. There is another nontidal factor that can contribute to either negative or positive acceleration of the Earth's rotation due to the change of mass distribution on the Earth (e.g. iceberg coverage variations and the Earth's mantle motion). At present, the nontidal factor is accelerating the rate of the Earth's rotation and countering the effect of tidal friction, but only at a rate of ~26% of the tidal frictional effect: the tidal frictional effect dominates, and so the Earth's rotation has been slowing down progressively as a net outcome.

As a result, a correction is sometimes made to adjust the time using a leap second to make civil time and UTC "compatible." Otherwise, over a long period of time, our civil time would not match UTC; for example, 12 noon might not be when the Sun is crossing the local meridian but rather in the morning or some other time. Imagine that the definition of noon would gradually shift in time; that would be inconvenient.

The idea of UTC is that we fix the length of 1 second, and all units smaller than 1 second (e.g. millisecond) have precise, unchangeable length defined by atomic oscillations. The length of a second will not be allowed to change with time. All the larger units of time, i.e., minute, hour, day, ... , can have a "slight" change in length by the introduction of a "leap second." This leap second cannot be predicted far ahead because of the irregularity of the rate of Earth's rotation. The correction has to rely on the measurements. This leap second occurs rarely, and most of minutes are still 60 seconds each.

However, the use of the leap second is still being debated. Sometimes, interruptions of apparatus or mistakes in data recording occur. An experienced GPS user may notice a problem using some of the old GPS receivers, which sometimes produce data that are messed up due to the leap second, especially if the data are saved at 1 second or shorter intervals. This is similar to the problem of the "leap hour" at 2:00 AM on the second Sunday of every March in the U.S., when the time is changed between standard time and daylight time.

3.8 The Angular Speed of the Earth's Rotation

We now examine the *Earth's angular speed of rotation.* The Earth rotates 360 degrees in less than 24 hours, so it would not be correct to use 24 hours (one solar day) as the amount of time the Earth rotates in space in a full circle. The fact is that the Earth rotates for a complete revolution in space (one sidereal day) in about 23 hours, 56 minutes, and 4.1 seconds (Fig. 3.1). The angular speed is then:

$$\omega = \frac{2\pi}{23 \text{ hr } 56 \text{ min } 4 \text{ s}} = 7.29 \times 10^{-5}$$

The unit is rad/s^{-1}. If one used 24 hours for the time of a full revolution of the Earth, the results would have been 7.27×10^{-5} s^{-1}, which may appear to be just a small difference, but conceptually, this is not correct. This mistake is occasionally seen in research articles and textbooks.

3.9 Julian Day Number

Because the length of the year is not really an integer, it is a little complicated to calculate time in continuous days. *Julian Day number* (JDN) is defined for that purpose. The question is which date to choose as the "first" day or *origin* from which the day starts to count from 0. The time should be a real number, i.e. it allows 1.5 day, 2.01 day, etc. It should also allow time before that origin which will be negative values.

The original Julian Day definition is a time of 0 at the *mean noon* of *Jan. 1, 4713 BC*. It is indeed a long time ago. Most oceanographers do not work on data of that long or old. It would be very inconvenient to use this as the time origin for present-day data, even though this does not matter as far as data analysis is concerned because all calculations are done by computers.

A modified Julian Day is defined at midnight of Nov. 17, 1858, or 2,400,000.5 (Julian Day time). In MATLAB, however, the very useful internal function `datenum` does not use any of these definitions. Like the MATLAB function `now` discussed earlier, the time origin for this function is also defined at 00:00, the "0th month of the year 0":

```
>> datenum(0,0,0,0,0,0)
ans =
     0
```

Note that here the "0th month" and the first month give the same result:

```
>> datenum(0,1,0,0,0,0)
ans =
     0
```

Or, equivalently, day 1 is defined at 00:00, Jan. 1 of the "year 0" (which is 1 BC):

```
>> datenum(0,1,1,0,0,0)
ans =
     1
```

We learned earlier that if one provides values for the hour, minute, and second of a day, e.g. 12:23:12, and the date, e.g. Mar. 3, 1923, the time value in days from 00:00:00, Jan. 1 of the year 0 can be calculated by:

```
>> datenum(1923,3,3,12,23,12)
ans =
  7.0242e+005
```

In addition, it should also be noted that the input values are fairly flexible, to the extent that it may provide confusing results. For example,

```
>> datenum(0,12,1,0,0,0)
ans =
     336
>> datenum(0,12,0,0,0,0)
ans =
     335
```

Obviously, there is no day 0 in December. This is equivalent to November 30:

```
>> datenum(0,11,30,0,0,0)
ans =
      335
```

Now compare the next two commands:

```
>> datenum(0,12,-10,0,0,0)
ans =
   325
>> datenum(0,11,20,0,0,0)
ans =
   325
```

Sometimes, invalid data are not always automatically excluded by a program, as in the case with this `datenum` function. Under this situation, it is the user's responsibility to use valid date vectors (year, month, day, hour, minute, second) with the function. Alternatively, a user-defined function can be written for validation of the date vectors before applying to the `datenum` function, which is left as an exercise at the end of the chapter.

The `datenum` function can also be used with an array of time stamps. Assume that we have an array of time stamps named *tstamp*, or we input:

```
>> tstamp = [ 2020 5 16 10 22 0; ...
             2020 5 16 10 23 0; ...
             2020 5 16 10 24 0; ...
             2020 5 16 10 25 0; ...
             2020 5 16 10 26 0; ...
             2020 5 16 10 27 0; ...
             2020 5 16 10 28 0; ...
             2020 5 16 10 29 0; ...
             2020 5 16 10 30 0]
```

We can use the following simple command to convert the array of time stamps into an array of real values for time:

```
format long
>> datenum(tstamps)
ans =
  1.0e+005 *
   7.379274319444444
   7.379274326388890
   7.379274333333333
   7.379274340277777
   7.379274347222223
```

```
7.379274354166666
7.379274361111111
7.379274368055556
7.379274375000000
```

3.10 Julian Century

A *Julian Century* is defined as exactly 36525 days. In reality, a century is not exactly this many days. There is a fraction: since one year is approximately 365.2425 days, 100 years is about 36524.25 days. The Julian Century is defined for convenience of having a unit with an integer number of days that is roughly 100 years.

In some applications of data analysis, particularly if a dataset has a time period within a year (it does not go from one year to the next or cross Dec. 31 of one year into Jan. 1 of the next), it is often convenient to use Jan. 1 of that year as the time origin. For that purpose, as an example, we do the following, assuming that the data have time stamps within the year 2018:

```
>> timeindays = datenum(year,month,day,hour,minute,second) - datenum
(2018,1,1);
```

assuming that the variables year, month, day, hour, minute, and second are those suggested by their names, respectively. In oceanography, such as in tidal analysis, we sometimes use 00:00:00, Jan. 1, 1900, as the time origin, in which case we can use the following command to define our time:

```
>> timeindays = datenum(year,month,day,hour,minute,second) - datenum
(1900,1,1);
```

Of course, for many applications, we can use any other time origin without changing the results. It should be noted that for tidal harmonic analysis, the change in time origin will result in a change in the phase of each tidal constituent. What is significant in dynamics is not an individual value of phase but relative phase differences among different tidal constituents and between elevation and velocity. These phase differences do not change with the change of time origin.

3.11 Time and Longitude

Because the Earth's surface is roughly a sphere and the angular speed of the Earth is fairly constant, it is obvious that in a given time period the Earth rotates the same amount in longitude. Since every 24 hours the Earth rotates 360 degrees relative to

Table 3.1. *Conversion between longitude and time*

Longitude ~ time	Time ~ longitude
360° ~ 24 hours	
1° ~ 4 minute in time	1 hour ~ 15°
1′ ~ 4 second in time	1 minute ~ 15′
1″ ~ 1/15 second in time	1 second ~ 15″

the mean Sun (or the fictitious mean solar position), each 15 degrees of longitude is equivalent to 1 hour in time. Thus, 1 degree in longitude is equivalent to 4 minutes in time; 1 minute in longitude is equivalent to 4 seconds in time; and 1 second in longitude is equivalent to 1/15 seconds in time (Table 3.1).

While Greenwich, England, has the origin of longitude (0° longitude), a line at roughly 180 degrees from the 0-degree longitude is defined as the *international date line*. This is a line defined in 1884 to separate two consecutive dates. When one crosses to the west of this line, the date is added by 1, and when one crosses to the east of the line, the date is subtracted by 1. The international date line happens to be in the middle of the Pacific Ocean. It is a line not exactly at 180 degrees longitude but runs from the north pole to south pole roughly along this longitude, zigzagging through the islands of Russia and the Aleutian Islands.

In comparison, latitude is used not to measure time, but rather distance. One degree in latitude is roughly 60 nautical miles or ~111 km, and so 1′ in latitude is 1 nautical mile. This conversion is not accurate but is good enough for a quick estimate in many applications. This is because the Earth is not a perfect sphere. It is more like an ellipsoid; the Earth's radius varies from ~6378 km at the equator to ~6357 km at the poles, with an average radius of ~6371 km. The error is thus ~0.3%. For better accuracy, the ellipsoidal effect (e.g. Vermeille, 2002) should be included in the calculation of distance, which is not discussed in this book.

3.12 GPS Data

As we already discussed, oceanographic data should always come with time stamps. They should also have position stamps: we need to know where the data are obtained. Present-day position measurements are often quite straightforward using Global Positioning System (US Government, 1996) satellite receivers. GPS provides worldwide public service for accurate *positioning, navigation, and timing* (PNT). The main part of the GPS consists of a constellation of at least 24 satellites circling the Earth twice a day at an altitude of ~20K km. There are six equally spaced orbital planes designated around the Earth at 20K km height, specifically for these satellites. This allows four satellites on each orbital plane. The number of satellites in orbit is usually more than 24 for enhanced performance.

The GPS satellites transmit continuous signals for their accurate positions and time. The user of a GPS signal should have a receiver, which is an electronic device available from the market. The GPS receiver receives the GPS satellite signals and calculates the three-dimensional position and time of the receiver. The GPS receiver will then output lines in formatted simple text showing the position and time at specified intervals. Though GPS units outputting at 5 Hz or even 10 Hz are available, for most normal oceanographic and environmental applications, the time interval of 1 second or longer will generally suffice.

For oceanographic observations, often the output of data measured from the sensor should be merged with the GPS information for time and position stamps. This can be done internally but many instruments need to have their data merged after the output of data. The output of a GPS unit is a series of lines of text. The following lines are from a GPS unit.

```
$GPRMC,030123.000,A,3023.0815,N,09103.6838,W,0.10,28.49,060112,,,A*4D
$GPRMC,030124.000,A,3023.0813,N,09103.6839,W,0.03,357.75,060112,,,A*7B
$GPRMC,030125.000,A,3023.0814,N,09103.6837,W,0.03,359.36,060112,,,A*7A
$GPRMC,030126.000,A,3023.0813,N,09103.6837,W,0.03,3.73,060112,,,A*73
$GPRMC,030127.000,A,3023.0813,N,09103.6836,W,0.04,333.44,060112,,,A*70
$GPRMC,030128.000,A,3023.0812,N,09103.6835,W,0.03,4.47,060112,,,A*7E
$GPRMC,030129.000,A,3023.0811,N,09103.6836,W,0.04,356.33,060112,,,A*7F
```

These are the *NMEA sentences.* Here NMEA stands for *National Marine Electronics Association* (SiRF Technology, 2005). The dollar sign at the beginning of the sentence indicates an NMEA sentence. The first two characters, GP, indicate that this line consists of Global Positioning System or GPS data. The characters that follow (RMC) indicate this line is the "recommended minimum data for GPS." The interpretation of the sentence is provided below.

```
       1          2 3       4 5       6 7  8   9   10  11 12
       |          | |       | |       | |  |   |   |   | |
$GPRMC,hhmmss.ss,A,llll.ll,a,yyyyy.yy,a,x.x,x.x,xxxx,x.x,a*hh
 1) Time (UTC)
 2) Status, A means valid data; V = Navigation receiver warning
 3) Latitude
 4) N or S (N is short for north; S is short for south)
 5) Longitude
 6) E or W (E is short for east; W is short for west)
 7) Speed over ground in knots
 8) Course over ground in degrees true
 9) Date, ddmmyy
```

```
10) Magnetic Variation, degrees E (east) or W (west)
11) Mode, A means autonomous; D means DGPS; E means DR (only applies to NMEA
version 2.3 and later)
12) Checksum
```

A typical GPS receiver module is available on the market for less than $30. Fig. 3.2 shows one such module used by the author to make a GPS recorder (Fig. 3.3). The module (Fig. 3.2) has an antenna (the light-colored square), a backup battery, an integrated circuit (IC), and four connectors: (1) 5-volt direct current (VDC); (2) ground; (3) reception; and (4) transmission.

A GPS module can be connected to a PC or integrated with a microcontroller for a standalone device (Fig. 3.3), and its output as NMEA sentences can be saved in the computer or on an SD card or be merged with a data string from a sensor through a com port.

With the ASCII file output in NMEA sentence, GPS data can be easily read, decoded, and analyzed. A free online tool can be used for a quick visualization of GPS data. The web address is www.gpsvisualizer.com/. It is a quite smart online app that can read the general NMEA sentence for GPS positions and plot the track of the GPS (e.g. from a boat or truck) on a web-based map.

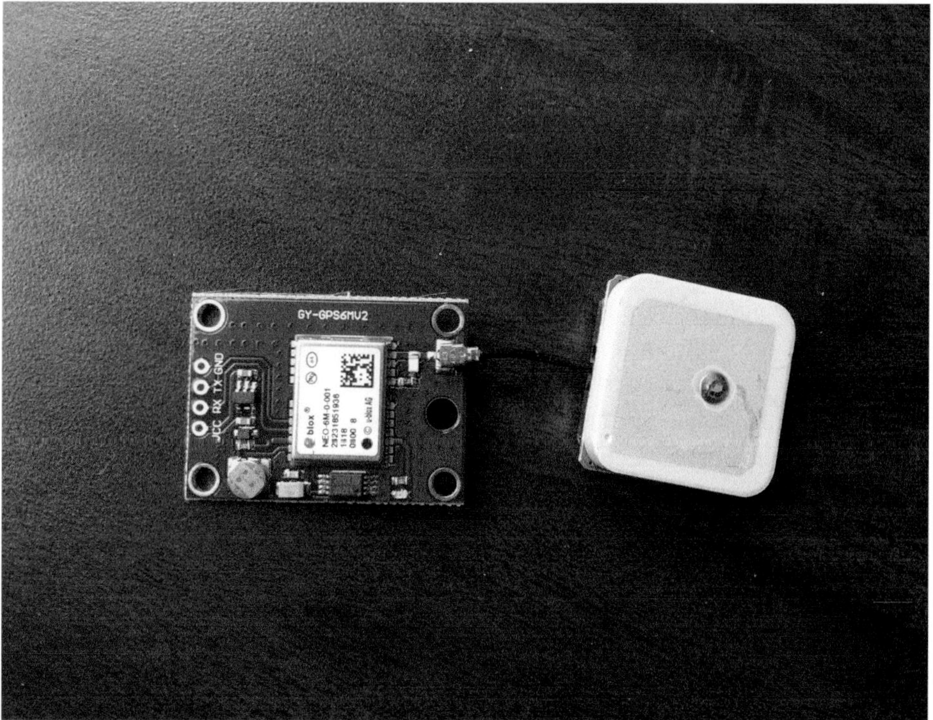

Figure 3.2 A GPS module.

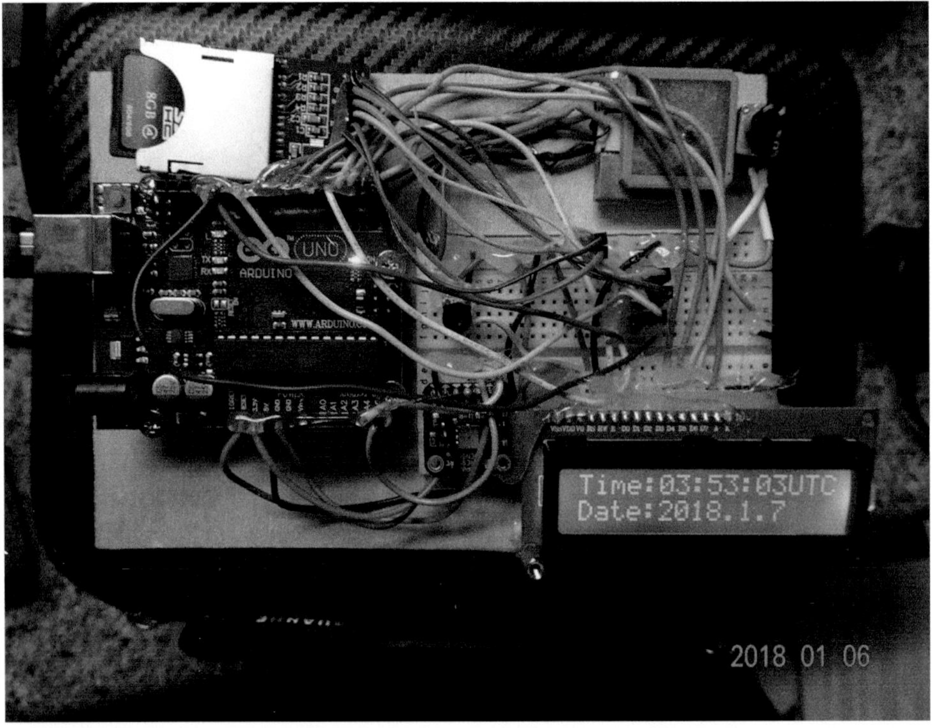

Figure 3.3 A homemade GPS receiver with display, SD card, and air pressure and temperature.

3.13 Interpolation

Time series data analysis often requires that the time intervals are constant, with some exceptions. In the harmonic analysis and a general least square method, time intervals are not required to be a constant. However, for Fourier analysis and filtering, the time intervals need to be a constant for a given time series. In real observations using different sensors, it is often programmed to sample at a constant rate. But sometimes a malfunction of sensors or instrument failure can occur due to various reasons. There may be a power problem or severe weather impact. As a result, it is common for real time series data have some gaps. If there are too many gaps, the data may not be usable. On the other hand, if there are only a few gaps, an interpolation should be done before any further analysis.

In MATLAB, we can use existing interpolation functions rather than building our own functions from scratch. Here we briefly discuss an example with some relevant MATLAB functions. The relevant MATLAB functions include `diff`, `mode`, `[ ]`, and `interp1`.

The `diff` function computes the first order difference of a given array by subtracting an element from its previous value, i.e. `z = diff(y)` means `z(i − 1) = y(i) − y(i − 1)`, with `i = 2,3, ... ,n`, assuming that the length of `y` is `n`. The resultant difference (`z`) is an array of length `n − 1`.

The `mode` function gets the value in an array with the most frequent occurrence. The pair of square brackets `[ ]` with nothing inside is used to delete element(s) of given array. The function `interp1` is a one-dimensional interpolation function allowing several options (different methods of interpolation). The usage is in the form of

```
yi = interp1(x,y,xi,METHOD)
```

in which METHOD can be any of the following:

- `'linear'` – is the default if not specified, using a linear interpolation.
- `'nearest'` – using the nearest neighbor value for the interpolation.
- `'next'` – using the next value for the interpolation.
- `'previous'` – using the previous value for the interpolation.
- `'spline'` – using the so-called piecewise cubic spline interpolation for smoothness.
- `'pchip'` – using a shape-preserving piecewise cubic interpolation.
- `'v5cubic'` – using a cubic interpolation from MATLAB 5, which does not extrapolate and uses `'spline'` – if the array is not equally spaced.
- `'akima'` – using a special method called modified Akima cubic interpolation.

```
%=====================================
% EXAMPLE - INTERPOLATION
%=====================================
%===============================
% DEFINE A TIME SERIES
%===============================

t = 1:100;          % an example time

x = exp(0.01 * t);  % an example time series
%===============================
% PLOT THE TIME SERIES
%===============================
figure; subplot(2,1,1) % Figure with 2 x 1 panels, 1st panel
plot(t,x)              % plot the time series
%===============================
% CALCULATE THE TIME INTERVALS
%===============================
dt = diff(t);        % calculate the time interval
%===============================
% PLOT dt
```

```
%===============================
subplot(2,1,2); plot(dt)  % show time interval is constant
%===================================================
% TAKE OUT A FEW DATA POINTS TO SIMULATE DATA GAPS
%===================================================
x([ 3 12 31 52 82 91]) =[];  % square brackets with nothing in
t([ 3 12 31 52 82 91]) =[];  % means to delete the element
%===============================
% PLOT THE GAPPY TIME SERIES
%===============================
figure; subplot(2,1,1)
plot(t,x,'o'); hold on; plot(t,x,'r')
%===============================
% CALCULATE THE TIME INTERVALS
%===============================
dt = diff(t); % check the time interval again
%===============================
% PLOT dt
%===============================
subplot(2,1,2); plot(dt)  % show the data gap
%==================================
% NOW TRY TO DO INTERPOLATION TO
% FILL THE GAPS OF THE TIME SERIES
%==================================
%==================================
% DEFINE A CONSTANT dt
%==================================
dt = mode(diff(t)); % mode function takes the most frequent
                    % occurrence of value
%==================================
% FIND THE LENGTH OF TIME
%==================================
T = t(end)-t(1); % the parameter end inside the parentheses
                 % accesses the last element of the array
%===================================
% FIND THE NUMBER OF TIME INTERVALS
%===================================
N = fix(T / dt);
%==========================================
% DEFINE NEW TIME WITH CONSTANT INTERVALS
%==========================================
t1 = t(1) + dt * (0:N - 1);
%===================================
% DO INTERPOLATIONS
%===================================
x1 = interp1(t,x,t1);
%===============================
% PLOT THE TIME SERIES
%===============================
figure; subplot(2,1,1); plot(t1,x1,'o')
```

```
hold on
%===============================
% PLOT dt - verify constant dt
%===============================
subplot(2,1,2); plot(t1(1:end - 1),diff(t1),'r')
```

3.14 Elapsed Time Computation

MATLAB has several other functions useful for time inquiry and elapsed time computations. The following provides more such time-related functions.

```
>>date
```

would give the present time in a string format, e.g.

```
>>08-Sep-2020
```

while

```
>>datetime
```

gives not only the present date but also present time information in a time string format, e.g.

```
>>08-Sep-2020 21:06:56
```

and

```
>>now
```

gives the present time as a real number, similar to what the `datenum` function provides, e.g.

```
ans =
  7.3804e+005
```

The `clock` commands would give a time vector, e.g.

```
>>format rat; clock
ans =
  Columns 1 through 4
    2020              9              25              11
  Columns 5 through 6
      18 4567/1000
```

`clock` can also be used as a function with a flag of daylight time, e.g.

```
[ T flag] = clock  % FLAG SHOWS IF DAYLIGHT SAVING TIME
                   % (TRUE OR FALSE)
T =
  1.0e+003 *
    2.0200    0.0090    0.0080    0.0200    0.0450    0.0527
flag =
 logical
 1
```

The output value of 1 for `flag` indicates that the time was daylight saving time. The following combination of commands can also be used to provide desired time data:

```
>>datestr(now)
ans =
08-Sep-2020 20:45:01

>>datevec(now)
ans =
  1.0e+003 *
    2.0200    0.0090    0.0080    0.0200    0.0450    0.0527

>>datenum(now)
ans =
  7.3804e+005 *
```

The following provides some examples of computation of elapsed time in MATLAB: method 1 uses the `tic` and `toc` pair; method 2 uses the function `etime` twice; method 3 uses the function `cputime` twice; method 4 uses the function `now` twice.

```
%=========================================
% ELAPSED TIME FUNCTION PAIRS
%=========================================
% METHOD #1
%=========================================
tic;                   % START TIME
for i = 1:100000;      % BLOCK OF COMPUTATIONS
for j = 1:100          % BLOCK OF COMPUTATIONS
9 * 9^0.009999;        % BLOCK OF COMPUTATIONS
end                    % BLOCK OF COMPUTATIONS
end                    % BLOCK OF COMPUTATIONS
toc                    % END TIME
%% Elapsed time is 1.691390 seconds.

%=========================================
% METHOD #2
%=========================================
etime(clock,[ 2020,1,1,0,0,0] )   % etime(t2,t1) gives elapsed
                                  % time between t2 and t1
```

```
ans =
  2.1762e+007 *% IN SECONDS
etime(clock,[ 2020,1,1,0,0,0] ) / 24 / 3600
ans =
  251.8734  % IN DAYS

%==========================================
% METHOD #3
%==========================================
t1 = cputime         % START TIME
for i = 1:100000;    % BLOCK OF COMPUTATIONS
for j = 1:100        % BLOCK OF COMPUTATIONS
9 * 9^0.009999;      % BLOCK OF COMPUTATIONS
end                  % BLOCK OF COMPUTATIONS
end                  % BLOCK OF COMPUTATIONS
t2 = cputime         % END TIME
dt = t2 - t1;        % ELAPSED TIME
dt                   % IN SECONDS

%==========================================
% METHOD #4
%==========================================
t1 = now;
for i = 1:100000;    % BLOCK OF COMPUTATIONS
for j = 1:100        % BLOCK OF COMPUTATIONS
9 * 9^0.009999;      % BLOCK OF COMPUTATIONS
end                  % BLOCK OF COMPUTATIONS
end                  % BLOCK OF COMPUTATIONS
t2 = now;
dt = t2 - t1;        % ELAPSED TIME (IN DAYS)
dt * 24 * 3600       % IN SECONDS
```

In addition to the above time related functions, MATLAB has several other functions useful for tick mark and tick format when making plots, for example, `datetick`, `xtickformat`, `ytickformat`, `ztickformat`, `thetatickformat`, and `rtickformat`.

Selected MATLAB Commands in This Chapter

`clock` – a function that returns the six number date vector (for the year, month, date, hour, minute, and second) of the present time.

`date` – a function that returns the current date string.

`datenum` – This is a very useful function that converts a time string or time vector to a real number in days counting from 00:00, Dec. 31, the year before year 0 (when day 0 starts). In other words, day 1 is 00:00, Jan. 1 of the year 0: datenum(0,1,1,0,0,0) = 1 or datenum(0,1,1) = 1.

`datestr` – This function converts a real number representing time in days counted from day 1 from Jan. 1 of the year 0 to a formatted string of date, e.g. datestr(50.9) = 19-Feb-0000 21:36:00. This function has more capability than the above example. Check MATLAB help for the entire functionality.

`datevec` – converts a real number representing time in days to a six-number vector representing the year, month, date, hour, minute, and second, e.g. datevec(50.9) = [0 2 19 21 36 0].

`now` – a function that returns a real number for the present time.

Review Questions for Chapter 3

(1) What is the difference between GMT and UTC in concept? Is there any difference between them in terms of oceanographic applications?

(2) Why do oceanographers prefer using GMT or UTC, not local time? Can you think of any advantage using local time instead?

Exercises for Chapter 3

(1) Conversion from time vector to time variable using MATLAB.

(a) Hurricane Katrina made its landfall at 1100 UTC on Aug 29, 2005. What is the date number in the year 2005 (the number of days counted from Jan. 1, 2005)?

(b) Hurricane Rita made its landfall at 0740 UTC on Sept 24, 2005. What is the date number in the year 2005?

(2) Working with GPS data for GPS recorded ship track.
Data file named "HollyBeach_GPS.TXT" is a GPS data file. Use the online GPS visualization tool at www.gpsvisualizer.com/ to upload this data file and make a map online.

(3) Working with GPS data for GPS recorded vehicle track.
Data file named "AK20140724.gpx" is a Garmin GPS data file. Use the online GPS visualization tool at www.gpsvisualizer.com/ to upload this data file and make a map online.

(4) Angular speed of a hypothetical Earth.
If the Earth's orbital motion were in the opposite direction of the real orbital motion, but its spin around its own axis and everything else were the same, what would be the angular speed of this hypothetical Earth? Write a MATLAB script to make that calculation.

(5) Work with MATLAB's time functions.

Some programs do not provide an automatic check on data validity. Try to convert some time vectors to time values using `datenum`. Let us execute the following commands in MATLAB: (a) `datenum(2020,7,5)`, (b) `datenum(2020,7,−5)`, (c) `datenum(−100,7,5)`, (d) `datenum(−100,−7,5)`, (e) `datenum(2020,7,50)`. Based on the results, can you provide some discussion and interpretation? What did you learn from these results?

(6) Load the MATLAB data file "adcp_data.mat" and check the time interval before doing interpolations for the variables u1, v1, w1, and temperature.

(7) (a) Define hypothetical time: t = i $*$ $i^{1/10}$ (i = 1, 2, . . . , 100); (b) define hypothetical data at time t: = sin(t/10); (c) calculate the difference of t using MATLAB command `diff`; (d) plot the difference of t to demonstrate that the time intervals are not constant; (e) now interpolate the data Y on to regular grids xi = [1, 2, 3, . . . 158] to find yi.

(8) (a) Define hypothetical time: t = i (i = 1, 2, . . . , 100); (b) take out data points at i = 20, 21, 22 (Hint: use "`t(20:22)=[]`"); (c) take out data points at i = 37, 38, 39; (d) define Y at these t values using: Y = sin(t/100); (e) interpolate the data yi onto xi = [1, 2, . . . , 100].

(9) Use the four different methods shown in this chapter to determine how long it takes for MATLAB to do the following computations:

(a) Start from 0 and add 1 part per million (i.e. 0.000001 or 10^{-6}) for 10^6 times using a for-loop (`ds = 1 / 1000000; S = S + ds`).

(b) Then subtract 2 parts per million (i.e. 0.000001 or 10^{-6}) for 10^6 times using another for-loop. The final result is S; what is the value of S? How long does it take for each of the four methods?

(Method 1: using `tic` & `toc`)

(Method 2: using `clock` twice and then using `etime` once)

(Method 3: using `cputime` twice)

(Method 4: using `now` twice)

(10) Write a MATLAB function to validate date string or vector arrays so that bad date strings or vectors can be excluded, i.e. the month value cannot be less than 1 or greater than 12; the day value cannot be less than 1 and greater than 31, etc. It should return an error message if the date string or vector does not pass the function.

4
Deterministic and Random Functions

About Chapter 4

This chapter discusses some basic concepts quantifying the characteristics of functions with random fluctuations. Deterministic functions are discussed first in order to introduce functions with randomness and ways to quantify them. The concepts of phase space, ensemble mean, ergodic process, moments, covariance functions, and correlation functions are discussed briefly.

In much of the development for time series analysis, the fundamentals are based on theories of mathematical operations for deterministic functions. Unfortunately, natural processes often have some uncertainties, and data always have random errors. Here we briefly discuss some basic concepts for when we deal with the uncertainties of time series data.

4.1 Deterministic Functions

4.1.1 Periodic and Quasi-Periodic Functions

Deterministic functions are those with no uncertainty. As an example, any mathematical function expressed analytically (using a conventional mathematical formula) with a uniquely defined value at any given time is deterministic. Among these deterministic functions, the simplest are perhaps periodic functions, each having a certain period over which the function repeats itself. An example of deterministic function is one with a fundamental frequency plus its harmonics (with frequencies being multiples of the fundamental frequency).

Sometimes, a function is only roughly periodic or *quasi-periodic*. Strictly speaking, it is not exactly a periodic function. Such a function has some oscillations linearly superimposed together, but the frequencies are not

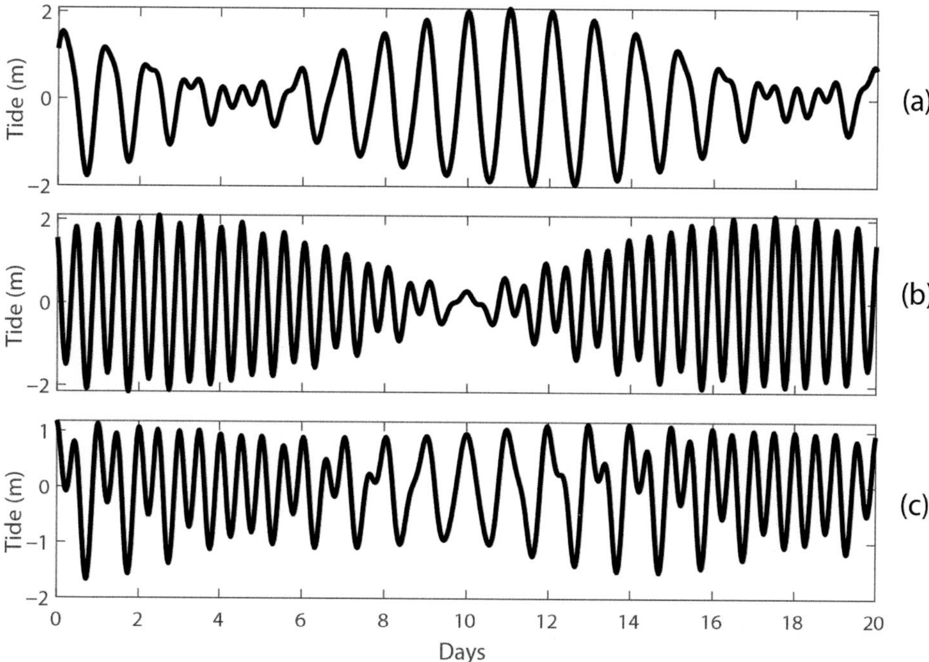

Figure 4.1 Quasi-periodic functions – different tides: (a) diurnal; (b) semi-diurnal; and (c) mixed tides.

commensurately related. A very good example is tidal signal. Tides often include multiple tidal constituents, each having a defined frequency; these can be roughly classified into *diurnal*, *semi-diurnal*, and *mixed tides* (Fig. 4.1). In most cases in coastal oceans and estuaries, there are some major tidal constituents that are dominant at a specific location such that it is considered either diurnal, semi-diurnal, or mixed type. A criterion often used by oceanographers to classify tides as one of these types is basically a ratio called tidal *form number* between the diurnal tidal amplitude and semi-diurnal tidal amplitude, for example:

$$F = \frac{K_1 + O_1}{M_2 + S_2}, \tag{4.1}$$

in which K_1 and O_1 are the amplitudes of two different lunar diurnal tides, while M_2 and S_2 are the amplitudes of principal lunar semi-diurnal tide and principal solar semi-diurnal tide, respectively. Slightly different form numbers can be found in literature, but for the one defined above, with a form number of 0.25 or less, tide is classified as semi-diurnal; if F is 3.0 or greater, it is classified as diurnal tide; otherwise, tide is of mixed type.

Most of the east coast of the United States has mainly semi-diurnal tides, while much of the northern Gulf of Mexico is dominated by either diurnal or mixed types. In any case, tides are *not* exactly periodic (Fig. 4.1) but are better called *quasi-periodic*.

4.1.2 Nonperiodic or Transient Functions

Another category of deterministic functions is nonperiodic or transient functions that do not have any obvious periodicity. These functions often have some trend or long-time scale variations in magnitude. A well-known example is the global atmospheric carbon dioxide (CO2) concentration, which has been increasing continuously since the mid-18th century from 280 to 415 ppm, which is the highest in more than 10 million years. This almost monotonic increase is not periodic and is considered an effect of anthropogenic activities, especially fossil fuel usage. Although the CO2 concentration is not exactly a deterministic function, the impact of human activity might have been an important factor in causing the overall "deterministic" trend of increase.

4.2 Random Process and Ergodic Process

4.2.1 Phase Space

Data or signals from the real world often have some characteristics of deterministic functions but also have some random fluctuations. As a result, observations can be considered as a function with certain intrinsic randomness, which may result from instrument error, measurement error, or simply the nature of dynamics such as chaos. The random functions can be classified into two main categories: *nonstationary random process* and *stationary random process*. For stationary random process, the statistical characteristics are not functions of time. Otherwise, it is nonstationary.

To help our discussion, we can use an example from statistical physics. In the microscopic world, individual atoms and molecules move rapidly in a random fashion. They also constantly interact with each other with frequent collisions and redistribution of momentum among them. Only a statistical mean of a quantity of them may yield a meaningful value. For example, the ocean temperature at one location is actually a statistical mean of a quantity proportional to the water molecular kinetic energy. Assume we have a system (e.g. a container) with N particles (e.g. atoms or molecules, such as water molecules in the ocean) with their positions and momentum denoted by $\mathbf{r}_i(t)$ and $\mathbf{p}_i(t)$, respectively, in which $i = 1, 2, \ldots, N$. These are all vectors, each having three components in a three-dimensional space. The state of the entire system can be described by these positions and momenta, or any quantity of the system H can be determined by these vectors:

$$H = H(\mathbf{r}_1, \mathbf{r}_2, \ldots, \mathbf{r}_N; \mathbf{p}_1, \mathbf{p}_2, \ldots \mathbf{p}_N). \tag{4.2}$$

The number N is extremely large for any macroscopic system. Mathematically, the quantify H is defined in a $6N$ dimensional "space," which is termed the *phase space*. At any instance, the system is a point in the phase space called *phase point*. As time advances, the system goes through different states and the phase point moves to form a *phase trajectory*. Because of the random motions of the microscopic particles, the state of the system depends on the phase trajectory and distribution of possible states in the phase space.

4.2.2 Ensemble Mean

Imagine there is a collection of *identical systems* as described above. Alternatively, we can consider that we have a collection of *identical experiments* with identical setup and conditions. Because of the random nature of motion in the microscopic world, even the systems and experiments are identical (in terms of initial setup and conditions, and their environmental or boundary conditions); the quantity H is not deterministic but has some random fluctuations. In reality, an ensemble of such identical systems is impossible. For instance, we do not have the luxury of a collection of Pacific Oceans. In this context, the word ensemble is useful only for concepts and thought experiments. However, this word is frequently used in observations for a collection of samples in *time* as well.

An ensemble mean of a given quantity $x(t)$ at a given time t is defined by an average over the entire collection of identical systems, a task impossible to realize in reality but which can be expressed mathematically below:

$$\mu_x(t) = \lim_{N\to\infty}\frac{1}{N}\sum_{k=1}^{N} x_k(t). \tag{4.3}$$

Here $x_k(t)$ is a measured value of the function $x(t)$ in the kth system or kth experiment at time t. Because of the randomness in function $x(t)$, the value $x_k(t)$ depends on the experiment or k. The ensemble mean is the mean across the systems or experiments and not over time – it is a "vertical" average, not a horizontal average; in Fig. 4.2, this is marked by vertical dashed lines that indicate the values have been averaged. The ensemble mean is also called the expected value $E[x(t)]$, which can be expressed using a *probability density function*:

$$\mu_x = E[x(t)] = \int_{-\infty}^{+\infty} xp\,dx, \tag{4.4}$$

in which ρ is the probability density function of the random variable x. Here time t is an implicit parameter. In the phase space case discussed above, the ensemble

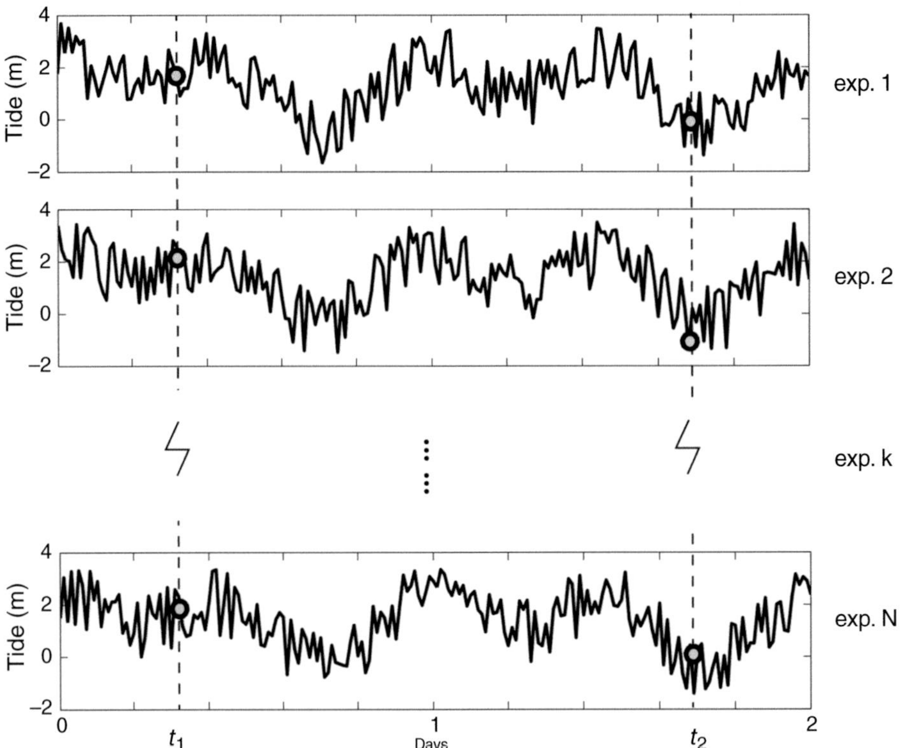

Figure 4.2 Example of an ensemble of experiments (experiment 1, experiment 2, . . . , experiment k, . . . , experiment N) and the function values used in the calculation of ensemble means at t_1 and t_2 (indicated by the dots).

mean at a given time is the average of all possible phase points for all the identical ensemble systems at a given time.

4.2.3 Autocorrelation Function

The ensemble mean is not enough to characterize a random process completely. Another function needed for further characterization is the *ensemble autocorrelation function* or *second order joint moment*:

$$R_{xx}(t_1, t_2) = \lim_{N\to\infty} \frac{1}{N} \sum_{k=1}^{N} x_k(t_1) x_k(t_2) \tag{4.5}$$

$$R_{xx}(t, \tau) = \lim_{N\to\infty} \frac{1}{N} \sum_{k=1}^{N} x_k(t) x_k(t+\tau). \tag{4.6}$$

This is a function of t_1 and t_2, or alternatively t and τ, in which $\tau = t_2 - t_1$ and $t = t_1$.

4.2.4 Weakly Stationary Process

If the ensemble mean and the ensemble autocorrelation do not vary with time,

$$\mu_x(t) = \mu_x \tag{4.7}$$

$$R_{xx}(t,\tau) = R_{xx}(\tau), \tag{4.8}$$

i.e. they are not functions of t; we call this random process x a *weakly stationary random process*. The reason we use the word weakly is because here only two statistical parameters are independent of time. This cannot guarantee that other, possibly important statistical parameters are also independent of time.

4.2.5 Higher-Order Moments

If the above conditions are further restricted, we may have a *strongly stationary random process*; if the above conditions are true and all other higher-order moments and joint moments are not dependent on time, then we refer to the process as strongly stationary. A second order de-meaned moment is the ensemble variance:

$$\sigma_x^2 = E[(x-\mu_x)^2] = \int_{-\infty}^{\infty} (x-\mu_x)^2 p(x)dx. \tag{4.9}$$

In general, the mth order moment is defined as

$$\lim_{N\to\infty} \frac{1}{N} \sum_{k=1}^{N} x_k^m(t) \quad \text{or} \quad E[x_k^m(t)].$$

An example of a higher order joint moment is

$$R_{xxx}(t_1,t_2,t_3) = \lim_{N\to\infty} \frac{1}{N} \sum_{k=1}^{N} x_k(t_1)x_k(t_2)x_k(t_3) \tag{4.10}$$

or

$$R_{xxx}(t_1,t_2,t_3) = E[x_k(t_1)x_k(t_2)x_k(t_3)]. \tag{4.11}$$

Obviously, there are infinite higher-order joint moments.

It is useful to introduce the ensemble mean but inconvenient to calculate the value for practical applications unless the probability density function is known. In practice, the time-averaged value is often used, as defined below:

$$\mu_x(k) = \lim_{T\to\infty} \frac{1}{T} \int_0^T x_k(t)\, dt. \tag{4.12}$$

Accordingly, the time autocorrelation function is defined as:

$$R_{xx}(k,\tau) = \lim_{T\to\infty} \frac{1}{T} \int_0^T x_k(t) x_k(t+\tau)\, dt. \tag{4.13}$$

Higher-order correlation functions can be defined likewise.

4.2.6 Ergodic Random Process

Now we return to the phase space concept. At any instant, the state of a system with a large number of particles (or large degrees of freedom) is described by a phase point of the system. The phase point draws a phase trajectory over time and eventually may occupy all possible states. In other words, after a long enough time, the system experiences all possible states in the phase space. This is understandable considering that the particles inside a container or certain region have occupied all possible positions and made all possible momentum combinations through random motions, collisions, and translations. If the random process is such that after a long enough time, it samples all possible values, then the temporal average is essentially the same as the ensemble average. In this case, we call the process *ergodic.*

More accurately, if $x(t)$ is stationary and its time-average and time autocorrelation do not depend on the ensemble number or identical experiment number (k), the random process that generates $x(t)$ is called ergodic. The time-averaged mean and autocorrelations under these conditions are the same as their ensemble counterparts.

Ensemble average is made over an infinite number of experiments. Time-average is made over time for one experiment. Ensemble averages are not doable in practice, but an ergodic process makes it possible because averaging over time is always doable. Ergodic assumption allows the replacement of the ensemble average by a temporal average. It should be noted that in instrument specification sheets, the phrase ensemble mean, if used, is often meant for temporal average.

4.2.7 Sample Mean and Sample Variance

In actual calculation of the mean value, the usual algorithm is to use the simple arithmetic average, which is also called the sample mean:

$$\bar{x} = \hat{\mu}_x = \frac{1}{N}\sum_{i=1}^{N} x_i. \tag{4.14}$$

The sample variance is defined as:

$$s^2 = \hat{\sigma}_x^2 = \frac{1}{N}\sum_{i=1}^{N} (x_i - \bar{x})^2. \tag{4.15}$$

The ensemble mean or expected value of the sample variance represented by (4.15), however, is not equal to the true variance. For that reason, (4.15) is called a biased estimate of the variance. The difference is only a factor of $\frac{N-1}{N}$. Sometimes, the factor $\frac{1}{N}$ in equation (4.15) is replaced by $\frac{1}{N-1}$ for an unbiased estimate of the variance. For very large N, $\frac{N-1}{N}$ is approximately 1 and there is essentially no difference between the biased and unbiased estimates.

4.3 Covariance Functions and Correlation Function

4.3.1 Covariance Functions

Now we look at the quantification of variability of random functions and statistical relationship between two random processes $x_k(t)$ and $y_k(t)$. Here again, k is the ensemble number. The ensemble means of them are, respectively

$$\bar{x}(t) = E[x_k(t)], \ \ \bar{y}(t) = E[y_k(t)], \tag{4.16}$$

which are all functions of t – they are not necessarily constants, as the average is done through the ensemble, not through the entire time. Equation (4.16) is the ensemble counterpart of the temporal average equation (4.12). The expression is slightly different from equation (4.4) but with no difference in meaning: the inclusion of the number k is only for emphasis on the ensemble mean being from the viewpoint of an identical experiment and not probability density function. The statistical *covariance functions* are used to describe their variabilities and relationship:

$$C_{xx}(t,\tau) = E\left[\left(x_k(t) - \bar{x}(t)\right)\left(x_k(t+\tau) - \bar{x}(t+\tau)\right)\right], \tag{4.17}$$

$$C_{yy}(t,\tau) = E\left[\left(y_k(t) - \bar{y}(t)\right)\left(y_k(t+\tau) - \bar{y}(t+\tau)\right)\right], \tag{4.18}$$

$$C_{xy}(t,\tau) = E\left[\left(x_k(t) - \bar{x}(t)\right)\left(y_k(t+\tau) - \bar{y}(t+\tau)\right)\right]. \tag{4.19}$$

The first two functions, C_{xx} and C_{yy}, are called *autocovariance functions* of x and y, respectively. The last function, C_{xy}, is called a *cross-covariance function* between x and y. These are all functions of t and τ. The autocovariance functions C_{xx} and C_{yy} are measures of variabilities of random functions x and y, respectively, at various time lags (τ). Likewise, the cross-covariance function C_{xy} tells us how much correlation they have at various time lags.

For the special case of no lag, i.e. $\tau = 0$, we have the *variances* of x and y, respectively:

$$\sigma_x^2 = C_{xx}(t) = E\left[\left(x_k(t) - \bar{x}(t)\right)^2\right], \tag{4.20}$$

$$\sigma_y^2 = C_{yy}(t) = E\left[\left(y_k(t) - \bar{y}(t)\right)^2\right], \tag{4.21}$$

and the covariance of x and y:

$$C_{xy}(t) = E\left[\left(x_k(t) - \bar{x}(t)\right)\left(y_k(t) - \bar{y}(t)\right)\right]. \tag{4.22}$$

These functions help to describe the random processes and their statistical relationship. The complete description, however, requires an infinite number of similar functions – e.g. the covariance calculated by fixing three or more times, e.g.:

$$C_{xxx}(t,\tau,\tau_1) = E\left[\left(x_k(t) - \bar{x}(t)\right)\left(x_k(t+\tau) - \bar{x}(t+\tau)\right)\left(x_k(t+\tau_1) - \bar{x}(t+\tau_1)\right)\right], \tag{4.23}$$

$$C_{yyy}(t,\tau,\tau_1) = E\left[\left(y_k(t) - \bar{y}(t)\right)\left(y_k(t+\tau) - \bar{y}(t+\tau)\right)\left(y_k(t+\tau_1) - \bar{y}(t+\tau_1)\right)\right], \tag{4.24}$$

$$C_{xxy}(t,\tau,\tau_1) = E\left[\left(x_k(t) - \bar{x}(t)\right)\left(x_k(t+\tau) - \bar{x}(t+\tau)\right)\left(y_k(t+\tau_1) - \bar{y}(t+\tau_1)\right)\right]. \tag{4.25}$$

......

If the functions (4.16)–(4.19) are independent of time t, then the random processes $x(t)$ and $y(t)$ are said to be weakly stationary. If all the other functions, including (4.23)–(4.25), are also independent of time, then the random processes are said to be strongly stationary. If the random processes x and y are Gaussian, then weak stationarity ensures that they are also strongly stationary.

4.3.2 Correlation Functions

The autocorrelation and cross-correlation functions of random processes $x(t)$ and $y(t)$ are defined by:

$$\sigma_x^2 = R_{xx}(t,\tau) = E[x_k(t)x_k(t+\tau)], \tag{4.26}$$

$$\sigma_y^2 = R_{yy}(t,\tau) = E[y_k(t)y_k(t+\tau)], \tag{4.27}$$

$$\sigma_{xy}^2 = R_{xy}(t,\tau) = E[x_k(t)y_k(t+\tau)]. \tag{4.28}$$

For stationary random process, the correlation and covariance functions are related by the following equations:

$$C_{xx} = R_{xx} - \bar{x}^2, \tag{4.29}$$

$$C_{yy} = R_{yy} - \bar{y}^2, \tag{4.30}$$

$$C_{xy} = R_{xy} - \bar{x}\bar{y}. \tag{4.31}$$

Therefore, if the mean values are zero, the correlation and covariance functions yield the same results.

4.3.3 Correlation Coefficient Functions

The cross-correlation function and the autocorrelation functions for stationary random processes $x(t)$ and $y(t)$ have the following relationship:

$$|R_{xy}(\tau)|^2 \leq R_{xx}(0)R_{yy}(0). \tag{4.32}$$

A similar relation is true for the cross-covariance function:

$$C_{xy}{}^2(\tau) \leq C_{xx}(0)C_{yy}(0) = \sigma_x^2\sigma_y^2. \tag{4.33}$$

This allows the definition of a correlation coefficient function (normalized cross-covariance function) by

$$\rho_{xy}(\tau) = \frac{C_{xy}(\tau)}{\sigma_x\sigma_y}, \tag{4.34}$$

with a magnitude between 0 and 1. This function is a measure of linear dependence between two random processes $x(t)$ and $y(t)$ for a time lag of τ.

4.3.4 MATLAB Functions for Covariance and Correlations

MATLAB has several built-in functions that compute the covariance and correlation coefficient. For example, `corr` and `corrcoef` output correlation coefficient. The latter has more functions, e.g. computing the coefficient of determination (see Chapter 8 for this concept) and 95% confidence interval, etc. The function `cov` computes the covariance values.

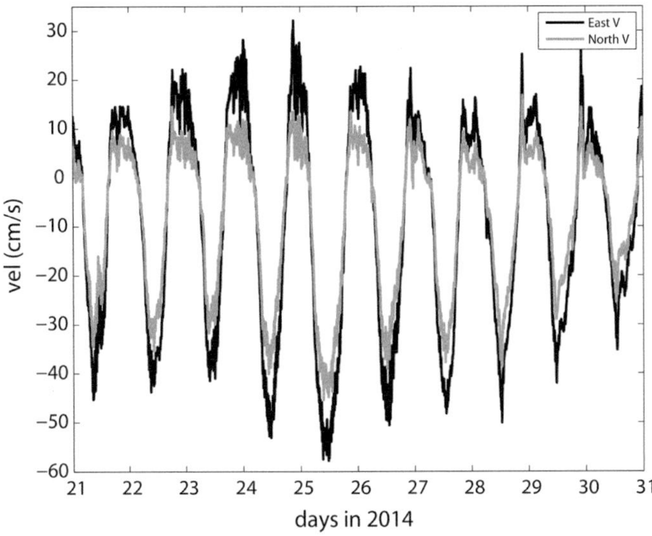

Figure 4.3 Part of the time series data of velocity measured by a horizontal acoustic Doppler current profiler at Caminada Pass, Barataria Bay. Data file name: CAMP-ADCP-Data.mat.

To demonstrate, we use a data file named "CAMP-ADCP-Data.mat," which is a MATLAB saved file for flow velocity data measured at Caminada Pass, Barataria Bay in 2014. There are three variables, timeCamF, CamFVE, and CamFVN, which are time in days in 2014 and the east and north velocity components, respectively, in cm/s (Fig. 4.3). The following is a few lines of MATLAB script for computing correlation coefficients and covariances.

```
load CAMP-ADCP-Data
figure;plot(timeCamF-timeCamF(1), CamFVE)
hold on;plot(timeCamF-timeCamF(1), CamFVN,'r')
axis([ 21 31 -60 35])
xlabel('days in 2014'); ylabel('vel (cm/s)')
legend('East V','North V')
corrVEVN = corr(CamFVN,CamFVE) % compute correlation coefficient
                               % which gives a value of 0.97
[ R,P,RLO,RUP] = corrcoef(CamFVN,CamFVE) % compute cross- and auto
                 % correlation coefficients
%========================================================
% R = 1.0000    0.9700   --- auto- and cross-correlation
%     0.9700    1.0000
% P =  1     0           --- p-value
%      0     1
% RLO = 1.0000    0.9693 --- lower limit of the estimate
%       0.9693    1.0000
% RUP = 1.0000    0.9706 --- upper limit of the estimate
%       0.9706    1.0000
%========================================================
```

```
cov(CamFVE)      % computes the autocovariance
% ans =   399.3421

cov(CamFVN,CamFVE) % computes the auto- and cross-covariances
% ans = 207.4168  279.1611
%       279.1611  399.3421
```

The command `corr(CamFVN,CamFVE)` computes the correlation coefficient between the north and east velocity components, which gives a value of 0.97, a high value as the two components are well correlated with each other (Fig. 4.3). In comparison, the command `[R,P,RLO,RUP] = corrcoef (CamFVN,CamFVE)` computes the auto- and cross-correlation coefficients, p-values, and the lower and upper limits of the correlation coefficient estimate with 95% confidence interval. In this example, the interval is [0.9693, 0.9706], with an (average) estimate of 0.97. The autocovariance for the east velocity using `cov(CamFVE)` gives a value of ~399.3; while `cov(CamFVN,CamFVE)` gives also the autocovariance of the north velocity (~207.4) and their covariance (~279.2).

Review Questions for Chapter 4

(1) What is an ensemble mean?
(2) What is an ergodic random process?

Exercises for Chapter 4

(1) Verify equations (4.29)–(4.31) for stationary random processes.
(2) Using the data file "CAMP-ADCP-Data.mat" and the commands shown in the example in Section 4.3.4, compute the correlation coefficient and covariance for the first half and second half of the time series and compare them. Are they similar? What if you divide the time series into 10-day segments and do the computation? [Compute and compare the results for each of the 10-day segments.]

5

Error and Variability Propagation

About Chapter 5

The objective of this chapter is to discuss the error propagation through computation from the error contained in the original data. We are not focusing on the measurement errors themselves, although the propagation of computational error is related to measurement errors. We also discuss the variability propagation with given mathematical relationships, or sensitivity of a system to a given parameter, which are all closely related to error propagation.

5.1 Source of Error

Measurements usually contain errors. They can be introduced during measurements, data transmission, data processing, or computations. The errors during measurements depend on the quality of instrument, conditions of the environment within which the measurements are made, and the skill of the person who uses the instrument. For example, when an oceanographer is trying to measure a vertical profile of water temperature, salinity, and density along a vertical water column, one usually uses equipment called a conductivity, temperature, and depth sensor, or *CTD* for short. If a CTD is manually lowered into the water column in shallow water, it needs to be lowered at a steady and relatively slow speed (not a free fall). This is to give the sensors enough response time to have smooth and reliable data and to avoid damage of sensors if they hit the bottom with a free fall. An inexperienced user without proper training sometimes lowers the instrument too fast. This may introduce significant errors, and the data might not be usable at the extreme. This is one type of error caused by mistakes in operation, and they are usually not of great interest in error analysis: the assumption is usually that the data are obtained with no major mistakes by the operator of the apparatus.

5.1.1 Systematic vs. Random Errors

Generally, errors other than those from the inexperience of the experimenter can be classified into (1) *systematic measurement errors* due to a repeatable systematic bias of the measuring device and (2) *random errors* due to uncontrollable and random change in the measurement environment and natural stochastic fluctuations. Although we generally discard the data with errors from an inexperienced experimenter, it is not always guaranteed that this type of error is excluded. An inexperienced experimenter can introduce either systematic or random error. For example, if a person using a ruler to measure the length of a fish tends to read the hatch marks with his head tilted, the reading will likely introduce some systematic offset. But since the tilt of his head is unlikely to be a constant, his readings may also contain additional random errors. Another example of the first kind of error is that produced in measurements when the device is off the scale or mis-calibrated with a constant offset. An example of the second kind of error is the random fluctuations of flow velocity in a turbulent ocean current. The systematic errors of measurement are associated with the inaccuracy of measurements, although they can be precise (highly repeatable given the same conditions). This brings up the concepts of *accuracy* and *precision*.

5.1.2 Accuracy vs. Precision

Accuracy is referred to the closeness of the measured values to the true value, while precision is referred to as the closeness of the measured values to the average value if the measurements are repeated numerous times. The averaged value from the samples might not be close to the real value, even though the precision may be high or the data scatter is within a small range. Fig. 5.1 shows that the measurements of the true value A in an experiment have relatively high accuracy, but with an offset from the true value A. The measurements are therefore relatively precise but inaccurate. In comparison, the measurements of the true value B in another hypothetical experiment have relatively low accuracy, but with an averaged value closer to the true value B. The measurements are therefore relatively accurate but with a relatively low precision.

5.1.3 Propagation of Error in Computations

In the following, we assume that certain measurements have been done and the data obtained have some errors. These errors are presumably the second kind – the random errors. In other words, we are dealing with the precision rather than the accuracy problem because it is a little more complicated to obtain the true

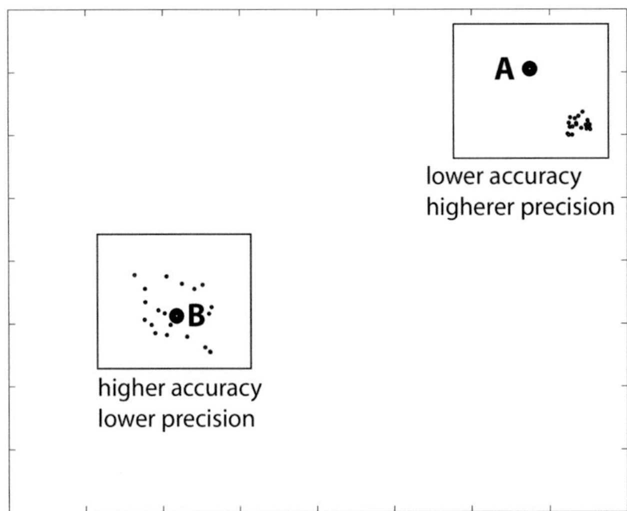

Figure 5.1 Precision and accuracy. The large circles marked as A and B are two "true" values. Points near A are the measured values of A with one hypothetical experiment. Points around B are the measured values of B with another hypothetical experiment.

value; what we usually do is use an average from a large number of samples to approximate a true value. If we knew the true value, there would be no need to do any measurements. In reality, it may of course include the first kind of error as well or even human errors because generally there is no way to exclude them for sure. Nevertheless, our interest is in examining how errors may propagate during data processing or computations using computers.

5.1.4 Computational Errors

How can computations introduce or propagate errors? In digital computations, there is potentially an error due to the limit in digital representation of numbers in computers, i.e. computers can only handle a finite number of bits (digits), and the calculations can have *round-off errors* – errors caused by finite bits. Computation using a computer can experience *overflow* when a number is too large to be represented by a finite number of bits in a given computer. It can also experience an *underflow* when a number is too small to be represented by a finite number of bits in a given computer. Even when a simple subtraction is performed, there can be the so-called *loss of significance* (or *catastrophic cancellation*) caused by a "bad subtraction," a subtraction of two very close numbers. There can be *error amplification* when errors are accumulated and/or amplified through computations. In addition to these, we may have *truncation errors* – errors caused by a truncation of an infinite series to a finite series, such as when we need to compute certain functions in a Fourier series (we will discuss Fourier series later).

5.2 An Example of Overflow Error for a Large Value

Imagine that we need to do some computations, e.g. calculate the first order derivative of a given theoretical function of a known form such as sin(x), e^x, and $\log_e(x)$. There should be no concern of error introduced in the calculation if only a theoretical function is being worked on. In this example, the theoretical results of the derivatives can be expressed analytically by cos(x), e^x, and $1/x$, respectively. Given any value of x except 0 for the last function $1/x$, the results for these functions are all finite. There is no issue about error introduced for this kind of derivation. If, on the other hand, the actual function values of a derivative need to be computed, we need to consider the round-off error introduced due to the use of a finite number of digits in the calculation.

If one is using a computer to do the computation digitally, one needs to know a little more about what may be expected. Even a simple calculation of e^{800} in MATLAB would produce an infinity. This is an overflow error. One may argue that e^{800} is really a large number (~2.7×10^{347}, which is too large for MATLAB to handle); isn't that close to infinity anyway, and why would we worry? If this is the only value or final step we are computing, saying that the result is infinity is probably not a big mistake – unless for certain problems we need to know the actual, albeit very large, value (e.g. for the distance of a star at the edge of the observable universe in astronomy). To calculate the actual value, one would have to be creative; there is usually not a one-size-fits-all type of solution. For the above example, we may do something like this:

```
>>format long
>>(exp(80)/10^34)^10
```

This will produce

```
ans =
    2.726374572112566e+007
```

This is a result for $e^{800}/10^{340}$, so the actual result for e^{800} is

2.726374572112566e+347 or $2.726374572112566\times10^{347}$,

which cannot be expressed explicitly in MATLAB. But, with a little maneuvering, this value can actually be calculated with the help of MATLAB.

Indeed, sometimes we can come up with a solution for an apparently intractable problem. This is particularly true for computations involving functions that can produce numbers too large or too small to be represented by a computer. This also depends on the capacity of a computer. With more powerful computers in the future, the largest and smallest numbers a computer or a software package like MATLAB can handle should be expected to further improve. This, however, will never change the fact that digital computers introduce and propagate errors.

5.3 An Example of Overflow Error

Sometimes, large numbers appear in the middle of the computation and the final answer is not necessarily an infinity. Any intermediate step encountering an overflow would ruin the entire computation, and a `NaN` (not a number) or an `Inf` may be the result in MATLAB. But with some attention and proper treatment, this can usually be avoided. To illustrate this, and before we really appreciate the importance of knowing the characteristics of digital computation, let us look at the following example related to the above example.

Let us assume that we are going to calculate digitally the value of $e^x/(e^x + x)$ for $x = 800$. With MATLAB, you are probably tempted to write the following code:

```
>>a = exp(800)/(exp(800) + 800)
```

From the above example in Section 5.2, we know that an overflow will occur. Indeed, if you use the above code, MATLAB will produce a `NaN` for the very reason that MATLAB cannot handle the large value `exp(800)`. A `NaN` is obviously not the right answer. The correct answer is 1, because the numerator and denominator are essentially the same. However, proper programming can avoid creating this error. You may notice that you can actually rewrite the expression with a mathematically equivalent form like this:

$$\frac{1}{1 + xe^{-x}}$$

and then do the calculation with MATLAB using this code:

```
>>a = 1 / (1 + 800 * exp(-800))
```

This will produce the correct answer: `a = 1`. Of course, we do not really need to use MATLAB to obtain the result because it can be easily concluded from the above equation. The point, however, is clear: sometimes we should take greater care of how we program in computation.

From this example, we can see that computations with a formula using a computer program may produce an incorrect or undesirable result for an unsuspecting person. Fortunately, this kind of pitfall is generally limited and less common in data analysis.

5.4 Absolute and Relative Errors

An *absolute error* e is defined by the absolute value of the difference between the true value X and its approximation x, i.e.

$$e = |X - x|. \tag{5.1}$$

A *relative error* ρ is defined by the ratio between the absolute error and the magnitude of the true value, i.e.

$$\rho = \frac{e}{|X|}, \qquad (X \neq 0). \tag{5.2}$$

These definitions are not very useful if we try to apply them literally in actual computations, because the true value of a quantity X is usually unknown. Should we know it, we would not have to approximate it by a less accurate value x. In real applications, the true value X has to be estimated from data. Often we use an averaged value obtained from numerous observations or measurements for an estimate of the true value. This, of course, implies that we have enough confidence that the measurements are reliable.

5.5 Relative Errors of Subtraction and Addition

Given the error resulted from observations in the original data, any calculations during data processing may result in an error in the final product. Intuitively, calculations cannot eliminate existing error. On the contrary, it would at least pass on some of the errors if not amplify them in the result of computations. When the computations are done with a digital computer, it should be anticipated that the accuracy of the output may worsen with an increased number of calculations. So, here the question to ask, given the fact that the original data always have an error one way or the other, is how will computations affect the error of the final result? This is a question about error propagation. We will first look at the simplest computation that one can think of: addition and subtraction.

Assume that we have two variables, x and y, both with some errors associated with them from less than perfect observations. The absolute errors of observations for x and y are, respectively,

$$e_x = |\delta x|, \quad e_y = |\delta y|. \tag{5.3}$$

Here $\delta x = X - x$, $\delta y = Y - y$. When these two variables x and y are added together, the errors will be added together. Chances are that there may be cancellation and superposition of the errors. However, the details are impossible to determine because of the random nature of the data and the unknown true values. For that reason, absolute errors should be used to obtain a "safe" estimate or the upper bound of the error as a result of addition. This is shown below mathematically:

$$e_{x\pm y} = |\delta(x \pm y)| = |\delta x \pm \delta y| \le |\delta x| + |\delta y| = e_x + e_y. \tag{5.4}$$

So, the error of an addition or a subtraction is the sum of the errors of x and y. The relative error is, according to the definition,

$$\rho_{x\pm y} = \frac{e_{x\pm y}}{|X \pm Y|} \le \frac{e_x + e_y}{|X \pm Y|} = \frac{\rho_x|X| + \rho_y|Y|}{|X \pm Y|}. \tag{5.5}$$

This equation tells us that the relative error after addition or subtraction is a weighted average. The problem is that the denominator is an addition or subtraction of two numbers, which can be positive or negative themselves. In a case when the two are close in magnitude with the same sign, any subtraction in the denominator can be problematic, i.e. increase the relative error unfavorably.

We now illustrate this by an example of subtraction: do the calculation using MATLAB with two mathematically equivalent equations for large values of x.

$$f_1 = x(x - \sqrt{x^2 - 2}) \tag{5.6}$$

$$f_2 = \frac{2x}{x + \sqrt{x^2 - 2}} \tag{5.7}$$

These two expressions are identical mathematically, regardless of the value x being large or small. Although in theory they are the same, when computations are being performed, they do not necessarily produce equivalent results, depending on the value of x and the capacity of the computer. As far as actual computations are concerned, for values of x that are not very large, the two equations (5.6) and (5.7) would give essentially equivalent results. When x is very large, however, the difference can be significant.

For very large x values, the first equation (5.6) involves a subtraction of two close numbers, while the second has mainly an addition of two large positive numbers in the denominator and a division. The subtraction $x^2 - 2$ is still a very large number because, for very large values of x, the value -2 is negligible. The second equation, (5.7), does not have a subtraction of two large and close numbers

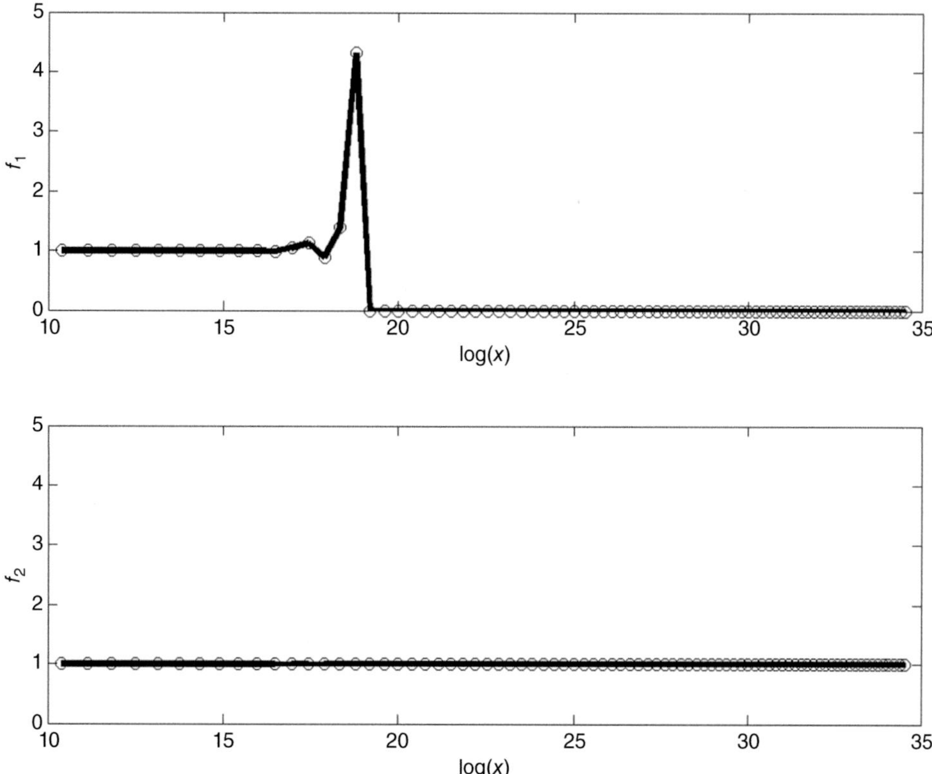

Figure 5.2 Subtraction of two very large and close numbers. The upper panel is obtained by using equation (5.6), while the lower panel is obtained by using equation (5.7).

and thus would not have an increased relative error in the computation. The following script is for this example and for producing Fig. 5.2.

```
clear
x = (2:0.1:10).^15;
f1 = x.*(x-sqrt(x.^2-2));
f2 = 2*x./(x+sqrt(x.^2-2));

figure
subplot(2,1,1)
plot(log(x),f1,'k','LineWidth',3)
hold on
plot(log(x),f1,'ko')
axis([ 10 35 0 5])
xlabel('log(x)')
ylabel('f1')
subplot(2,1,2)
```

```
plot(log(x),f2,'ko')
hold on
plot(log(x),f2,'k','LineWidth',3)
axis([ 10 35 0 5])
xlabel('log(x)')
ylabel('f2')
print -dpng chapter5.1.png
```

Computations with large x values using these equations with MATLAB show a distinct difference for very large x values (Fig. 5.2). The deviation of the first equation (5.6), at some point followed by an abrupt drop to zero, is what we call a loss of significance by the catastrophic cancellation that occurs when a subtraction is performed for two very close large numbers.

5.6 Relative Errors of Multiplication and Division

Now we examine multiplication of two numbers. For convenience of discussion, let us consider a logarithm of the product xy as shown below:

$$\ln(xy) = \ln(x) + \ln(y). \tag{5.8}$$

By looking at the departures of the random variables x and y from their true values X and Y, respectively, we do a derivative to (5.8) and obtain

$$\frac{\delta(xy)}{|XY|} = \frac{\delta(x)}{|X|} + \frac{\delta(y)}{|Y|}. \tag{5.9}$$

Here, δ is a difference operator.

The absolute value of the above is estimated to have the upper bound expressed as

$$\frac{|\delta(xy)|}{|XY|} = \left|\frac{\delta(x)}{|X|} + \frac{\delta(y)}{|Y|}\right| \leq \frac{|\delta(x)|}{|X|} + \frac{|\delta(y)|}{|Y|}. \tag{5.10}$$

Therefore, we have

$$\frac{|\delta(xy)|}{|XY|} \leq \frac{e_x}{|X|} + \frac{e_y}{|Y|} \tag{5.11}$$

or

$$\rho_{xy} \leq \rho_x + \rho_y. \tag{5.12}$$

Unlike error propagation in addition and subtraction, multiplication will produce a product with a relative error as the simple superposition of the relative errors of x

and y. Therefore, this operation will not cause abrupt change such as the catastrophic cancellation demonstrated above with subtraction of two very close and large numbers.

For a division of two numbers, we can evaluate in the same way:

$$\ln\left(\frac{x}{y}\right) = \ln(x) - \ln(y). \tag{5.13}$$

By looking at the departures of the random variables x and y away from their true values X and Y, respectively, we do a derivative to (5.13) and obtain

$$\frac{\delta(x/y)}{|X/Y|} = \frac{\delta(x)}{|X|} - \frac{\delta(y)}{|Y|}. \tag{5.14}$$

The only difference between (5.14) and (5.9) is the sign of the second term. Now we look at the absolute value:

$$\frac{|\delta(x/y)|}{|X/Y|} = \left|\frac{\delta(x)}{|X|} - \frac{\delta(y)}{|Y|}\right| \leq \frac{|\delta(x)|}{|X|} + \frac{|\delta(y)|}{|Y|}. \tag{5.15}$$

This is the same as (5.10) on the right-hand side, or

$$\frac{|\delta(x/y)|}{|X/Y|} \leq \frac{e_x}{|X|} + \frac{e_y}{|Y|}. \tag{5.16}$$

The relative error of division is expressed in a similar format as that for multiplication – it is an addition of the two relative errors:

$$\rho_{x/y} \leq \rho_x + \rho_y. \tag{5.17}$$

From the above discussion, we conclude that, for an addition of two positive variables, there is no major problem, as the relative error is bounded by a weighted average, while a subtraction, when the two numbers are close, tends to produce large relative errors. For multiplication and division, the relative errors are bounded by the addition of their respective relative errors – so usually there is no major concern here. Obviously, a multiplication involving N variables will produce a relative error bounded by the superposition of the respective relative errors of the N factors.

5.7 Variability Propagation

In the discussion above, we have examined the propagation of errors with addition, subtraction, multiplication, and division. The derivation uses calculus, and it is clear that the difference-operator δ or derivatives, e.g. $\frac{\partial}{\partial t}$, are in reference to the variations in the variables in general. The variations are considered to be caused by random errors in observations and subsequent computations. In other words, the above conclusions are not necessarily only applicable to error propagations. They are also useful for *variability propagations* in computations, assuming that the variabilities are relatively small (so the difference-operator can be used). For example, if variable z is a product of variables x and y, then

$$z = xy. \tag{5.18}$$

If the variables x and y have standard deviations due to the variabilities (not necessarily due to random errors) of δx and δy, respectively, then from (5.11), the variability of z would be

$$|\delta z| \leq |\bar{z}| \left| \frac{\delta x}{\bar{x}} + \frac{\delta y}{\bar{y}} \right|, \tag{5.19}$$

in which the variable with an overbar denotes the average value of the variable. We can also use the relative variability expression

$$\left| \frac{\delta z}{\bar{z}} \right| \leq \left| \frac{\delta x}{\bar{x}} \right| + \left| \frac{\delta y}{\bar{y}} \right|. \tag{5.20}$$

In other words, the percentage of variability of z is the sum of the percentage variability of x and y. If the variability of x and y are ~10% and ~5%, respectively, the variability of z would be bounded by ~15%. The condition of the above equation is also important: the variabilities should be "small." If the variability is not "small," we may be able to do a similar estimate by the inclusion of higher-order terms. The discussion is omitted here.

5.8 Examples

5.8.1 Example 1 – Error Propagation

At the terminal fall velocity of a particle in a quiescent water column, the drag on the particle depends on a few factors. It is described by the following equation (Yang, 2003):

$$F = \frac{C_D \rho S w^2}{2}, \tag{5.21}$$

in which F is the drag force, C_D the drag coefficient, ρ the density of water, S the cross-section area of the particle perpendicular to the direction of fall, and w the fall velocity. If one needs to estimate the drag force based on this equation, the relative error should be the sum of the relative errors of the drag coefficient, water density, and cross-section area, plus two times the relative error of the fall velocity. The denominator 2, like any constant factor, does not affect the error in any way.

5.8.2 Example 2 – Variability Propagation

The water surface wave dispersion relationship is given by (Landau and Lifshitz, 1987):

$$\omega^2 = gk \tanh kh, \tag{5.22}$$

in which ω, g, k, and h are the angular frequency, gravitational acceleration, wavenumber, and undisturbed water depth, respectively. The wavenumber is related to wavelength by

$$k = \frac{2\pi}{\lambda}. \tag{5.23}$$

With a wave maker, a large wave tank can be used to do experiments with waves and related processes. The frequency of the wave can be controlled by the wave maker. The wavelength is determined by the dispersion relationship under idealized conditions.

If a lab technician would like to know what range of wavelength would be available for experiments with a range of frequency values (ω) that the wave maker can provide, given the water depth (h), he/she can estimate it by a derivative to equation (5.22):

$$2\omega\,\delta\omega = g\,\delta k \tanh kh + gk\,\delta\left(\frac{e^{kh} - e^{-kh}}{e^{kh} + e^{-kh}}\right), \tag{5.24}$$

which leads to

$$\delta\lambda = -\frac{\lambda^2}{2\pi}\frac{2\omega\,\delta\omega}{g\{\tanh kh + k[1 - (\tanh kh)^2]\}} \tag{5.25}$$

or

$$\frac{\delta\lambda}{\lambda} = -\frac{\lambda}{2\pi}\frac{2\omega^2}{g\{\tanh kh + k[1 - (\tanh kh)^2]\}}\frac{\delta\omega}{\omega}. \tag{5.26}$$

The above equation can be used for both error propagation estimates and variability estimates: with a relative change of frequency, i.e. $\delta\omega/\omega$, the relative change in wavelength is described by the above equation. Likewise, given a relative error estimate of the frequency, the relative error of wavelength obtained from the given dispersion relationship can be derived from the following equation:

$$e_\lambda = \frac{\lambda}{2\pi} \frac{2\omega^2}{g\{\tanh kh + k[1 - (\tanh kh)^2]\}} e_\omega, \tag{5.27}$$

in which e_λ and e_ω are the relative errors of the wavelength and angular frequency, respectively, i.e.

$$e_\lambda = \left|\frac{\delta\lambda}{\lambda}\right|, \quad e_\omega = \left|\frac{\delta\omega}{\omega}\right|. \tag{5.28}$$

This example is a little different from that in Section 5.6, in which the error propagation of a simple multiplication is considered. Here the dispersion relationship is more complicated than a simple multiplication. The hyperbolic tangent function tanhkh is not linearly related to the wavenumber k. In addition, in this example, the water depth is assumed constant. If the water depth measurement has an error or the depth is allowed to have a variability, the resultant error or variability of the wavelength can be estimated but the derivation will be a little more complicated (omitted here but left as an exercise).

5.8.3 Example 3 – Application of Error Estimation

Here is an example to estimate relative errors of a calculation with a given equation involving observational data and several variables. More specifically, the example is about the calculation of bottom drag coefficient based on observations and an equation relating the drag coefficient to water depth, velocity amplitude, and the phase difference between tidal velocity and tidal surface elevation gradient.

In a frictional tidal channel, tidal velocity amplitude variation in a cross-channel transect is found to be dependent on water depth. Mathematically, the momentum balance requires that (Li, 2003)

$$U = -\frac{gh}{i\sigma h + \beta} \frac{\partial A}{\partial x}, \tag{5.29}$$

in which $U, g, \sigma, h, \beta, A, x$ are tidal velocity amplitude, gravitational acceleration (9.8 m/s^2), angular frequency of the major tidal constituent, water depth, linear friction coefficient, water level amplitude, and along-channel distance, respectively. The variables U and A are all complex functions (with real and imaginary parts) and i is the unit imaginary number $\sqrt{-1}$. The linear friction coefficient is expressed as

$$\beta = \frac{8C_D U_0}{3\pi}. \tag{5.30}$$

Here C_D is the bottom drag coefficient and U_0 is the estimated order of magnitude of the tidal velocity. The tidal velocity amplitude as a function of water depth is

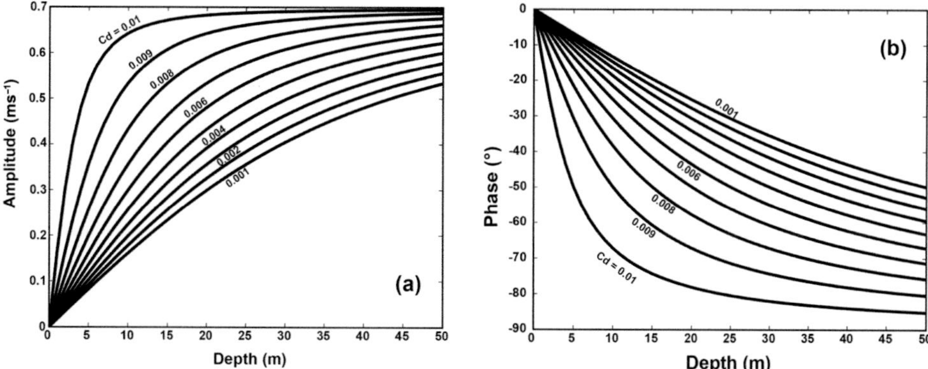

Figure 5.3 Functions from theory of frictional tide in a channel. (a) Velocity amplitude as a function of depth and drag coefficient across a tidal channel. (b) Phase as a function of depth and drag coefficient across a tidal channel. These are all for semi-diurnal tides.

shown in Fig. 5.3a. The phase of the tidal velocity as a function of water depth is shown in Fig. 5.3b.

From the momentum relationship equation (5.29), the bottom drag coefficient C_D is found to be related to h, σ, U_0, and the phase difference $\delta\varphi$ between tidal velocity and tidal water elevation gradient by the following equation (Li, 2003):

$$C_D = \frac{3\pi\sigma h}{8U_0 \tan(\delta\varphi)}. \tag{5.31}$$

The question is, given the estimated observational errors in h, U_0, and $\delta\varphi$, what is the error for the estimated drag coefficient C_D? This is a question of error propagation with multiplication and division. Taking a logarithm and a derivative on both sides of (5.31), it can be shown that

$$\left|\frac{\delta C_D}{C_D}\right| \le \frac{|\delta h|}{h} + \frac{|\delta U_0|}{U_0} + \frac{\left|\delta\Big(\tan(\delta\varphi)\Big)\right|}{|\tan(\delta\varphi)|}. \tag{5.32}$$

If observations are made along a line across the channel to obtain the water depth, tidal velocity amplitude, and phase difference at various locations, the bottom drag coefficient can be calculated by equation (5.31), with the relative error estimated by (5.32) at various water depths across the channel. More specifically, by vessel-based and repeated observations along a cross-channel transect using an acoustic Doppler current profiler, the along-channel tidal velocity amplitude at different locations along the transect can be obtained using a harmonic analysis (harmonic analysis will be discussed in Chapter 11).

Based on this theory, the drag coefficient can be estimated if relevant variables can be obtained from observations. The method was applied to data measured by a

600 kHz ADCP across two tidal channels in Sapelo Sound (31°32′N, 81°11′W). The observations in each channel lasted for ~12 hours. Twenty repetitions were completed. The length of the cross-channel transect was 1.1 km. The velocity profiles were first vertically integrated and divided by the mean water depth. A harmonic analysis was then applied to the depth-averaged velocity to obtain the semi-diurnal tidal velocity amplitude (Fig. 5.4a) and phase (Fig. 5.4b). Using equation (5.31), the drag coefficient was estimated (Fig. 5.5a).

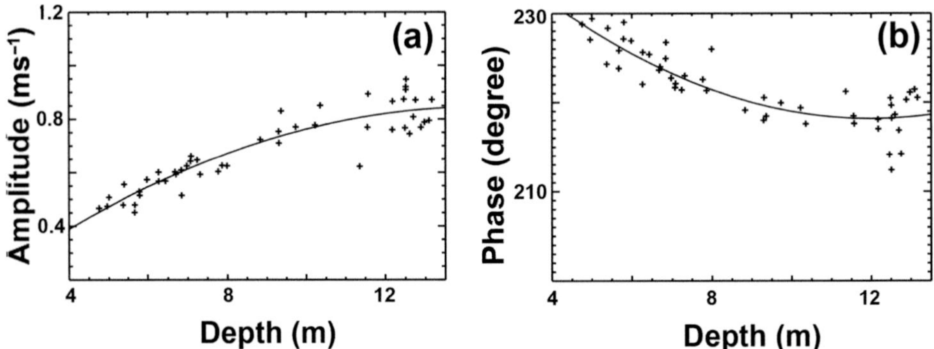

Figure 5.4 Observed semi-diurnal tidal velocity amplitude (a) and phase (b) from a cross-channel transect at Sapelo Sound, Georgia, USA. The crosses are from observations and the solid lines are from a least square fit to a Taylor series expansion or a polynomial to the second order.

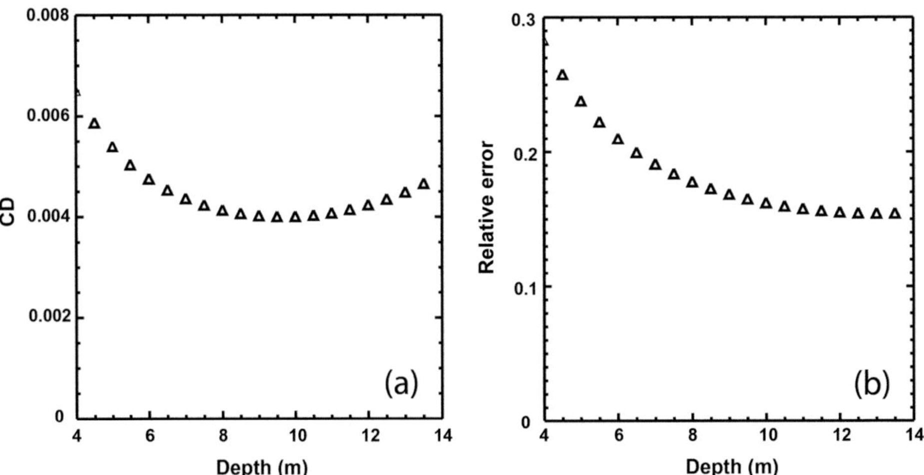

Figure 5.5 (a) Estimated function (drag coefficient) from a tidal channel (Sapelo Sound, Li, 2013). (b) Estimated relative error of the drag coefficient calculation at different depths in two tidal channels.

Equation (5.32) was then applied to estimate the error of drag coefficient. Omitting the details of the procedures of calculations, the estimated error for the drag coefficient is shown in Fig. 5.5b.

5.9 Note on Relative Error

The use of relative error is helpful for a "normalized" quantification of error. It is, however, not always the best choice. This is particularly true when the true value, or the ensemble-mean value of the observed quantity, is close to zero. According to the definition of relative error,

$$\rho = \frac{e}{|X|}, \quad (X \neq 0).$$

The relative error can be very large if the actual value of the parameter is very small. This can be demonstrated by Fig. 5.6. Here some hypothetical observations are assumed (the thick line) with some error bars. The error bars are shorter at some points and longer at other points in time. In this specific case, the error bars are shorter when the values are small and longer when the values are larger. The error bars are proportional to the absolute errors (e in the above equation). Equation (5.2) is undefined at $X = 0$. The value of X may not be exactly zero, but the relative error can be enlarged dramatically. As shown in Fig. 5.6, the

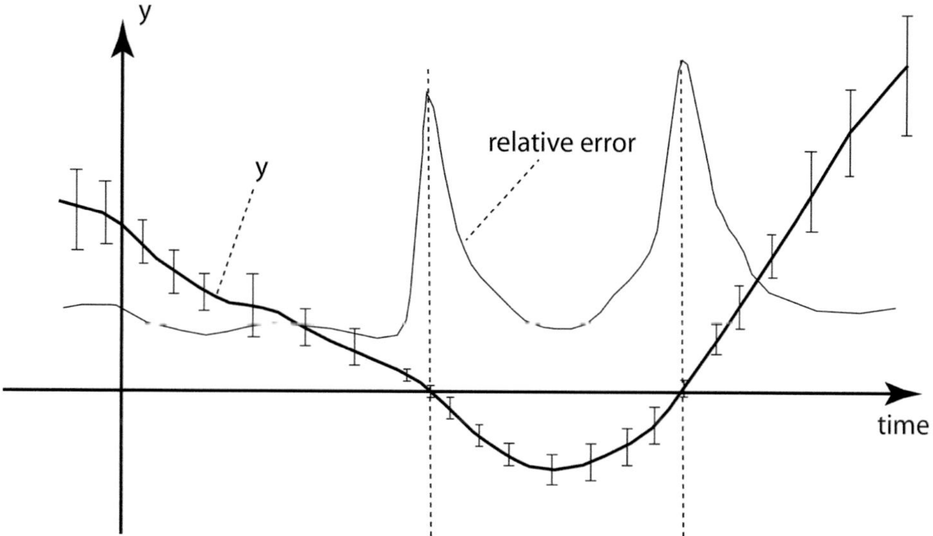

Figure 5.6 The pitfall when using relative error. The thick line is a hypothetical time series with error bars. The think line is the relative error. The scales and units are arbitrary and just for demonstration purposes.

relative errors have two peaks around the time when the observed value y is close to zero.

There are two important points to note. First, as far as data analysis is concerned, it is quite subjective to decide how small X is when we say it is zero. The second point is that when the function is zero, the magnitude of the relative error may be unimportant. To give an example, consider the flux of water through a tidal pass. Because of tidal oscillations, the flow alternates between flood and ebb in a quasi-periodic manner. At the turn of the tide, velocity is the minimum and there is virtually no transport through the pass. The relative error of measurements at this turning point is usually relatively large, but the absolute error may not matter that much for the total transport because the values observed are relatively small anyway. The reason for having a large relative error is simply because the denominator in equation (5.2) is small.

Because of these points, it is advised that the relative error equation (5.2) be used with caution, especially around the time of small values, assuming a common-sense order of magnitude is established to decide how small is small. A large relative error may not always suggest inadequate data quality.

Review Questions for Chapter 5

(1) Compare accuracy and precision; compare systematic error and random error.

(2) Compare the absolute error with the relative error. Are they the same in terms of the usefulness? Do you have a preference for which one to use for your work?

(3) Try to think of some real examples for measurements which are (a) accurate but not precise; (b) inaccurate and not precise; (c) accurate and precise; or (d) inaccurate but precise.

(4) Can you think of a real-life example with known true value based on which the accuracy can be determined, not estimated?

(5) Suppose you are going to measure current velocity in a stream with a handheld mechanical rotary current meter. You are going to hold it still next to a stream either on an anchored boat or at a dock, or from a bridge. List all possible sources of errors in each case (from a boat, a dock, or a bridge) that you can think of.

(6) With a 16-point compass scale for wind direction (Fig. 4.2), if we digitize these 16 directions (using numbers instead of words) as defined in this chapter, what is the error in direction (degrees)? When this error is present, and if the wind velocity components are calculated using equations (4.2) and (4.3), what is the error in the east and north components contributed by the error in direction? What if a 32-point compass scale is used?

Exercises for Chapter 5

(1) Assume that you are going to write a MATLAB script to compute

$$y = \frac{e^{900}}{e^{900} + e^{902}}.$$

The result is a finite number. Try to think of an algorithm (method) to do the computation using MATLAB.

(2) The shallow water wave propagation speed has a theoretical relationship with depth: $c = \sqrt{gh}$, in which h is the undisturbed water depth and g is the gravitational acceleration. This relationship, of course, is only correct under idealized conditions (e.g. frictionless, constant water depth, and hydrostatic condition with negligible vertical acceleration). A lab experiment in a long flume is done using certain apparatus to measure the speed c and compare with the theoretical value. The error of measurement of the speed c has a relative error of $e_{cm} = 3.5\%$. If the water depth measurement has a relative error of $e_{cm} = 2.0\%$, which will result in an error in c from the theoretical relationship e_{ct}, is this error e_{ct} smaller or larger than the measurement error e_{cm}? (This may be used to test if the hydrostatic approximation is satisfied with the given lab setup, including the water depth. This can be done with different water depth values to examine at what depth value the hydrostatic approximation may be broken.)

(3) Wind stress is determined by $\tau = C_D w^2$, in which C_D and w are the drag coefficient and wind speed. The measurement of w has an error δw and the drag coefficient also has an uncertainty or variability δC_D. Determine what the relative error in wind stress e_τ is because of the uncertainties in wind speed and drag coefficient. Write a MATLAB function and use some example values to work out some MATLAB script for computing the relative error.

(4) Suppose we have obtained a time series (an array) of water salinity s, density ρ, and velocity u through measurements. The salt flux per unit time per unit area f_s is determined by the simple equation:

$$f_S = \rho s u.$$

If we know that the salinity has an observational error of 5%, density has an error of 10%, and velocity has an error of 10%, how much error would we anticipate for the calculated quantity f_s?

(5) If the total precipitation measured at a location for a given day has 1% error of measurement, then when one examines the total precipitation in a month (i.e. a 30-day period), what is the error for the monthly estimate?

(6) Standard observational wind data are often expressed as wind speed w_{spd} and wind direction w_{dir} in degrees. In analysis and modeling, however, the usage of wind data may need to decompose the wind into the east and north components. This can be done by (consult Chapter 4 for these equations):

$$\begin{aligned} w_e &= w_{spd} \cos(270 - w_{dir}) \\ w_n &= w_{spd} \sin(270 - w_{dir}) \end{aligned}.$$

Here w_e and w_n are the east and north component of the wind velocity, respectively. The wind direction is sometimes expressed as accurate to 10 degrees, e.g. $w_{dir} = 253°$ is actually shown in the data file as $w_{dir} = 250°$. With this error ($\delta w_{dir} = 10$), what are the errors of the decomposed wind velocity components, assuming that the error of wind speed can be estimated as δw_{spd}? Write a MATLAB function and use some example values to work out some MATLAB script for computing the errors.

(7) For the example discussed in Section 5.8.1, write a MATLAB function for computing the error of the drag force. Write a MATLAB script to test the function.

(8) For the example discussed in Section 5.8.2, write a MATLAB function for computing the variability of the wavenumber based on the variability of the wave maker's frequency range. Write a MATLAB script to test the function.

(9) Consider the example in Section 5.8.2 and the water surface wave dispersion relationship again. If the wave maker can allow a variability of frequency of $\delta\omega$ and depth variation in the wave tank or flume of δh, what range of wavelength would one anticipate according to the equation (5.22)? Write a MATLAB function and use some example values to work out some MATLAB script for computing the relative error.

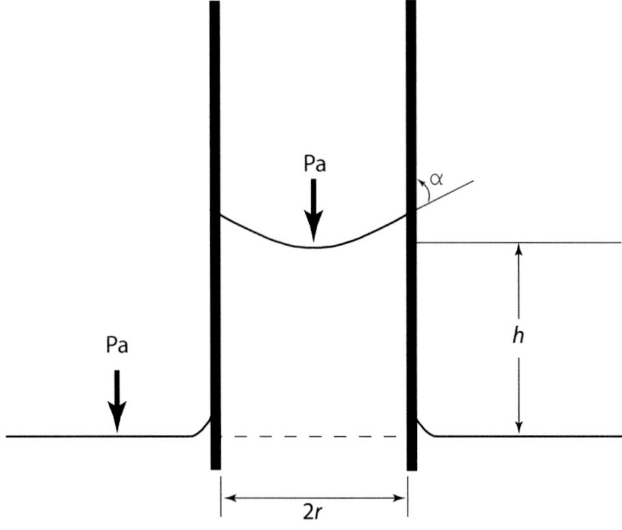

Figure 5.E1 Diagram for exercise 10. Liquid level in a thin tube. Pa is the atmospheric pressure.

(10) A liquid is placed in a large container. A thin tube is inserted into the container (Fig. 5.E1). Because of surface tension, the liquid in the thin tube is higher than the liquid level outside of the tube by h, which is related to the radius of the tube r, density of the liquid ρ, surface tension σ, and contact angle α by the following equation (Kundu and Cohen, 2004):

$$h = \frac{2\sigma \cos\alpha}{\rho g r}.$$

If one can measure h with an error δh, radius r with an error δr, and density ρ with an error $\delta\rho$ and obtain surface tension value σ from temperature measurement with an error of $\delta\sigma$, what is the error estimate of the contact angle $|\delta\alpha|$? Use your result and write a MATLAB function for the computation. Test it with an example.

6

Taylor Series Expansion and Application in Error Estimate

About Chapter 6

The objective of this chapter is to review the Taylor series expansion and discuss its usage in error estimation. The unique value of Taylor series expansion is often neglected. The major assumption is that a function must be infinitely differentiable to use the Taylor series expansion. In real applications in oceanography, however, hardly there is a need to worry about a derivative higher than the 3rd order, although one may think of some exceptions. The point is, there is rarely a need in oceanography and other environmental sciences to actually consider calculating a very high order derivative, unless for theoretical investigations or under special situations. So the application of Taylor series expansion usually only involves the first two derivatives. In this chapter, some simple examples are included for a better understanding of the applications.

6.1 Taylor Series Expansion

Taylor series expansion is a formula which we use to approximate a differentiable function with a *polynomial function* in a neighborhood region of a point where the function value and derivatives are somehow known. This can be useful in error estimations in laboratory experiments, data analysis, and stability analysis for computational schemes in numerical models. For instance, it can be used for error estimates for interpolation or extrapolation. Often we do not know the form of a function, but we may know the function value, its slope, and its higher-order derivatives at a given point based on measurements or computer model output.

The question is: can we approximate an unknown function within a certain range of the independent variable with a simple function based on knowledge of the values and derivatives of the function at a given point?

The answer is yes, with Taylor series expansion, which provides the solution. It is an approximation valid within a certain range of the independent variable. In theory, it is an infinite series of terms in a *polynomial form*. The polynomial has the standard form, with increasing power of the independent variable and with coefficients to be determined by available data. The determination of coefficients is done in terms of various orders of the rate of change of the function. The Taylor series usually converges in the neighborhood of a given point or value of the independent variable. Sometimes, however, this limitation is relaxed so that the series is convergent for wider intervals or even all possible values of the independent variable.

6.1.1 The 0th-Order Approximation

The simplest function is a straight line. If you need to use a straight line to approximate a function at a given point, how should we choose this line to approximate the function passing the point $(x_0, f(x_0))$? The simplest function that passes this point is a constant:

$$P(x) = f(x_0). \tag{6.1}$$

This can be considered as the 0th order approximation: it is a horizontal straight line or a constant for x within an unspecified range (range of x values). This function (6.1), $P(x)$, depends on the independent variable x, but the function value is a constant that equals the value of the original function $f(x)$ at x_0.

6.1.2 The 1st Order Approximation

If we further require that the approximate function has not only the same function value but also the same slope (the first order derivative) at x_0, we can use the modified equation:

$$P(x) = f(x_0) + f'(x_0)(x - x_0). \tag{6.2}$$

Here $f'(x_0)$ is the first order derivative of the function $f(x_0)$ at x_0. This is again a polynomial of $x - x_0$. The above equation is geometrically a straight line passing the point $(x_0, f(x_0))$ with a slope equal to that of the tangent line of the function at that point (Fig. 6.1).

This is the 1st order approximation. It can be easily verified that $P(x_0) = f(x_0)$ and $P'(x_0) = f'(x_0)$. Again, this is an approximation good at the neighborhood of x_0. The further away from this point, the less likely the polynomial (6.2) would be a good approximation of the original function $f(x)$. But we can see that (6.2) must be a better approximation of $f(x)$ than (6.1).

6.1.3 The 2nd Order Approximation

Based on the above discussion, we can further require the derivative of the approximate function to be expressed by a function similar to (6.2), i.e.

$$P'(x) = f'(x_0) + f''(x_0)(x - x_0), \tag{6.3}$$

which satisfies $P'(x_0) = f'(x_0)$, and $P''(x_0) = f''(x_0)$.

An integration of (6.3) over x yields

$$P(x) = f(x_0) + f'(x_0)(x - x_0) + \frac{f''(x_0)}{2}(x - x_0)^2. \tag{6.4}$$

This more complicated polynomial function satisfies $P(x_0) = f(x_0)$, $P'(x_0) = f'(x_0)$, and $P''(x_0) = f''(x_0)$. Obviously, (6.4) is a better approximation than (6.2). Geometrically, this is shown by Fig. 6.2: the 2nd order polynomial approximation in the neighborhood of x_0 has approximately the same *curvature* of the original function. In comparison, if a 2nd order polynomial does not have the same curvature, it would have more error. As an example, the curved dashed line in Fig. 6.2 does not have the same curvature of the original function at x_0.

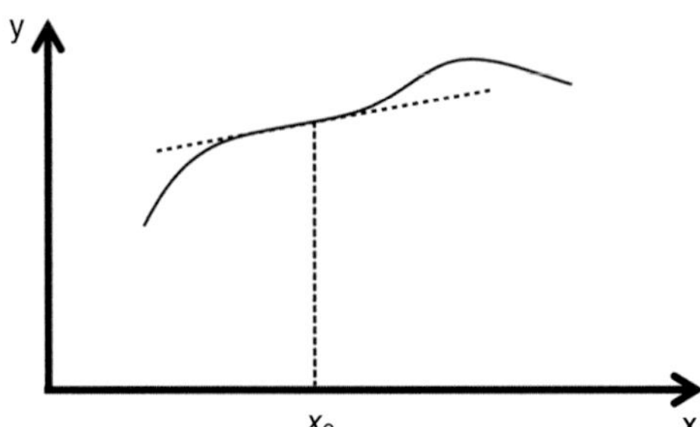

Figure 6.1 First order Taylor series expansion.

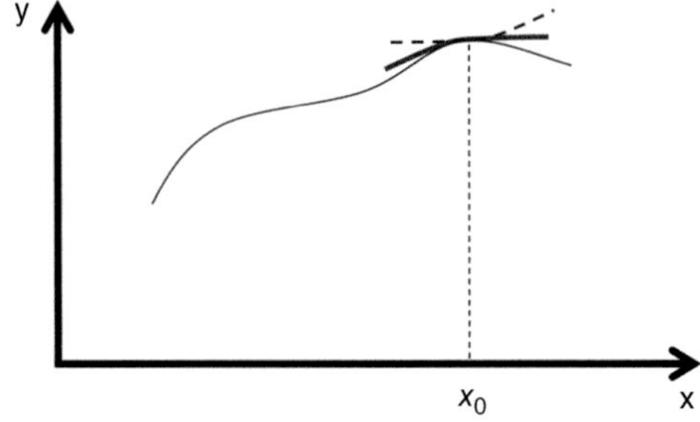

Figure 6.2 The second order Taylor series expansion. The thick curved line satisfies the curvature requirement, while the dashed curve line does not. The vertical dashed line indicates the location x_0.

6.1.4 Taylor Series Expansion

The above reasoning can be iteratively done an arbitrary number of times, at least in theory. Here, for convenience of discussion, we assume that this *iteration* has been done for $n-1$ times and a similar equation is obtained:

$$P^{(k-1)}(x)=f^{(k-1)}(x_0)+f^{(k)}(x_0)(x-x_0), k=1,2,3,\ldots,n-1, \tag{6.5}$$

which satisfies $P^{(k-1)}(x_0)=f^{(k-1)}(x_0)$ and $P^{(k)}(x_0)=f^{(k)}(x_0)$. Here the integer within parentheses on the upper right corner of the function name or polynomial function name, e.g. k, is the order of derivative, with zeroth order derivative being defined as the function itself.

If we integrate (6.5) over x, we have

$$P^{(k-2)}(x)=f^{(k-2)}(x_0)+f^{(k-1)}(x_0)(x-x_0)+\frac{f^{(k)}(x_0)}{2}(x-x_0)^2. \tag{6.6}$$

Here in (6.6) we have used a relation $P^{(k-2)}(x_0)=f^{(k-2)}(x_0)$. If we integrate the above equation again, we will get (6.7):

$$P^{(k-3)}(x)=f^{(k-3)}(x_0)+f^{(k-2)}(x_0)(x-x_0)+\frac{f^{(k-1)}(x_0)}{2}(x-x_0)^2+\frac{f^{(k)}(x_0)}{2\cdot 3}(x-x_0)^3, \tag{6.7}$$

after making use of the requirement that $P^{(k-3)}(x_0)=f^{(k-3)}(x_0)$.

Repeating this process, we can see that we will end up with the following equation if the maximum k is n:

$$P_n(x)=f(x_0)+f'(x_0)(x-x_0)+\frac{f''(x_0)}{2!}(x-x_0)^2+\frac{f'''(x_0)}{3!}(x-x_0)^3+\cdots+\frac{f^{(n)}(x_0)}{n!}(x-x_0)^n \tag{6.8}$$

The assumption for (6.8) equation to hold is that the function $f(x)$ is continuous and differentiable for any number of times at x_0. The first term is the function evaluated at the fixed position x_0. The terms with primes $f'(x_0), f''(x_0), \ldots, f^{(n)}(x_0)$ are, respectively, the first, second, ... , and nth order derivatives evaluated at the single position x_0. The only variable here is x, in the form of $(x-x_0)^k$, $k=1,2,3,\ldots,n$. The above equation is the so-called Taylor series expansion. A nice property is that the estimated *error of Taylor series expansion* can be given by,

$$f(x)-P_n(x)=R_n(x)=\frac{f^{(n+1)}(c)}{(n+1)!}(x-x_0)^{n+1}, \quad x_0 \text{ is point of expansion} \tag{6.9}$$

in which c is a point between x and x_0. Even though c is unknown, the above equation usually provides enough information for the estimate of an upper bound of error of the Taylor series expansion. In theory, Taylor series expansion can have an infinite series. In reality, any computation must involve only a finite number of terms. The infinite series must be truncated at some point. The error estimate, equation (6.9), says that if the Taylor series expansion is truncated at the $(n+1)$th term, the error is proportional to $(x - x_0)^{n+1}$. Therefore, if $x - x_0 < 1$, $(x - x_0)^{n+1}$ will approach 0 as n approaches infinity.

6.2 The Essence of Taylor Series Expansion

The idea of Taylor series expansion is to approximate a target function at a point x within a region near a given point x_0 with a polynomial of the distance from x_0, i.e. a polynomial of $x - x_0$. The coefficients of the polynomial can be uniquely determined by requiring that the polynomial has the same value and the same first order, second order, … , and nth order derivatives of the target function at the point x_0.

In other words, with Taylor series expansion, we require the approximate function to have the same value as the target function being represented. Further, we require that the approximate function must have the same slope as the target function. We require also that the two have the same rate of change of the slope (or same curvature) in the neighborhood of position x_0, and so on for all higher-order derivatives.

A polynomial function is simple in form, and the Taylor series expansion is quite generic. The beauty of this is that we are using the simplest function that we can think of, i.e. a polynomial function, to express any smooth or differentiable function. The condition that the target function has any order of derivative is a strict requirement. In real applications, however, we rarely need to deal with higher-order derivatives except for theoretical investigations.

6.3 Some Examples

6.3.1 Interpolation

Assume that a survey across a channel is conducted by measuring certain concentration in the water. The survey is somehow successful only partially so that there is a gap along the transect where the data are missing, as shown in Fig. 6.3. A common remedy is to fill the gap with an interpolation. If a linear interpolation is used, we are essentially using the first two terms of the Taylor series expansion: equation (6.2).

If only one point is inserted into the gap, the linear interpolation may or may not be smooth. This depends on the extent of the gap. The interpolation depends on

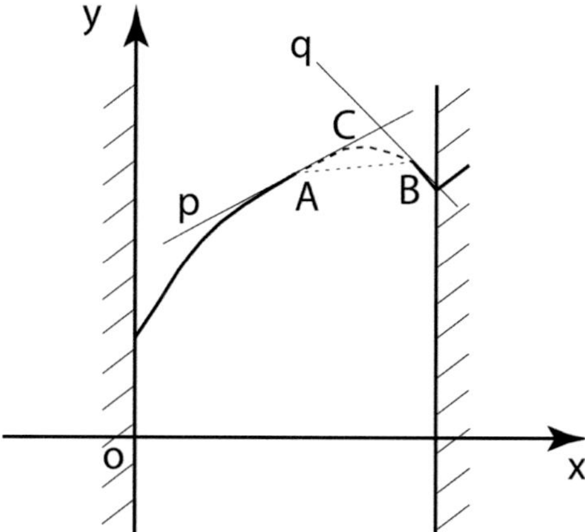

Figure 6.3 Gap of data across a channel. The x axis is the distance across the channel and y is the value of the measured quantity. Here y also indicates the direction of the channel. The dashed curve C between A and B is the hypothetical data that would have been measured. The solid curves are measured. The thin dashed line between A and B is a straight-line segment. The lines p and q are tangent lines at A and B, respectively.

how the first order derivative is estimated. If values at points A and B are used for the averaged derivative, the interpolated point would be on the straight line between A and B (the thin dashed line in Fig. 6.3). If multiple points on the left-hand (or right-hand) side of the gap were used to estimate the slope at A (or B) and an extrapolation were done, the filled data would be along line p (or q). In general, we would like to avoid using extrapolation as much as possible. The absolute error of the interpolation would be limited by

$$R_1 \leq \left|\frac{f''(\xi)}{2}\right| \overline{AB}^2. \tag{6.10}$$

Here ξ is a value between A and B. $\overline{AB}$ is the distance between A and B.

For a better approximation, a second order Taylor series expansion may be used, which would need to estimate the second order derivative. The equation to use would be (6.4). The actual implementation should be done according to the actual problems, e.g. how many data points are available, how large the gap is, and what accuracy is needed for the interpolation. The absolute error of the interpolation would be limited by

$$R_2 \leq \left|\frac{f'''(\xi)}{6}\right| \overline{AB}^3. \tag{6.11}$$

Here ξ again is a value between A and B.

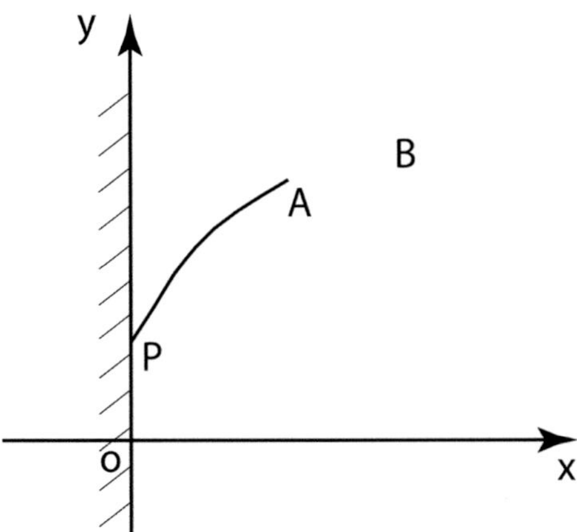

Figure 6.4 Data extrapolation from A to B. The x axis is the distance from shoreline toward offshore and y is the value of the measured quantity.

6.3.2 Extrapolation

Sometimes, due to a lack of sufficient data, an extrapolation is the only choice based on limited information. Now imagine that a model needs input data at point B (Fig. 6.4) in the ocean where there is no measurement. But there are measurements between the shoreline at P to an offshore location A. An extrapolation is the only way to achieve the goal. In this case, all the derivatives required in using the Taylor series expansion, including the error estimate, will have to be done using data between P and A – all one-sided derivatives. The error estimate equations (6.10) and (6.11) are still applicable, depending on whether (6.2) or (6.4) or a higher-order polynomial is used for the Taylor series expansion. Another example of extrapolation is when working with ADCP data for total transport through a water column. The velocity profiles measured by an ADCP have a blanking distance and side lobe effect which will result in a reduced valid data range. Extrapolations have to be done to allow a correct integration through the water column for the total transport.

6.4 Some Examples of Taylor Series Expansion

The following offers a few examples of Taylor series expansion of some well-known functions. The expansion is all made at $x_0 = 0$. Usually, the closer x is to x_0, the smaller the error of the approximation will be; the larger the number n is, the smaller the error of the approximation will be, based on equation (6.9). To get the more general form of Taylor series expansions at $x - x_0$, simply replace x in the following equations with $x - x_0$. The exponential function's Taylor expansion is

$$e^x = 1 + x + \frac{x^2}{2!} + \frac{x^3}{3!} + \cdots + \frac{x^n}{n!} + \cdots. \tag{6.12}$$

Taylor series expansion for the sine function is

$$\sin x = x - \frac{x^3}{3!} + \frac{x^5}{5!} - \frac{x^7}{7!} + \cdots + (-1)^{n-1}\frac{x^{2n-1}}{(2n-1)!} + \cdots. \tag{6.13}$$

Taylor series expansion for the cosine function is

$$\cos x = 1 - \frac{x^2}{2!} + \frac{x^4}{4!} - \frac{x^6}{6!} + \cdots + (-1)^{n-1}\frac{x^{2n-2}}{(2n-2)!} + \cdots. \tag{6.14}$$

Taylor series expansion for the simple rational function $1/(1 \pm x)$ is

$$\frac{1}{1 \pm x} = 1 \mp x + x^2 \mp x^3 + \cdots \quad (|x| < 1). \tag{6.15}$$

Taylor series expansion for the natural logarithm function $\ln(1 + x)$ is

$$\ln(1 + x) = x - \frac{x^2}{2} + \frac{x^3}{3} - \frac{x^4}{4} + \cdots + (-1)^{n-1}\frac{x^n}{n} + \cdots \quad (|x| < 1). \tag{6.16}$$

Taylor series expansion for the tangent function is

$$\tan x = x + \frac{x^3}{3} + \frac{2x^5}{15} + \frac{17x^7}{315} + \cdots. \tag{6.17}$$

Sometimes, the above equations (Taylor series expansion for $x_0 = 0$) is also called *the Maclaurin Series*.

6.5 An Example Using Taylor Series Expansion for Error Estimate

6.5.1 The Problem and Solution

Time series of water level and current velocity profiles from the Rigolets, one of Lake Pontchartrain's narrow tidal passes, were obtained during Hurricanes Gustav and Ike (Fig. 6.5) in 2008. The water level data (Li et al., 2010) obtained from a pressure sensor on an ADCP recorded 3953 data points at 5-minute intervals. The instrument was deployed at the bottom of the water, which was not all flat. The pressure sensor measured the water depth from the sensor to the surface of water. To have reliable measurements, the sensor should not move. Should it move over the undulating bottom, the total water depth due to water level change would be contaminated by the movement over variable bottom. This usually is not a concern as instruments are installed on a heavy metal frame and the frame does not move unless under extreme weather conditions, when the water flows at an abnormally high speed.

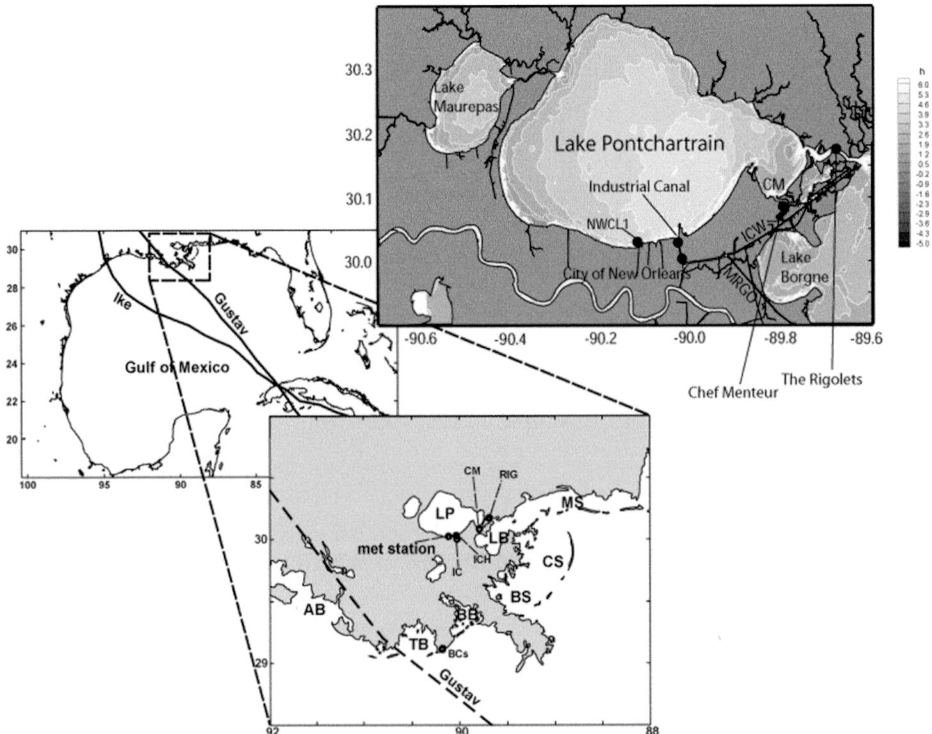

Figure 6.5 Study site, hurricane tracks, and mooring locations (circles) where CM is Chef Menteur, ICW is the Intracoastal Waterway, MRGO is the Mississippi River Gulf Outlet, NWCL1 is a NOAA NOS station, LP is Lake Pontchartrain, LB is Lake Borgne, MS is the Mississippi Sound, CS is the Chandeleur Sound, BS is the Breton Sound, BB is Barataria Bay, TB is Terrebonne Bay, RIG is for Rigolets, IC is the Industrial Canal and ICH is the location of the horizontal ADP, AB is Atchafalaya Bay, and BC is Bay Champaign.

Among the 3953 data points, most of the pressure (or water level) data had smooth changes, but there were 23 "sharp" jumps, corresponding to ~0.10–2.5 m depth changes over the 5-minute intervals in the Rigolets (or at a rate of 1.2–30.0 m/h) during the peak of Gustav's storm surge, suggesting abrupt movements of the deployed instrument under extreme flow conditions. With 3953 total records of data from the Rigolets up to September 10, the 23 jumps form about 0.58% of the total record. The platform was found to have moved about 50 m downstream from where it was originally deployed. The data after redeployment on September 10 were retrieved on October 8 and did not appear to have moved at all during Hurricane Ike.

Except during the 23 jumps, the maximum rate of increase of the surge (abnormal water level change) from Hurricane Gustav was about 0.35 m/h, much

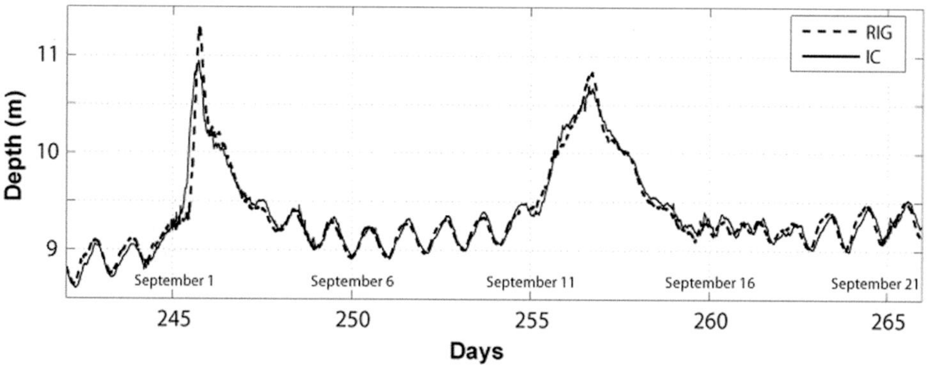

Figure 6.6 Corrected water level time series compared to that from a nearby location in the Industrial Canal.

smaller than the rate of the sharp jumps. The maximum velocity reached 2.3–2.5 m/s during the storm surge, according to the current meter data. The recorded pressures were a combination of the storm surge and the depth changes caused by the horizontal movement of the platform along a nonuniform bottom topography due to the swift currents during the storm surge.

To correct these errors, the times of the sharp pressure (or water level) changes were identified and corresponding data excluded, and each of these "bad" data was then replaced by a value consistent with the slope or rate of change of the adjacent (pre-jump time) observations. Accordingly, an adjustment was made to all data points after each jump. By doing so, the resultant storm surge curves were recovered. The final results produce a time series in agreement with data from nearby NOAA water level measurements and those from one of the other stations at the Industrial Canal (Fig. 6.6). This operation is not without error because the pressure changes induced by the movements of the two platforms cannot be 100% removed. An analysis of the errors introduced by this procedure is discussed below.

6.5.2 The Error Estimate

The 23 problematic data points in the time series during Hurricane Gustav were produced when the platform apparently moved. Excluding these "bad" data points, the averaged depth change and the associated standard deviation over the 5-minute time intervals were estimated to be 23.4 and 24.6 mm, respectively. If we use the standard deviation as the maximum error, the 23 corrections would result to a maximum error of 0.57 m. Although this is possible, it is most likely an overestimate. This is because within the 23 points, the correction may be larger or smaller than the actual water level change over the 5-minute interval. The positive and negative errors should cancel each other to some extent and reduce the overall error. This can be further justified by the following. In the correction, we used the slope of the surge to estimate the water level value at the bad data point. This is equivalent to using the Taylor series expansion:

$$f(t_1) = f(t_0) + f'(t_0)(t_1 - t_0) + \frac{f''(t_0)}{2!}(t_1 - t_0)^2 + o\left((t_1 - t_0)^3\right), \tag{6.18}$$

in which t_0 and t_1 are adjacent time instances when measurements were made in our application and $t_1 - t_0 = dt = 0.0035$ days (5 min). In our application, t_0 and t_1 are also the times with and without a valid data point, respectively. In the analysis, we used the adjacent "correct" slope of the data before the jump in value to correct the value at time t_1 when there was a jump in depth value. Therefore, we are using the first two terms of the above equation to approximate the depth value at t_1. The error of this operation is thus of the order of

$$E(t_1) = \frac{f''(t_0)}{2!}(t_1 - t_0)^2. \tag{6.19}$$

Here $f(t)$ is the depth time series. Since the data variations have certain random natures, the derivatives of the depth time series have large errors. We therefore used a 5-point smoothing function (a moving average) to smooth the time series before calculating the first and second order time derivatives, from which $E(t_1)$ can be calculated, as shown by Fig. 6.7. It is found that each of the corrections resulted in an error of a few millimeters, and the maximum accumulated error would be less than 0.23 m even if all errors had the same sign (so no cancellations of positive and negative errors). A further verification of this is the fact that the water level time series from the Rigolets matched that of a nearby station at the Industrial Canal after the Hurricane Ike-induced surge dissipated (Fig. 6.6). Based on this analysis and reasoning, we conclude that the algorithm of correction applied to the "bad data" due to apparent movement of the platform has a reasonably small total of accumulated errors (less than 0.23 m with the worst-case scenario).

6.6 Taylor Series Expansion for Two Variables

The beauty of Taylor series expansion does not stop at single-variable functions. Taylor series expansion can be extended to two variables and even multiple variables. This can be useful for multivariable problems in data analysis, experiments, and numerical modeling. Assuming that we have two independent variables x and y of a function f, the *two-variable Taylor series expansion* for $f(x, y)$ is expressed as

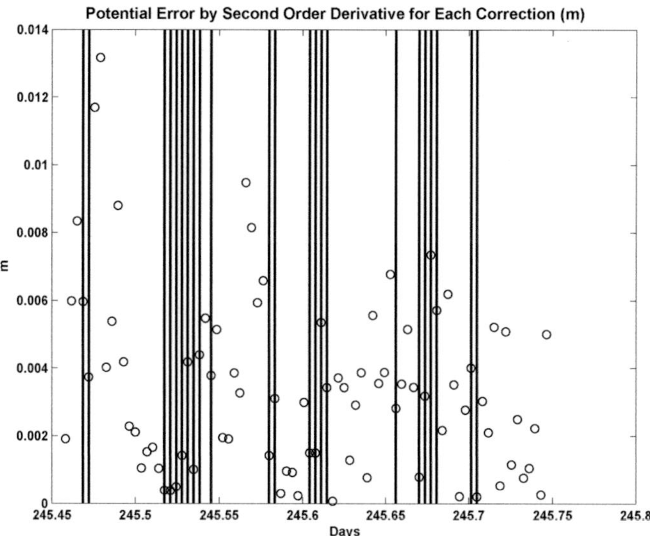

Figure 6.7 Potential error introduced when the water elevation value is approximated by the Taylor series expansion to the first order (using the function value and the slope value). The vertical bars show the actual times when the mooring apparently moved with a jump in depth values and where corrections were made on September 1, 2008 (Day 245 in 2008 was September 1).

$$
\begin{aligned}
f(x,y) =& f(x_0,y_0) + \left.\frac{\partial f}{\partial x}\right|_{(x_0,y_0)} (x-x_0) + \left.\frac{\partial f}{\partial y}\right|_{(x_0,y_0)} (y-y_0) \\
&+ \frac{1}{2!}\left[\left.\frac{\partial^2 f}{\partial x^2}\right|_{(x_0,y_0)} (x-x_0)^2 + 2\left.\frac{\partial^2 f}{\partial x \partial y}\right|_{(x_0,y_0)} (x-x_0)(y-y_0) + \left.\frac{\partial^2 f}{\partial y^2}\right|_{(x_0,y_0)} (y-y_0)^2 \right] \\
&+ \frac{1}{3!}\left[\left.\frac{\partial^3 f}{\partial x^3}\right|_{(x_0,y_0)} (x-x_0)^3 + 3\left.\frac{\partial^3 f}{\partial x^2 \partial y}\right|_{(x_0,y_0)} (x-x_0)^2(y-y_0) \right. \\
&\left. + 3\left.\frac{\partial^3 f}{\partial x \partial y^2}\right|_{(x_0,y_0)} (x-x_0)(y-y_0)^2 + \left.\frac{\partial^3 f}{\partial y^3}\right|_{(x_0,y_0)} (y-y_0)^3 \right] + \cdots \\
&+ \frac{1}{n!}\left[(x-x_0)\frac{\partial}{\partial x} + (y-y_0)\frac{\partial}{\partial y} \right]^n f|_{(x_0,y_0)} + R_n.
\end{aligned}
\tag{6.20}
$$

The error R_n is

$$
R_n = \frac{1}{(n+1)!}\left[(x-x_0)\frac{\partial}{\partial x} + (y-y_0)\frac{\partial}{\partial y} \right]^{n+1} f|_{(\xi,\eta)}, \quad (x_0 \le \xi \le x, y_0 \le \eta \le y).
\tag{6.21}
$$

Here the vertical bar with the coordinate (x_0, y_0) or (ξ, η) at its lower right corner specifies at which point the value of the function or the value of the derivative of the function is evaluated AFTER the operators have been applied to the function. Note that the above equation for R_n means that after the term within the square brackets with the $(n+1)$th power is expanded into individual terms of operators (of various order derivatives) and the partial derivatives have been done to the function f, the result is evaluated for the derivatives at (ξ, η). The factors $(x - x_0)$ and $(y - y_0)$, however, are not involved in this operation, i.e.

$$\left[(x - x_0)\frac{\partial}{\partial x} + (y - y_0)\frac{\partial}{\partial y}\right] f|_{(x_0, y_0)} = (x - x_0)\left.\frac{\partial f}{\partial x}\right|_{(x - x_0)} + (y - y_0)\left.\frac{\partial f}{\partial y}\right|_{(y - y_0)}. \tag{6.22}$$

In the operation, $(x - x_0)$ and $(y - y_0)$ are kept untouched.

In other words, in general, the symbol

$$\left[(x - x_0)\frac{\partial}{\partial x} + (y - y_0)\frac{\partial}{\partial y}\right]^n$$

is an operator that treats the partial derivatives $\frac{\partial}{\partial x}$ and $\frac{\partial}{\partial y}$ as variables when doing the expansion into a number of terms with the partial derivative operators, which are then applied to a function on the right (e.g. function $f(x, y)$, not shown in the above expression). It only has meaning when it is combined with a function on its right. If we select $n = 2$, the above operator applied to function $f(x, y)$ is

$$\begin{aligned}
&\left[(x - x_0)\frac{\partial}{\partial x} + (y - y_0)\frac{\partial}{\partial y}\right]^2 f|_{(x_0, y_0)} \\
&\quad = \left[(x - x_0)^2\frac{\partial^2}{\partial x^2} + 2(x - x_0)(y - y_0)\frac{\partial^2}{\partial x \partial y} + (y - y_0)^2\frac{\partial^2}{\partial y^2}\right] f|_{(x_0, y_0)} \\
&\quad = (x - x_0)^2\left.\frac{\partial^2 f}{\partial x^2}\right|_{(x_0, y_0)} + 2(x - x_0)(y - y_0)\left.\frac{\partial^2 f}{\partial x \partial y}\right|_{(x_0, y_0)} + (y - y_0)^2\left.\frac{\partial^2 f}{\partial y^2}\right|_{(x_0, y_0)}.
\end{aligned} \tag{6.23}$$

Again, the above operation keeps terms like $(x - x_0)^2$, $(x - x_0)(y - y_0)$, and $(y - y_0)^2$ untouched. If instead these terms were evaluated at e.g. $x = x_0$, they would be all zeros. Symbols like these are used to simplify the mathematical expressions.

6.7 Taylor Series Expansion for Multiple Variables

The above theory can be extended to a generic *Taylor series expansion for multiple variables*. For convenience of discussion, we now change our notation slightly. We assume that we have a function f which depends on m independent variables $x_1, x_2, \ldots, x_m$. We now use the following list of variables to denote the small change of these independent variables, respectively: $\delta x_1, \delta x_2, \ldots, \delta x_m$. The Taylor series expansion for $f(x_1, x_2, \ldots, x_m)$ around $(x_1, x_2, \ldots, x_m)$ is then

$$f(x_1+\delta x_1, x_2+\delta x_2, \ldots, x_m+\delta x_m) = f(x_1, x_2, \ldots, x_m) + \sum_{i=1}^{n}\frac{1}{i!}\left(\sum_{j=1}^{m}\delta x_j\frac{\partial}{\partial x_j}\right)^i f(x_1, x_2, \ldots, x_m) + R_n \cdot \tag{6.24}$$

The error is estimated by

$$R_n\frac{1}{(n+1)!}\left(\sum_{j=1}^{m}\delta x_j\frac{\partial}{\partial x_j}\right)^{n+1} f(x_1+\theta_1\delta x_1, x_2+\theta_2\delta x_2, \ldots, x_m+\theta_m\delta x_m), \tag{6.25}$$

in which $\theta_1, \theta_2, \ldots, \theta_m$ are all real numbers between 0 and 1.

An example of a multivariable function is salinity in the ocean as a function of the coordinate and time, i.e. $s = s(x, y, z, t)$, in which x, y, and z are the three-dimensional coordinates and t is time.

Equation (6.24) can be further generalized to *Taylor series expansion for an array of functions with multiple variables*, as in the following:

$$\begin{pmatrix} f_1(x_1+\delta x_1, x_2+\delta x_2, \ldots, x_m+\delta x_m) \\ f_2(x_1+\delta x_1, x_2+\delta x_2, \ldots, x_m+\delta x_m) \\ \vdots \\ f_q(x_1+\delta x_1, x_2+\delta x_2, \ldots, x_m+\delta x_m) \end{pmatrix} = \begin{pmatrix} f_1(x_1, x_2, \ldots, x_m) \\ f_2(x_1, x_2, \ldots, x_m) \\ \vdots \\ f_q(x_1, x_2, \ldots, x_m) \end{pmatrix} + \sum_{i=1}^{n}\frac{1}{i!}\left(\sum_{j=1}^{m}\delta x_j\frac{\partial}{\partial x_j}\right)^i \begin{pmatrix} f_1(x_1, x_2, \ldots, x_m) \\ f_2(x_1, x_2, \ldots, x_m) \\ \vdots \\ f_q(x_1, x_2, \ldots, x_m) \end{pmatrix} + R_n, \tag{6.26}$$

with an error estimate of

$$R_n = \frac{1}{(n+1)!}\left(\sum_{j=1}^{m}\delta x_j\frac{\partial}{\partial x_j}\right)^{n+1} \begin{pmatrix} f_1(x_1+\theta_1\delta x_1, x_2+\theta_2\delta x_2, \ldots, x_m+\theta_m\delta x_m) \\ f_2(x_1+\theta_1\delta x_1, x_2+\theta_2\delta x_2, \ldots, x_m+\theta_m\delta x_m) \\ \vdots \\ f_q(x_1+\theta_1\delta x_1, x_2+\theta_2\delta x_2, \ldots, x_m+\theta_m\delta x_m) \end{pmatrix}. \tag{6.27}$$

Again, $\theta_1, \theta_2, \ldots, \theta_m$ are all real numbers between 0 and 1. Here, $f_1, f_2, \ldots, f_q$ are functions. An example of this is for an array of parameters of ocean dynamics, e.g. the velocity components (u, v, w), water temperature (t), salinity (s), density (ρ), and wind speed components (w_x, w_y), so that

$$\begin{pmatrix} f_1(x_1, x_2, \ldots, x_m) \\ f_2(x_1, x_2, \ldots, x_m) \\ \vdots \\ f_8(x_1, x_2, \ldots, x_m) \end{pmatrix} = \begin{pmatrix} u \\ v \\ w \\ t \\ s \\ \rho \\ w_x \\ w_y \end{pmatrix}. \tag{6.28}$$

6.8 An Example of Taylor Series Expansion for an Array of Functions with Multiple Variables

The applications of Taylor series expansion related to data analysis in oceanography, especially those for multivariables, are not common in the literature. The value of this tool may have been underappreciated. Here we discuss one example to provide some ideas (Li et al., 2019a).

We know that the depth-averaged linearized momentum and continuity equations are usually given in the following form:

$$\frac{\partial u}{\partial t} - fv = -g\frac{\partial \zeta}{\partial x} + \frac{\tau_{ax}}{\rho h} - \frac{\tau_{bx}}{\rho h} - \frac{1}{\rho}\frac{\partial p_a}{\partial x}, \tag{6.29}$$

$$\frac{\partial v}{\partial t} + fu = -g\frac{\partial \zeta}{\partial y} + \frac{\tau_{ay}}{\rho h} - \frac{\tau_{by}}{\rho h} - \frac{1}{\rho}\frac{\partial p_a}{\partial y}, \tag{6.30}$$

$$\frac{\partial \zeta}{\partial t} + \frac{\partial hu}{\partial x} + \frac{\partial hv}{\partial y} = 0. \tag{6.31}$$

Here u and v are the depth-averaged horizontal velocity components in the x and y directions, respectively, as functions of time t; τ_{ax} and τ_{ay} are the surface (wind) stress components in the x and y directions, respectively, as functions of time t; while τ_{ax} and τ_{ay} are the bottom stress components; and ρ, h, ζ, and p_a are water density, depth, and surface elevation, respectively. Note that we are not trying to solve these equations, nor do they have a generic solution for an arbitrary situation. We are going to start from this point and use the Taylor series expansion to come up with a statistical model for empirical data. This will demonstrate the usefulness of Taylor series expansion for functions with multiple variables.

Although these momentum and continuity equations (6.29)–(6.31) are dependent on time, the *low-pass filtered functions* (after filtering out the tidal oscillations) sometimes satisfy a quasi-steady state balance in which the temporal derivatives are much smaller than the other terms in the above equations, and the following equations are approximately true for some weather-driven motions:

$$-f\bar{v} = -g\frac{\partial\bar{\zeta}}{\partial x} + \frac{\bar{\tau}_{ax}}{\rho h} - \frac{\bar{\tau}_{bx}}{\rho h} - \frac{1}{\rho}\frac{\partial\bar{p}_a}{\partial x}, \tag{6.32}$$

$$f\bar{u} = -g\frac{\partial\bar{\zeta}}{\partial y} + \frac{\bar{\tau}_{ay}}{\rho h} - \frac{\bar{\tau}_{by}}{\rho h} - \frac{1}{\rho}\frac{\partial\bar{p}_a}{\partial y}, \tag{6.33}$$

$$\frac{\partial h\bar{u}}{\partial x} + \frac{\partial h\bar{v}}{\partial y} = 0. \tag{6.34}$$

In the above equations, the bar over a variable means a low-pass filtered (Chapter 20) or subtidal signal (or a signal after taking out the tidal variations from the original time series). For the subtidal motion, the wind stress and air pressure are also low-pass filtered. The temporal variations of such problems are implied in the forcing factors, i.e. the wind stress and air pressure terms. The temporal derivatives are neglected simply because they are too small compared to the other terms. The wind stress $(\bar{\tau}_{ax}, \bar{\tau}_{ay})$ and air pressure $\bar{p}_a$ are slowly varying with time in the sense that they change at a rate slower than tides (timescale is longer than 24 hours).

The low-pass filtered wind stress is expressed by the quadradic law:

$$\bar{\boldsymbol{\tau}}_a = \rho C_{Da}\bar{\mathbf{w}}|\bar{\mathbf{w}}|, \tag{6.35}$$

in which $\bar{\boldsymbol{\tau}}_a = (\bar{\tau}_{ax}, \bar{\tau}_{ay})$ is the low-pass filtered wind stress vector, C_{Da} is the drag coefficient in the atmosphere on the ocean surface, and $\bar{\mathbf{w}}$ is the low-pass filtered wind velocity vector. For convenience, we omit the subscript a in the components so that $\bar{\boldsymbol{\tau}}_a = (\bar{\tau}_x, \bar{\tau}_y)$.

The three equations (6.32), (6.33), and (6.34) have three unknowns $(\bar{u}, \bar{v}, \bar{\zeta})$. Because of the complicated bathymetry, coastlines, and temporally and spatially varying forcing factors, it is impossible to obtain a general (or even a special) solution in an analytic form explicitly. However, with a proper boundary condition, a solution can be obtained for the unknowns, at least in theory. Assume that a solution for $(\bar{u}, \bar{v}, \bar{\zeta})$ is obtained. It must be a function of the forcing factors, i.e. the pressure change and wind stress $(\Delta\bar{p}_a, \bar{\tau}_x, \bar{\tau}_y)$. Here the air pressure change $\Delta\bar{p}_a$ instead of the air pressure $\bar{p}_a$ is used because what is important is the spatial gradient of the air pressure, not the absolute value. In theory, the solution can be expressed by the general format:

$$\begin{pmatrix} \bar{u} \\ \bar{v} \\ \bar{\zeta} \end{pmatrix} = \begin{pmatrix} \bar{u}(\Delta\bar{p}_a, \bar{\tau}_x, \bar{\tau}_y) \\ \bar{v}(\Delta\bar{p}_a, \bar{\tau}_x, \bar{\tau}_y) \\ \bar{\zeta}(\Delta\bar{p}_a, \bar{\tau}_x, \bar{\tau}_y) \end{pmatrix}, \tag{6.36}$$

in which the low frequency air pressure anomaly $\Delta\bar{p}_a$ is $\bar{p}_a - \bar{p}_{a0}$ and $\bar{p}_{a0}$ is the background or environmental air pressure or the average air pressure at sea level.

Now we can do a Taylor series expansion for the multiple functions $\bar{u}$, $\bar{v}$, and $\bar{\zeta}$ in terms of the variables $\Delta\bar{p}_a$, $\bar{\tau}_x$, and $\bar{\tau}_y$ to the first order:

$$\begin{pmatrix} \bar{u} \\ \bar{v} \\ \bar{\zeta} \end{pmatrix} = \begin{pmatrix} \bar{u}(0,0,0) + \dfrac{\partial \bar{u}}{\partial p_a}\Delta p_a + \dfrac{\partial \bar{u}}{\partial \bar{\tau}_x}\bar{\tau}_x + \dfrac{\partial \bar{u}}{\partial \bar{\tau}_y}\bar{\tau}_y \\ \bar{v}(0,0,0) + \dfrac{\partial \bar{v}}{\partial p_a}\Delta p_a + \dfrac{\partial \bar{v}}{\partial \bar{\tau}_x}\bar{\tau}_x + \dfrac{\partial \bar{v}}{\partial \bar{\tau}_y}\bar{\tau}_y \\ \bar{\zeta}(0,0,0) + \dfrac{\partial \bar{\zeta}}{\partial p_a}\Delta p_a + \dfrac{\partial \bar{\zeta}}{\partial \bar{\tau}_x}\bar{\tau}_x + \dfrac{\partial \bar{\zeta}}{\partial \bar{\tau}_y}\bar{\tau}_y \end{pmatrix}. \tag{6.37}$$

All the derivatives in equation (6.37) are evaluated at (0,0,0) for $(\Delta\bar{p}_a, \bar{\tau}_x, \bar{\tau}_y)$. Note that, for the quadratic wind stress, another Taylor series expansion can be applied to the quadratic velocity law to the second order or with a linear term and another term proportional to velocity squared, i.e.

$$\bar{\tau}_x = \alpha\bar{w}_x + \beta\bar{w}_x^2 + \text{higher order terms}, \tag{6.38}$$

$$\bar{\tau}_y = \gamma\bar{w}_y + \delta\bar{w}_y^2 + \text{higher order terms}, \tag{6.39}$$

where $\alpha, \beta, \gamma, \delta$ are all constants of the Taylor series expansions. The zeroth order terms are zero because when there is no wind, there is no stress, i.e. $\bar{\tau}_x = 0$, when $\bar{w}_x = 0$; and $\bar{\tau}_y = 0$, when $\bar{w}_y = 0$. Substituting (6.38) and (6.39) into (6.37) and neglect the higher order terms, we obtain:

$$\begin{pmatrix} \bar{u} \\ \bar{v} \\ \bar{\zeta} \end{pmatrix} = \begin{pmatrix} \bar{u}(0,0,0) + \dfrac{\partial \bar{u}}{\partial p_a}\Delta p_a + \dfrac{\partial \bar{u}}{\partial \bar{\tau}_x}(\alpha\bar{w}_x + \beta\bar{w}_x^2) + \dfrac{\partial \bar{u}}{\partial \bar{\tau}_y}(\gamma\bar{w}_y + \delta\bar{w}_y^2) \\ \bar{v}(0,0,0) + \dfrac{\partial \bar{v}}{\partial p_a}\Delta p_a + \dfrac{\partial \bar{v}}{\partial \bar{\tau}_x}(\alpha\bar{w}_x + \beta\bar{w}_x^2) + \dfrac{\partial \bar{v}}{\partial \bar{\tau}_y}(\gamma\bar{w}_y + \delta\bar{w}_y^2) \\ \bar{\zeta}(0,0,0) + \dfrac{\partial \bar{\zeta}}{\partial p_a}\Delta p_a + \dfrac{\partial \bar{\zeta}}{\partial \bar{\tau}_x}(\alpha\bar{w}_x + \beta\bar{w}_x^2) + \dfrac{\partial \bar{\zeta}}{\partial \bar{\tau}_y}(\gamma\bar{w}_y + \delta\bar{w}_y^2) \end{pmatrix}. \tag{6.40}$$

This can be rewritten as:

$$\begin{pmatrix} \bar{u} \\ \bar{v} \\ \bar{\zeta} \end{pmatrix} = \begin{pmatrix} A\bar{w}_x + B\bar{w}_x^2 + C\bar{w}_y + D\bar{w}_y^2 + E\Delta\bar{p}_a + F \\ G\bar{w}_x + H\bar{w}_x^2 + I\bar{w}_y + J\bar{w}_y^2 + K\Delta\bar{p}_a + L \\ M\bar{w}_x + N\bar{w}_x^2 + O\bar{w}_y + P\bar{w}_y^2 + Q\Delta\bar{p}_a + R \end{pmatrix}, \tag{6.41}$$

in which,

$$A = \alpha \frac{\partial \bar{u}}{\partial \tau_x},\ B = \beta \frac{\partial \bar{u}}{\partial \tau_x},\ C = \gamma \frac{\partial \bar{u}}{\partial \tau_y},\ D = \delta \frac{\partial \bar{u}}{\partial \tau_y},\ E = \frac{\partial \bar{u}}{\partial p_a},\ F = \bar{u}(0,0,0),$$

$$G = \alpha \frac{\partial \bar{v}}{\partial \tau_x},\ H = \beta \frac{\partial \bar{v}}{\partial \tau_x},\ I = \gamma \frac{\partial \bar{v}}{\partial \tau_y},\ J = \delta \frac{\partial \bar{v}}{\partial \tau_y},\ K = \frac{\partial \bar{v}}{\partial p_a},\ L = \bar{v}(0,0,0),$$

$$M = \alpha \frac{\partial \bar{\zeta}}{\partial \tau_x},\ N = \beta \frac{\partial \bar{\zeta}}{\partial \tau_x},\ O = \gamma \frac{\partial \bar{\zeta}}{\partial \tau_y},\ P = \delta \frac{\partial \bar{\zeta}}{\partial \tau_y},\ Q = \frac{\partial \bar{\zeta}}{\partial p_a},\ R = \bar{\zeta}(0,0,0).$$

Taking the first equation of (6.41) as an example,

$$\bar{u} = A\bar{w}_x + B\bar{w}_x^2 + C\bar{w}_y + D\bar{w}_y^2 + E\Delta\bar{p}_a + F. \tag{6.42}$$

This equation is not an explicit solution, and the coefficients are all unknowns that can be determined with a regression using observational data. It has been used for data from several estuaries to statistically calculate the estuarine flows caused by weather, especially by atmospheric frontal passages (Li et al., 2019a,b). The statistical regression is a topic discussed in Chapter 8.

Now we discuss an example, applying the above theory for the Barataria Bay, Louisiana. The observations were done between Dec. 2014 and Mar. 2015. A Sontek Argonaut DP SL 500 kHz horizontal acoustic Doppler current profiler (ADCP) was used at the center of the Caminada Pass. The ADCP was deployed at ~2 m below the surface, looking horizontally, with the signal beam perpendicular to the flow, covering a distance of ~100 m with five data points along the beam of the acoustic signal – about one point every 20 m. The instrument was powered by a battery recharged during the day through a solar panel mounted on site. The data were saved at 15-minute ensemble intervals. The wind data were from the Grand Isle station of the National Ocean Service located at 29.265° N 89.958° W and the South Timbalier Block 52 station or CSI 6. The wind data have 6-minute ensemble average intervals and are consistent in magnitude and temporal variations. The wind data from CSI 6 are only used to fill the gaps in the Grand Isle data, mostly in 2015. The coordinate system is rotated first to be aligned with the principal component axis (which is basically the same as the axis of the main channel) of the data before analysis. This way, the major component will be along the channel because of the geometric constraint of a very narrow channel. A 40-hour *low-pass filter* (Chapters 20 and 21) was used for the along-channel velocity u to filter out all tidal and higher frequency components of the flow. The same filter was used for the wind vector time series $\boldsymbol{w}$. A regression is then done for equation (6.42) to determine the coefficients A, B, C, D, E, and F, which are:

$A = 0.886$ with a confidence interval of $C_A = [0.855, 0.917]$,
$B = 0.049$ with a confidence interval of $C_B = [0.044, 0.054]$,
$C = 1.623$ with a confidence interval of $C_C = [1.599, 1.647]$,

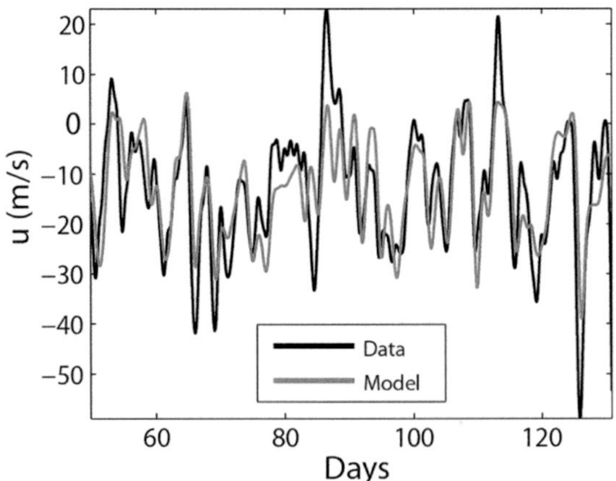

Figure 6.8 Comparison between low-pass filtered observed along-channel velocity measured at Caminada Pass of the Barataria Bay, Louisiana. The time axis starts from the first date of observations.

$D = -0.045$ with a confidence interval of $C_D = [-0.048, -0.042]$,
$E = 0.409$ with a confidence interval of $C_E = [0.385, 0.433]$,
$F = -5.364$ with a confidence interval of $C_F = [-5.555, -5.173]$.

The overall goodness of fit is $R^2 = 0.584$. In this specific example, the unit for wind speed is m/s, the unit for air pressure is millibar, and the final unit for the low-pass filtered velocity is cm/s. Fig. 6.8 shows the comparison between the regression model from (6.42) and the observations.

Review Questions for Chapter 6

(1) If a continuous function has a sharp point, does it have a Taylor series expansion?
(2) Can you think of any limitation of Taylor series expansion?

Exercises for Chapter 6

(1) Verify Taylor series expansions in (6.12)–(6.17), using (6.8).
(2) Do a Taylor series expansion for a two-variable function $f(x) = e^x \cos x$ up to the third order using equation (6.8).
(3) Use Taylor series expansion for a single variable function e^x from (6.12) and that for $\cos x$ from (6.14) and multiply the results. Rearrange and order the

terms in ascending order. Compare this result with that from the direct Taylor series expansion from the above question. Are they the same?

(4) Do a Taylor series expansion for a two-variable function $f(x, y) = e^x \cos y$ up to the third order using equation (6.20).

(5) Use Taylor series expansion for a single variable function e^x and that for another $\cos y$ and multiply the results. Compare this result with that from the direct Taylor series expansion for two variables. Are they equivalent?

(6) We have learned that for a variable that is a product of several other variables, the relative error after the multiplication is the sum of the relative error of each factor. For example, the volume of a cuboid (V) is the multiplication of its length (L), width (W), and height (H), $V = LWH$. If a measurement is done for $L, W,$ and H with absolute random errors of δL, δW, and δH, respectively, the relative error of V is

$$e_V \leq e_L + e_W + e_H,$$

then, use a Taylor series expansion of the function V in terms of the multiple variables L, W, and H to the first order (neglecting the second order) and verify the above error estimate. What happens if you keep the second order?

(7) The limit of $f(x) = \frac{e^x - 1}{x}$ at $x \to 0$ is 1. A direct computation using MATLAB, however, can be problematic; while using Taylor series expansion, the problem can be resolved. Do the following to verify:

(a) Define a variable that is approaching 0: $x = \left[0.1, 0.01, \ldots, 0.1^{18}\right]$.

(b) Write a MATLAB script and compute $f(x) = \frac{e^x - 1}{x}$.

(c) Plot $f(x)$ (does the result approach to 1?).

(d) Use the Taylor series expansion of $f(x) = \frac{e^x - 1}{x} \approx 1 + \frac{1}{2!}x + \frac{1}{3!}x^2 + \frac{1}{4!}x^3$ and write a MATLAB script.

(e) Plot the result from Taylor series expansion (does the result approach 1?).

(6) Compare the results from these two methods and explain why there is a difference. [Note: do not use MATLAB's Symbolic Math Toolbox for computing functions.]

7

Spherical Trigonometry and Distance Computation

About Chapter 7

This chapter discusses some basic spherical trigonometry applicable to distance computations between points on the surface of the Earth, including in the ocean, particularly for large-scale problems in oceanography. Because of the curvature of the Earth, the plane geometry is not applicable.

7.1 Need for Spherical Trigonometry

For most environmental data, it is necessary to know the position from where the data are obtained. The position of measurements may be used to calculate distance traveled, speed, acceleration, etc. On the spherical Earth, the computation is done with given information of the geolocations (i.e. latitude, longitude, and height in the air or depth in the water with a given vertical reference). Since the surface of the Earth is approximately a sphere, the measurement of distance over the surface of the Earth can use a mathematical tool based on the theory of spherical trigonometry if the accuracy requirement is not greater than 0.3%. Spherical trigonometry is most often used in astronomy and navigation in the ocean. For higher resolution applications, a more accurate theory based on an ellipsoidal Earth may be needed, which is not discussed here.

With the objective of understanding the basic spherical trigonometry for the calculation of the distance on the surface of the globe, we will present some equations derived from spherical trigonometry. We will also discuss an approximation in a *local Cartesian coordinate system* if the points of interest are all very close (much less than the Earth's radius). In the following, we introduce some concepts before we present the equations for the distance calculations.

7.2 A Few Definitions

7.2.1 The Sphere

A sphere centered at O with a radius r is a collection of points satisfying the following equation in a Euclidean Space Cartesian Coordinate System centered at O:

$$x^2 + y^2 + z^2 = r^2. \tag{7.1}$$

The area of a sphere is $4\pi r^2$, while the volume enclosed by the sphere is $4\pi r^3/3$.

To get a flavor of what spherical trigonometry is about or can do, we present the following interesting theorem: it can be proven that, for horizontal strips of a sphere with equal difference in z coordinate (Fig. 7.1), their areas on the sphere are equal.

This theorem can be verified by the integration of the area over a strip of the sphere defined between two z coordinates (Fig. 7.1).

$$S_{1,2} = \int_{z_1}^{z_2} ds = \int_{z_1}^{z_2} 2\pi R dl = \int_{z_1}^{z_2} 2\pi R\sqrt{dR^2 + dz^2}, \tag{7.2}$$

where $R = \sqrt{x^2 + y^2} = \sqrt{r^2 - z^2}$, which leads to

$$dR = -\frac{z}{R} dz. \tag{7.3}$$

We obtain

$$S_{1,2} = \int_{z_1}^{z_2} 2\pi R\sqrt{\left(\frac{z}{R}\right)^2 + 1}\, dz = \int_{z_1}^{z_2} 2\pi r\, dz = 2\pi r(z_2 - z_1). \tag{7.4}$$

The theorem is thus proven.

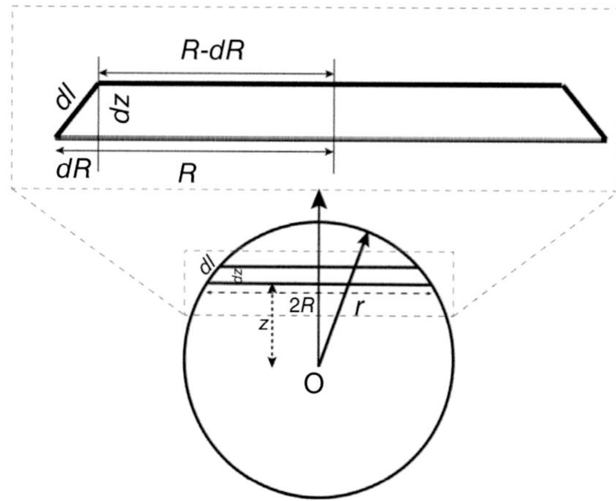

Figure 7.1 Proof of equation (7.4). z is the vertical coordinate.

Obviously, the result says that the area of the strip only depends on the difference in the z coordinate and the radius of the sphere. It does not depend on the latitude or the position of the strip. This is a generic and interesting conclusion but may not be obvious.

7.2.2 Radian

In plane geometry, an angle can be measured in degrees, but it is also measured in *radian*, which is defined by the length of an arc of a unit circle (Fig. 7.2a). So, a 360° circle is an angle of 2π in radian and a right angle (90°) is an angle of $\pi/2$ in radian.

7.2.3 Solid Angle and Steradian

Similar to the use of radian, an angle in space (or *solid angle*) is defined by a collection of rays coming from a given origin (e.g. the center of a sphere), measured by the area of A on the *unit sphere* (with a unit radius) centered at the origin. This area A is a collection of all points of the rays intersecting with the unit sphere (Fig. 7.2b). The unit of a solid angle is *steradian*. Using equation (7.4), for the area between $z = 0$ and $z = r$ (half of the sphere), the area is $S = 2\pi r(z_2 - z_1) = 2\pi r^2$. This verifies that it is indeed half of the surface area of a sphere. For a unit sphere, $r = 1$ and $S = 2\pi$. This is also the solid angle in steradian for half of the sphere. The entire sphere has a solid angle of 4π.

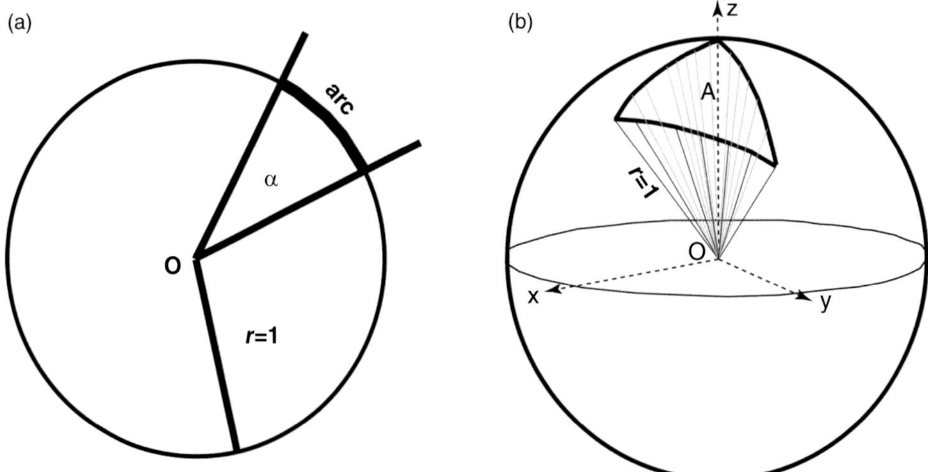

Figure 7.2 The left graph shows the definition of radian (the length of the thick line arc) while the right graph shows an example of the *steradian* (the area enclosed by the thin rays, assuming the sphere has a unit length radius).

7.3 The Measure of "Distance" in Spherical Trigonometry

In the traditional spherical trigonometry, we often use angles to measure a "distance." For the actual distance or length of lines on a sphere, we must also use the radius in the calculation.

7.3.1 Geodesic Line

A *geodesic line* is a line on a surface with the shortest distance between two points. For instance, on a two-dimensional plane, a geodesic line is a straight line between two points. Generally, for a curved surface, the geodesic line can be complicated to find. The geodesic line on a sphere, however, is simpler to find, which is discussed below.

7.3.2 The Great Circle, Small Circle, and Poles

A geodesic line on a sphere is a line on a *great circle*. A great circle is a circle on a sphere with the center of the sphere as the center of the circle. A longitude line is half of a great circle. The equator is a great circle. Except the equator, all latitudinal circles are not great circles. Any circle on a sphere other than great circles is a small circle.

The *poles* of a great circle are two points on the sphere, each of which is 90° away from all points on the great circle. An *arc* is a segment of a circle. For convenience, the *north pole* of a great circle is at $(0, 0, r)$, while the *south pole* is defined at $(0, 0, -r)$; here r is the radius of the sphere.

7.3.3 Spherical Lune or Spherical Biangle

A *spherical lune* or a *spherical biangle* on a sphere is defined by two half great circles sharing the same poles (Fig. 7.3). This is like a "polygon" on a sphere with only two sides, which are all half great circles. The angle α at one of the vertices A is defined by the angle between the two half lines tangent to the two half-great-circle edges (or sides) e and e′ at point A (Fig. 7.3), respectively. Similarly, the angle α' at the vertex A′ is the angle between the two half lines tangent to e and e′ at point A′, respectively. It can be seen that the lines tangent to e at A and A′ are parallel and those to e′ are also parallel. Thus, it can be seen that the two angles α and α' must be the same. For that reason, we can just call this angle the angle of the biangle. It can be shown that, for a unit sphere, the area of a biangle is twice the biangle α. For any sphere with a given radius, the area of a biangle is twice the angle α multiplied by the square of the radius of the sphere.

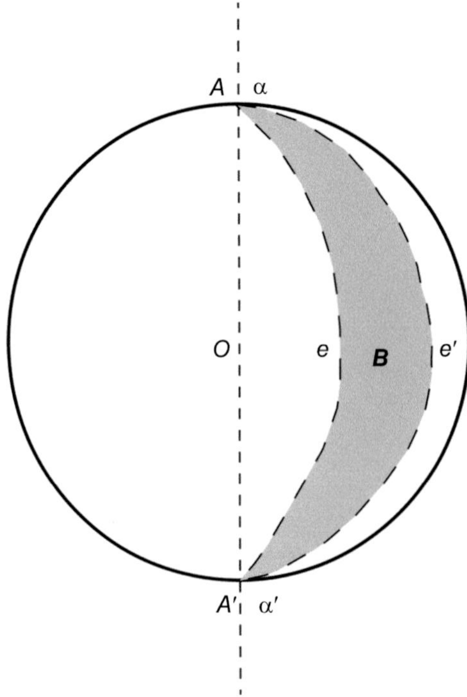

Figure 7.3 Biangle definition.

7.3.4 Spherical Triangle

A *spherical triangle* is a "polygon" on a sphere formed by the arcs of three great circles. When distance is measured on a sphere, we use a great circle, which is similar to a straight line in plane geometry as both are geodesic lines on their given respective surfaces. Between two points on a sphere, the great circle line segment gives the shortest distance. The three sides of a spherical triangle are measured by angles (in degrees or radians), not by length. The three angles of a spherical triangle are also measured by angles (in degrees or radians). Therefore, the description of a spherical triangle can be given in terms of six angles in degrees or radians. All theorems in spherical trigonometry are presented with respect to these six angles. This is a little different from the convention in plane geometry in which the lengths of sides of a triangle often appear as quantities in the formulas and theorems.

For convenience, it is customary that we use upper case letters, e.g. A, B, C, to denote both the vertices and the angles of a spherical triangle at the three vertices, whereas we use lower case letters, e.g. a, b, c, to denote the angles of the arcs or sides of the spherical triangle (Fig. 7.4).

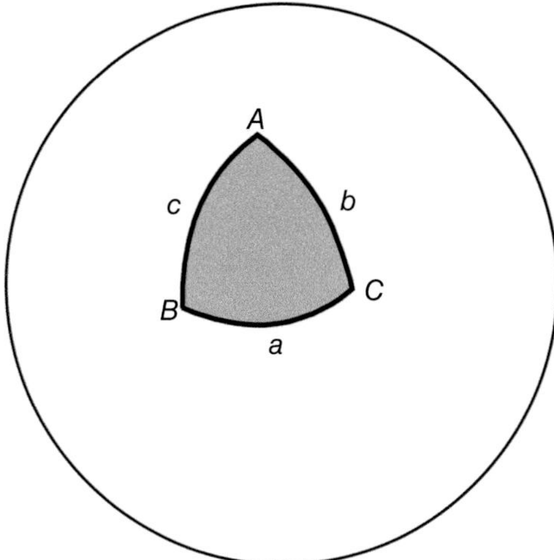

Figure 7.4 Spherical triangle.

7.3.5 Properties of Spherical Triangles

There are some basic properties of a spherical triangle. Assume the sides of a spherical triangle are denoted by a, b, and c, respectively, and the three angles by A, B, and C, respectively (Fig. 7.4); side a is on the opposite side of A, side b is on the opposite side of B, and side c is on the opposite side of C. The main properties of a general spherical triangle include the following:

(1) **Limit on each angle:** each spherical angle is $< 180^\circ$.

(2) **Sum of three sides:** the sum of the three *sides* is between 0° and 360°, i.e.

$$0^\circ < a + b + c < 360^\circ. \tag{7.5}$$

(3) **Sum of two sides:** any two *sides* of a spherical triangle added together are greater than the third, i.e.

$$a + b > c. \tag{7.6}$$

(4) **Difference of two sides:** the difference of two *sides* is smaller than the third side, i.e.

$$|a - b| < c. \tag{7.7}$$

(5) **Side-angle relationship**: the side opposite a greater angle is greater than the side opposite a smaller angle, i.e.

$$\text{If } A > B, \text{ then } a > b. \tag{7.8}$$

(6) **Sum of two angles:** the sum of any two of the three angles is less than the third angle plus 180°, i.e.

$$A + B < C + 180^\circ. \tag{7.9}$$

(7) **Sum of three angles:** the sum of all three *angles* is greater than 180° but smaller than 540°

$$180^\circ < A + B + C < 540^\circ. \tag{7.10}$$

Because the sum of the three angles of any spherical triangle is greater than 180°, the difference between the sum of the three angles and 180° is called the *spherical excess*, i.e.

$$\delta = A + B + C - 180^\circ. \tag{7.11}$$

(8) **Area of spherical triangle:** the area of a spherical triangle is related to the spherical excess as the following:

$$S = r^2 \frac{\pi}{180}\delta, \tag{7.12}$$

in which r is the radius of the sphere and δ is defined in (7.11).

7.4 Laws in Spherical Trigonometry

7.4.1 The Cosine Law for Sides

In plane geometry, there is a cosine law that relates the sides of a triangle and one of the three angles. More specifically, it states that one side squared is equal to the sum of the squares of the other two sides, minus twice the product of the two sides, multiplied by the cosine of the angle formed by these two sides (Fig. 7.5). In mathematics, it is:

$$c^2 = a^2 + b^2 - 2ab\cos\alpha. \tag{7.13}$$

Here α is the angle C or $\angle C$. The main message is that, given two sides of a triangle and the angle formed by these two sides, the third side can be calculated with this cosine law, which is a generalized *Pythagorean theorem*: if $\alpha = \pi/2$, (7.13) becomes the Pythagorean theorem.

In spherical trigonometry, the "distance" is the angle measured by an arc on a great circle. A similar *cosine law* exists relating the angles defined on a sphere:

$$\cos a = \cos b \cos c + \sin b \sin c \cos A. \tag{7.14}$$

This is a well-known relation in spherical trigonometry. Even though equation (7.14) looks quite different from equation (7.13), the idea is similar: given two sides (b and c) of a spherical triangle and the angle formed by the two sides (A), the third side (a) can be calculated with the cosine law of spherical trigonometry.

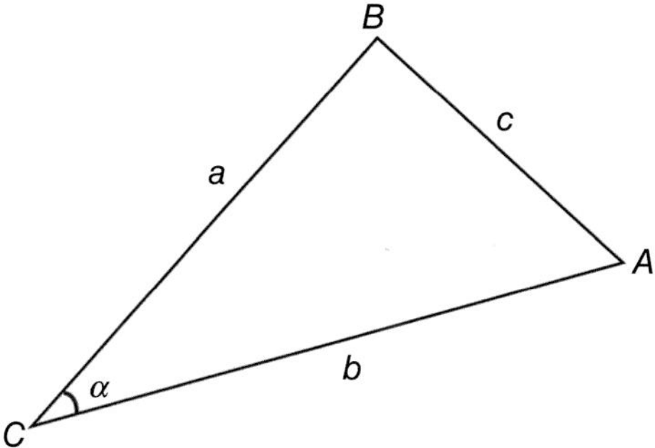

Figure 7.5 Plane triangle and cosine law.

The cosine law for sides has other two forms:

$$\cos b = \cos c \cos a + \sin c \sin a \cos B, \tag{7.15}$$

$$\cos c = \cos a \cos b + \sin a \sin b \cos C. \tag{7.16}$$

7.4.2 Other Laws

In addition to the cosine law, we also have the *sine law*:

$$\frac{\sin A}{\sin a} = \frac{\sin B}{\sin b} = \frac{\sin C}{\sin c}, \tag{7.17}$$

and the *five-element law*:

$$\sin a \cos B = \cos b \sin c - \sin b \cos c \cos A, \tag{7.18}$$

$$\sin a \cos C = \cos c \sin b - \sin c \cos b \cos A, \tag{7.19}$$

$$\sin A \cos b = \cos B \sin C + \sin B \cos C \cos a, \tag{7.20}$$

$$\sin A \cos c = \cos C \sin B + \sin C \cos B \cos a. \tag{7.21}$$

In addition, spherical trigonometry has cosine laws for angles, sine–cosine laws, polar sine–cosine laws, and many useful formulas, which are all omitted. As far as our applications are concerned, the most useful is probably the cosine law for sides in calculating the distance between two points on the surface of the Earth.

7.4.3 Examples

Example 1 Given the latitude and longitude of two points on the surface of the Earth, P1 (C) and P2 (B) (Fig. 7.6), find the shortest distance between them.

If we use λ and φ to denote longitude and latitude, respectively, the coordinates of the two points are expressed as:

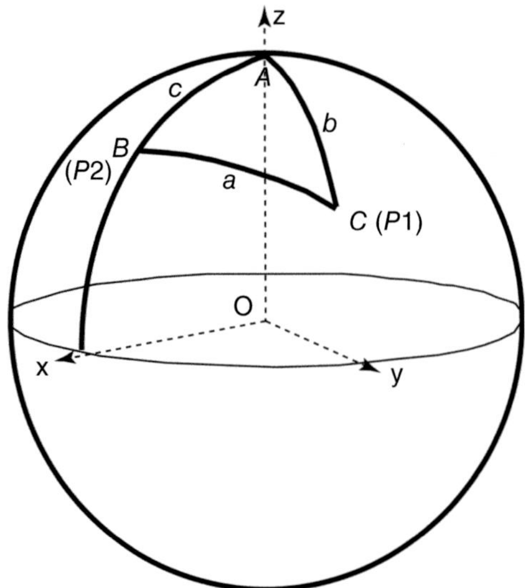

Figure 7.6 Calculation of distance on a sphere.

$$P1 = (\lambda_1, \varphi_1), P2 = (\lambda_2, \varphi_2). \tag{7.22}$$

From the definition of latitude, we have the measure in radian of the arcs b and c:

$$b = \frac{\pi}{2} - \varphi_1, c = \frac{\pi}{2} - \varphi_2. \tag{7.23}$$

Now note that

$$\begin{aligned}
\cos b &= \cos\left(\frac{\pi}{2} - \varphi_1\right) = \sin\varphi_1 \\
\cos c &= \cos\left(\frac{\pi}{2} - \varphi_2\right) = \sin\varphi_2 \\
\sin b &= \cos\varphi_1, \quad \sin c = \cos\varphi_2 \\
\cos A &= \cos(\lambda_2 - \lambda_1).
\end{aligned}$$

Applying these to the cosine law or equation (7.14), we have

$$\cos a = \sin\varphi_1 \sin\varphi_2 + \cos\varphi_1 \cos\varphi_2 \cos(\lambda_2 - \lambda_1), \tag{7.24}$$

$$a = \arccos\big(\sin\varphi_1 \sin\varphi_2 + \cos\varphi_1 \cos\varphi_2 \cos(\lambda_2 - \lambda_1)\big), \tag{7.25}$$

$$\overline{P1P2} = Ra, \tag{7.26}$$

in which R in (7.26) is the radius of the Earth, which has an average value of 6371 km.

Example 1 Calculate the shortest distance between New York and Los Angeles. New York = [40° 46′N, 73° 57′W], Los Angeles = [34° 2′N, 118° 19′W].

Answer: the latitudes and longitudes are

$$\varphi_1 = 40 + 46/60, \quad \varphi_2 = 34 + 2/60$$
$$\lambda_1 = 73 + 57/60, \quad \lambda_2 = 118 + 19/60.$$

Using equations (7.24)–(7.26),

$$\begin{aligned} \cos a &= \sin\varphi_1 \sin\varphi_2 + \cos\varphi_1 \cos\varphi_2 \cos(\lambda_2 - \lambda_1) \\ &= 0.8141 \\ a &= \arccos(0.8141) = 0.6195 \\ \overline{P1P2} &= Ra = 6371 \times 0.6195 = 3947(\text{km}). \end{aligned}$$

7.5 Calculation When the Angle Is Small

Sometimes, two points on a sphere are "very close," and the calculation of the distance can use an approximation of the cosine law in a local x-y Cartesian coordinate system (Fig. 7.7).

Given P1 and P2 in Fig. 7.6 are very close, with $a \ll 1$, i.e.

$$a \ll 1; \varphi_1 \approx \varphi_2 \approx \varphi = (\varphi_1 + \varphi_2)/2; \lambda_2 - \lambda_1 \ll 1. \tag{7.27}$$

We now seek an approximation of equations (7.24)–(7.26). From Taylor series expansion, we have

$$\cos a = \sqrt{1 - \sin^2 a} \approx 1 - \frac{a^2}{2},$$

$$\cos(\delta\lambda) = \sqrt{1 - \sin^2(\delta\lambda)} \approx 1 - \frac{(\delta\lambda)^2}{2},$$

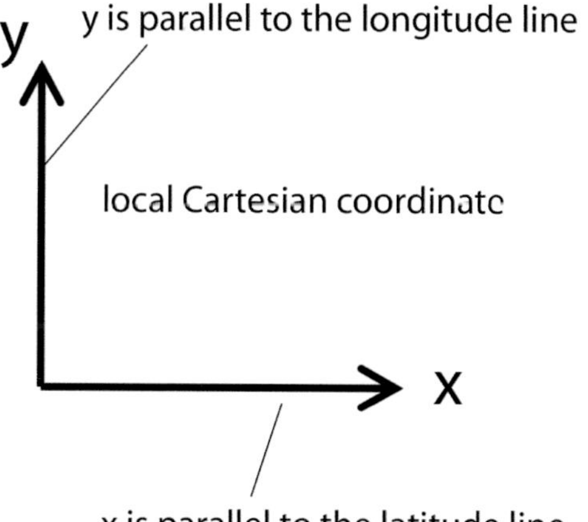

Figure 7.7 Local Cartesian coordinate.

$$\delta\lambda = \lambda_2 - \lambda_1, \tag{7.28}$$

$$\cos(\delta\varphi) = \sqrt{1 - \sin^2(\delta\varphi)} \approx 1 - \frac{(\delta\varphi)^2}{2},$$

$$\delta\varphi = \varphi_2 - \varphi_1. \tag{7.29}$$

Therefore, we have

$$\begin{aligned} 1 - \frac{a^2}{2} &\approx \sin\varphi_1 \sin\varphi_2 + \cos\varphi_1 \cos\varphi_2 \left(1 - \frac{\delta\lambda^2}{2}\right) \\ &= \cos(\varphi_2 - \varphi_1) - \frac{\delta\lambda^2}{2} \cos\varphi_1 \cos\varphi_2 \\ &\approx 1 - \frac{\delta\varphi^2}{2} - \frac{\delta\lambda^2}{2} \cos^2\varphi, \quad \left(\varphi = \frac{\varphi_1 + \varphi_2}{2}\right) \end{aligned}$$

This leads to

$$\begin{aligned} & a \approx \sqrt{\delta\varphi^2 + \delta\lambda^2 \cos^2\varphi} \\ & \therefore \text{distance} = Ra = \sqrt{(R\delta\varphi)^2 + \delta\lambda^2 (R\cos\varphi)^2} = \sqrt{dy^2 + dx^2}. \end{aligned} \tag{7.30}$$

The key is to find dx (Fig. 7.8) and dy (Fig. 7.9):

$$dx = \delta\lambda R \cos\varphi, \quad dy = R\delta\varphi, \tag{7.31}$$

in which φ is the averaged latitude using the middle equation of (7.27).

In oceanography, like in meteorology, the use of local Cartesian coordinates is for "small scale" problems. Hereby small scale, it is for problems with spatial scales small enough so that we can ignore the effect of the Earth's curvature. In such a coordinate system, the positive x-axis is usually in the east direction, while the positive y-axis is in the north direction, and the z-axis is positive upward. For

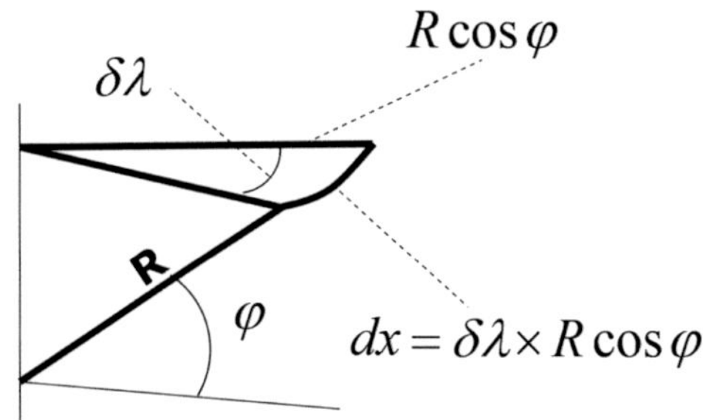

Figure 7.8 Calculation of dx in a local Cartesian coordinate system.

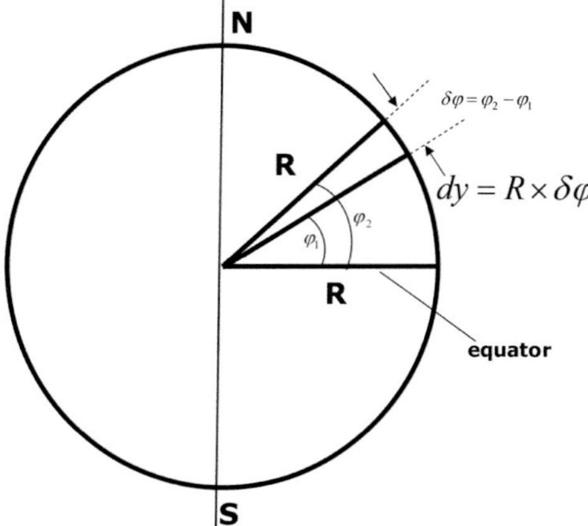

Figure 7.9 Calculation of dy in a local Cartesian coordinate system.

example, for problems in an estuary or a small bay of tens of kilometers, the use of local coordinates is usually sufficient.

Review Questions for Chapter 7

(1) What is the minimum angle of a biangle?
(2) What is the maximum angle of a biangle?
(3) Why do we use an angle to describe a side of a spherical triangle, not length?
(4) If we have to consider the topographic features of the Earth's surface, will a geodesic line between two points on the Earth be along a great circle?

Exercises for Chapter 7

(1) Using results from equation (7.4), verify that the surface area of a sphere is indeed the well-known $A = 4\pi r^2$.
(2) Write a function that calculates the distance on Earth's surface given the latitudes and longitudes of two points, and then use this function to calculate the distance between the following two cities: Baton Rouge (30°27′N, 91°09′W) and Shanghai (31°08′, 121°49′E).
(3) Using Google Earth, find the latitudes and longitudes of Washington, DC, and Norfolk, Virginia. Use spherical trigonometry calculate their distance, and compare with the results using the local Cartesian coordinate approximation.
(4) A ship (A) in the Atlantic Ocean is in trouble and sends out an SOS signal with its location at 31°53′6″N, 66°09′22″W. The signal is received by a nearby ship (B) at 29°51′11.7″N, 73°19′56.38″W. Assuming that ship A is anchored (no

movement) and ship B has a maximum speed of 12 knots, how long will it take for B to reach A, assuming the ship is moving along a great circle?

(5) Calculate speed and acceleration of a boat using GPS data: a GPS data file named "GPS20140729a.txt" is provided for this problem. The data include the time and position (in longitude and latitude) of a boat doing a survey in an arctic lagoon. Write a MATLAB script to calculate the boat's speed and acceleration as a function of time.

(6) Use the last property presented by equation (7.12) to verify that the area of half of a sphere is $2\pi r^2$, in which r is the radius of the sphere.

(7) Prove that the spherical cosine law, or

$$\cos a = \cos b \cos c + \sin b \sin c \cos A$$

leads to the plane geometry cosine law, or equation

$$a^2 = b^2 + c^2 - 2bc \cos(\alpha)$$

when the radius of the sphere approaches infinity or the curvature of the sphere becomes negligible.

(8) Prove that for a unit sphere, the area of a biangle is twice the angle of the biangle α.

(9) Prove that for any sphere with a given radius, the area of a biangle is twice its angle α multiplied by the square of the radius of the sphere.

8

A System of Linear Equations and Least Squares Method

About Chapter 8

The objective of this chapter is to prepare for the use of MATLAB to find solutions for a system of linear equations. The important concepts include the inverse matrix, solution of a system of linear equations, least squares method, and MATLAB's "left-division," which is essentially an implementation of the least squares method. MATLAB functions are very useful and efficient in the job. For real problems, the equations are often highly overdetermined, which occurs when there are many more equations (or measurements) than the number of unknowns, and thus it requires the use of the least squares method to solve the solution in a statistical sense. The contents in this chapter lay out a foundation for several later chapters because many theories and methods are related in one way or the other to the solution of a system of linear equations, e.g. Fourier analysis and harmonic analysis.

8.1 Linear Equations and Matrix Operation

A large part of time series data analysis involves computations using *matrix operations*. These mainly include *matrix multiplication* and *matrix inversion*. In addition, we should also know the general operations of matrices such as matrix addition, scalar multiplication, and some properties of matrix operations. The matrix operations are needed because several important analyses essentially solve a set of linear equations that are best presented with matrix notations. The linear equations expressed in matrix format are concise and helpful in understanding the problems in an abstract way and convenient in theoretical deductions.

For this reason, in time series data analysis, we frequently encounter matrices or arrays and related mathematical operations and computations. With matrix operations, the calculations involving large numbers of equations can be expressed in clear and short formats. This makes the presentation of solving a large set of equations appear to be easier. Obviously, finding the actual solution for a set of equations and doing the computation manually can be tedious. For data analysis, the number of equations can be prohibitive for any attempt to manually solve the equations. With computers, it is an easy job if a *computational algorithm* is properly designed for a given computer program. With MATLAB, computation with matrix operations is particularly fast and convenient because it is designed to work with matrices easily. In addition, most matrix operations expressed in MATLAB scripts are visually consistent with conventions in mathematics. This makes it very easy to transfer the mathematical equations to MATLAB scripts for actual computations.

8.1.1 Matrix Multiplication

Before we look at the linear system of equations, it is necessary to review matrix multiplication. Matrix multiplication is done with some specific rules mentioned earlier. It is order dependent and may or may not exist depending on the dimensions of the matrices under discussion.

A matrix A with dimensions $n \times l$ is sometimes denoted as $A_{n \times l}$ or simply A. A matrix $A_{n \times l}$ can multiply another matrix $B_{l \times m}$ to yield a third matrix of $C_{n \times m}$, i.e.

$$A_{n \times l} B_{l \times m} = C_{n \times m}. \tag{8.1}$$

This equation shows that the number of columns (i.e. l) of the first matrix is the same as the number of rows of the second matrix. Otherwise, matrix multiplication does not exist. The resultant matrix has a dimension of $n \times m$. For convenience, we usually do not use the above expression, but rather use

$$AB = C. \tag{8.2}$$

Here the subscripts are all omitted, implying that the number of columns of the first matrix on the left-hand side (A) is the same as the number of rows of the second matrix (B). The structures of the matrices are shown in the following equation:

$$\begin{pmatrix} a_{11} & \cdots & a_{1k} & \cdots & a_{1l} \\ \vdots & & \vdots & \vdots & \\ a_{i1} & \cdots & a_{ik} & \cdots & a_{il} \\ \vdots & & \vdots & \vdots & \\ a_{n1} & \cdots & a_{nk} & \cdots & a_{nl} \end{pmatrix} \begin{pmatrix} b_{11} & \cdots & b_{1j} & \cdots & b_{1m} \\ \vdots & & \vdots & \vdots & \\ b_{k1} & \cdots & b_{kj} & \cdots & b_{km} \\ \vdots & & \vdots & \vdots & \\ b_{l1} & \cdots & b_{lj} & \cdots & b_{lm} \end{pmatrix}$$
$$= \begin{pmatrix} c_{11} & \cdots & c_{1j} & \cdots & c_{1m} \\ \vdots & & \vdots & \vdots & \\ c_{i1} & \cdots & c_{ij} & \cdots & c_{im} \\ \vdots & & \vdots & \vdots & \\ c_{n1} & \cdots & c_{nj} & \cdots & c_{nm} \end{pmatrix}. \tag{8.3}$$

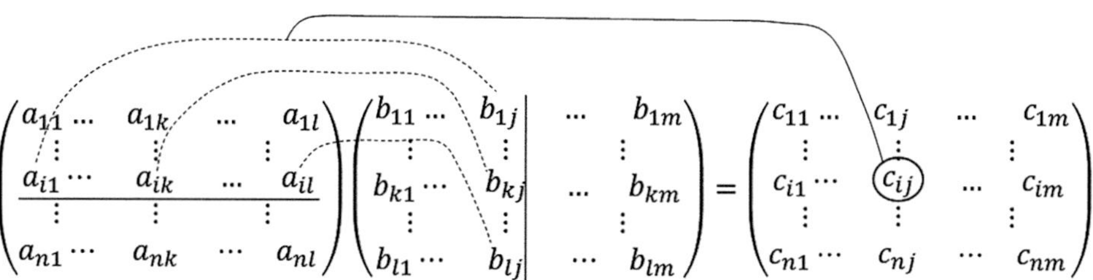

Figure 8.1 Diagram showing how matrix multiplication is done.

The actual computation is done by *multiplication and addition* of *a row of the first matrix and a column of the second matrix* of the left-hand side of the equation; this yields *a single element* of the resultant matrix on the right-hand side of the equation. This is shown in Fig. 8.1.

More specifically, the element c_{ij} at the position of the ith row and jth column in the matrix C is obtained by (Fig. 8.1)

$$a_{i1}b_{1j} + a_{i2}b_{2j} + \cdots + a_{ik}b_{kj} + \cdots + a_{il}b_{lj} = c_{ij} \tag{8.4}$$

or

$$c_{ij} = \sum_{k=1}^{l} a_{ik}b_{kj}, \quad (i = 1, 2, \ldots, n;\ j = 1, 2, \ldots, m). \tag{8.5}$$

It can be seen from this definition that matrix multiplication is order dependent, i.e., if A and B are two matrices, generally, $AB \neq BA$, even if both matrix multiplications exist.

8.1.2 Other Matrix Operations

Matrices can have other mathematical operations in addition to multiplication. These mathematical operations are not possible without the definitions of identity matrix and zero matrix.

Identity Matrix. An identity matrix is a square matrix where the total number of rows equals the total number of columns; all elements are zero except those along the diagonal, which are all 1s. A 3×3 identity matrix is

$$I = \begin{pmatrix} 1 & 0 & 0 \\ 0 & 1 & 0 \\ 0 & 0 & 1 \end{pmatrix}.$$

In general, an $n \times n$ identity matrix is

$$I = \begin{pmatrix} 1 & 0 & \cdots & 0 \\ 0 & 1 & \cdots & 0 \\ \vdots & \vdots & \ddots & \vdots \\ 0 & 0 & \cdots & 1 \end{pmatrix}.$$

A property of an identity matrix is that the multiplication of any matrix with an identity matrix, or the multiplication of an identity matrix with another matrix, yields the same matrix (the order of multiplication does not matter here) as long as these multiplications exist. The role of an identity matrix in matrix multiplication is very much like the role of number 1 in number multiplication.

Zero Matrix. A zero matrix is one in which all elements are zero. Unlike the identity matrix, it does not have to be a square matrix. It is sometimes denoted just by the number 0; the following is a short expression of a zero matrix with a dimension of $n \times m$:

$$0 = (0)_{n \times m}.$$

This simply means that the matrix is $n \times m$ with all elements being 0.

Matrix Addition and Subtraction. Matrix addition or subtraction is meaningful only for those matrices with the same dimension. If A and B are all $n \times m$ matrices, the addition (subtraction) of them ($A \pm B$) is defined as the addition (subtraction) of the elements at the same positions:

$$\begin{pmatrix} a_{11} & \cdots & a_{1j} & \cdots & a_{1l} \\ \vdots & & \vdots & \vdots & \\ a_{i1} & \cdots & a_{ij} & \cdots & a_{il} \\ \vdots & & \vdots & \vdots & \\ a_{n1} & \cdots & a_{nj} & \cdots & a_{nl} \end{pmatrix} \pm \begin{pmatrix} b_{11} & \cdots & b_{1j} & \cdots & b_{1l} \\ \vdots & & \vdots & \vdots & \\ b_{i1} & \cdots & b_{ij} & \cdots & b_{il} \\ \vdots & & \vdots & \vdots & \\ b_{n1} & \cdots & b_{nj} & \cdots & b_{nl} \end{pmatrix}$$

$$= \begin{pmatrix} a_{11} \pm b_{11} & \cdots & a_{1j} \pm b_{1j} & \cdots & a_{1l} \pm b_{1l} \\ \vdots & & \vdots & & \vdots \\ a_{i1} \pm b_{i1} & \cdots & a_{ij} \pm b_{ij} & \cdots & a_{il} \pm b_{il} \\ \vdots & & \vdots & & \vdots \\ a_{n1} \pm b_{n1} & \cdots & a_{nj} \pm b_{nj} & \cdots & a_{nl} \pm b_{nl} \end{pmatrix}.$$

Scalar Multiplication. A scalar multiplication to a matrix is a multiplication of a scalar with all elements of a matrix:

$$\alpha A = \alpha \begin{pmatrix} a_{11} & \cdots & a_{1j} & \cdots & a_{1l} \\ \vdots & & \vdots & \vdots & \\ a_{i1} & \cdots & a_{ij} & \cdots & a_{il} \\ \vdots & & \vdots & \vdots & \\ a_{n1} & \cdots & a_{nj} & \cdots & a_{nl} \end{pmatrix} = \begin{pmatrix} \alpha a_{11} & \cdots & \alpha a_{1j} & \cdots & \alpha a_{1l} \\ \vdots & & \vdots & \vdots & \\ \alpha a_{i1} & \cdots & \alpha a_{ij} & \cdots & \alpha a_{il} \\ \vdots & & \vdots & \vdots & \\ \alpha a_{n1} & \cdots & \alpha a_{nj} & \cdots & \alpha a_{nl} \end{pmatrix}.$$

With these definitions, we have the following properties of matrix operations:

Commutative property: $A + B = B + A$, given that A and B have the same dimensions (e.g. $n \times m$).

Additive property: $0 + A = A + 0$, if both matrices have the same dimension.

Associative property: $(A + B) + C = A + (B + C)$, if all matrices have the same dimensions.

Distributive property for scalars: $(\alpha + \beta)A = \alpha A + \beta A$.

Distributive property for matrices: $\alpha(A + B) = \alpha A + \alpha B$, if both matrices have the same dimensions.

Associative property for scalars: $\alpha(\beta A) = (\alpha\beta)A$.

Identity matrix multiplication property: $IA = AI = A$, if A and I are square matrices of the same dimensions, and I is an identity matrix.

Associativity of matrix multiplication of several matrices: $(AB)C = A(BC)$, if the number of columns of A is the same as the number of rows of B, and the number of columns of B is the same as the number of rows of C.

Left distributive property of multiplication: $A(B + C) = AB + AC$, if the number of columns of A is the same as the number of rows of B; B and C have the same dimensions.

Right distributive property of multiplication: $(A + B)C = AC + BC$, if the number of columns of A is the same as the number of rows of C, and A and B have the same dimensions.

Scalar associative property: $\alpha(AB) = (\alpha A)B = A(\alpha B)$, if the number of columns of A is the same as the number of rows of B.

8.1.3 Linear Equations – A 3-Equation Example

A *linear system of equations* can be expressed in a matrix format using the definition of matrix multiplication. Here we use a set of three linear equations as an example to demonstrate this. Suppose we have the following three equations:

$$\begin{aligned} f_1 &= a_1x + a_2y + a_3z \\ f_2 &= b_1x + b_2y + b_3z \\ f_3 &= c_1x + c_2y + c_3z \end{aligned} \tag{8.6}$$

in which the *a*s, *b*s, and *c*s are coefficients of the variables or unknowns x, y, and z; and f_1, f_2, and f_3 are usually known values (e.g. from measurements).

In matrix form, (8.6) becomes

$$\begin{pmatrix} f_1 \\ f_2 \\ f_3 \end{pmatrix} = \begin{pmatrix} a_1 & a_2 & a_3 \\ b_1 & b_2 & b_3 \\ c_1 & c_2 & c_3 \end{pmatrix} \begin{pmatrix} x \\ y \\ z \end{pmatrix}. \tag{8.7}$$

Here the matrix multiplication is used on the right-hand side. The above equation can be written in a simpler matrix form:

$$F = AX, \tag{8.8}$$

in which

$$F = \begin{pmatrix} f_1 \\ f_2 \\ f_3 \end{pmatrix}; \quad A = \begin{pmatrix} a_1 & a_2 & a_3 \\ b_1 & b_2 & b_3 \\ c_1 & c_2 & c_3 \end{pmatrix}; \quad X = \begin{pmatrix} x \\ y \\ z \end{pmatrix}. \tag{8.9}$$

Each equation in (8.6) can be obtained by the matrix multiplication in (8.7). For the right-hand side of (8.7) or (8.8), it is a matrix multiplication resulting in a 3-element column array, corresponding to the left-hand side 3-element array F as defined in (8.9). Each of these three elements is obtained by multiplying each element of a row of the matrix A by each corresponding element in the column array X and adding them together. For instance, for the first equation, the right-hand side is: $a_1x + a_2y + a_3z$. Note that AX is equal to F, but XA does not exist because each row in X has only one element or one column, while each column of A has three rows. The above discussion can be easily generalized to an n-equation linear system, with n being an arbitrary positive integer.

8.2 A MATLAB Example to Solve an Equation

8.2.1 Doing Matrix Multiplication in MATLAB

Now we are going to work with MATLAB to see how to do the matrix multiplications. It is rather straightforward. As an example, say the matrix A and array X are known, e.g.

$$A = \begin{pmatrix} 1 & -2 & 3 \\ -4 & -5 & 6 \\ 7 & -8 & 9 \end{pmatrix}; \quad X = \begin{pmatrix} 10 \\ 11 \\ 12 \end{pmatrix}. \tag{8.10}$$

In MATLAB, A and X can be defined using two lines of command:

```
>>A = [ 1 -2 3; -4 -5 6; 7 -8 9] ;
>>X = [ 10;11;12] ;
```

The calculation of F in (8.8) using MATLAB is done by the simple line:

```
>>F = A * X
```

MATLAB will output the following:

```
>>F =
    24
   -23
    90
```

8.2.2 Solving an Equation Using MATLAB

Once relevant information about the matrices is given, solving a linear set of equations is straightforward with a line of MATLAB command.

To demonstrate, suppose that in equation (8.8), the array X is unknown, while A and F are known by (8.11):

$$A = \begin{pmatrix} 1 & -2 & 3 \\ -4 & -5 & 6 \\ 7 & -8 & 9 \end{pmatrix}; \quad F = \begin{pmatrix} 24 \\ -23 \\ 90 \end{pmatrix}. \tag{8.11}$$

We can use the inverse of A to solve X:

```
>>X = inv(A) * F
```

which yields

```
>>X =
   10.0000
   11.0000
   12.0000
```

This is consistent with the values of X (8.10).

The MATLAB function `inv` with an argument A, i.e. `inv(A)`, is to compute the *inverse matrix* of A. An inverse matrix is a matrix which, if it exists, can be used to multiply the left-hand side array F of (8.8) to yield the solution X. This is not a formal definition of an inverse matrix; nor is it limited to a 3×3 matrix. The following section will discussion the inverse matrix a little more.

8.3 Inverse Matrix

The concept of inverse matrix is crucial to the understanding of how to solve a system of linear equations. The definition of an inverse matrix can be made with the aid of an identity matrix. If two matrices A and B satisfy the condition that both AB and BA are equal to the identity matrix, we say A and B are an inverse matrix of each other. Using 3×3 matrices as examples, if we have

$$AB = \begin{pmatrix} 1 & 0 & 0 \\ 0 & 1 & 0 \\ 0 & 0 & 1 \end{pmatrix}; \quad BA = \begin{pmatrix} 1 & 0 & 0 \\ 0 & 1 & 0 \\ 0 & 0 & 1 \end{pmatrix}, \tag{8.12}$$

then A and B are inverse of each other.

For $n \times n$ matrices A and B, it is similar, i.e. if

$$AB = BA = \begin{pmatrix} 1 & 0 & \cdots & 0 \\ 0 & 1 & \cdots & 0 \\ \vdots & \vdots & \ddots & \vdots \\ 0 & 0 & \cdots & 1 \end{pmatrix}, \tag{8.13}$$

then A and B are inverse of each other.

We write the inverse of A, which is B, as

$$B = A^{-1}. \tag{8.14}$$

Or the inverse of B, which is A, as

$$A = B^{-1}.$$

This is much like a reciprocal of a number, e.g. $5^{-1} = 1/5, 5^{-1} \times 5 = 1$. In matrix world, a matrix multiplied by its inverse matrix is equal to an identity matrix.

An inverse matrix is only meaningful for square matrices. A rectangular matrix with different numbers of rows and columns does not have a definition of inverse matrix. The inverse matrix of a given square matrix is not guaranteed to exist, either.

With the inverse matrix, it is easier (at least conceptually) to solve a system of linear equations. Assume that we are trying to solve an equation like (8.8). If the inverse matrix exists, we can multiply both sides of (8.8) by the inverse matrix to obtain the solution:

$$A^{-1}F = A^{-1}AX = \left(A^{-1}A\right)X = IX = X.$$

Here we have used the associativity property of matrix multiplication involving more than two matrices.

8.4 A Different Approach of MATLAB

Although using an inverse matrix is a standard way to solve a system of linear equations, MATLAB allows the use of a different approach by a "left-division" to solve the equation. Let us look at an example first:

```
>> A = [ 1 -2 3;  -4 -5 6;  7 -8 9] ;
>> F = [ 24;  -23;  90]
F =

      24
     -23
      90
>>X = A \ F
```

```
X =
        10
        11
        12
```

Here the operator \ is the "left-division."

The MATLAB output is the same as that using the inverse matrix. This is a command equivalent to `X = inv(A) * F` for this problem. This shows that either `X = inv(A) * F` or `X = A\ F` will yield the solution `X` defined in (8.10). The left-division approach is actually a least squares method, which is much more commonly used in data analysis.

Not all linear equations will have a solution. To illustrate this, let us look at an example with MATLAB. Suppose in equation (8.8) we have

$$A = \begin{pmatrix} 1 & 2 & 3 \\ 4 & 5 & 6 \\ 7 & 8 & 9 \end{pmatrix}; \quad X = \begin{pmatrix} 10 \\ 11 \\ 12 \end{pmatrix}. \tag{8.15}$$

```
>>A = [ 1 2 3; 4 5 6; 7 8 9] ;
>>X = [ 10;11;12] ;
```

If we define F to be AX:

```
>>F = A * X
F =
     68
    167
    266
```

This would establish the following equation:

$$\begin{pmatrix} 1 & 2 & 3 \\ 4 & 5 & 6 \\ 7 & 8 & 9 \end{pmatrix} \begin{pmatrix} 10 \\ 11 \\ 12 \end{pmatrix} = \begin{pmatrix} 68 \\ 167 \\ 266 \end{pmatrix}.$$

Is it true? In other words, if we know A and F, can we solve the equation and verify that indeed

$$X = \begin{pmatrix} 10 \\ 11 \\ 12 \end{pmatrix}?$$

This can be easily checked by `inv(A) * F`, which leads to

```
>> inv(A) * F
Warning: Matrix is close to singular or badly scaled.
Results may be inaccurate. RCOND = 1.541976e-018.
```

```
ans =
     0
     0
   256
```

Obviously, this is not equal to $X = \begin{pmatrix} 10 \\ 11 \\ 12 \end{pmatrix}$.

If we use left-division, we get

```
>> A\F
Warning: Matrix is close to singular or badly scaled.
Results may be inaccurate. RCOND = 1.541976e-018.
ans =
   -66
   163
   -64
```

The left-division also does not give the "correct results."

In both cases, MATLAB gives some warnings. To explain this, we need to discuss an important concept of the *rank of a matrix.*

8.5 A Concept of Rank for Matrix

The inconsistency of these two methods is not because either of these methods is wrong; it is a problem with matrix A. In essence, this is because not all equations represented by matrix A are independent: at least one of the equations is the same as one of the others. They might not appear to be exactly the same, but at least one of them can be deduced from the rest of the equations. In other words, the degree of freedom of the n equations represented by the $n \times n$ matrix A is less than n. Or, we have n unknowns but only m independent equations, and m is smaller than n. Therefore, there is not enough information to derive a unique solution for the n unknowns. Here the number of independent equations is called the *rank of matrix A*.

The rank can be checked by a MATLAB function `rank`:

```
>> rank(A)
ans =
     2
```

This is smaller than the number of unknowns (which is three, in this case). This means that with the given matrix A, only two of the three equations are really different from each other; the third equation is a disguised form of one of the other two. Another way to check is to use a MATLAB function `det(A)`, i.e. *determinant of matrix A*:

```
>> det(A)
ans =
  6.6613e-016
```

Here the determinant is defined as a number calculated by:

$$\det A = \begin{vmatrix} a_{11} & a_{12} & a_{13} \\ a_{21} & a_{22} & a_{23} \\ a_{31} & a_{32} & a_{33} \end{vmatrix} = a_{11}a_{22}a_{33} - a_{11}a_{23}a_{32} - a_{12}a_{21}a_{33} + a_{12}a_{23}a_{31} + a_{13}a_{21}a_{32} - a_{13}a_{22}a_{31} \quad . \tag{8.16}$$

The determinant in this case, `6.6613e-016`, is very small – essentially zero. If a square matrix has a zero determinant, its inverse matrix does not exist. In other words, in this case with the given matrix A, the three equations (8.6) only have two of them that are independent of each other. The third equation is effectively a repetition of one of the other two equations. For example, if we have

$$\begin{aligned} f_1 &= 1x + 2y + 3z \\ f_2 &= 4x + 5y + 6z \, . \\ f_3 &= 7x + 8y + 9z \end{aligned} \tag{8.17}$$

Subtract the first equation from the second and subtract the second equation from the third; we get two identical equations, and therefore only two of the three equations are independent.

The rank of the matrix is 2. There is not enough information for a unique solution.

8.6 A General Discussion on the Solution of Linear Equations

8.6.1 Determinant

The definition of determinant for an $N \times N$ matrix is a number:

$$\det(A) = \begin{vmatrix} a_{11} & a_{12} & \cdots & a_{1N} \\ a_{21} & a_{22} & \cdots & a_{2N} \\ \vdots & & & \vdots \\ a_{N1} & a_{N2} & \cdots & a_{NN} \end{vmatrix} = \sum_{(\lambda_1\lambda_2,\ldots,\lambda_N)} v(\lambda_1, \lambda_2, \ldots, \lambda_N) a_{1\lambda_1} a_{2\lambda_2} \cdots a_{N\lambda_N}, \tag{8.18}$$

in which $\lambda_1, \lambda_2, \ldots, \lambda_N$ are an *arrangement of natural numbers* 1, 2, ... , N: an arrangement is a sequence of a given collection of numbers aligned in a certain order (for example, (1,5,3,4,2), (5,3,2,1,4), (3,1,2,4,5), ... , are arrangements of (1,2,3,4,5)); $v(\lambda_1, \lambda_2, \ldots, \lambda_N)$ is a very simple function of the arrangement $\lambda_1, \lambda_2, \ldots, \lambda_N$ of the natural numbers $1, 2, \ldots, N$: it is either 1 or -1, depending on the order of $(\lambda_1, \lambda_2, \ldots, \lambda_N)$, i.e.

$$v(\lambda_1, \lambda_2, \ldots, \lambda_N) = \mathrm{sgn} \prod_{1 \le r < s \le N} (\lambda_s - \lambda_r). \tag{8.19}$$

Here the symbol Π is a product of all the terms, similar to the symbol Σ for addition of the terms. If all of these numbers have an ascending order, the function value is 1. For example: $v(1, 2, 3, 4, 5) = 1$. For each pair of numbers with a descending order, a negative sign is multiplied, e.g. $v(1, 2, 3, 5, 4) = -1$, $v(1, 3, 2, 5, 4) = 1$.

8.6.2 The Linear Equations

In order to understand a large portion of the time series data analysis techniques, we first need to understand the solution of linear equations. We now look at a general set of linear equations. For a linear set of M equations with N unknowns:

$$\begin{cases} a_{11}x_1 + a_{12}x_2 + \cdots + a_{1N}x_N = y_1 \\ a_{21}x_1 + a_{22}x_2 + \cdots + a_{2N}x_N = y_2 \\ \vdots \\ a_{M1}x_1 + a_{M2}x_2 + \cdots + a_{MN}x_N = y_M \end{cases}, \tag{8.20}$$

in which $x_k (k = 1, 2, \ldots, N)$ are generally the unknowns; the coefficients a_{ij} are prescribed, and the y's on the right-hand side are usually from direct observations. In matrix format, it is

$$Ax = y, \tag{8.21}$$

in which

$$A = \begin{pmatrix} a_{11} & a_{12} & \cdots & a_{1N} \\ a_{21} & a_{22} & \cdots & a_{2N} \\ \vdots & & & \vdots \\ a_{M1} & a_{M2} & \cdots & a_{MN} \end{pmatrix} \tag{8.22}$$

$$x = \begin{pmatrix} x_1 \\ x_2 \\ \vdots \\ x_N \end{pmatrix}, \quad y = \begin{pmatrix} y_1 \\ y_2 \\ \vdots \\ y_M \end{pmatrix} \tag{8.23}$$

or

$$\begin{pmatrix} a_{11} & a_{12} & \cdots & a_{1N} \\ a_{21} & a_{22} & \cdots & a_{2N} \\ \vdots & \vdots & \vdots & \vdots \\ a_{M1} & a_{M2} & \cdots & a_{MN} \end{pmatrix} \begin{pmatrix} x_1 \\ x_2 \\ \vdots \\ x_N \end{pmatrix} = \begin{pmatrix} y_1 \\ y_2 \\ \vdots \\ y_M \end{pmatrix}. \tag{8.24}$$

Note that the length of x is different from that of y, unless $N = M$. There are three scenarios for the above equation, in terms of the nature of the solution, if any. These correspond to the situations when $N = M$, $N > M$, and $N < M$, respectively. The following sections will discuss them.

8.6.3 The First Case, $N = M$

For the first case, if $N = M$, A is a square matrix, and the unknowns in x can be solved if the inverse matrix exists. This is the same as saying that the rank of A is n; or the matrix A is full-rank; or the determinant of A is non-zero. Under this condition, the solution is

$$x = By = A^{-1}y. \tag{8.25}$$

Here the inverse matrix B (denoted as A^{-1}) of A is defined as another matrix that, when multiplied by the original matrix A, it yields an identity matrix as shown in (8.13) or, in more detail:

$$AB = \begin{pmatrix} a_{11} & a_{12} & \cdots & a_{1N} \\ a_{21} & a_{22} & \cdots & a_{2N} \\ \vdots & \vdots & \vdots & \vdots \\ a_{N1} & a_{N2} & \cdots & a_{NN} \end{pmatrix} \begin{pmatrix} b_{11} & b_{12} & \cdots & b_{1N} \\ b_{21} & b_{22} & \cdots & b_{2N} \\ \vdots & \vdots & \vdots & \vdots \\ b_{N1} & b_{N2} & \cdots & b_{NN} \end{pmatrix} = \begin{pmatrix} 1 & 0 & \cdots & 0 \\ 0 & 1 & \cdots & 0 \\ \vdots & & & \vdots \\ 0 & 0 & \cdots & 1 \end{pmatrix}, \tag{8.26}$$

$$BA = \begin{pmatrix} b_{11} & b_{12} & \cdots & b_{1N} \\ b_{21} & b_{22} & \cdots & b_{2N} \\ \vdots & \vdots & \vdots & \vdots \\ b_{N1} & b_{N2} & \cdots & b_{NN} \end{pmatrix} \begin{pmatrix} a_{11} & a_{12} & \cdots & a_{1N} \\ a_{21} & a_{22} & \cdots & a_{2N} \\ \vdots & \vdots & \vdots & \vdots \\ a_{N1} & a_{N2} & \cdots & a_{NN} \end{pmatrix} = \begin{pmatrix} 1 & 0 & \cdots & 0 \\ 0 & 1 & \cdots & 0 \\ \vdots & \vdots & \vdots & \vdots \\ 0 & 0 & \cdots & 1 \end{pmatrix}. \tag{8.27}$$

As discussed earlier, an inverse matrix is like a reciprocal of a number.

If the determinant of A is 0 or very close to zero, the inverse of A is essentially nonexistent. In this situation, we call the matrix A *ill-conditioned.*

In summary, in this case (when $N = M$), unless the matrix A is ill-conditioned, there is a unique solution for the linear system of equations (8.31). MATLAB allows for the use of an inverse function (`inv(A)`) or a left-division (\) to do the calculation (as in the command `A\x` shown earlier).

8.6.4 The Second Case: $N > M$

For the second case, if $N > M$, there are more unknowns than the number of equations, and thus there are an infinite number of solutions. We will not deal with this situation with MATLAB because of its arbitrary nature: there are too many degrees of freedom to have a unique or meaningful solution. This is an *underdetermined problem*. When this situation is encountered, it means that more conditions are needed to better constrain the problem.

8.6.5 The Third Case: $N < M$

For the third case, if $N < M$, there are more equations than the number of unknowns. As a result, this is *an overdetermined problem*. It usually does not have such a solution to satisfy all equations precisely at the same time. However, we can expect that these equations hold "statistically" (if $M \gg N$). This can be very useful, because observations have inevitable random errors one way or the other. As a result, to resolve N unknowns, N observations are not enough because of statistical fluctuations. We need to have more observations so we can obtain a statistically meaningful estimate of the unknowns. With more equations than the number of unknowns, the equations cannot be satisfied precisely. Instead, we can require that the equations hold in a statistical sense to have the least error possible with the given set of observations.

In theory, equation (8.21) should hold, but the existence of random error suggests that

$$Ax - y = \varepsilon \neq 0. \tag{8.28}$$

Here ε can be considered as the "error." It is an array with the same length as y. The TOTAL error squared is then:

$$\Delta = \varepsilon_1^2 + \varepsilon_2^2 + \varepsilon_3^2 + \cdots \varepsilon_M^2 = (\varepsilon_1 \quad \varepsilon_2 \quad \varepsilon_3 \cdots \varepsilon_M) \begin{pmatrix} \varepsilon_1 \\ \varepsilon_2 \\ \vdots \\ \varepsilon_M \end{pmatrix}. \tag{8.29}$$

This is also, in matrix form:

$$\Delta = (Ax - y)^T (Ax - y). \tag{8.30}$$

Here the superscript T means a transpose of the matrix within the parentheses. We are trying to find the solution $x = (x_1, x_2, \ldots, x_M)^T$, so that the error squared is minimized. In other words, based on the theorem on extreme values in calculus, we require that

$$\frac{\partial \Delta}{\partial x} = 0. \tag{8.31}$$

Or, equivalently,

$$\frac{\partial \Delta}{\partial x_k} = 0, \quad k = 1, 2, \ldots, M. \tag{8.32}$$

It can be shown that this is equivalent to

$$A^T(Ax - y) = 0, \tag{8.33}$$

which leads to

$$A^T A x = A^T y. \tag{8.34}$$

The solution is

$$x = \left(A^T A\right)^{-1} A^T y. \tag{8.35}$$

This is a generic least squares method solution of an overdetermined system of linear equations. Note that in this case, M is larger than N and matrix A is not a square matrix, which means it does not have an inverse defined. But the product of the transposed of A with A itself is guaranteed a square matrix. In MATLAB, one may choose to use equation (8.35) to find the solution for x of (8.21). However, using the left-division of MATLAB for equation (8.21) is much easier. We can also use the pseudoinverse function `pinv(A)`. The pseudoinverse function is applicable to a non-square matrix. The following commands are essentially the same:

```
>> x = pinv(A) * y
>> x = A\y
>> x =(A'*A)^(−1)*A'*y
```

In summary, in MATLAB, the solution of a set of overdetermined linear equations is straightforward: it needs only one line of command once the matrix A and the array y are provided.

8.7 Examples of Least Squares Method

8.7.1 Example 1

We now use an example to compare the conventional least squares method with that using MATLAB left-division. Suppose that we have obtained a set of observations of 300 pairs of data points. The data are known to represent a linear model like this:

$$y = ax + b, \tag{8.36}$$

in which $x = \sin\left(\frac{\pi t}{180}\right), t = 1 : 300$. Because of random errors, this equation is only approximately correct. For the 300 observations, we should have 300 equations, as the following:

$$\begin{aligned} ax_1 + b &= y_1 \\ ax_2 + b &= y_2 \\ \cdots \\ ax_M + b &= y_M \end{aligned} \quad . \tag{8.37}$$

Or, equivalently,

$$y_i = ax_i + b, \quad i = 1, 2, \ldots, M, \tag{8.38}$$

in which $M = 300$. These equations, however, are not necessarily accurate, i.e. the following difference or error is usually non-zero:

$$\delta_i = y_i - (ax_i + b). \tag{8.39}$$

The error squared is defined as

$$\delta_i^2 = [y_i - (ax_i + b)]^2. \tag{8.40}$$

The total error squared is then

$$\Psi = \sum_{i=1}^{M} \delta_i^2 = \sum_{i=1}^{M} [y_i - (ax_i + b)]^2. \tag{8.41}$$

Note that the total error squared Ψ depends on the choice of a and b. So, it is a function of the pair of parameters (a, b). To minimize the total error squared, we require that

$$\frac{\partial \Psi}{\partial a} = 0, \quad \frac{\partial \Psi}{\partial b} = 0, \tag{8.42}$$

from which we can obtain the optimal values for a and b. The above equation can be shown as

$$\frac{\partial \Psi}{\partial a} = \sum_{i=1}^{M} 2[y_i - (ax_i + b)](-x_i) = 0; \frac{\partial \Psi}{\partial b} = \sum_{i=1}^{M} 2[y_i - (ax_i + b)](-1) = 0, \tag{8.43}$$

which leads to

$$\begin{aligned} a\sum_{i=1}^{M} x_i + bM &= \sum_{i=1}^{M} y_i \\ a\sum_{i=1}^{M} x_i^2 + b\sum_{i=1}^{M} x_i &= \sum_{i=1}^{M} x_i y_i. \end{aligned} \tag{8.44}$$

The above two equations have two unknowns (a and b) and can be solved as

$$a = \frac{T_1 T_2 - M T_3}{T_1^2 - M T_5}, \quad b = \frac{T_1 T_3 - T_2 T_5}{T_1^2 - M T_5} \tag{8.45}$$

in which

$$T_1 = \sum_{i=1}^{M} x_i, \; T_2 = \sum_{i=1}^{M} y_i, \; T_3 = \sum_{i=1}^{M} x_i y_i, \; T_5 = \sum_{i=1}^{M} x_i^2. \tag{8.46}$$

In real applications, the y values on the right-hand side of equation (8.37) are obtained from observations or provided in some format. Here, for convenience, we will construct some simulated y values. We can simulate the y values by the following with an addition of some randomness:

$$y = ax + 0.2 \times \text{rand}(1, M), \tag{8.47}$$

Here `rand(1,M)` is a one row and M column of uniformly distributed pseudorandom numbers between 0 and 1. After the generation of simulated observations for y, we can use (8.45) to solve a and b in a statistical sense (i.e. the optimal estimate). The following MATLAB code is used for this task:

```
clear
t = 1:300;
X = sin(t * pi / 180);
N = size(X,2);          % length of the data array
a0=1;
Y = a0 * X + 0.2 * rand(1,N);
T1 = sum(X);
T2 = sum(Y);
T3 = sum(X.*Y);
T4 = sum(X)^2;
T5 = sum(X.*X);
a = (T1 * T2 - N * T3) / (T4 - N * T5)    % LEAST SQUARES METHOD
b = (T1 * T3 - T2 * T5) / (T4 - N * T5)
Yfit = a * X + b;                % FIT TO A LINE
plot(t, Y, 'ro')                 % THE ORIGINAL DATA
hold on
plot(t, Yfit, 'LineWidth', 2)              % THE BEST-FIT REGRESSION LINE
title('Data & Their Best Fit Line Using Least Squares Methods')
xlabel('t (arbitrary Unit)')
ylabel('Y (Data & Fit)')
```

MATLAB outputs the following result for a and b:

```
a =

    0.9974
b =

    0.1002
```

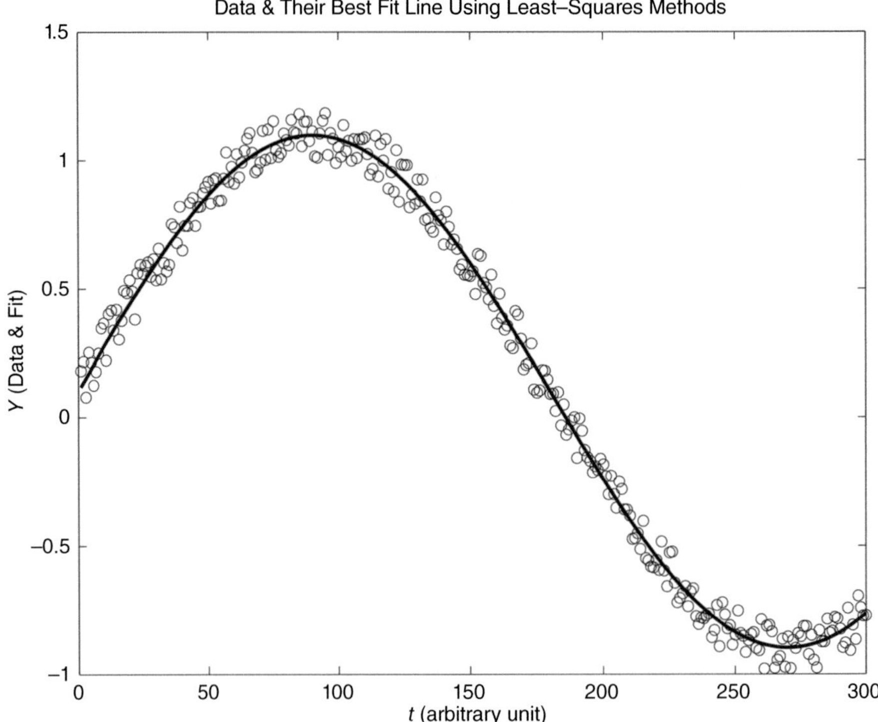

Figure 8.2 An example for best-fitting observational data using the least squares method.

Here the theoretical value for a is 1 (`a0` was defined to be 1) and that for b is 0.1 because the mean of `rand(1,N)` is 0.5. The least squares method apparently provided a consistent result (Fig. 8.2).

An alternative and usually easier method using MATLAB is to use left-division. To do that, we should first write the equation in matrix format:

$$\begin{pmatrix} x_1 & 1 \\ x_2 & 1 \\ \vdots & \vdots \\ x_M & 1 \end{pmatrix} \begin{pmatrix} a \\ b \end{pmatrix} = \begin{pmatrix} y_1 \\ y_1 \\ \vdots \\ y_M \end{pmatrix}, \tag{8.48}$$

$$Ax = y, \tag{8.49}$$

in which

$$A = \begin{pmatrix} x_1 & 1 \\ x_2 & 1 \\ \vdots & \vdots \\ x_M & 1 \end{pmatrix}; \quad x = \begin{pmatrix} a \\ b \end{pmatrix}; \quad y = \begin{pmatrix} y_1 \\ y_1 \\ \vdots \\ y_M \end{pmatrix}. \tag{8.50}$$

In this case, A is a two-column matrix. To use the left-division feature of MATLAB, we first define the variables using the same first few lines:

```
clear
t = (1:300)';          % transpose to get a column-vector
X = sin(t * pi / 180);
N = length(X);         % length of the data array
Y = 0.2 * rand(N,1) + X;
```

and then we can construct the matrix A using these commands:

```
I = ones(N,1);
A = [ X I] ;
```

The MATLAB function `ones(N,1)` gives a column array of N 1s, after which we can apply left-division to solve a and b:

```
x = A\Y  % NOTE THAT THE Y IS A COLUMN-VECTOR
```

which yields something like this (each run may have slightly different values due to the random function):

```
x =
     0.9960
     0.1064
```

The first value of x is a and the second is b. The results are also consistent with the theoretical values of a and b. We can also use one of the following two commands to achieve the same goal:

```
x1 = pinv(A) * Y'
x2 = (A' * A)^-1 * A' * Y'
```

They all produce the same results.

8.7.2 Example 2

Assume that we have a proposed statistical model:

$$y = ax^2 + bx + c. \tag{8.51}$$

We now omit the derivation of the exact solution for a least squares method but only use MATLAB's left-division to do a quick least squares fitting of some simulated data to the model and then compare the estimated coefficients a, b, and c with those "theoretical values." For each of the M observations, the following equation should hold in a statistical sense:

$$y_i = ax_i^2 + bx_i + c, \quad i = 1, 2, \ldots, M. \tag{8.52}$$

In matrix format, the equations are combined into a single equation:

$$\begin{pmatrix} x_1^2 & x_1 & 1 \\ x_2^2 & x_2 & 1 \\ \vdots & \vdots & \vdots \\ x_M^2 & x_M & 1 \end{pmatrix} \begin{pmatrix} a \\ b \\ c \end{pmatrix} = \begin{pmatrix} y_1 \\ y_2 \\ \vdots \\ y_M \end{pmatrix} \tag{8.53}$$

or

$$AX = Y, \tag{8.54}$$

in which

$$A = \begin{pmatrix} x_1^2 & x_1 & 1 \\ x_2^2 & x_2 & 1 \\ \vdots & \vdots & \vdots \\ x_M^2 & x_M & 1 \end{pmatrix}, \quad X = \begin{pmatrix} a \\ b \\ c \end{pmatrix}, \quad Y = \begin{pmatrix} y_1 \\ y_2 \\ \vdots \\ y_M \end{pmatrix}. \tag{8.55}$$

The statistical solution using MATLAB can be sought by the left-division `x = A\Y`. The following is a sample script to demonstrate how to implement. For convenience, we assume some "theoretical values" of a, b, and c, and add some random errors when we construct some simulated "data." In real applications, the observational data will replace the few lines making the simulated data.

```
clear
X = (1:300)';           % transpose to get a column-vector
N = length(X);          % length of the data array
a0 = 2.02;      % The "theoretical value for a"
b0 = 101.31;    % The "theoretical value for b"
c0 = −52.8;     % The "theoretical value for c"
Y = a0 * X.^2 + b0 * X + c0 + 10 * rand(N,1); % This is simulated
                  % "data." In real applications, Y should be
                  % replaced by data to be used for the fitting
                  % to the statistical model
I = ones(N,1);
A = [ X.^2 X I];
x = A \ Y
```

The output is

```
x =

    2.0200
  101.3184
  −47.8480
```

Compared with the "theoretical" values of `a`, `b`, and `c`:

```
a0 = 2.02; b0 = 101.31; c0 = −52.8
```

the MATLAB output gives consistent results.

8.8 Coefficient of Determination of Linear Models

From the above examples, we see that the least squares method for solving the linear equations in (8.20) can be used for linear models of observed time series. Although (8.55) is nonlinear in terms of the variable x, it is linear in terms of the coefficients a, b, and c, which are the unknowns of the linear equations. In this case, the total error squared of the model is

$$es = (Ax - y)^T (Ax - y). \tag{8.56}$$

The root-mean-squared error is

$$\sigma = \sqrt{\frac{es}{M - N}}. \tag{8.57}$$

The *coefficient of determination* or R-squared value is

$$R^2 = 1 - \frac{es}{R_y}, \tag{8.58}$$

in which

$$R_y = (y - \bar{y})^T (y - \bar{y}) = \sum_{i=1}^{M} (y_i - \bar{y})^2, \quad \bar{y} = \frac{1}{M} \sum_{i=1}^{M} y_i. \tag{8.59}$$

The coefficient of determination R^2 has a maximum of 1 and is often used in assessing the closeness of fit between the model and the time series data. When the model has a perfect fit, the total error squared is zero ($es = 0$) and the coefficient of determination is $R^2 = 1$. If the model is inaccurate, but only predicts the mean value $\bar{y}$, the coefficient of determination is $R^2 = 0$.

Review Questions for Chapter 8

(1) Can a matrix with a dimension of 3×7 have an inverse matrix?

(2) Can any 3×3 matrix have an inverse matrix?

(3) Compare the least squares method and the "left-division" in MATLAB. How are they related?

Exercises for Chapter 8

(1) Given a statistical model

$$y = ae^{t/\delta} + r,$$

in which a is a constant; $t = [1, 300]$; δ is the growth constant; r is a random number between 0 and 50. If we know the growth constant is 50, then, using $a = 1$ to simulate the data with $r_{\max} = 50$, generate 300 "data" points and use MATLAB commands to solve the parameters a and r using the least squares method. Calculate the coefficient of determination.

(2) Assume that we have a statistical model

$$y = a\sin(\omega t) + b\cos(\omega t) + c,$$

in which $\omega = 12.1414$ (rad day^{-1}), $t = 0$ to 5 days. Assuming that the theoretical values of a and b are 1.2 and -0.7, respectively, the observational data for y can be simulated by

$$y = a\sin(\omega t) + b\cos(\omega t) + \text{noise}$$

or by using this MATLAB command: `Y = 1.2 * sin(omega * t) -0.7 * cos(omega * t) + 0.2 * rand(N,1)`. Write a MATLAB script and use left-division to find the parameters a, b, and c from the "simulated data," and then calculate the coefficient of determination.

(3) The data file named "time-depth1.txt" has two columns. The first is time in days and the second is water depth in mm. It is a time series of water depth with diurnal tide (with a period of ~24 hours or 1 day). So, we can use a model with sine and cosine functions $\sin(2\pi ft)$ and $\cos(2\pi ft)$ as base functions to fit to the time series, in which $f = 1$ cycle per day. In other words, the model is expressed as

$$y(t) = a\sin(2\pi ft) + b\cos(2\pi ft) + c.$$

Construct matrix A by these two sinusoidal functions plus another base function 1. Use MATLAB left-division to solve the coefficients of the model a, b, and c. Compare the model with data by a plot of both time series curves on the same figure, and then calculate the coefficient of determination.

9

Base Functions and Linear Independence

About Chapter 9

The objective of this chapter is to discuss the concept of base functions and the basics of using some simple base functions to represent other functions. These base functions are needed in many commonly used analyses. An example of using base functions to approximate an almost arbitrary target function is the Taylor series expansion we have discussed. Here we say that an "almost arbitrary function" is not really arbitrary because, in theory, the target function must be differentiable an arbitrary number of times for the Taylor series expansion to be valid. The concept of linear independence of functions is important in understanding the selection of base functions.

9.1 Base Functions

9.1.1 Base Functions Used in Taylor Series Expansion

Taylor series expansion, discussed in Chapter 6, is a useful tool – with certain conditions, a (smooth) target function can be expressed by a series of simple polynomial functions even if we may not know exactly what the function looks like. The target function being represented is otherwise arbitrary. The polynomial functions are

$$1, x, x^2, x^3, \ldots, x^n, \ldots$$

or, more generally,

$$1,\ x - x_0,\ (x - x_0)^2,\ (x - x_0)^3,\ \ldots,\ (x - x_0)^n,\ \ldots,$$

in which x_0 is a given value of the independent variable x.

These polynomial functions are called *base functions*, because they are the basic elements on which the target function is approximated.

9.1.2 A Generalized Question

Base functions are not limited to polynomials. They can be many other functions. The question, then, is whether we can use any combination of functions to linearly combine them together to represent or approximate a target function. For example, can we use a combination of

$$\sin x, x, x^{1/3}, \ln x$$

to represent another function, for example

$$\frac{x \tan x}{1 + x^2}?$$

Here, we are seeking a relationship like this:

$$\frac{x \tan x}{1 + x^2} = a \sin x + bx + cx^{\frac{1}{3}} + d \ln x + e.$$

How useful can this relationship be, or how large can the error be? While the question of the error of such representation is not to be discussed in this chapter, in essence, it is a question about the *completeness of the base functions*, i.e. are they sufficient? Consider using linear functions 1 and x to represent a nonlinear function x^2, which seems impossible or at least inconsistent. We will defer the discussion of completeness to the Fourier analysis chapter. Here we would like to point out what may already appear obvious to some – that there is no guarantee that a collection of functions can be used to correctly represent another function.

The above hypothetical relationship involves different kinds of functions (i.e. sine function, polynomial function, and logarithmic function) and may not have any real applications. Fortunately, the real problems are somewhat simpler, e.g. like the polynomial Taylor series expansion, which only involves one type of function (polynomial functions).

To answer the question of whether we can use any linear combination of functions to represent a target function, and to better understand linear regression, Fourier analysis, harmonic analysis, and other methods involving a system of linear equations, we should first have a clear understanding of base functions and the *linear independence of functions*.

9.2 Linear Dependence and Linear Independence

9.2.1 Linearly Dependent Functions

Here we first define what linearly dependent functions are.

Definition 1 *Suppose there are continuous functions*

$$f_1(x), f_2(x), \ldots, f_n(x).$$

They are called *linearly dependent* if there exists a series of scalars $a_1, a_2, \ldots,$ and a_n; not all of them are zeros, such that

$$a_1 f_1(x) + a_2 f_2(x) + \cdots + a_n f_n(x) = 0 \tag{9.1}$$

holds for all x values. In other words, these functions are not all unrelated to each other. They have some type of linear relationship, and at least one of them can be expressed by some of the rest in the group. This relationship will at least reduce the *degrees of freedom* by 1, and thus among the n functions there are at most $n - 1$ independent functions.

For example, if $f_1(x) = \sin x, f_2(x) = \sin x/10$, then

$$f_1(x) = 10f_2(x)$$

or

$$f_1(x) - 10f_2(x) = 0$$

and

$$a_1 = 1,\ a_2 = -10.$$

In this example, functions $f_1(x)$ and $f_2(x)$ are essentially the same function except with different factors. If these two were used as base functions to represent another function, there would be essentially a duplicate and an error will occur when doing matrix computations. To select base functions, we really need to avoid using those that are linearly related. We need to use linearly independent functions, which are defined below.

9.2.2 Linearly Independent Functions

Definition 2 *Suppose there are continuous functions*

$$f_1(x), f_2(x), \ldots, f_n(x).$$

They are *linearly independent* if there exist no scalars $a_1, a_2, \ldots,$ and a_n; not all of them are zeros, to allow

$$a_1 f_1(x) + a_2 f_2(x) + \cdots + a_n f_n(x) = 0 \tag{9.2}$$

to be true for all values of the independent variable x. That is, if those functions ($f_1, f_2, \ldots, f_n$) are linearly independent, and if the above equation holds, then all the coefficients in the above equation must be zero for all values of x. Conceptually, the linearly independent functions are all different functions because any subset of them cannot be used to linearly express one of the rest: all of the functions are independent of each other. *Only linearly independent functions can be used as base functions.* Otherwise, the matrix inversion can have problems.

For example, if $f_1(x) = \sin x$, $f_2(x) = \cos x$, then there is no way to have $a_1 f_1(x) + a_2 f_2(x) = 0$ for all values of x, unless $a_1 = 0$, $a_2 = 0$.

9.2.3 Linear Model

Linearly independent functions can be chosen as a set of base functions, with which other functions can be uniquely expressed. We can choose to form a minimum number of functions to express all other functions within certain *functional space*: a collection of functions (usually with an infinite number of them) depends on a few base functions by a linear combination. Generally, a linear combination of a few base functions can only *approximate* a target function. In the event that a series of base functions can represent a collection of target functions (e.g. all continuous real functions) with no error at all, we say the base functions form a complete set. Once a complete set of base functions is selected, there is only a unique way to express any member function and thus no confusion is possible.

Physical processes are often expressed by different *components* in a linear model. The linear model is expressed, in general, as

$$y(x) = a_1 f_1(x) + a_2 f_2(x) + \cdots + a_n f_n(x), \tag{9.3}$$

in which, $f_1, f_2, \ldots$, and f_n are the base functions – equivalent to *i*, *j*, and *k* in a coordinate system, but here we are not limited to two or three dimensions. In this example, we have n-dimensions, or n degrees of freedom because there are n different ways we can change the model by changing the values of coefficients. Note also that here the word *linear* is in terms of the coefficients $a_1, a_2, \ldots$, and a_n, but not the base functions $f_1, f_2, \ldots$, and f_n, which can be nonlinear functions of x.

9.3 An Example of Simple Tide

9.3.1 Simple Tide

Now let us look at the modeling of a tidal signal with a single major frequency. Assume that observations of water level at a coastal station contain a major tidal frequency f. The tidal oscillations of the water level with a single frequency are sinusoidal functions, and thus they can be simulated by:

$$z_{obs} = a \sin(2\pi f t) + b \cos(2\pi f t) + 2c\,\mathrm{rand}(t), \tag{9.4}$$

in which rand is a function of t, of uniformly distributed *pseudorandom real numbers* between 0 and 1, t is time, f is a given tidal frequency (if we assume that tide is a semi-diurnal M_2 tide, the period is $T = 12.42$ hours and the frequency is $f = 1/T$), and a, b, and c are constants (independent of time t). The reason we use a factor 2 in front of the coefficient c is because rand(t) has an average of 0.5, so that $2c$ rand(t) has an average of c. Note that we are using (9.4) to generate some pseudo-observational data. In data analysis, there is no need to produce pseudo-observational data – data are already given. Here we use simulated data just for convenience. We could have used real data for the discussion.

9.3.2 The Matrix Expression

In the above example of simulated data, the mathematical model that we are going to use is obviously

$$z = a \sin(2\pi f t) + b \cos(2\pi f t) + c, \tag{9.5}$$

in which the parameters a, b, and c are to be determined with a "best" estimate.

From this model, for the ith observation, we have the following equation:

$$z_i = a \sin(2\pi f t_i) + b \cos(2\pi f t_i) + c, \quad i = 1, 2, \ldots, n. \tag{9.6}$$

For all the observations, if we write all the equations:

$$\begin{aligned} z_1 &= a \sin(2\pi f t_1) + b \cos(2\pi f t_1) + c \\ z_2 &= a \sin(2\pi f t_2) + b \cos(2\pi f t_2) + c \\ &\vdots \\ z_n &= a \sin(2\pi f t_n) + b \cos(2\pi f t_n) + c \end{aligned} . \tag{9.7}$$

Or in a matrix format:

$$\begin{pmatrix} z_1 \\ z_2 \\ \vdots \\ z_n \end{pmatrix} = \begin{pmatrix} \sin(2\pi f t_1) & \cos(2\pi f t_1) & 1 \\ \sin(2\pi f t_2) & \cos(2\pi f t_2) & 1 \\ \vdots & \vdots & \vdots \\ \sin(2\pi f t_n) & \cos(2\pi f t_n) & 1 \end{pmatrix} \begin{pmatrix} a \\ b \\ c \end{pmatrix}. \tag{9.8}$$

The left-hand side of the above equation is a column-vector for the observed water level with noise, which in this case is given by (9.4). Again, in real applications, the results from equation (9.4) must be replaced by an actual array of observed values. The right-hand side is a matrix multiplication involving a matrix of three columns: the first and second columns are the sine and cosine functions (the functions for the model), while the last column is a constant 1, which can also be considered as part of the model for the mean of the entire time series. Here, $\sin(2\pi f t)$, $\cos(2\pi f t)$, and 1 are the base functions. The parameters a, b, and c form the three-column-vector on the right. In a more concise matrix format, the above equation becomes

$$z = Ax, \tag{9.9}$$

in which

$$z = \begin{pmatrix} z_1 \\ z_2 \\ \vdots \\ z_n \end{pmatrix}, \tag{9.10}$$

$$A = \begin{pmatrix} \sin(2\pi f t_1) & \cos(2\pi f t_1) & 1 \\ \sin(2\pi f t_2) & \cos(2\pi f t_2) & 1 \\ \vdots & \vdots & \vdots \\ \sin(2\pi f t_n) & \cos(2\pi f t_n) & 1 \end{pmatrix}, \tag{9.11}$$

$$x = \begin{pmatrix} a \\ b \\ c \end{pmatrix}. \tag{9.12}$$

This is essentially an overdetermined problem because there are three unknowns (a, b, and c), and as long as the number of observations is more than three, which should always be the case, there will be more equations than the number of unknowns. The solution then must be in the "least square" sense. This can be achieved in MATLAB by the left-division.

9.3.3 A Special Case

Suppose that, in this case, the data were obtained every 20 minutes for 10 days. For the purpose of data simulation, we assume that the constants a, b, and c have values 2, 1, and 0.3, respectively. In real applications, these are the variables to be determined, and the data are measured, not simulated with an equation. The following MATLAB code defines the "data" (simulated observations), creates the three-column matrix A, and uses the left-division command of MATLAB to find the solution for x (including a, b, and c).

```
t = (0:1 / 24 / 3:10)';     % DAYS
a = 2; b = 1; c = 0.3; N = length(t);
f = 1 /(12.42 / 24);
z = a * sin(2 * pi * f * t) + b * cos(2 * pi * f * t) + 2 * c * rand(N,1);
A = [ sin(2 * pi * f * t) cos(2 * pi * f * t) ones(N,1)] ;
x = A \ z
zm = A * x;
figure; plot(t,z,'*k');
hold on
plot(t,zm,'k')
```

Because of the random function included in the calculation, each time the above scripts are run (or applied) in the MATLAB command window, the results are a little different, but overall the results are consistent. One run shows the following output in MATLAB:

```
x =
     2.0201
     1.0074
     0.3093
```

The first element of x is $a = 2.0201$, which is very close to the exact value of 2; the second element of x is $b = 1.0074$, also very close to the theoretical value of 1; the last element of x is $c = 0.3093$, not much different from the theoretical value of 0.3, either. The small errors are expected because a random number is introduced that made them be not exactly reproduced but still quite close to the "true" values. In this example, the model works well (Fig. 9.1).

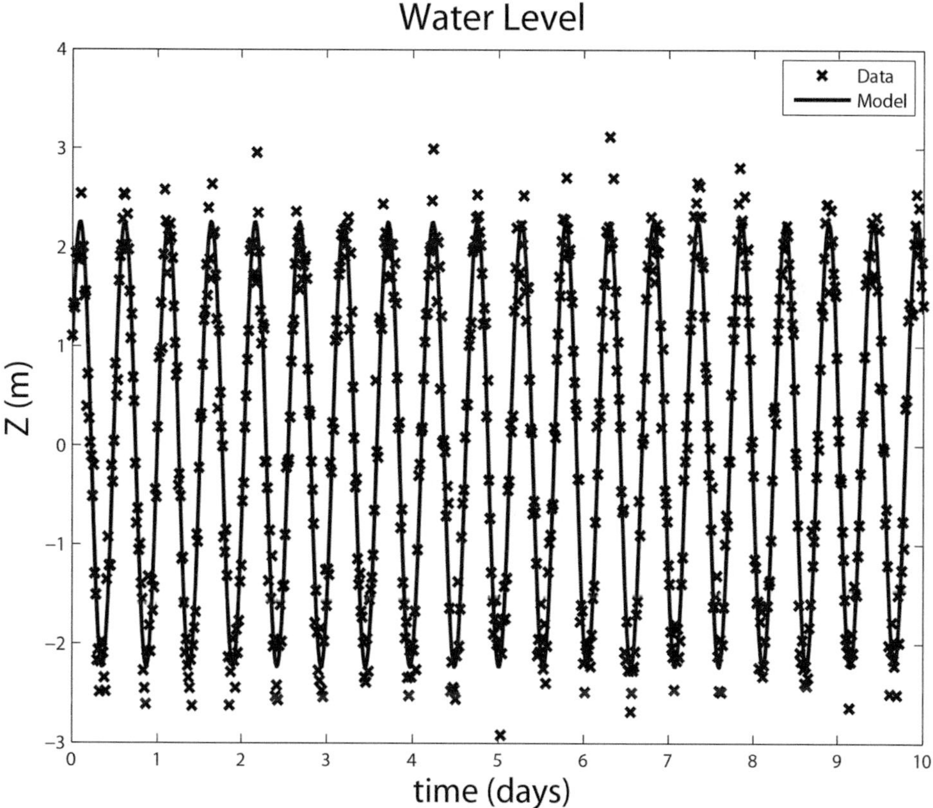

Figure 9.1 Simulated data and model results from left-division. * represents the results from the simulated data with added randomness (equation 9.4). The solid line represents the model from equation 9.5.

9.4 Examples of Improper Base Functions

9.4.1 Example 1

As discussed earlier, equation (9.5) is a model using the three base functions $f_1(t) = \sin(2\pi ft)$, $f_2(t) = \cos(2\pi ft)$, and $f_3(t) = 1$. These three functions are linearly independent of each other. What happens if linearly dependent functions are used as the base functions? Let us try a different model:

$$z = a[\sin(2\pi ft) + 30] + b\sin(2\pi ft) + c. \quad (9.13)$$

In this case, the base functions are $f_1(t) = \sin(2\pi ft) + 30$, $f_2(t) = \sin(2\pi ft)$, and $f_3(t) = 1$. The first two base functions are linearly related. With the same definition of time, the following MATLAB code is used to create a simulated time series and to recover the parameters a, b, and c:

```
t = (0:1/24/3:10)';   % DAYS
a = 2; b = 1; c = 0.3; N = length(t);
f = 1/(12.42/24);
z = a*(sin(2*pi*f*t)+30)+b*sin(2*pi*f*t)+2*c*rand(N,1);
A = [ sin(2*pi*f*t)+30   sin(2*pi*f*t)   ones(N,1)];
x = A\z
```

MATLAB output the following warning together with the results:

```
Warning: Rank deficient, rank = 2, tol = 1.290524e-010.

x =
    2.0101
    0.9844
         0
```

The results are consistent with the first two parameters of the model. However, the third parameter was not correctly recovered. More importantly, MATLAB does not like it: the warning basically says that matrix A has some problems. It has a rank of 2, meaning that two of the three base functions are linearly related and there are only two, instead of three, degrees of freedom, which we noted at the beginning of this chapter.

9.4.2 Example 2

If this is not enough to worry you about the improper selection of base functions, let us look at one more example. Can the following model be used?

$$z = at + b(t + 1) + c(t + 2) + d. \quad (9.14)$$

In this case, we have the following base functions: $f_1(t) = t$, $f_2(t) = t + 1$, $f_3(t) = t + 2$, and $f_4(t) = 1$. Obviously, they are linearly dependent. Using the same definition of time, the MATLAB code is

```
t = (0:1/24/3:10)';  % DAYS
a = 2; b = 2; c = 1; d = 0.3; N = length(t);
z = a*t+b*(t+1)+c*(t+2)+2*d*rand(N,1);
A = [ t t+1  t+2  ones(N,1)];
x = A\z
```

MATLAB produces the following:

```
Warning: Rank deficient, rank = 2, tol =   3.2556e-011.
x =
    2.8749
         0
    2.1324
         0
```

In this case, MATLAB provided a warning, and the results are not consistent with what was provided as parameters for the model: the first and second parameters have a value of 2, the third is 1, and the last is 0.3. Again, the four-column matrix has only a rank of 2, meaning only two of the four base functions are not "related" to each other. Indeed, the first three base functions are all similar – different only by a constant.

These examples demonstrate that base functions cannot be arbitrarily chosen. They need to be linearly independent of each other. Otherwise, there is a reduction in degree of freedom or the rank of the matrix A will be smaller than the total number of base functions. This is similar to representing a three-dimensional statue by a two-dimensional surface – it may only be a projection of the 3-D structure on a 2-D surface, and the representation cannot be 100% accurate: there is a loss of one dimension.

9.5 More Discussion

9.5.1 A Geometry Point of View

Conceptually, the selection of base functions is very much the same as the selection of the axes of a coordinate system. A vector can be expressed by a linear combination of other vectors. In a two-dimensional space, any vector can be expressed by a linear combination of any two other vectors that are not parallel to each other. In a three-dimensional space, any vector can be expressed by a linear combination of any three other vectors that are not on a plane.

We are accustomed to using Cartesian coordinate systems in which vectors are usually expressed by the three perpendicular unit vectors, **i**, **j**, and **k** in the x, y, and z directions, respectively. For example, we can express $\mathbf{v} = (2,7)$ by the unit vectors in the x and y directions **i** and **j**: $\mathbf{v} = 2\mathbf{i} + 7\mathbf{j}$. Obviously, this is a linear superposition of **i** and **j** with the coefficients of 2 and 7. We can also try to express $\mathbf{v} = (2,7)$ by any two other vectors, for instance $\mathbf{v}_1 = (1,1)$ and $\mathbf{v}_2 = (-1,1)$:

$$\mathbf{v} = a\mathbf{v}_1 + b\mathbf{v}_2 \text{ or } (2,7) = a(1,1) + b(-1,1), \tag{9.15}$$

which yields two equations for a and b: $2 = a - b$; and $7 = a + b$, which leads to $a = 4.5$ and $b = 2.5$, or

$$\mathbf{v} = 4.5\mathbf{v}_1 + 2.5\mathbf{v}_2. \tag{9.16}$$

The result is shown in Fig. 9.2.

If we try to express $\mathbf{v} = (2,7)$ by $\mathbf{v}_1 = (1,1)$ alone, it would not be possible. Likewise, if we were to express $\mathbf{v} = (2,7)$ by two vectors that are parallel to each other, e.g. $\mathbf{v}_1 = (1,1)$ and $\mathbf{v}_2 = 3\mathbf{v}_1 = (3,3)$, it can be seen that this is no different from using $\mathbf{v}_1 = (1,1)$ alone to express **v**.

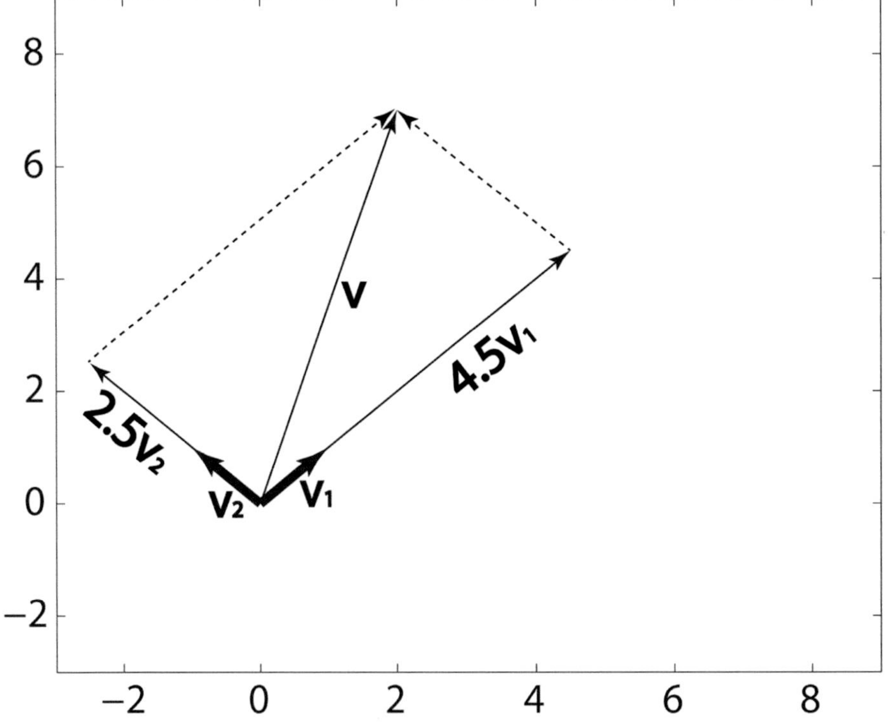

Figure 9.2 Example of linear combination of two vectors to express the third vector: $\mathbf{v} = 4.5\mathbf{v}_1 + 2.5\mathbf{v}_2$.

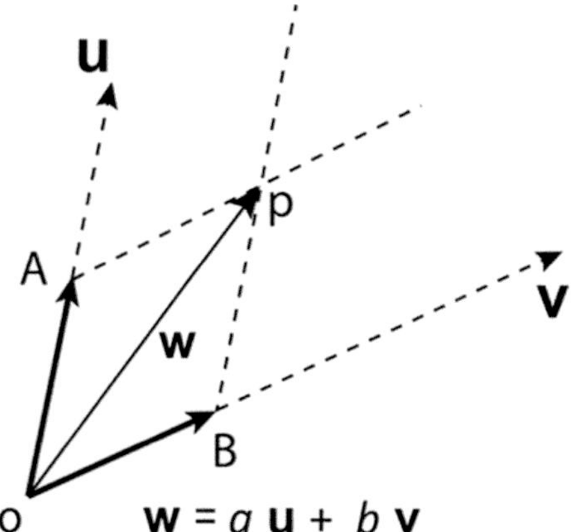

Figure 9.3 Vector decomposition by other vectors.

In general, in a two-dimensional space, to express a vector **w** in terms of **u** and **v** (Fig. 9.3), one has to make sure that **u** and **v** are not parallel to each other. Otherwise, unless **w** happens to be parallel to **u**, it is not possible.

From a geometric point of view, a group of n linearly independent functions can form a n-dimensional functional space to represent another group of functions. If any two of the n functions are linearly dependent, they are said to be "parallel" to each other. This effectively reduces the dimension of the functional space that they can represent by 1. This makes them impossible to correctly represent the n-dimensional functional space but only $n - 1$ dimensional space at most. This would make the matrix inversion using the left-division problematic because the rank of the matrix is reduced at least by 1. This is important in understanding the proper selection and use of base functions for regression analysis.

9.5.2 About Nonlinear Models

The discussion on the linear independence of base functions is only in the context of linear models. As indicated earlier, a linear model is linear in the sense of the coefficients, not the base functions. The base functions are almost always nonlinear, except a special function $ax + b$ (in which x is an independent variable and a, b are constants). There are two questions that may be asked: (1) is a linear model too simple; and (2) what about *nonlinear models* – are they worthy of discussion?

For the first question, the answer is no: a linear model is not necessarily "simple." In fact, a few widely used theories, which we are going to discuss in later chapters, use linear models. These include the Fourier series analysis, Fourier Transform or spectrum analysis, harmonic analysis, wavelet analysis, and empirical orthogonal function analysis. These encompass the backbones of the most significant advances in signal processing in modern practical mathematics. We can never overestimate their roles in the development of science and technology in almost all fields: physics, astronomy, medical science, image processing, voice recognition, stock market analysis, earthquake prediction, etc. Of course, this is also true for oceanography and atmospheric sciences.

For the second question, while nonlinear models may be of value, depending on specific questions, it is less generic and more difficult, with a few exceptions, to generalize. Of course, there are far more nonlinear models than linear ones. Nonlinear models are those with the coefficients of certain (base) functions being nonlinearly related to the target function being approximated. For example, the following model,

$$y = ae^{bx}, \tag{9.17}$$

is a nonlinear model, with the coefficients a and b to be determined (statistically). The coefficient b is what makes the model nonlinear. In this specific example, the nonlinear model can be converted to a linear model. This can be seen by taking a natural logarithm on both sides of equation (9.17):

$$\ln y = \ln a + bx. \tag{9.18}$$

If we let

$$z = \ln y; \quad c = \ln a, \tag{9.19}$$

$$z = bx + c. \tag{9.20}$$

This is a linear model. If we know observed values of y and therefore z as a function of y and thus of x, and if we have enough repetitions of measurements (many z values corresponding to x values), we can use the least squares method to estimate the coefficients b and c and therefore a and b. This kind of model is commonly seen in biological and chemical processes such as *growth models* of various kinds, or a *radioactive material decay model*.

Most nonlinear models cannot be converted to a linear model, e.g.

$$y = a \sin(bx), \tag{9.21}$$

$$y = \log \frac{x}{ax + b} + x^c, \tag{9.22}$$

$$y = \frac{ax^b}{e^c + \sin(dx)}. \tag{9.23}$$

Here a, b, c, and d are all coefficients to be determined. We cannot even clearly define what a base function is in some of these cases. In later chapters we will only discuss applications related to linear models with the subjects mentioned earlier (e.g. Fourier series).

Review Questions for Chapter 9

(1) If two sets of functions are linearly independent within each set, when the two sets of functions are put together into one larger set of functions, are they still linearly independent? Why?

(2) If we know a function looks like this:

$$y = a + bx + cx^2 + dx^3,$$

given base functions $f_1 = 1, f_2 = x$, and $f_3 = x^2$ to approximate the function, what do you anticipate and why?

(3) Are these functions linearly independent?

$$1, x^{1/2}, x^{1/3}, \ldots, x^{1/n}.$$

(4) Are these functions linearly independent?

$$1, \ln(x^{1/2}), \ln(x^{1/3}), x, \ldots, x^n, \ldots.$$

Exercises for Chapter 9

(1) The following model is nonlinear; can you convert it to a linear model? If you can, show the converted linear model.

$$y = a \sin(bx).$$

(2) The following model is nonlinear; can you convert it to a linear model? If you can, show the converted linear model.

$$y = ae^{bx^2+cx}.$$

10

Generic Least Squares Method and Orthogonal Functions

About Chapter 10

This chapter discusses a generic least squares method and a special situation when the base functions are orthogonal to each other, which makes the solution explicit; in addition, we learn that the essence of the least squares method can be viewed as a way to project the target function in a higher dimension onto a lower dimension formed by the base functions. The least squares method ensures that the error vector is "perpendicular" to the projected (or approximate) vector in the base function dimension (a lower dimension) and thus has the shortest "length" or minimized error. Although this chapter does not have much computation involved, it is very important for a good understanding of the meaning of many techniques and methods in the subsequent chapters.

10.1 Inner (Dot) Product

To understand the error of a model represented by a linear combination of base functions, we start with the discussion of projection of a 3-D vector onto a 2-D surface or a 1-D line. This may sound irrelevant, but indeed, there is an intrinsic relationship between the approximation of a function and the projection of a vector. Our discussion starts with n-dimensional vectors. We then use knowledge in plain geometry of 3-D and 2-D vectors for the definition of an angle in a higher dimension.

In multidimensional functional space, as we discussed in Chapter 9, the projection of a vector in a n-dimensional space onto a lower dimension has something to do with the error of a linear (regression) model. To demonstrate this, we can start with a quick review of the *inner product* or *dot product* of vectors.

10.1.1 Inner (Dot) Product

We first review some of the geometric meanings of vector/array calculations. For two vectors (arrays) of the same length $\mathbf{u} = (u_1, u_2, \ldots, u_n)$ and $\mathbf{v} = (v_1, v_2, \ldots, v_n)$, the inner product or dot product of them is defined as

$$ip = \mathbf{u} \cdot \mathbf{v} = \sum_{i=1}^{n} u_i v_i, \tag{10.1}$$

i.e. it is the sum of the element-by-element multiplication of the two vectors. The reason it is called a dot product is apparently because of the conventional mathematical symbol of using a dot between the vectors. In MATLAB, the sum of the element-by-element product of two vectors can be computed by the following command, assuming that both u and v have been defined and have the same length:

```
>>ip = dot(u,v);
```

For the self-inner product, i.e. if $\mathbf{u} = \mathbf{v}$,

$$ip = \mathbf{u} \cdot \mathbf{u} = \sum_{i=1}^{n} u_i^2, \tag{10.2}$$

which is sometimes expressed by the square of the vector within two vertical bars:

$$ip = |\mathbf{u}|^2. \tag{10.3}$$

Here $|\mathbf{u}|$ is called the *module (length) of the vector* $\mathbf{u}$.

Of course, this implies that the module is defined to be the square root of the sum of the squared elements, combining (10.2) and (10.3):

$$|\mathbf{u}| = \sqrt{\sum_{i=1}^{n} u_i^2}. \tag{10.4}$$

10.1.2 The Projection of a Vector

In two-dimensional problems, $n = 2$, according to the *Pythagorean theorem*, (10.4) gives the length of the hypotenuse of a right-angle triangle with the lengths of the other two sides to be u_1 and u_2, respectively. For $n = 3$, (10.4) gives the diagonal of the cuboid with lengths of u_1, u_2, and u_3, respectively. For $n > 3$, this is apparently an extension of the Pythagorean theorem to dimensions higher than 3 that cannot be visualized. The self-inner product of (10.2) thus has the geometric meaning of the "length" squared of the vector $\mathbf{u}$. For dimensions higher than 3, we can only imagine and cannot draw the length in an n-dimensional space.

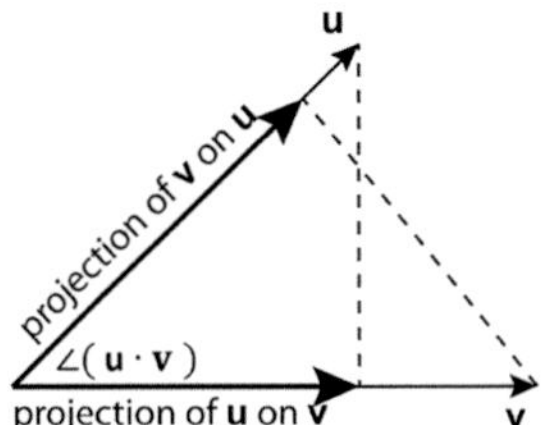

Figure 10.1 The projection of a vector onto another and its relation to a dot product.

For 2-D and 3-D vectors, a dot product of two vectors is the geometric projection of the first vector onto the second vector, multiplied by the length of the second vector (Fig. 10.1):

$$\mathbf{u} \cdot \mathbf{v} = |\mathbf{u}||\mathbf{v}| \cos\left(\angle(\mathbf{u}, \mathbf{v})\right), \tag{10.5}$$

in which $\angle(\mathbf{u}, \mathbf{v})$ is the angle between the two vectors $\mathbf{u}$ and $\mathbf{v}$. This operation is obviously symmetric, i.e.

$$\mathbf{u} \cdot \mathbf{v} = \mathbf{v} \cdot \mathbf{u}.$$

However, the projection of one vector onto the other does not necessarily equal the projection of the other vector onto the first vector, i.e. except for cases when the length of the two vectors is the same, in general:

$$|\mathbf{v}| \cos\left(\angle(\mathbf{u}, \mathbf{v})\right) \neq |\mathbf{u}| \cos\left(\angle(\mathbf{u}, \mathbf{v})\right). \tag{10.6}$$

The inner product or dot product of two vectors can also be viewed as a measure of similarity or correlation between the two vectors. Obviously, when they are parallel to each other, the inner product is maximized because $|\cos x| \leq 1$ and $\cos 0 = 1$. When they are perpendicular to each other, the inner product is zero because $\cos(\pi/2) = 1$, so that from (10.5), $\mathbf{u} \cdot \mathbf{v} = 0$.

10.1.3 The Angle between Vectors

The concepts discussed above can be extended to an arbitrary n-dimensional functional space. Here n is equal to the number of base functions. The n-dimensional space has n *spatial directions*, much like $\mathbf{i}$, $\mathbf{j}$, and $\mathbf{k}$ in a Cartesian coordinate, that determine the directions of x, y, and z. We can define the "angle" between two vectors $\mathbf{u}$ and $\mathbf{v}$ $(\angle(\mathbf{u}, \mathbf{v}))$ in an n-dimensional space by:

$$\cos\left(\angle(\mathbf{u}, \mathbf{v})\right) = \frac{\mathbf{u} \cdot \mathbf{v}}{|\mathbf{u}||\mathbf{v}|}. \tag{10.7}$$

Since the dot product and the modules of $\mathbf{u}$ and $\mathbf{v}$ are all defined, the angle, albeit abstract and purely mathematical, can be defined. It can be proven that

$$|\mathbf{u} \cdot \mathbf{v}| \leq |\mathbf{u}||\mathbf{v}| \tag{10.8}$$

or

$$u_1 v_1 + u_2 v_2 + \cdots u_n v_n \leq \sqrt{(u_1^2 + u_2^2 + \cdots u_n^2)(v_1^2 + v_2^2 + \cdots v_n^2)}.$$

This is the so-called Schwarz Inequality for vectors. From (10.8), we see that the magnitude of the right-hand side of (10.7) cannot be greater than 1 and is thus within the range of values of a cosine function. The angle therefore is meaningfully defined mathematically.

Perhaps the usefulness of this angle is more than what it appears to be when we look at the dot product $\mathbf{u} \cdot \mathbf{v}$ as a correlation between $\mathbf{u}$ and $\mathbf{v}$. Given fixed lengths for $\mathbf{u}$ and $\mathbf{v}$, when they are "similar" (or close to being parallel), the magnitude of the dot product is relatively large. When they are linearly related (or parallel to each other), the dot product reaches the maximum. When they are "uncorrelated" or not similar at all (opposite of being parallel to each other), the dot product is zero and we say the two vectors are perpendicular to each other.

10.2 Generic Least Squares Method

10.2.1 Selection of Base Functions

Based on the discussion on linear independence of functions, it has been established that we should only select linearly independent functions as base functions when constructing a regression model. Fortunately, unless using the same function more than twice or with a different factor, almost all functions are linearly independent, for example, polynomial functions ($1, x, x^2, \ldots, x^n$), trigonometric functions ($\sin x$, $\cos x$), exponential functions (e^x, e^{2x}, $\ldots$), and logarithm functions ($\log x$, $\log^2 x, \ldots, \log^n x$), and their reciprocals. These functions are linearly independent not only within their own groups but also among different groups. With these, there can be a combination of many different functions to form linearly independent groups. For example, 1, $x+1$, $(x+1)^2, \ldots, (x+1)^n$ are linearly independent; $x \sin x$, $x \cos x, \ldots$, are also linearly independent.

As long as we choose linearly independent functions, there will be no problem with matrix inversion; there is no *deficiency in matrix rank* when the rank is not "full," i.e. less than the dimension of the square matrix: the rank of a $N \times N$ matrix A is less than N. We say that the matrix is from an *ill-posed problem* when the determinant of the matrix is 0 or near zero. A matrix with zero determinant does not have a full rank, and vice versa.

10.2.2 An Example of Tidal Signal

In ocean data analysis, we frequently encounter problems with tidal signals, those with oscillations at known tidal frequencies. Often, we do not know the amplitude and phase of such signals. The least squares method allows us to resolve these amplitudes and phases through a best estimate by the minimization of errors. This mathematical process, i.e. given tidal frequencies and trying to determine the amplitude and phase at these frequencies, is called *harmonic analysis of tide*. Though more detailed discussion on harmonic analysis will be provided later, here we can examine the base functions for harmonic analysis. Suppose we are examining water level with tidal oscillations; the "model" is then:

$$\begin{aligned}\zeta &= \zeta_0 + \sum_{i=1}^{N} A_i \cos(\omega_i t - \phi_i) \\ &= B_1 g_1(t) + B_2 g_2(t) + \cdots + B_N g_N(t) + B_{N+1} g_{N+1}(t) + B_{N+2} g_{N+2}(t) + \cdots + B_{2N} g_{2N}(t).\end{aligned} \tag{10.9}$$

in which ζ_0 is a constant and

$$\begin{aligned} g_1 &= \cos(\omega_1 t),\ g_2 = \cos(\omega_2 t),\ g_N = \cos(\omega_N t) \\ g_{N+1} &= \sin(\omega_1 t),\ g_{N+2} = \sin(\omega_2 t),\ g_{2N} = \sin(\omega_N t)\end{aligned}. \tag{10.10}$$

These cosine and sine functions are the base functions. They are obviously linearly independent. The frequencies $(\omega_1, \omega_2, \ldots, \omega_n)$ are already determined from astronomy. What the harmonic analysis does is find the amplitude A_i and phase φ_i. With N tidal constituents (N frequencies), there will be $2N$ base functions. Usually, a mean value can be included unless the data are de-meaned first – meaning the mean value is subtracted from the original data. So, an additional base function $g_0 = 1$ should be included that corresponds to the first term ζ_0.That gives a total of $2N + 1$ base functions. It can be verified that these functions are linearly independent.

In the following, we will first discuss a generic least squares method. The generic method encompasses the harmonic analysis in principle, and the general mathematics is very much the same, except some details (e.g. the phase determination is specifically needed in the harmonic analysis but not for most other regression analysis).

10.3 Generic Least Squares Method

10.3.1 For Any Linearly Independent Base Functions

Now we will do a little mathematical work on the derivation of a generic least squares method with arbitrary linearly independent base functions in an N-dimensional space. The linear model with time t as the independent variable is

$$y(t) = a_1 g_1(t) + a_2 g_2(t) + \cdots + a_N g_N(t), \tag{10.11}$$

which is often an approximation of a more complex system. For example, if a system has either (1) nonlinear relationships with the coefficients, or (2) errors in actual observations, or (3) errors in the model (i.e. processes that have been neglected in the model), the above model will not be exactly correct and the equations not accurate.

Even if the above complications did not exist and there were no observational errors, the model might still not be accurate. Often, a complete list of base functions is infinite in number, and we must truncate the infinite series at some point. This means that (10.11) is an approximation of a higher dimensional space "vector" (here it is a realization of the target function, e.g. the "observations" in our case) by a N-dimensional space using the N base functions. In any case, the above model has some errors, and we aim to find the best approximation in the lower N-dimensional space.

With M observations, the above model is applied M times, i.e. for $t = t_1, \ldots, t_M$, and the equations become

$$\begin{cases} a_1 g_1(t_1) + a_2 g_2(t_1) + \cdots + a_N g_N(t_1) = y_1 \\ a_1 g_1(t_2) + a_2 g_2(t_2) + \cdots + a_N g_N(t_2) = y_2 \\ \vdots \\ a_1 g_1(t_M) + a_2 g_2(t_M) + \cdots + a_N g_N(t_M) = y_M \end{cases}. \tag{10.12}$$

For each given set of coefficients, $a_1, a_2, \ldots, a_N$ and a given t value t_i, the error

$$e_i = y(t_i) - [a_1 g_1(t_i) + a_2 g_2(t_i) + \cdots + a_N g_N(t_i)], i = 1, 2, \ldots, M \tag{10.13}$$

is usually not zero. The total error squared is defined by

$$\Psi = \sum_{i=1}^{M} e_i^2 = \sum_{i=1}^{M} \left(y(t_i) - \sum_{j=1}^{N} a_j g_j(t_i) \right)^2. \tag{10.14}$$

The purpose is usually to obtain the coefficients, a_1, a_2, ... , a_N from data, assuming that the model is correct. However, the model can never be accurate and errors do exist, as we discussed. Thus, the M equations from observations cannot be true at the same time in a strict sense. For each given set of coefficients, the "total error" can be calculated. If this error can be minimized, we can obtain *a best possible* solution for the coefficients – the idea of the general least squares method. In matrix format, the above equation is

$$\Psi = \sum_{i=1}^{M} e_i^2 = \sum_{i=1}^{M} \left(\sum_{j=1}^{N} a_j g_j(t_i) - y(t_i) \right)^2 = (Ax - y)^T (Ax - y), \tag{10.15}$$

in which $x = (a_1, a_2, \ldots, a_N)^T$. Here T means a transpose operation to change the row-vector to a column-vector. Equation (10.15) is the error squared sum and is a number. But this number depends on the parameters $a_1, a_2, \ldots, a_N$ or x. So Ψ is a function of a_1, a_2, … , a_N. We can now minimize the error squared sum Ψ by finding the solution for a_j, satisfying the conditions that the partial derivatives with respect to a_j are all zeros, which leads to

$$\frac{\partial \Psi}{\partial a_j} = 2\sum_{i=1}^{M}\left(y_i - \sum_{k=1}^{N} a_k g_k(t_i)\right)(-g_j(t_i)) = -2\left[\sum_{i=1}^{M} y_i g_j(t_i) - \sum_{i=1}^{M}\sum_{k=1}^{N} a_k g_k(t_i) g_j(t_i)\right]$$
$$= -2\sum_{i=1}^{M} y_i g_j(t_i) + 2\sum_{i=1}^{M}\sum_{k=1}^{N} a_k g_k(t_i) g_j(t_i) = 0. \tag{10.16}$$

Therefore,

$$\sum_{k=1}^{N} a_k \sum_{i=1}^{M} g_k(t_i) g_j(t_i) = \sum_{i=1}^{M} y_i g_j(t_i), \quad j = 1, 2, \ldots, N. \tag{10.17}$$

This is equivalent to solving a set of N linear equations for a_1, a_2, … , and a_N. In matrix format, the above equation is

$$(A^T A)x = A^T y \tag{10.18}$$

or

$$Bx = d, \tag{10.19}$$

in which

$$B = \begin{pmatrix} \sum_{i=1}^{M} g_1(t_i) g_1(t_i) & \sum_{i=1}^{M} g_1(t_i) g_2(t_i) & \cdots & \sum_{i=1}^{M} g_1(t_i) g_N(t_i) \\ \sum_{i=1}^{M} g_2(t_i) g_1(t_i) & \sum_{i=1}^{M} g_2(t_i) g_2(t_i) & \cdots & \sum_{i=1}^{M} g_2(t_i) g_N(t_i) \\ \vdots & \vdots & \vdots & \vdots \\ \sum_{i=1}^{M} g_N(t_i) g_1(t_i) & \sum_{i=1}^{M} g_N(t_i) g_2(t_i) & \cdots & \sum_{i=1}^{M} g_N(t_i) g_N(t_i) \end{pmatrix} \tag{10.20}$$

$$x = \begin{pmatrix} a_1 \\ a_2 \\ \vdots \\ a_N \end{pmatrix}, d = \begin{pmatrix} \sum_{i=1}^{M} y_i g_1(t_i) \\ \sum_{i=1}^{M} y_i g_2(t_i) \\ \vdots \\ \sum_{i=1}^{M} y_i g_N(t_i) \end{pmatrix}. \tag{10.21}$$

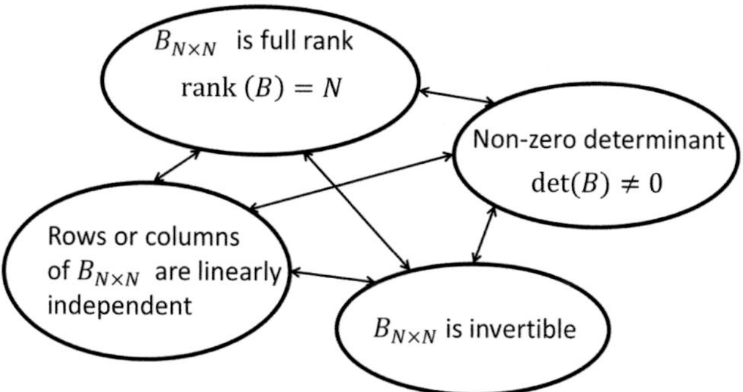

Figure 10.2 Matrix characteristics: determinant, rank, linear independent, and invertibility.

It is very interesting to see that matrix B is basically an array of *correlations of the base functions*. For example:

$$b_{kj} = \sum_{i=1}^{M} g_k(t_i) g_j(t_i) \tag{10.22}$$

is a dot product between $g_k(t)$ and $g_j(t)$. As we discussed in 10.1 on the dot product, equation (10.22) is also a correlation between vectors $(g_k(t_1), g_k(t_2), \ldots, g_k(t_M))^T$ and $(g_j(t_1), g_j(t_2), \ldots, g_j(t_M))^T$. Here T again means a transpose of the array.

If the base functions $g_1(t), g_2(t), \ldots, g_N(t)$ are linearly independent, then matrix B has a *full rank*, or the rank of B is N. Otherwise, if the kth and jth base functions are linearly dependent, the kth and jth rows are essentially the same (linearly dependent of each other). The matrix is full rank if its determinant is nonzeros and vice versa (Fig. 10.2). Otherwise, no inverse matrix exists. In general, the rows or columns of an $N \times N$ matrix are linearly independent, equivalent to the matrix being invertible, or the determinant nonzeros or the matrix has a full rank. Fig. 10.2 gives a diagram showing the relationships among the determinant, rank, linear independency, and invertibility.

If the base functions are all linearly independent from (10.18), we have

$$x = (A^T A)^{-1} A^T y. \tag{10.23}$$

10.3.2 For Orthogonal Base Functions

In a special case when all the base functions are completely uncorrelated – i.e. b_{kj} from (10.22) is zero unless $k = j$ – the matrix B is simplified to one where only the diagonal elements are non-zero, i.e.

$$b_{kj} = \sum_{i=1}^{M} g_k(t_i)g_j(t_i) = \begin{cases} \sum\limits_{i=1}^{M} \left(g_k(t_i)\right)^2 & (k = j; k = 1, 2, \ldots, N) \\ 0 & (i \neq j) \end{cases} \tag{10.24}$$

or

$$B = \begin{pmatrix} \sum\limits_{i=1}^{M} g_1(t_i)g_1(t_i) & 0 & \cdots & 0 \\ 0 & \sum\limits_{i=1}^{M} g_2(t_i)g_2(t_i) & \cdots & 0 \\ \vdots & \vdots & \vdots & \vdots \\ 0 & 0 & \cdots & \sum\limits_{i=1}^{M} g_N(t_i)g_N(t_i) \end{pmatrix}. \tag{10.25}$$

These base functions are called *orthogonal base functions*. An analogy in geometry is when a Cartesian coordinate system is used and the unit vectors in x, y, and z directions, i.e. **i**, **j**, and **k**, are perpendicular to each other and so their dot products are all zero: $\mathbf{i} \cdot \mathbf{j} = 0, \mathbf{i} \cdot \mathbf{k} = 0, \mathbf{k} \cdot \mathbf{j} = 0$. The beauty of using orthogonal base functions is that it simplifies the mathematics significantly. More specifically, the solution for $x = (a_1, a_2, \ldots, a_N)^T$ becomes explicit and can be immediately shown as

$$a_k = \frac{\sum\limits_{i=1}^{M} y_i g_k(t_i)}{\sum\limits_{i=1}^{M} \left(g_k(t_i)\right)^2}. \tag{10.26}$$

There is no need to do the left-division or a matrix inversion.

10.4 Geometric Meaning of Least Squares Method

Linear equations in matrix format

$$Ax = y$$

often represent a relationship involving a physical process (or a mathematical model). Within this model, the unknown parameters are the N-dimensional vector x. Once the array x is determined, the model is fully defined. Here, what is practical is the case when the number of data points (or observations) M is much greater than the length of x (an overdetermined situation). For the above relationship, more specifically, it includes:

(1) An input, or information of time (t), which is implied in the matrix A; this independent variable t (t_i – note that it is different from x here) is an M-dimensional vector (M is an integer).
(2) A linear system or a physical process or a relationship, which can also be viewed as a constraint or a mathematical model, implied in the matrix A; the linear system has finite number of parameters that determine the system.
(3) An output (the vector y); which is often the observed variable, and more complicated than what the model represents because it at least contains random errors of observations.

For time series problems, the input often includes some information of time, i.e., time intervals, and length of time (for real problems, no time series can be infinite in length). The time intervals can be a constant or variable in time. As stated above, the output of the system y is more complicated than the model of the actual physical process. The model has simplified the actual process to some extent. So, the actual physical process should be viewed as having variations in a *higher dimensional space*, while the model represents this higher dimensional space with a lower dimension (N-dimensions formed by the N base functions).

Conceptually, this is very much like the approximation of a three-dimensional space vector using a two-dimensional space vector. If on a two-dimensional plane, any vector is allowed to approximate a given three-dimensional vector (Fig. 10.3), we can use the distance between the original vector and the approximate 2-D vector as the error of approximation. In Fig. 10.3, OA is the original 3-D vector; the vectors OB, OC, and OD are all on the 2-D plane. Obviously, some of these two-dimensional vectors have more errors than the others if each is used to approximate the three-dimensional vector OA. It does not take long before one realizes that the best approximation within the 2-D plane is the vector that is a projection of the 3-D vector onto the plane, i.e. OB. The difference vector AB is perpendicular to the plane and has the shortest distance between point A and point

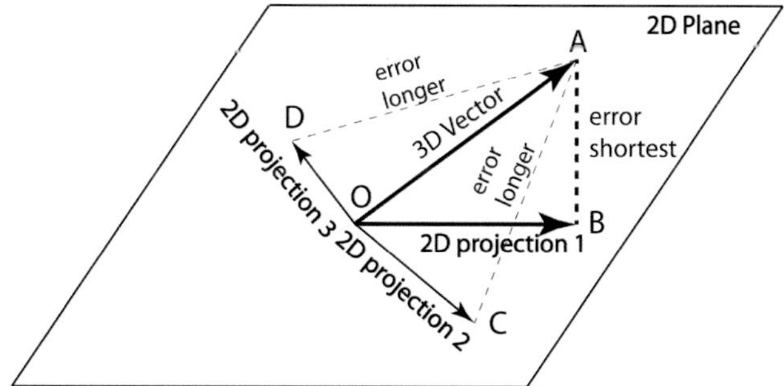

Figure 10.3 Projection of a 3-D vector on a 2-D plane.

B (AB is shorter than AC and AD). This difference (AB) is actually the error of the approximation.

Now, to come back to the problem at hand, a physical process is approximated by a model using N base functions $\big(g_1(t), g_2(t), \ldots, g_N(t)\big)$, i.e. equation (10.11):

$$y(t) = a_1 g_1(t) + a_2 g_2(t) + \cdots + a_N g_N(t).$$

The right-hand side of the above equation can be viewed as a function of the functions or a function in a functional space, which means it depends on the base functions $\big(g_1(t), g_2(t), \ldots, g_N(t)\big)$. For this reason, the RHS can be viewed as a vector in an n-dimensional space. The coefficients $(a_1, a_2, \ldots, a_N)$ can be viewed as the coordinates in this space.

Since the RHS is only an approximation of the LHS $\big(y(t)\big)$, and the output $y(t)$ usually represents more complicated processes, we can view $y(t)$ as a vector in a higher dimension. The equation is only approximate. The total error squared of this approximation is given by equation (10.15). This total error squared is a dot product or a distance squared, which is analogous to the error vector AB in Fig. 10.3. To make this analogy even clearer, recall equation (10.18): it can be rewritten as

$$A^T(Ax - y) = 0. \tag{10.27}$$

Here $Ax - y$ is the error vector. The transpose of matrix A is simply the base functions expressed in row-vectors. Equation (10.22) therefore means that each base function is perpendicular to the error vector! In other words, the error vector must be perpendicular to the approximate vector because it is formed by a linear superposition of the base functions so they must be in the same dimensional "plane." To further see consistency with the idea of the projection of a 3-D vector onto a 2-D plane as an analogy for higher dimensions, the original vector is observations or y, and the projection of y onto the lower dimension formed by the base functions is Ax, while the error vector is $Ax - y$. So, the requirement that the error vector is perpendicular to the projected vector can be translated to: the inner product between them must be zero, i.e.

$$(Ax)^T(Ax - y) = 0, \tag{10.28}$$

which is equivalent to

$$x^T A^T(Ax - y) = 0 \tag{10.29}$$

or

$$x^T[A^T(Ax - y)] = 0. \tag{10.30}$$

From (10.27), we see that this must be true. This is such an esthetic result!

Review Questions for Chapter 10

(1) On a two-dimensional flat plane, vector **a** is perpendicular to vectors **b** and **c**. What is the relationship between vectors **b** and **c**?

(2) In a three-dimensional space, vector **a** is perpendicular to vectors **b** and **c**. What is the relationship between vectors **b** and **c**?

(3) On a n-dimensional space, vector **a** is perpendicular to vectors **b** and **c**. What is the relationship between vectors **b** and **c**?

(4) In equation (10.9), the tidal oscillation is expressed as

$$\zeta = \zeta_0 + \sum_{i=1}^{N} A_i \cos(\omega_i t - \phi_i)$$

Can we use a plus sign in front of the phase, i.e.

$$\zeta = \zeta_0 + \sum_{i=1}^{N} A_i \cos(\omega_i t + \phi_i)$$

Can we use sine function instead, i.e.

$$\zeta = \zeta_0 + \sum_{i=1}^{N} A_i \sin(\omega_i t - \phi_i)?$$

(5) What is the advantage of using orthogonal base functions in a linear model?

Exercises for Chapter 10

(1) Two two-dimensional vectors in a Cartesian coordinate system are perpendicular to each other (Fig. 10.E1) and their slopes are k_1 and k_2, respectively. Prove, using dot product, that the two slopes are negative reciprocal of each other, i.e. $k_1 = -1/k_2$.

(2) Use dot product to verify that the vector (2, 3) is perpendicular to (−3, 2) and (3, −2). Are these related to the problem above?

(3) Given vectors (a, b), (12, −22), (0, 20), and (111, $\sqrt{3}$), find a perpendicular vector for each of these vectors, respectively. Is each perpendicular vector uniquely defined or does it have alternative answers?

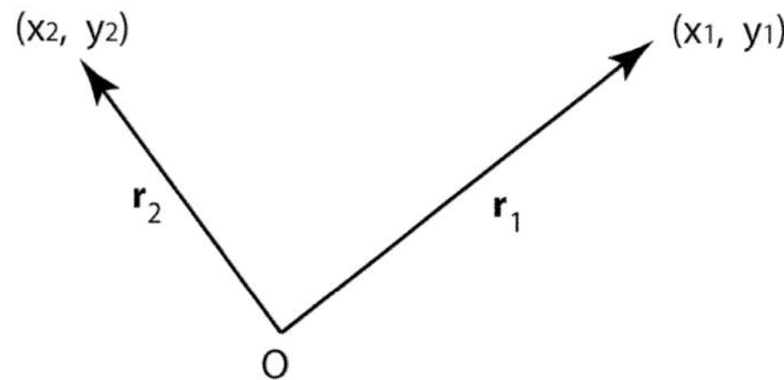

Figure 10.E1 Relationship of slopes of vectors perpendicular to each other.

11

Harmonic Analysis of Tide

About Chapter 11

The objective of this chapter is to discuss some background information of tides and the idea, purpose, and method of harmonic analysis of tides. Harmonic analysis is a special application of the least squares method to tidal signals. A list of 37 major tidal frequencies is provided. The basic theory and an example for the analysis is presented. The time origin of expression of tidal time series and longer-term variation of tidal constituents are discussed. A concise equilibrium tidal theory is included at the end of the chapter for reference.

11.1 About Tides

11.1.1 The Quasi-Periodic Tide

In ocean data analysis, we unavoidably encounter astronomical tides and need to know how to extract tidal information from data. This is especially true for coastal waters, where tides are more prominent than in deep oceans. But coastal waters also have lots of variabilities that are not related to tides. These include wind effect, river discharge effect, and the effects of heat flux, including latent heat flux, which all contribute to the dynamics of the atmosphere and ocean. Observed data include all contributions of various factors. In addition, nonlinear interactions in shallow water of different tidal constituents as well as the interactions between tidal and nontidal motions further complicate and enrich the spectrum.

Sometimes, we need to extract tidal signal from the data, while other times we need to exclude tidal signal and focus on nontidal signals such as wind-driven flows or density currents. There are several different ways to accomplish these goals, but one of the conventional ways is to use the least squares method for tidal analysis. Another way is to use filters (Chapter 21).

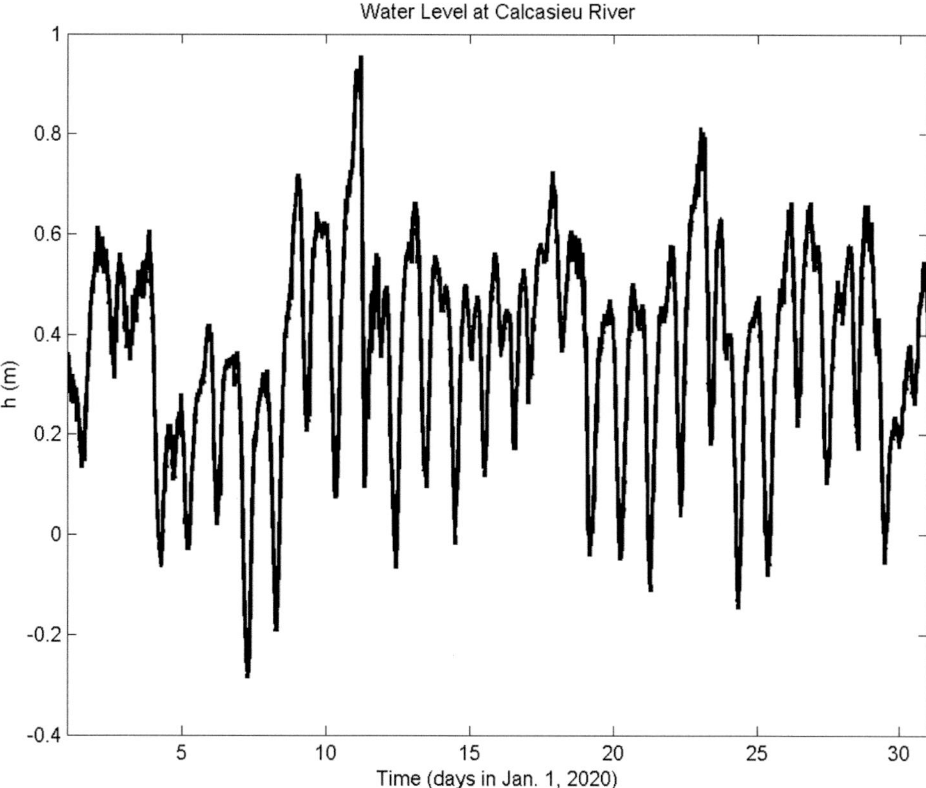

Figure 11.1 Water level at an USGS station in Cameron, Louisiana (USGS 08017118 Calcasieu River at Cameron, LA), showing the quasi-periodic tidal signal.

When using the least squares method, the first question is: what base functions should we use? This has something to do with the nature of tide; that is, tidal motion is oscillatory or periodical or, more accurately, *quasi-periodical* (Chapter 4). The reason we say that tidal oscillations are quasi-periodical is because tides have many different periodicities superimposed on each other in addition to the nontidal, especially weather-induced variations. That makes tide not a single frequency periodical oscillation but a combination of many periodical oscillations.

Although tide is generally not a single-period oscillation, it often has a dominant tidal period. This can be verified by looking at the water level records from many stations around the world. Fig. 11.1 shows an example of water level recorded at the Calcasieu River Estuary, Louisiana. The diurnal tides have periods of around 24 hours, while the semi-diurnal tides have periods around 12 hours, but they are essentially different. When these oscillations with different periods are added, they are not exactly sinusoidal nor periodical.

11.1.2 The Base Functions for Tidal Analysis

To predict tides empirically, one needs to know the periods of tidal oscillations and a time series of observations. Here we have implied that tidal variations are a superposition of a series of sinusoidal functions, i.e. our base functions will be the sine and cosine functions with given frequencies – *the tidal frequencies* plus the mean value:

$$1, \sin(\omega_1 t), \cos(\omega_1 t), \sin(\omega_2 t), \cos(\omega_2 t), \ldots, \sin(\omega_n t), \cos(\omega_n t).$$

Here t is time, and $\omega_1, \omega_2, \ldots, \omega_n$ are the angular frequencies of tide. The first base function is a constant 1, which represents the mean component.

As demonstrated earlier, the least squares method can be applied to obtain the coefficients of these base functions. For n frequencies, there are $2n + 1$ base functions and thus $2n + 1$ unknown coefficients to be determined. Once these coefficients are obtained, we can use them to predict future tidal oscillations.

11.1.3 Tidal Constituents

The question is, what frequencies should we use? The quasi-periodic water level variations of tide are caused by the tide-generating force of the Sun and the Earth's moon. The periodicities of tide can be lumped into several groups: *semi-diurnal tides*, *diurnal tides*, *long-term tides*, and short-term *overtides*. Each group has many different frequencies, which are determined by an expansion of the tide-generating potential into a superposition of simple sinusoidal functions. The tidal motion at each frequency is also called a *tidal constituent*.

The last group (overtides) is one with frequencies generated from nonlinear dynamics of tidal wave propagations. Usually, a *nonlinear parameter* a/h is used to characterize the importance of this effect. Here a and h are the tidal amplitude and water depth, respectively. The larger this parameter, the stronger the nonlinear effect. In shallow water, the nonlinear effect is stronger. The components in this group are sometimes called the *shallow water constituents*.

The exact periods or frequencies for all of the above tidal constituents have already been accurately determined from analysis of astronomical *tide-generating force*, a long time ago. Most of them have astronomical origins. The rest are from nonlinear interactions among the astronomical frequencies – the abovementioned shallow water constituents. Additional information about tide-generating force is included at the end of this chapter for the reader's reference.

Each tidal constituent has a name and commonly used symbol (Table 11.1). The semi-diurnal tidal constituents are highlighted in Table 11.1 in boldface. The shallow water tidal constituents are indicated in Table 11.1 in italic font. The tidal

Table 11.1. *Tidal frequencies*

#	Name	Speed (degree/hour)	Description
1	**M2**	28.9841042	Principal lunar semi-diurnal
2	**S2**	30	Principal solar semi-diurnal
3	**N2**	28.4397295	Larger lunar elliptic semi-diurnal
4	K1	15.0410686	Lunar diurnal
5	*M4*	*57.9682084*	*Shallow water overtides of principal lunar*
6	O1	13.9430356	Lunar diurnal
7	*M6*	*86.9523127*	*Shallow water overtides of principal lunar*
8	*MK3*	*44.0251729*	*Shallow water terdiurnal*
9	*S4*	*60*	*Shallow water overtides of principal solar*
10	*MN4*	*57.4238337*	*Shallow water quarter diurnal*
11	**NU2**	28.5125831	Larger lunar evectional
12	*S6*	*90*	*Shallow water overtides of principal solar*
13	**MU2**	27.9682084	Variational
14	**2N2**	27.8953548	Lunar elliptical semi-diurnal second-order
15	OO1	16.1391017	Lunar diurnal
16	**LAM2**	29.4556253	Smaller lunar evectional
17	S1	15	Solar diurnal
18	M1	14.4966939	Smaller lunar elliptic diurnal constituent
19	J1	15.5854433	Smaller lunar elliptic diurnal constituent
20	MM	0.5443747	Lunar monthly constituent
21	SSA	0.0821373	Solar semiannual constituent
22	SA	0.0410686	Solar annual constituent
23	MSF	1.0158958	Lunisolar synodic fortnightly constituent
24	MF	1.0980331	Lunisolar fortnightly constituent
25	RHO	13.4715145	Larger lunar evectional diurnal constituent
26	Q1	13.3986609	Larger lunar elliptic diurnal constituent
27	**T2**	29.9589333	Larger solar elliptic constituent
28	**R2**	30.0410667	Smaller solar elliptic constituent
29	2Q1	12.8542862	Larger elliptic diurnal
30	P1	14.9589314	Solar diurnal constituent
31	*2SM2*	*31.0158958*	*Shallow water semi-diurnal constituent*
32	M3	43.4761563	Lunar terdiurnal constituent
33	**L2**	29.5284789	Smaller lunar elliptic semi-diurnal constituent
34	*2MK3*	*42.9271398*	*Shallow water terdiurnal constituent*
35	**K2**	30.0821373	Lunisolar semi-diurnal constituent
36	*M8*	*115.9364166*	*Shallow water eighth diurnal constituent*
37	*MS4*	*58.9841042*	*Shallow water quarter diurnal constituent*

frequencies are provided in a different way, using "speed" rather than frequency. The speed has a unit of degrees per hour. For instance, the S2 tide (the *principal solar semi-diurnal constituent*) has a speed of 30 degrees per hour, so 12 hours correspond to 360 degrees (a complete cycle). The M2 tide (*the principal lunar constituent*) has a speed of 28.9841042 degrees per hour, so in ~12.42 hours, it results in ~360 degrees, completing a full cycle.

To convert the speed to frequency in *cycles per day* (CPD), multiply the speed by 24/360. For instance, M2, S2, N2, K1, O1, S1, M1, and P1 tides (Table 11.1) have speed of 28.9841, 30.0000, 28.4397, 15.0411, 13.9430, 15.0000, 14.4967, and 14.9589 degrees per hour, respectively. Multiplying the speed by 24/360 yields frequencies of 1.9323, 2.0000, 1.8960, 1.0027, 0.9295, 1.0000, 0.9664, and 0.9973 cycles per day, respectively. To convert the tidal frequencies to periods in hours, multiply the reciprocal of the above frequencies in CPD by 24. For the above tidal constituents, these translate to periods of 12.4204, 12.0000, 12.6582, 23.9354, 25.8203, 24.0000, 24.8344, and 24.0650 hours, respectively.

11.2 Harmonic Analysis of Tides

11.2.1 Solution of the Tidal Constituents

Tide is generated in deep ocean and propagated into shallower water where tides are significantly amplified. This is mainly because tide-generating force (see Section 11.3) is a body force which has a magnitude proportional to the mass. Deep ocean has more water mass than shallower water, and thus tide-generating force has more effect on deep ocean. Tide in deep ocean, however, is still very weak despite the large amount of ocean mass. The tide-generating force is one order of magnitude smaller than the gravitational force.

The propagation of tide from deep ocean to shallower water generates *tidal waves*. Because of the complex bathymetry of the water, complex geometry of coastlines, and greater nonlinearity, tidal wave propagation in shallower water is a complex process. For the analysis or empirical prediction of tides at individual stations, we usually do not have to consider tidal wave propagation dynamics. Because of the quasi-periodic nature and perpetual motion involving definitive astronomical tidal frequencies in the process, a regression analysis can do the job. We assume that the tidal variation is a linear superposition of all the tidal constituents, including the shallow water constituents from nonlinear interactions:

$$\zeta = \zeta_0 + \sum_{i=1}^{K} a_i \cos(\omega_i t - \varphi_i). \tag{11.1}$$

Here K is the total number of tidal constituents selected. The first term on the right-hand side, ζ_0, is the mean value of ζ. The frequencies ω_i ($i = 1, 2, \ldots, K$) are given, and the coefficients a_i and phase φ_i are to be determined. The least squares method cannot be applied directly to the above equation until we expand the cosine function to cosine and sine functions. It can be shown that the above equation is equivalent to

$$\zeta = \zeta_0 + \sum_{i=1}^{K} [A_i \cos(\omega_i t) + B_i \sin(\omega_i t)]. \tag{11.2}$$

This is based on the trigonometric relation:

$$\cos(\omega_i t - \varphi_i) = \cos(\omega_i t)\cos(\varphi_i) + \sin(\omega_i t)\sin(\varphi_i). \tag{11.3}$$

This gives

$$A_i = a_i \cos(\varphi_i),\ B_i = a_i \sin(\varphi_i),\ \varphi_i = \arctan\left(\frac{B_i}{A_i}\right),\ a_i = \sqrt{A_i^2 + B_i^2}. \tag{11.4}$$

We can see that the base functions are the sine and cosine functions with known frequencies from Table 11.1. As we learned earlier, the sine and cosine functions are linearly independent and can be used as proper base functions for the least squares method to obtain the best estimate of the amplitude and phase of each tidal constituent.

How many tidal constituents to choose? It depends. At the very least, one might want to choose one major tidal frequency. For example, if data are obtained in a period of only 25 hours, and we know the area has mainly diurnal tide, then it would be reasonable to choose a 24-hour diurnal tide as the only constituent for the analysis. In that case, there will be three base functions: $g_1 = 1$, $g_2 = \cos(\omega t)$, and $g_3 = \sin(\omega t)$, in which ω is the angular frequency of tide corresponding to the 24-hour period. We may also add a semi-diurnal tidal constituent such as M_2 tide, which would make five base functions. Obviously, for K tidal constituents, we would have $2K + 1$ base functions:

$$g_1 = 1,\ g_{2i} = \cos(\omega_i t),\ g_{2i+1} = \sin(\omega_i t),\ i = 1, 2, \ldots, K. \tag{11.5}$$

If there are M observations, the matrix A from the base functions will be

$$A = \begin{pmatrix} 1 & g_2(t_1) & \cdots & g_{2K+1}(t_1) \\ 1 & g_2(t_2) & \cdots & g_{2K+1}(t_2) \\ \vdots & \vdots & \vdots & \vdots \\ 1 & g_2(t_M) & \cdots & g_{2K+1}(t_M) \end{pmatrix}. \tag{11.6}$$

If the array of observations of ζ in MATLAB is y, then the solutions for coefficients ζ_0, A_i, and B_i are obtained by the least squares solution:

$$x = (A^T A)^{-1} A^T y. \tag{11.7}$$

Here $x = (\zeta_0, A_1, B_1, \ldots, A_M, B_M)$. Again, with MATLAB, the above equation may be replaced by left-division. The last two equations in (11.4) will then be used to calculate the amplitude (a_i) and phase (φ_i) of each constituent. It should be noted that there are programs for tidal analysis out there, including those available online. For example, a commonly used tidal analysis program, t-tide, has a MATLAB version that can be easily applied.

11.2.2 The Error Estimation

The total error squared can be calculated by

$$R_S = (y - Ax)^T (y - Ax). \tag{11.8}$$

Here x is the estimated array of coefficients from the least squares method or equation (11.7). R_S is a number because $y - Ax$ is a column-vector, and thus R_S is an inner product of the vector with itself, which is a single real number.

The root mean-squared error of the model (11.1) is

$$\sigma = \sqrt{\frac{R_S}{M - (2K + 1)}}. \tag{11.9}$$

This is also a real number. The subtraction of $2K + 1$ in the denominator is because the $2K + 1$ coefficients are determined and the M data points have $M - (2K + 1)$ degrees of freedom.

Another statistical parameter – the coefficient of determination discussed in Chapter 4 or the percentage of explained variability by the model – the R^2 value, is

$$R^2 = 1 - \frac{R_S}{R_V}, \quad R_V = (y - \bar{y})^T (y - \bar{y}), \tag{11.10}$$

in which $\bar{y}$ is the mean of the observations y. R_S/R_V is the unexplained variability.

11.2.3 The Procedure

Given the above discussion, the procedure of harmonic analysis should be as follows:

(1) Prepare the data from observations or model output, i.e. the left-hand-side time series of ζ in equation (11.2), and define that as the MATLAB variable y of length M.
(2) Determine how many tidal frequencies to use, i.e. the number K and ω_1, ω_2, ... , ω_K.
(3) Know the M time instances, i.e. t_1, t_2, ... , t_M.

(4) Define the matrix A by defining each of its columns (the base functions); do not forget the column of ones.
(5) Do the MATLAB left-division or use equation (11.7) directly to resolve the coefficient array x.
(6) Convert the coefficients of the base functions to the tidal constituents for each of the K tidal frequencies, i.e. the amplitude and phase of each tidal frequency using (11.4).
(7) Calculate the total error squared and the root-mean squared error of the analysis using (11.8) and (11.9).
(8) Compute the R^2 from (11.10).

For example, if we choose the M2, S2, N2, K1, O1, S1, M1, and P1 tidal frequencies, we can define them in MATLAB like the following:

```
Omega   = [ 28.9841042    % M2 angular freq. in degrees per hour
            30            % S2 angular freq. in degrees per hour
            28.4397295    % N2 angular freq. in degrees per hour
            15.0410686    % K1 angular freq. in degrees per hour
            13.9430356    % O1 angular freq. in degrees per hour
            15            % S1 angular freq. in degrees per hour
            14.4966939    % M1 angular freq. in degrees per hour
            14.9589314] ; % P1 angular freq. in degrees per hour
```

If we have also defined the time variable `time` as a column array with a unit of days, which is the case if we use the MATLAB command `datenum` to generate the variable `time`, we can then define the base function matrix by:

```
M   = length(omega);
G1  = ones(size(time));
A   = [];
for i = 1:M
 A = [A sind(omega(i)*time*24) cosd(omega(i)*time*24)];
end
A = [ G1 A];
```

The next step of left division for the solution of the coefficients of base functions is almost trivial:

```
x = A \ y;
```

in which `y` is the data array or the observed time series of tidal data.

These commands can be either issued separately in a sequence or built into a function that can be used for a more automated processing. It is always a good practice to build a function for repeated applications because it will be more efficient and worry-free once it is reliably tested and validated.

11.2.4 Selection of Time Origin

In oceanography, a standard practice is to avoid using local time and instead only use UTC. This is also true for harmonic analysis for tide. Aside from this, it is important to know that the time origin must also be properly selected. There are several choices, though. Depending on the problem, there may be some flexibility – but with some caveats. In theory, the time origin does not really matter for conducting harmonic analysis. However, the phase depends on the time origin we use. The absolute value of the phase is not necessarily significant. Rather, it is the phase difference or the phase in reference to a specific time origin that is of significance. A common practice is to use Jan. 1, 1900, as the time origin, e.g. for NOAA's tidal constituents. Another choice is to use Jan. 1 of the year in which the first data were collected.

11.2.5 Long-Term Variations

Equation (11.2) is fairly accurate for tidal prediction unless we are working on a long-term prediction. By long term, we mean more than 10 years. This is because among all the tidal periodicities, there is another period of roughly 18.6 years caused by the precession of the Moon, which is the change in orientation of the Moon's orbit. It turns out that this motion will affect most of the amplitude and phase of the tidal constituents except solar constituents S2, S4, annual Sa, and semi-annual Ssa. To account for this complication, equation (11.2) is sometimes empirically modified as

$$\zeta = \zeta_0 + \sum_{i=1}^{K} f_i a_i \cos(\omega_i t - \varphi_i + u_i), \tag{11.11}$$

in which f_i and u_i are the so-called *lunar node factors*, which have some formulas varying at 18.6 years periodically. In practice, for most of our applications of short-term analysis and predictions, we can lump the factors with the amplitude and phase, respectively; and understand them as changing slowly over the 18.6-year period.

In actual applications of harmonic analysis and predictions, lunar node factors in fast-changing coastal environments may not be needed or useful. This is because in such an environment, erosion, deposition, and coastal engineering such as harbor construction etc. may alter the coastlines and bathymetry significantly over long-time scales of 18.6 years. In other words, the harmonic "constants" in this situation are no longer really constants. The changing coastlines and bathymetry will likely affect tidal wave propagation and tidal currents significantly, so that the amplitude and phase of any given tidal constituent are also slowly changing over time. This change may overwhelm the variation due to the lunar nodal variations, and thus those factors are not very useful in accurately describing the tidal constituents, should a time scale longer than 18 years be of interest. The anthropogenic effect obviously may play a more important role than the astronomical factors in this case.

11.3 Appendix: Tide-Generating Force

11.3.1 Tide-Generating Force

The cause of ocean tide was not correctly explained until Isaac Newton. The prediction of tidal oscillations of water surface and water flow velocity was not established until after Newton's discovery of *gravitational force* and his explanation of tide-generating force. Tide-generating force is closely related to gravitational force, but they are not the same. The only meaningful tide-generating forces on the Earth are from either the Sun or the Moon, with that from the Moon a little larger than from the Sun – perhaps a little counterintuitive to some. Any other celestial body produces only negligible tide-generating force to the Earth.

So, what is a tide-generating force? It is the difference of gravitational forces between that exerted on the center of the Earth and that on the place of consideration – in the ocean, in our case. In concept, it is a *tearing force*: it is the slight difference in gravitational force that tends to "rip the Earth apart," although the magnitude of the force is tiny on Earth and would not really tear Earth apart.

Newton's gravitational law applied for the gravitational force of the Moon (or Sun) on a unit mass at a given point on the Earth in mathematics is:

$$\mathbf{f}_1 = \frac{GM}{r^2}\hat{\mathbf{r}}. \tag{11.12}$$

Here M is the mass of the Moon (or the Sun); G is the gravitational constant, which is $6.67408 \times 10^{-11}\ \mathrm{m^3\,kg^{-1}\,s^{-2}}$; r is the distance between the center of the Moon and the point of consideration P on Earth; and $\hat{\mathbf{r}}$ is the unit vector in the $\mathbf{r}$ direction (Fig. 11.2). The gravitational pull to a unit mass at the center of the Earth is

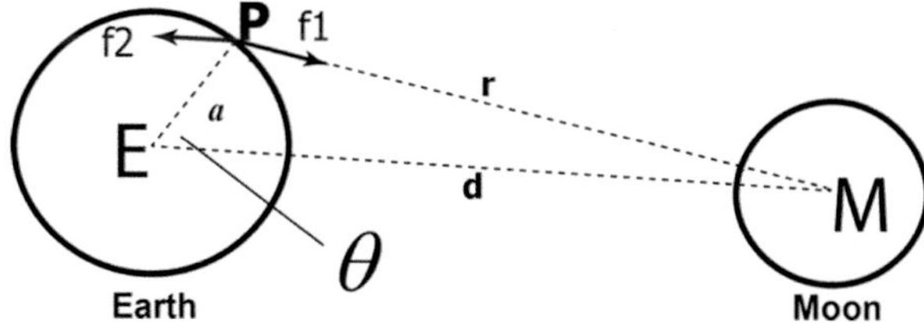

Figure 11.2 Tide-generating force.

$$-\mathbf{f}_2 = G\frac{M}{d^2}\mathbf{d}. \tag{11.13}$$

Here we use a negative sign because we want to use $\mathbf{f}_2$ to denote the centrifugal force used to make the Earth–Moon rotate around their center of mass. The tide-generating force is the sum of $\mathbf{f}_1$ and $-\mathbf{f}_2$:

$$\mathbf{f}_{tide} = \mathbf{f}_1 - \mathbf{f}_2 = G\frac{M}{r^2}\hat{\mathbf{r}} - G\frac{M}{d^2}\mathbf{d}. \tag{11.14}$$

With the vector relationship

$$\overrightarrow{PM} = \overrightarrow{PE} + \overrightarrow{EM}, \tag{11.15}$$

the vector $\mathbf{f}_1$ becomes

$$\mathbf{f}_1 = G\frac{M}{r^3}\overrightarrow{PM} = G\frac{M}{r^3}\left(\overrightarrow{PE} + \overrightarrow{EM}\right) = \mathbf{f}_{11} + \mathbf{f}_{12} \tag{11.16}$$

in which $\mathbf{f}_{11}$ is along the local vertical at P and $\mathbf{f}_{12}$ is parallel to $\mathbf{f}_2$ (Fig. 11.3), or

$$\mathbf{f}_{11} = \frac{GM}{r^3}\overrightarrow{PE}; \quad \mathbf{f}_{12} = \frac{GM}{r^3}\overrightarrow{EM}.$$

The vertical component of the tide-generating force is too small compared to gravity, so the effect from $\mathbf{f}_{11}$ is negligible and the tide-generating force is thus the horizontal components of $\mathbf{f}_2$ and $\mathbf{f}_{12}$. This can be shown, respectively, as

$$-f_2 \sin(\theta) = -G\frac{M}{d^2}\sin(\theta), \tag{11.17}$$

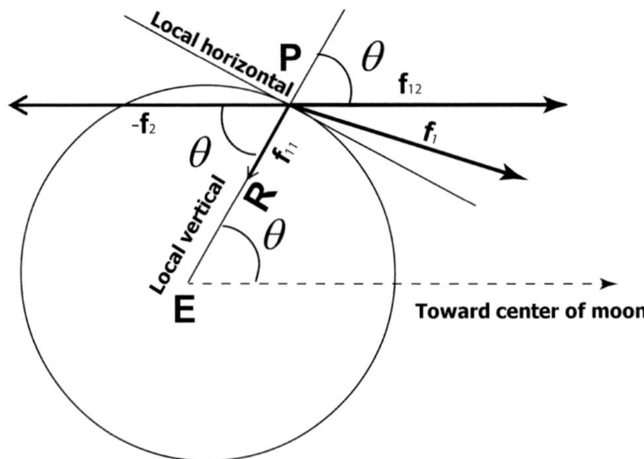

Figure 11.3 Derivation of tide-generating force.

$$f_{12}\sin(\theta) = G\frac{M}{r^3}d\sin(\theta). \quad (11.18)$$

The tide-generating force is then

$$f_{tide} = GM\left(\frac{d}{r^3} - \frac{1}{d^2}\right)\sin(\theta). \quad (11.19)$$

This can be further reduced to (Proudman, 1953)

$$f_{tide} = 3\frac{M}{E}\left(\frac{R}{d}\right)^3 g\cos(\theta)\sin(\theta), \quad (11.20)$$

in which E is the mass of the Earth, R is the radius of Earth, and g is the gravitational acceleration, which has an averaged value of $g = 9.8\ \mathrm{m/s^2}$ on Earth's surface.

From this result, we can see that the tide-generating force is *inversely proportional to distance cubed*, in contrast to the gravitational pull which is inversely proportional to distance squared. Mathematically, this makes sense because the tearing force is a difference of gravitational force, and we know that the derivative of $1/r^2$ is proportional to $1/r^3$. The magnitude also depends on the angle θ in a complex way. In any case, this means that tide-generating force is one order of magnitude smaller than the gravitational force.

The above equation works for the tide-generating force from both the Sun and the Moon. Since the mass of the Sun is ~1.9885×10^{30} kg, while that of the Moon is ~7.342×10^{22} kg, the mean distance between the Sun and Earth is ~1.496×10^8 km, and the mean distance between the Earth and Moon is ~3.84×10^5 km, it is easy to verify that the tide-generating force from the Sun is only ~46% of that from the Moon. Even though the Sun is significantly more massive than the Moon, it is further away, and since tide-generating force is inversely proportional to the distance cubed but proportional to mass, the effect of distance is much greater.

11.3.2 The Equilibrium Tide

The above discussion provides the tide-generating force. To examine the response of the ocean surface, we need to look at the dynamics. The following is the momentum equation for this problem:

$$\frac{d\mathbf{v}}{dt} + f\mathbf{k}\times\mathbf{v} = -g\nabla\zeta + \mathbf{g} + \mathbf{F}_{tide} + \mathbf{F}_{friction}, \quad (11.21)$$

where $\mathbf{V}$, f, $\mathbf{k}$, ζ, $\mathbf{F}_{tide}$, and $\mathbf{F}_{friction}$ are the ocean velocity vector, Coriolis parameter, unit vector along the local vertical with positive being upward, water surface elevation, tide-generating force vector in the local horizontal, and bottom frictional force vector, respectively. The first term on the right-hand-side, $-g\nabla\zeta$, is the pressure gradient force due to surface slope. The tide-generating force at equilibrium (when $\mathbf{v} = 0$) will deform the water surface to $\tilde{\zeta}$ so that the pressure gradient force caused by the gradient of $\tilde{\zeta}$ is balanced by $\mathbf{F}_{tide}$ or

$$g\frac{\partial\tilde{\zeta}}{\partial y} = -3g\frac{M}{E}\left(\frac{R}{d}\right)^3 \sin\theta\cos\theta. \tag{11.22}$$

By an integration and considering the volume conservation of the ocean (omitted), we can get the following equation for the *equilibrium tide*:

$$\tilde{\zeta} = R\frac{M}{E}\left(\frac{R}{d}\right)^3\left(\frac{3}{2}\cos^2\theta - \frac{1}{2}\right). \tag{11.23}$$

The angle θ is dependent on the point of consideration on the Earth and is changing all the time. We can express it by several astronomical or geographic parameters (Fig. 11.4), i.e., the *declination of the Moon* (δ), *hour angle of the Moon* (ψ), and latitude (φ), with the formula in spherical trigonometry we learned earlier:

$$\cos\theta = \cos\delta\cos\psi\cos\varphi + \sin\delta\sin\varphi. \tag{11.24}$$

The equilibrium tide is then expressed in terms of the parameters δ, ψ, and φ by (Proudman, 1953):

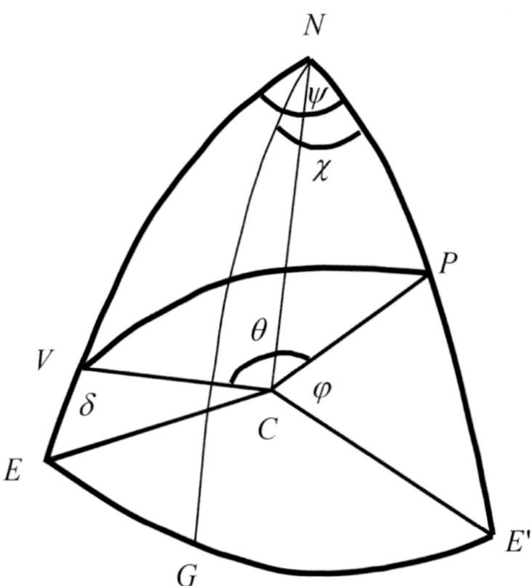

Figure 11.4 Relation of point of interest on Earth, declination of the Moon (δ), the hour-angle of the Moon (ψ) in reference to observer (at P) with its latitude (φ) and longitude (χ).

$$\tilde{\zeta} = R\frac{M}{E}\left(\frac{R}{d}\right)^3\left(\frac{3}{4}\cos^2\delta\cos^2\varphi\cos 2\psi + \frac{3}{4}\sin 2\delta\sin 2\varphi\cos\psi \right.$$
$$\left. + \frac{3}{4}\cos^2\delta\cos^2\varphi + \frac{3}{2}\sin^2\delta\sin^2\varphi - \frac{3}{10}\right), \tag{11.25}$$

In the above equation, the variables are d, δ, and ψ. The distance between the Earth and Moon has a monthly periodicity. The declination of the Moon δ has a monthly variation and thus $\cos^2\delta$ has a half-month period. Likewise, ψ has a diurnal variability, so $\cos 2\psi$ has a semi-diurnal period; $\sin(2\delta)$ has a half-monthly variation; and $\sin^2\delta$ has a half-monthly variation as well.

Without getting into more details, the above results give us different groups of tidal constituents. These groups include (1) diurnal, (2) semi-diurnal, (3) biweekly, and (4) monthly constituents for those related to the Moon. For those related to the Sun, it is similar: (1) diurnal, (2) semi-diurnal, (3) semi-annual, and (4) annual constituents.

These tidal constituents represent the force of astronomical origin. The ocean under such astronomical force will respond with waves propagating in the ocean basin. The tidal waves are small in the deep ocean where they are generated but are amplified significantly as they approach coastal regions, estuaries, and bays. As the water gets shallower, the nonlinear effect will kick in, producing new frequencies based on the original tidal frequencies. These new tidal constituents, due to the dynamics of ocean movement, are called shallow water constituents. They are produced not in (11.25) but through the advective acceleration in the momentum equation. Sometimes we call them overtides, as they mostly have higher frequencies. In addition, there are also lower frequencies generated by nonlinear effect. These, however, are within the low frequency band overlapping with the effect of weather.

11.3.3 Another Perspective from an Inertial System

Another way to understand the nature of tide-generating force is to examine the relationship between the acceleration in an inertial coordinate system with that in the rotational Earth system. Assume that the vector based on an inertial coordinate system from the center of mass of the Sun–Earth system (which is almost at the center of mass of the Sun) to the center of mass of the Earth is $\mathbf{r}_0$; the vector from the center of Earth to the position of interest in the ocean in the rotational system is $\mathbf{r}'$; and the vector from the center of mass of the Sun–Earth system to the position of interest in the ocean in the inertial system is $\mathbf{r}$. They have an obvious relation (Fig. 11.5):

$$\mathbf{r} = \mathbf{r}_0 + \mathbf{r}'. \tag{11.26}$$

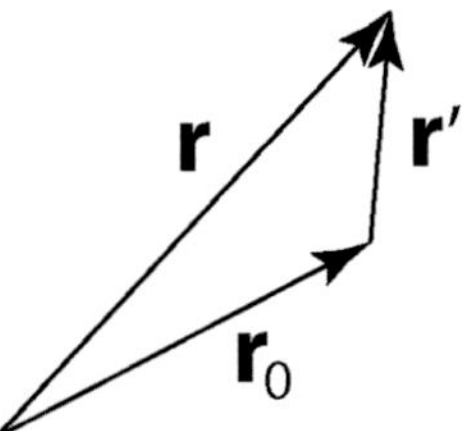

Figure 11.5 Vector relation: inertial system and rotational system.

The accelerations will have a similar relationship as

$$\mathbf{a} = \mathbf{a}_0 + \mathbf{a}'. \tag{11.27}$$

Recall the second of Newton's laws, $\mathbf{F} = \mathrm{m}\mathbf{a}$; the above relation leads to

$$\mathbf{a} = \mathbf{a}_0 + \mathbf{a}' = \mathbf{f}, \tag{11.28}$$

in which the vector $\mathbf{f}$ represents the total force per unit mass. One force acted on a unit mass is the gravitational pull from the Sun, which is

$$\mathbf{f}_{Sun} = G\frac{M_S}{r^2}\hat{\mathbf{r}} = \mathbf{a}, \tag{11.29}$$

in which M_S is the mass of Sun, and $\hat{\mathbf{r}}$ is the unit vector of $\mathbf{r}$. The center of mass of the Earth is moving around the Sun under the gravitational pull between the two celestial bodies, and the acceleration $\mathbf{a}_0$ is a result of that pull. This acceleration allows the Earth to rotate around the Sun, and this force applied to a unit mass is

$$\mathbf{f}_{Sun0} = \mathbf{a}_0 = G\frac{M_S}{r_0^2}\hat{\mathbf{r}}_0. \tag{11.30}$$

Here r_0 is the distance between the center of mass of the Sun–Earth system and the center of mass of the Earth; $\mathbf{r}_0$ is the vector between these two centers pointing toward the center of mass of the Earth to that of the Sun–Earth system; and $\hat{\mathbf{r}}_0$ is the unit vector of $\mathbf{r}_0$. Substituting equations (11–29) and (11–28) into (11–26) yields

$$\mathbf{a}' = G\frac{M_S}{r^2}\hat{\mathbf{r}} - G\frac{M_S}{r_0^2}\hat{\mathbf{r}}_0. \tag{11.31}$$

Equation (11.31) says that the acceleration of ocean particles relative to the Earth is determined by the tide-generating force that we defined earlier, or the difference of gravitational pulls between that at the center of the Earth and that on the ocean particle of consideration. For the situation of the Earth–Moon system, the discussion is the same, and we should get the same equation as above, except that the mass of the Sun is replaced by that of the Moon, and the distances to the Sun are replaced by those to the Moon.

Review Questions for Chapter 11

(1) All tidal oscillations are sinusoidal at given frequencies. Why, when all these oscillations at different frequencies are added together, does this not lead to an exactly periodic motion?

(2) What is the difference between gravitational force and tide-generating force?

(3) Tide is usually referred to as a phenomenon in the ocean, but it also occurs in the atmosphere with a much less obvious signal. Do you think the atmospheric tide should have the same frequencies as in the ocean (Table 11.1)?

(4) Suppose tide at a given location has five major frequencies, and in a harmonic analysis one should choose these five frequencies. What if you add 10 additional tidal frequencies: do you anticipate the results will be much different (i.e. do you anticipate a significant difference in predicting tide between using 5 and using 15 tidal frequencies)?

Exercises for Chapter 11

(1) Write a MATLAB script to convert the unit of tidal frequencies in Table 11.1 from degree per hour to cycle per day.

(2) Write a MATLAB script to convert the unit of tidal frequencies in Table 11.1 from degree per hour to hours per cycle.

(3) The data file named "S6-depth-time.mat" has two variables; the first is time in days and the second is water depth in m. It is a time series of water depth with both diurnal and semi-diurnal tides measured from the Louisiana shelf in the spring of 2020. The base functions are sine and cosine functions at eight different frequencies (cycle per day):
f1 = 1.9323; f2 = 2.0000; f3 = 1.8960; f4 = 2.0055; f5 = 1.0027; f6 = 0.9295; f7 = 1; f8 = 0.9973

(a) Construct 16 base functions $g_1, g_2, \ldots, g_{16}$ based on the above frequencies:

$$\cos(2\pi f_i t),\ \sin(2\pi f_i t), \quad i = 1, 2, 3, 4, 5, 6, 7, 8,$$

and a last base function of ones:

```
g17 = ones(length(time),1);
```

(b) Construct the A matrix, `A = [g1 g2 g3 g4 g5 g6 g7 ... g17]`.

(c) Find the coefficients `x` by left-division.

(d) Reconstruct the model depth, `depthm = A * x`.

(e) Plot the original data and compare with the model depth.

(4) Write a function to do harmonic analysis, including the eight constituents M2, S2, N2, K1, O1, S1, M1, and P1, and use data file "CLAKE-WaterLevel0.txt" to test the function. Note:

(a) The data file has seven columns, representing year, month, day, hour, minute, water pressure (db), and water level (m).

(b) Use `datenum(year,month,day,hour,minute,second)` to define the time; you may choose the time origin to be Jan. 1, 2010, so the time can be defined as `time = datenum(year,month,day,hour,minute,second) - datenum(2010,1,1) + 1,` i.e. day 1 to be starting from 00:00 UTC, Jan. 1, 2010]. The output should include the amplitude and phase for each constituent. It should also output the root-mean squared error.

(5) Modify the above function by adding the shallow water tidal constituents from Table 11.1 and write a new function. Apply the new function to the same data. Compare the results. How important are the nonlinear shallow water tides in the given data?

(6) Repeat problem 3 but adding the root- mean squared error estimate.

12

Fourier Series

About Chapter 12

The objective of this chapter is to extend the ad hoc least squares method of somewhat arbitrarily selected base functions to a more generic method applicable to a broad range of functions – the Fourier series, which is an expansion of a relatively arbitrary function (with certain smoothness requirement and finite jumps at worst) with a series of sinusoidal functions. An important mathematical reason for using Fourier series is its "completeness" and almost guaranteed convergence. Here "completeness" means that the error goes to zero when the whole Fourier series with infinite base function is used. In other words, the Fourier series formed by the selected sinusoidal functions is sufficient to linearly combine into a function that converges to an arbitrary continuous function. This chapter on Fourier series will lay out a foundation that will lead to Fourier Transform and spectrum analysis. In this sense, this chapter is important as it provides background information and theoretical preparation.

12.1 Linear Combination of Sinusoidal Functions

In physics, simple sinusoidal and periodic functions are quite common. For example, (1) an idealized wave; (2) the motion of a frictionless pendulum; and (3) the motion of a perfect spring with a mass attached at one end under gravity are all described by sine functions. These simple sinusoidal functions can be superimposed to construct more complicated functions.

From the least squares method, we know that many functions can be approximated by a linear combination of a series of base functions. A good example is harmonic analysis for ocean tides. It is a powerful method that has been used for many decades.

12.1.1 A Periodic Function

Now let us look at the problem in a slightly different way. We select some sinusoidal functions as base functions. Given a few sinusoidal time series functions with frequencies based on a fundamental frequency $1/T$, the resultant time series can be a rather complicated function but periodic nevertheless. As an example, we choose four such base functions and add them together. The definitions of the functions are given below, and the graph is shown in Fig. 12.1.

$$x_1 = \cos(2\pi f t), \quad x_2 = \cos(4\pi f t)/2, \quad x_3 = \cos(6\pi f t)/3, \quad x_4 = \cos(8\pi f t)/4$$
$$x = x_1 + x_2 + x_3 + x_4, \quad \text{here } f = 1/T, \quad T = 1$$

Here the four constituent functions are sinusoidal with frequencies f (which is $1/T$), $2f$, $3f$, and $4f$, respectively. The amplitudes of the functions at these frequencies are 1, 1/2, 1/3, and 1/4, respectively. The frequencies $2f$, $3f$, and $4f$ are two, three, and four times the lowest (fundamental) frequency f and are called the harmonics of f. These parameters are arbitrarily chosen for demonstration purposes and, of course, different values can be used. The linear superposition of

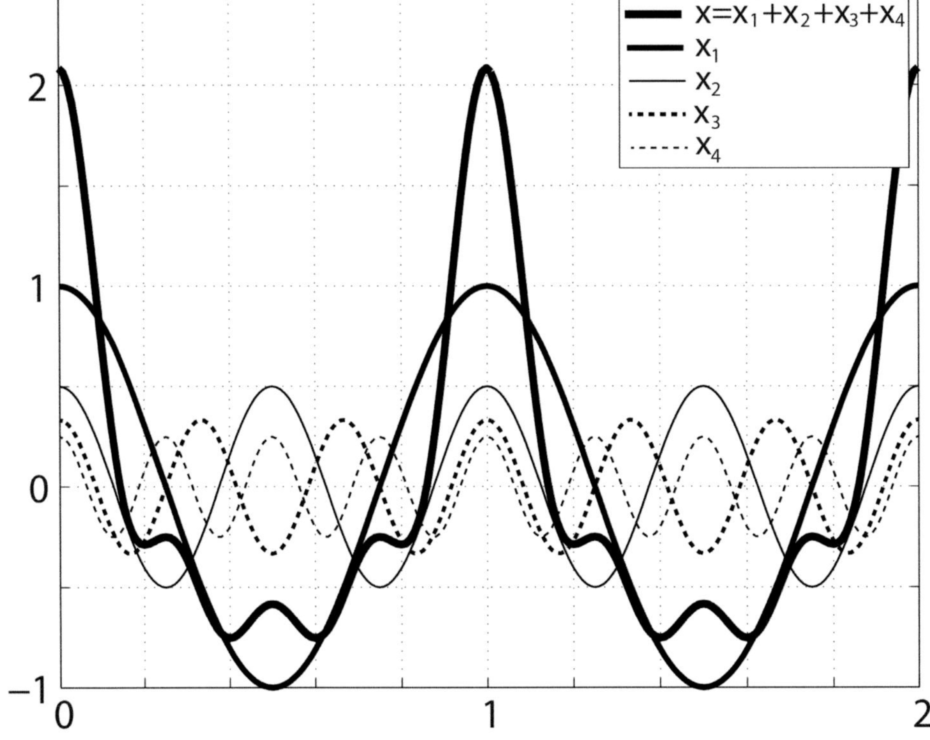

Figure 12.1 Superposition of in-phase sinusoidal functions.

oscillations at the fundamental frequencies and those of its harmonics has the same period corresponding to the fundamental frequency f, or the period is $T = 1/f$.

The resultant function, which is the thick line in Fig. 12.1, is periodic as a superposition of the other periodic functions. It just appears to be a little more complicated and nothing seems to be out of the ordinary.

12.1.2 A Second Example

If we change the base functions just a little bit – we keep the same frequencies in the base functions but add a phase to each:

$$x_1 = \cos(2\pi f t), \quad x_2 = \cos\left(4\pi f t + \frac{\pi}{2}\right)\Big/2, \quad x_3 = \cos\left(6\pi f t + \frac{\pi}{3}\right)\Big/3$$

$$x_4 = \cos\left(8\pi f t + \frac{\pi}{4}\right)\Big/4, \quad x = x_1 + x_2 + x_3 + x_4, \quad \text{here } f = 1/T, \quad T = 1$$

The values for the phases, i.e. 0, $\pi/2$, $\pi/3$, and $\pi/4$, are also arbitrarily chosen for demonstration purposes.

The new function, by adding all of the base functions with the arbitrary phase shifts, appears to have lost certain symmetry compared to the earlier example, and it is less easily seen as a periodic function even though it is one (Fig. 12.2). The point is, when a series of periodical functions are linearly superimposed, this yields a more complicated function, even beyond recognition. No one can easily tell that the resultant function (the thick line) is four perfect sinusoidal functions added together.

12.1.3 A Third Example

Let us keep the momentum and go one more step: we use eight functions to generate a new function. Again, the coefficients and phase shifts are arbitrarily given. The equations are as the following:

$$x_1(t) = \cos(2\pi f t); \; x_2(t) = \frac{1}{4}\cos\left(4\pi f t + \frac{\pi}{2}\right); \; x_3(t) = \frac{1}{3}\cos\left(6\pi f t + \frac{\pi}{3}\right)$$

$$x_4(t) = \frac{1}{2}\cos\left(8\pi f t + \frac{\pi}{4}\right); \; x_5(t) = \cos(3\pi f t); \; x_6(t) = \frac{1}{2}\cos\left(3.5\pi f t + \frac{2\pi}{3}\right);$$

$$x_7(t) = \frac{1}{2.5}\cos\left(11\pi f t - \frac{\pi}{3}\right); \; x_8(t) = \frac{1}{2.7}\cos\left(13\pi f - \frac{5\pi}{7}\right)$$

Now the frequencies are f, $2f$, $3f$, $4f$, $1.5f$, $1.75f$, $5.5f$, and $6.5f$, respectively. The amplitudes are 1, 1/4, 1/3, 1/2, 1, 1/2, 1/2.5, and 1/2.7, respectively; and the phases are 0, $\pi/2$, $\pi/3$, $\pi/4$, 0, $2\pi/3$, $-\pi/3$, and $-5\pi/7$, respectively. Again, these

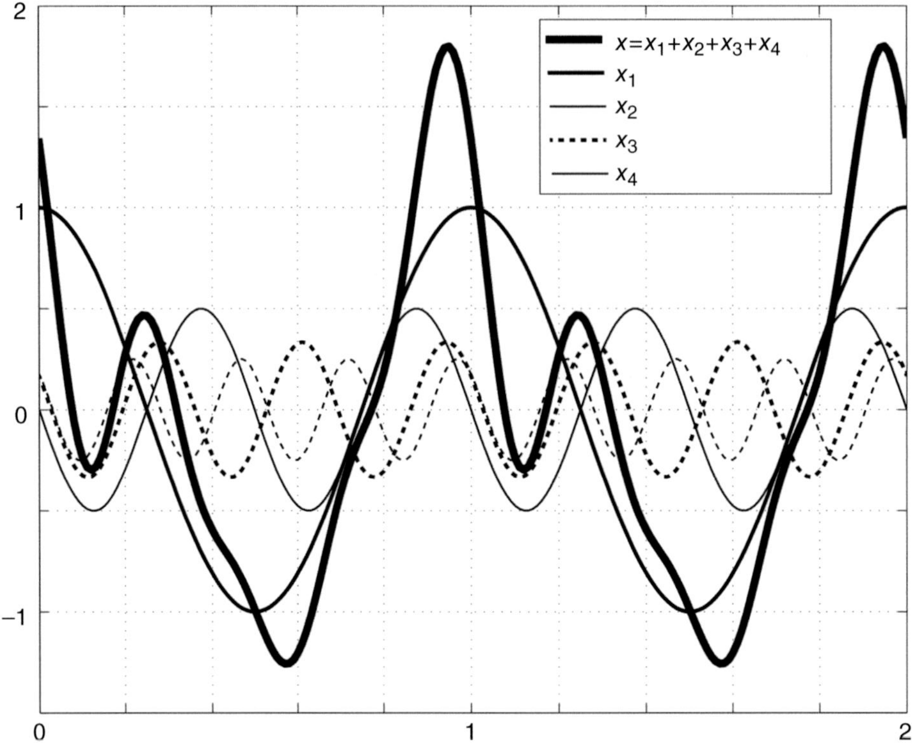

Figure 12.2 Superposition of four sinusoidal functions with phase shifts.

parameters are arbitrarily selected for demonstration purposes and can be modified in many different ways.

In this example, with eight sinusoidal functions, the resultant function is now hardly recognizable as one obtained by a linear combination of a few simple sinusoidal functions (Fig. 12.3). It is more like a randomly changing quasi-periodic function than a deterministic and quasi-periodic function.

12.2 Least Squares Method for Continuous Functions

12.2.1 Asking a Reversed Question

By the above three examples, it is not too much of a stretch to imagine that we can construct what appears to be a very complicated function by a simple linear combination just of sinusoidal functions. In the first example, the resultant function from four sinusoidal functions already shows certain complication, and it is barely recognizable as an addition of four cosine functions. In the second example, the complication seems to be enhanced by simply shifting the phases of the same

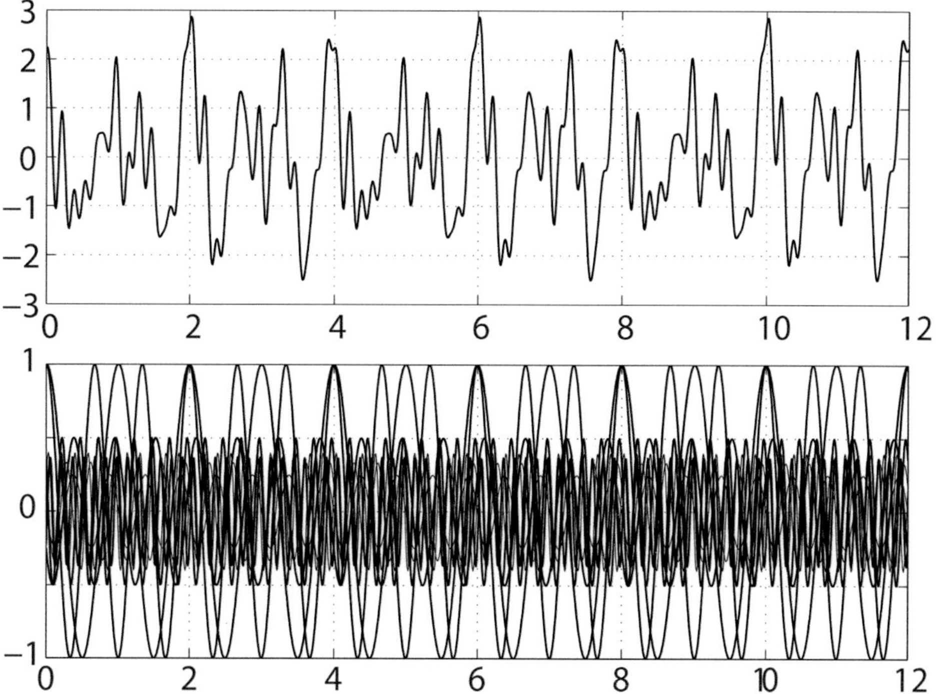

Figure 12.3 Superposition of eight sinusoidal functions with phase shifts. The upper panel is the result of the superposition. The lower panel shows each of the eight constituent functions.

functions of example 1. The third example further shows that when a good number of simple sinusoidal functions are added together, the function seems to be quite arbitrary, even resembling that of a random function.

Up to this point, it is all empirical and nothing fundamental. But now we can ask a reversed question, which is of more fundamental significance: given any periodic function, or even any continuous function, can we use a series of sinusoidal functions to represent this function? If the answer is yes, even a conditional yes, it would be of tremendous benefit to us, as we can simplify the expression of complicated functions by simple ones in a general sense. To put this question in a slightly different way: can we generally use just sine and cosine functions to represent a general function? In theory, we are interested in using an infinite series of (sine and cosine) functions, very much like Taylor series expansion. Here the functions are no longer polynomial as in the case of Taylor series expansion. A more interesting question is: will an infinite sine and cosine function series be convergent to a finite function? In the context of time function, the question can be translated into this: can a rather arbitrary time function be represented by the superposition of simple (sinusoidal) oscillations with different frequencies?

12.2.2 A Generic Method for Continuous Functions

Before we answer this question, let us examine a generic least squares method again, but in a slightly different way. We are still using a linear combination of selected base functions to approximate another function:

$$y(t) \approx g(t) = a_1 g_1(t) + a_2 g_2(t) + \cdots + a_N g_N(t). \tag{12.1}$$

Instead of calculating the total error squared at discrete data points, we evaluate it within a continuous interval $[t_1, t_2]$:

$$\Psi = \int_{t_1}^{t_2} (y - g(t))^2 dt \tag{12.2}$$

This is similar to (10.14), but the sigma over discrete data points is replaced by integration over a continuous interval. Mathematically, the difference between (10.14) and (12.2) is that they use different mathematical "operators" on the right-hand side: one is a summation and the other is an integration over time t. The following is a similar procedure but a little different in mathematical details. The least squares method requires that the total error squared must be minimized with the necessary condition that each of the following derivatives is zero:

$$\frac{\partial \Psi}{\theta a_j} = 2\int_{t_1}^{t_2} (y - g(t))\big(- g_j(t)\big)dt = -2\int_{t_1}^{t_2} (y - g(t))g_j(t)dt = 0, \quad j = 1, 2, \ldots, N \tag{12.3}$$

or

$$\int_{t_1}^{t_2} \left(\sum_{k=1}^{N} a_k g_k(t) \right) g_j(t)dt = \int_{t_1}^{t_2} y g_j(t)dt. \tag{12.4}$$

The summation here is for the base functions according to (12.1), not a summation over t (t is continuous).

On the left-hand side, we can switch the order of summation and integration to yield

$$\sum_{k=1}^{N} a_k \int_{t_1}^{t_2} g_k(t)g_j(t)dt = \int_{t_1}^{t_2} y g_j(t)dt. \tag{12.5}$$

The above equation is again an expression for N equations ($j = 1, 2, \ldots, N$).

Equation (12.5) is also a matrix equation when all j equations are combined:

$$Bz = b, \tag{12.6}$$

in which

$$B = \begin{pmatrix} \int g_1^2 dt & \int g_1 g_2 dt & \cdots & \int g_1 g_N dt \\ \int g_2 g_1 dt & \int g_2^2 dt & \cdots & \int g_2 g_N dt \\ \vdots & \vdots & \vdots & \vdots \\ \int g_N g_1 dt & \int g_N g_2 dt & \cdots & \int g_N^2 dt \end{pmatrix}, \tag{12.7}$$

$$z = \begin{pmatrix} a_1 \\ a_2 \\ \vdots \\ a_N \end{pmatrix}, \quad b = \begin{pmatrix} \int g_1 y dt \\ \int g_2 y dt \\ \vdots \\ \int g_N y dt \end{pmatrix}. \tag{12.8}$$

Now we consider a special set of orthogonal base functions which satisfy

$$\int_{t_1}^{t_2} g_j g_k dt = 0, \quad j \neq k. \tag{12.9}$$

Similar to earlier discussions, these functions are "perpendicular" to each other. With orthogonal base functions, we can greatly simplify the equations and results:

$$B = \begin{pmatrix} \int g_1^2 dt & 0 & \cdots & 0 \\ 0 & \int g_2^2 dt & \cdots & 0 \\ \vdots & \vdots & \ddots & \vdots \\ 0 & 0 & \cdots & \int g_N^2 dt \end{pmatrix}. \tag{12.10}$$

Therefore,

$$\begin{pmatrix} \int g_1^2 dt & 0 & \cdots & 0 \\ 0 & \int g_2^2 dt & \cdots & 0 \\ \vdots & \vdots & \ddots & \vdots \\ 0 & 0 & \cdots & \int_N^2 dt \end{pmatrix} \begin{pmatrix} a_1 \\ a_2 \\ \vdots \\ a_N \end{pmatrix} = \begin{pmatrix} \int g_1 y dt \\ \int g_2 y dt \\ \vdots \\ \int g_N y dt \end{pmatrix}. \tag{12.11}$$

Now we do not need to use the matrix operations to solve the equations because we have already got the explicit solution from the above equation:

$$a_k = \frac{\int_{t_1}^{t_k} g_k y dt}{\int_{t_1}^{t_2} g_k^2 dt} = \frac{(g_k, y)}{\|g_k\|^2}, \quad (k = 1, 2, \ldots, N). \tag{12.12}$$

In the above equation,

$$(g_k, y) \equiv \int_{t_1}^{t_2} g_k y dt$$

is a dot product between g_k and y in an integral sense, i.e., instead of using the summation

$$\sum_{i=1}^{M} g_k(t_i) y(t_i),$$

it is an integration. The denominator of (12.12) is the square of modulus, or magnitude, or length of the vector g_k:

$$\|g_k\| \equiv \sqrt{\int_{t_1}^{t_2} g_k^2 dt}.$$

The a_ks in (12.12) are called *generalized Fourier coefficients*. The geometric meaning of generalized Fourier coefficients are the "coordinates" of the vector approximating the "true" vector y. Or they are the projections of the vector y on each of the base functions, measured by the magnitude of the base functions. This can be seen by

$$a_k = \frac{(g_k, y)}{\|g_k{}^2\|} = (\tilde{g}_k, \tilde{y}), \tag{12.13}$$

in which

$$\tilde{g}_k = \frac{g_k}{\|g_k\|}, \quad \tilde{y} = \frac{y}{\|g_k\|}, \quad k = 1, 2, \ldots, N. \tag{12.14}$$

In other words, $\tilde{g}_k$ is g_k normalized by its own length (modulus $\|g_k\|$), and $\tilde{y}$ is y normalized by $\|g_k\|$. Obviously, the magnitude squared of each of the base functions is given by the integration or self-dot product of the base function g_k:

$$\|g_k\|^2 = \int_{t_1}^{t_2} g_k^2 dt. \tag{12.15}$$

So, the base functions normalized by the magnitude of itself are unit vectors in each of the dimensions of this n-dimensional space, while the vector y normalized by the magnitude of these vectors ($\|g_k\|$) can be seen as the measure of this vector in each of the dimensions. The a_k coefficients are therefore projections of the normalized vector $\tilde{y}$ onto the normalized vectors $\tilde{g}_k$ or the coordinates. The $\tilde{g}_k$s form a functional space. The vector formed by these projections onto this functional space is the best approximation of the original vector y in this space. The linearly independent g_k vectors and $\tilde{g}_k$ unit vectors are all perpendicular to the error vector, and

$$g(x) = a_1 g_1(x) + a_2 g_2(x) + \cdots + a_N g_N(x)$$

is a projection of y onto the lower dimensional space $\{g_1, g_2, \ldots, g_N\}$.

12.2.3 Discrete vs. Continuous Methods

The least squares method for discrete data can be viewed as a special case of the above theory, if we use the summation to replace the integral. For analysis of data, it is almost always for discrete and not continuous functions, as observations are done at discrete times. Even though some analog sensors may be recording continuously, they must have certain temporal resolutions. So, when the data are digitized for analysis, they have to be presented in finite time intervals. Therefore, it is natural to use the summation rather than the integration. The theoretical development, however, has its own value because it is convenient to work with analytical expressions of continuous functions.

We know that the crudest approximation of an integration is the so-called *midpoint rule*, which is a simple summation of the function at discrete points multiplied by the interval of the variable (in this context, it is the time interval dt). There are better approximations of integration, such as the *trapezoidal rule*, *Simpson's rule*, etc. The accuracy of these rules of digitally computing an integral depends on the size of the intervals. So, here the integration without any approximation is more general and more accurate. In mathematics, using summation or integration means using different functional space: the former is a finite dimensional space (M-dimensional for M observations), while the latter is an infinite dimensional space with continuous functions. From another point of view, the midpoint rule, trapezoidal rule, and Simpson's rule correspond to different "mathematical operators." When a least squares method is used to find the best solution with minimum errors, it means the minimum within relevant functional space corresponding to the associated mathematical operator. The minimum total error squared within one functional space using one operator (e.g. summation) is usually different from that within another functional space using a different operator (e.g. integration).

12.2.4 The Fourier Series

With all the preparations above, now we come back to this question: given any periodic function, can we use a series of sinusoidal functions to represent a periodic function? (Note: this periodic function is not necessarily sinusoidal. In other words, it can be a quite different periodic function, such as a "square wave.")

To put it in mathematics, for any periodic function $y(t)$, with a period of T, can we use the following base functions – $\cos(2\pi kft)$, $\sin(2\pi kft)$, $(k = 1, 2, \ldots)$ – to express it exactly? Here $f = 1/T$ is the *fundamental frequency*. Or, in mathematics, we are asking if

$$y(t) = a_0 + \sum_{k=1}^{+\infty}[a_k \cos(2\pi kft) + b_k \sin(2\pi kft)] \tag{12.16}$$

holds or not. Note that here we have a higher demand; we are seeking an exact relationship, not an approximation as is the case in the least squares method. This is equivalent to the question: does the expansion on the right-hand side of (12.16) converge to function $y(t)$?

Let us first ignore the convergence issue (assume it is convergent) and examine the least squares method anyway. We have to truncate the infinite series into that expressed by a finite number of terms. So, we replace infinity with an integer, $2N + 1$, in which N is the number of non-zero frequencies. When N approaches infinity, we have equation (12.16). For a given N, the base functions are

$$1, \cos(2\pi kft), \sin(2\pi kft), \quad (k = 1, 2, \ldots, N). \tag{12.17}$$

For this given N, there are $2N + 1$ base functions:

$$g_1 = 1, g_2 = \cos(2\pi ft), g_3 = \sin(2\pi ft), \ldots, g_{2N} = \cos(2N\pi ft), g_{2N+1} = \sin(2N\pi ft). \tag{12.18}$$

It can be verified that these base functions are indeed orthogonal in the interval $t = [0, T]$, which means they are mutually uncorrelated or their mutual inner products are all zeros, or they are all "perpendicular" to each other (for $j \neq k$):

$$\int_0^T \sin(2\pi jft)\cos(2\pi kft)dt = 0, \tag{12.19}$$

$$\int_0^T \sin(2\pi jft)\sin(2\pi kft)dt = 0, \tag{12.20}$$

$$\int_0^T \cos(2\pi jft)\cos(2\pi kft)dt = 0, \tag{12.21}$$

$$\int_0^T \sin(2\pi jft)dt = 0, \tag{12.22}$$

$$\int_0^T \cos(2\pi jft)dt = 0. \tag{12.23}$$

Therefore, we can use the results shown earlier for the generalized Fourier coefficients and obtain:

$$a_0 = \frac{\int_0^T ydt}{\int_0^T dt} = \frac{1}{T}\int_0^T ydt. \tag{12.24}$$

This is simply the mean value – the "direct current" or *DC component* in electronic circuit theory. The coefficients for the sine and cosine functions are, respectively,

$$a_k = \frac{\int_0^T \cos(2\pi kft)ydt}{\int_0^T \cos^2(2\pi kft)dt} = \frac{2}{T}\int_0^T \cos(2\pi kft)ydt, \tag{12.25}$$

$$b_k = \frac{\int_0^T \sin(2\pi kft)ydt}{\int_0^T \sin^2(2\pi kft)dt} = \frac{2}{T}\int_0^T \sin(2\pi kft)ydt. \tag{12.26}$$

These a_ks and b_ks are the Fourier coefficients in equation (12.16). When k is approaching infinity, (12.16) is called the *Fourier series expansion* of function y

within the interval $t = [0, T]$. The next question is: will the infinite series (12.16) converge to the function $y(t)$ at all time t?

12.3 The Convergence of Fourier Series

12.3.1 Theory

Now we examine the convergence of Fourier series. In the Fourier series equation (12.16), let

$$x = 2\pi f t, \quad \left(f = \frac{1}{T}\right), \tag{12.27}$$

after which the period of the new variable x becomes 2π and the Fourier series becomes

$$y(x) = \frac{a_0}{2} + \sum_{k=1}^{\infty} [a_k \cos(kx) + b_k \sin(kx)]. \tag{12.28}$$

Note that here we have changed the constant term (the first term) on the right-hand side by dividing it by 2. This is to unify the expression for the a_ks. So, the new a_0 is not the same a_0 as in (12.16). This is only for convenience of expression. The corresponding Fourier coefficients are then

$$a_k = \frac{1}{\pi} \int_{-\pi}^{\pi} y(x) \cos(kx) dx, \ k = 0, 1, 2, \ldots, \tag{12.29}$$

$$b_k = \frac{1}{\pi} \int_{-\pi}^{\pi} y(x) \sin(kx) dx, \ k = 1, 2, \ldots. \tag{12.30}$$

Note that here the coefficient for a_k allows $k = 0$. For finite n frequencies ($2n + 1$ terms), the truncated expansion at a given point x_0 becomes

$$S_n(x_0) = \frac{a_0}{2} + \sum_{k=1}^{n} [a_k \cos(kx_0) + b_k \sin(kx_0)]. \tag{12.31}$$

Substituting the Fourier coefficients into the above equation gives

$$\begin{aligned} S_n(x_0) = &\frac{1}{2\pi} \int_{-\pi}^{\pi} y(x) dx \\ &+ \sum_{k=1}^{n} \left\{ \left[\frac{1}{\pi} \int_{-\pi}^{\pi} y(x) \cos(kx) dx \right] \cos(kx_0) + \left[\frac{1}{\pi} \int_{-\pi}^{\pi} y(x) \sin(kx) dx \right] \sin(kx_0) \right\}. \end{aligned} \tag{12.32}$$

This leads to

$$
\begin{aligned}
S_n(x_0) &= \frac{1}{2\pi}\int_{-\pi}^{\pi} y(x)dx + \sum_{k=1}^{n}\left\{\frac{1}{\pi}\int_{-\pi}^{\pi} y(x)[\cos(kx)\cos(kx_0) + \sin(kx)\sin(kx_0)]dx\right\} \\
&= \frac{1}{2\pi}\int_{-\pi}^{\pi} y(x)dx + \sum_{k=1}^{n}\left[\frac{1}{\pi}\int_{-\pi}^{\pi} y(x)\cos k(x-x_0)dx\right] \\
&= \frac{1}{\pi}\int_{-\pi}^{\pi} y(x)\left[\frac{1}{2} + \sum_{k=1}^{n}\cos k(x-x_0)\right]dx.
\end{aligned}
\tag{12.33}
$$

Now we need to use a mathematical identity, without a lengthy proof, as the following:

$$
\frac{1}{2} + \sum_{k=1}^{n}\cos k(x-x_0) = \frac{\sin\left[(2n+1)\dfrac{x-x_0}{2}\right]}{2\sin\dfrac{x-x_0}{2}}. \tag{12.34}
$$

If we let $\tau = x - x_0$, it leads to

$$
S_n(x_0) = \frac{1}{\pi}\int_{0}^{\pi}[y(x_0+\tau) + y(x_0-\tau)]\frac{\sin\left[\left(n+\dfrac{1}{2}\right)\tau\right]}{2\sin\dfrac{\tau}{2}}d\tau. \tag{12.35}
$$

Now we need to use another mathematical theorem, Dirichlet's Theorem, also without a lengthy proof:

$$
\lim_{n\to\infty}\frac{1}{\pi}\int_{-\delta}^{\delta} y(x+t)\frac{\sin nt}{t}dt = \frac{1}{2}[y(x+0) + y(x-0)], (\delta > 0). \tag{12.36}
$$

Here $y(x+0)$ means the value of function $y(x)$ as a limit when x is approached from the right of x (larger than x), or the "right limit"; while $y(x-0)$ means the value of function $y(x)$ as a limit when x is approached from the left of x (smaller than x), or the "left limit."

Applying this theorem to equation (12.35), we can see that

$$
\lim_{n\to\infty} S_n(x_0) = \frac{1}{2}[y(x_0+0) + y(x_0-0)]. \tag{12.37}
$$

The conclusion, then, is that the limit (as $n \to +\infty$) of the truncated Fourier series is the average of the right limit and left limit of the function at the point of consideration.

12.3.2 Interpretation

The above result given by (12.37) means that the Fourier series converges to the original function in an average sense. Since the function value at where it is continuous will have equal limits from the right and left, i.e.

$$y(x+0) = y(x-0) = y(x)$$

we have $\lim_{n\to\infty} S_n(x_0) = y(x_0)$, or the Fourier series converges to the function precisely.

However, at a discontinuous point, the two limits are different. The Fourier series converges to neither point but instead to their average value (the two limits added together and divided by 2).

The most important conclusion is that the Fourier series can be used to represent pretty much any practical functions: that is, if the function is continuous. In contrast, the Taylor series expansion is valid only for functions that can be differentiated an infinite number of times, which is a much stricter condition. If a function is not continuous at a finite number of points, where the function has a finite jump, the Fourier series will converge to the average of the jump. Elsewhere (at the continuous points), the Fourier series will converge to the original function. This is a very useful property of the Fourier series because it allows us to use the Fourier series extensively for practical applications involving continuous functions or piecewise continuous functions.

12.4 Some Comments on Fourier Series

12.4.1 Important Properties

Before we look at some real examples of Fourier series expansion, in the following we first look at some important properties: why do we care about Fourier series expansion, which just appears to be the result of another least squares method? It is because of the nice properties of Fourier series. Here is a list of these properties:

(1) Fourier series has orthogonal base functions and so the solution is explicit: in practice this is very useful – at least there is no need to do the matrix inversion in computation of the coefficients and, in the case of using MATLAB, there is no need to do the left-division.

(2) The base functions of a Fourier series are COMPLETE: in other words, if all base functions (an infinite series) are used, the least squares error approaches zero as the number of terms approaches infinity if the function being expanded is continuous. This renders an EXACT rather than an approximate relationship, given the condition that the function is continuous.

We can relax the continuous function requirement: as long as the original function $y(t)$ is piecewise continuous, the Fourier series will work at the points of continuity. For functions with *finite discontinuity*, the Fourier series converges in an "average" sense – the relationship may not be exact where the function has a finite jump, but the series converges to the average value of the jump. There is an effect of finite jump of the Fourier series. It is called the *Gibbs Effect*. At discontinuity points, the Fourier series oscillates no matter how many terms one chooses. At the discontinuity, the Gibbs Effect causes the approximated function to *overshoot* and *undershoot* (an oscillation). The overshooting and undershooting cannot be eliminated by adding more terms, but the magnitude of these oscillations approach to a constant limit, which cannot be reduced.

12.4.2 Fundamental Frequency

It is crucial to understand that Fourier series expansion is done with an assumption that the function is periodic, i.e. it repeats itself at fixed time intervals (or space intervals). The period we choose for the Fourier series expansion, T, is called the fundamental period, as we know, and $1/T$ the fundamental frequency. If the original function is truly periodic and continuous everywhere, the Fourier series converges relatively fast. Otherwise, at any point with a discontinuity, there is the Gibbs Effect so that oscillations exist and the Fourier series only converges to the original function in an averaged sense where there is discontinuity. This repeats itself at the fundamental frequency $1/T$.

For a continuous periodic function, in which the beginning and ending of the period have the same value, the continuity holds to every point along the time axis. For a periodic function which is continuous inside the period, but with different values at the beginning and ending points, the discontinuity points have a finite jump (the discontinuity of the first kind). A discontinuity of the first kind can be inside the period as well.

For observational data, the time series is rarely truly periodic. In that case, the entire time series is treated as one period, and the length of record in time (T) is treated as the fundamental period. As a result, we can view the observed time series as a complete cycle that repeats itself beyond the observational period, forming an infinitely long "periodic" function. The fundamental frequency is $f = 1/T$.

12.5 Examples of Fourier Series and Gibbs Effect

The following are some examples showing the convergence of Fourier series for different functions. For smooth functions with no jumps anywhere, the Fourier series does not need too many terms for a good representation of the original function.

Here we discuss three examples of a Fourier series of three simple periodic functions with finite jumps in function value (for the first two examples) or with finite jumps in the first order derivative (the last example). The theoretical solutions of the Fourier series are used and computed using MATLAB to compare the original function and to examine the convergence of the Fourier series. At the same time, we can see with real examples the characteristics of Fourier series, especially the convergence where there is discontinuity in the function value or the derivative. More importantly, we can observe with these examples about the Gibbs Effect.

The first periodic function is

(1) $y(x) = x, x = [0, T]$. Here, in order to make the graph, we choose $T = 10$.

This is a periodic function, but within the interval $[0, T]$, it is a linear function. The definition was given only within the interval $[0, T]$ but should be viewed as repeating and extending infinitely in both directions. This function has a finite jump at $x = 0$ and $x = T, 2T, 3T, \ldots$ as well as at $x = -T, -2T, -3T, \ldots$

The second function is

(2) $y(x) = x^2, x = [0, T]$. For demonstration purposes, we choose $T = 10$.

This is a slightly different function compared to the first one. The only difference is that the function within each period, e.g. $[0, T]$, is quadratic instead of linear. Likewise, the function should be viewed as repeating and extending infinitely in both directions. This function also has a finite jump at $x = 0$ and $x = T, 2T, 3T, \ldots$ as well as at $x = -T, -2T, -3T, \ldots$

The third function is continuous everywhere but has a sharp point at both ends of each period. Or there is a discontinuity in the first order derivative:

(3) $y(x) = x^2, x = [-T/2, T/2]$. Again, for demonstration purposes, we choose $T = 20$.

Using the results presented earlier, i.e. equations (12.24)–(12.26), the Fourier series expansion can be analytically solved. The following equations are the Fourier series expansions for the above three periodic functions, respectively. The verification of these expansions is left as exercises at the end of the chapter.

(1) $y(x) = \frac{T}{2} + \sum_{k=1}^{\infty} \left[-\frac{T}{\pi k} \sin\left(\frac{2\pi k}{T} x\right) \right]$.

(2) $y(x) = \frac{T^2}{3} + \sum_{k=1}^{\infty} \left[-\frac{T^2}{\pi k} \sin\left(\frac{2\pi k}{T} x\right) + \frac{4}{(2\pi k/T)^2} \cos\left(\frac{2\pi k}{T} x\right) \right]$.

(3) $y(x) = \frac{T^2}{12} + \frac{4}{T} \sum_{k=1}^{\infty} \left[\frac{20(-1)^k}{(2\pi k/T)^2} \cos\left(\frac{2\pi k}{T} x\right) \right]$.

12.5.1 MATLAB Scripts 1

The MATLAB scripts for the calculations of the Fourier series expansion and comparison with the original function of the first example are:

```
clear
T = 10;
x = 0:0.01:T;
y1 = x;
a0 = T/2;
MM = 100;
y = a0;
for i = 1:MM
y = y-T/(pi*i)*sin(2*pi*i/T*x);
end
figure; plot(x,y,'LineWidth',1); hold on;
plot(x,y1,'r','LineWidth',1)
```

Here we use MM = 100 for the number of terms to use in the computation. The remaining infinite terms are truncated. Fig. 12.4 shows results in which only one period is presented – it should be considered as periodically repeating in both

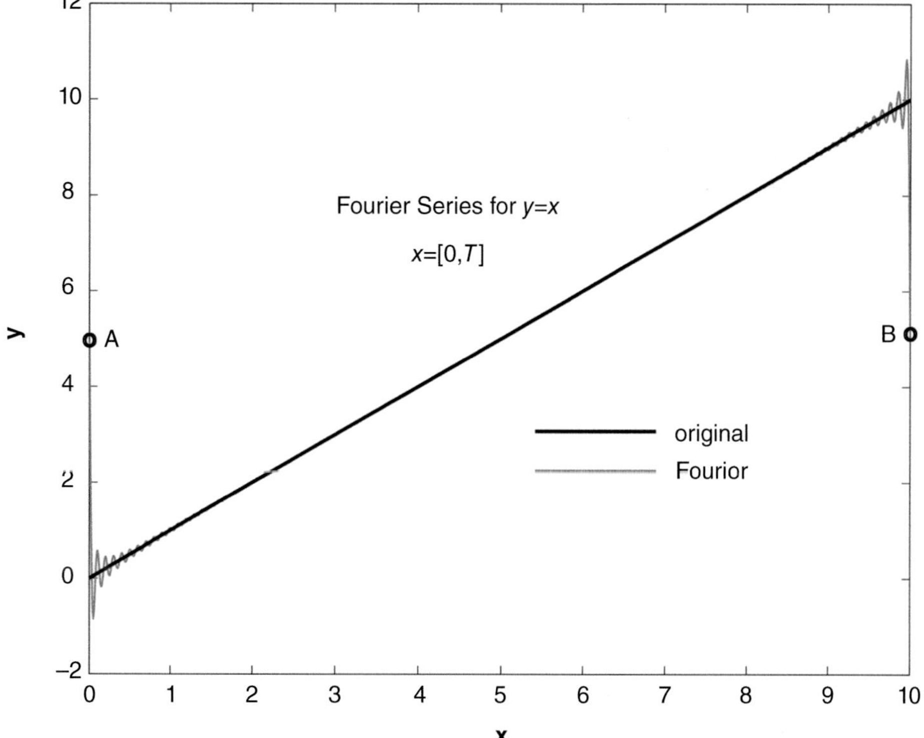

Figure 12.4 The first example. Fourier series and the original function. M = 100.

directions of the axis. Obviously, the Fourier series converges to the original function quite well away from the two ends. At both ends of each period, oscillations can be seen, and the convergence is to A and B for the beginning and end, respectively. These oscillations are the Gibbs Effects we mentioned earlier.

12.5.2 MATLAB Scripts 2

For the second example, the MATLAB scripts for the calculations of the Fourier series expansion and comparison with the original function are:

```
clear
T = 10;
x = 0:0.01:T;
y1 = x.^2;
a0 = T^2 / 3;
MM = 100;
y = a0;
for i = 1:MM
y = y-2*T/(2*pi*i*T)*sin(2*pi*i/T*x)+4/(2*pi*i/T)^2*cos(2*pi*i/T*x);
end
figure; plot(x,y,'LineWidth',1); hold on;
plot(x,y1,'r','LineWidth',1)
```

Here we still use MM = 100. Fig. 12.5 shows the results. Again, the Fourier series converges to the original function quite well away from both ends. At both ends of each period, the Gibbs Effect can be seen, and the convergence is to A and B for the beginning and end of each period, respectively.

12.5.3 MATLAB Scripts 3

For the last example, the MATLAB scripts for the calculations of the Fourier series expansion and comparison with the original function are:

```
clear
T = 20;
x = -T/2:0.01:T/2;
y1 = x.^2;
a0 = T^2/12;
MM = 10;
y = a0;
for i = 1:MM
y = y+(-1)^i*20/(2*pi*i/T)^2*cos(2*pi*i/T*x)*4/T;
end
figure; plot(x,y,'LineWidth',1); hold on;
plot(x,y1,'r','LineWidth',1)
```

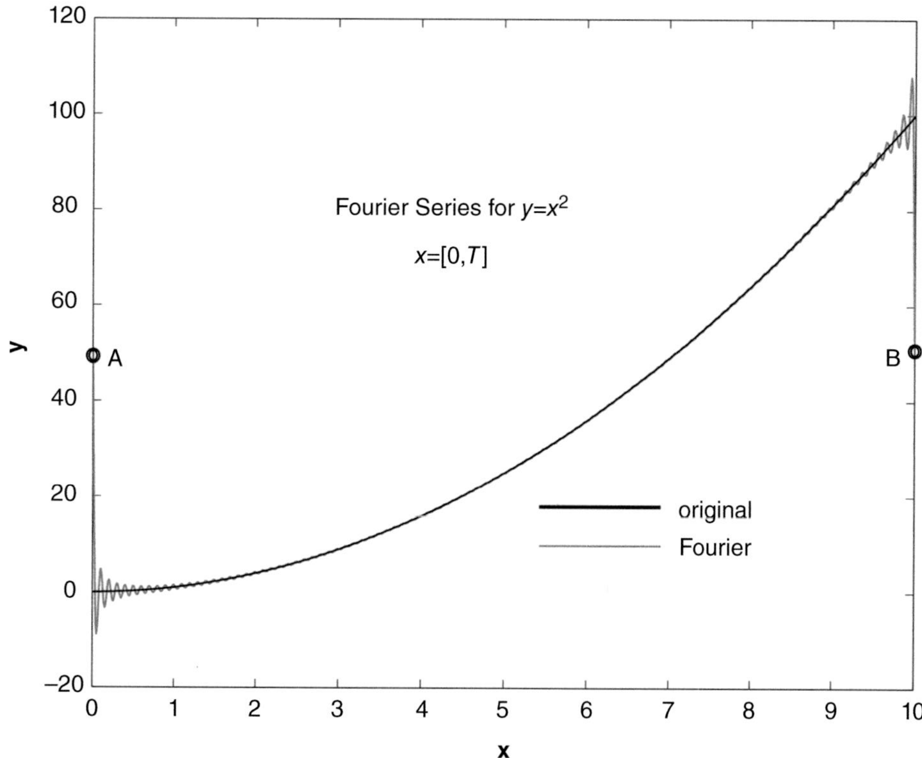

Figure 12.5 The second example. Fourier series and the original function. M = 100.

Note that here we use a much smaller MM = 10. Fig. 12.6 shows the results. Even with only 10 terms in the summation of the solution of the Fourier series expansion, the convergence is obvious. The error appears to be enlarged a little at both ends at $-T/2$ and $T/2$, but there is no Gibbs Effect. The lack of Gibbs Effect is because the function is continuous. The enlarged error at both ends is because the first order derivative is discontinuous.

Let us replace `MM = 10` with `MM = 100`, just for comparison purposes, and use the following scripts:

```
clear
T = 20;
x = −T / 2:0.01:T / 2;
y1 = x.^2;
a0 = T^2/12
MM = 100;
y = a0;
for i = 1:MM
y = y+(−1)^i*20/(2*pi*i/T)^2*cos(2*pi*i/T*x)*4/T;
```

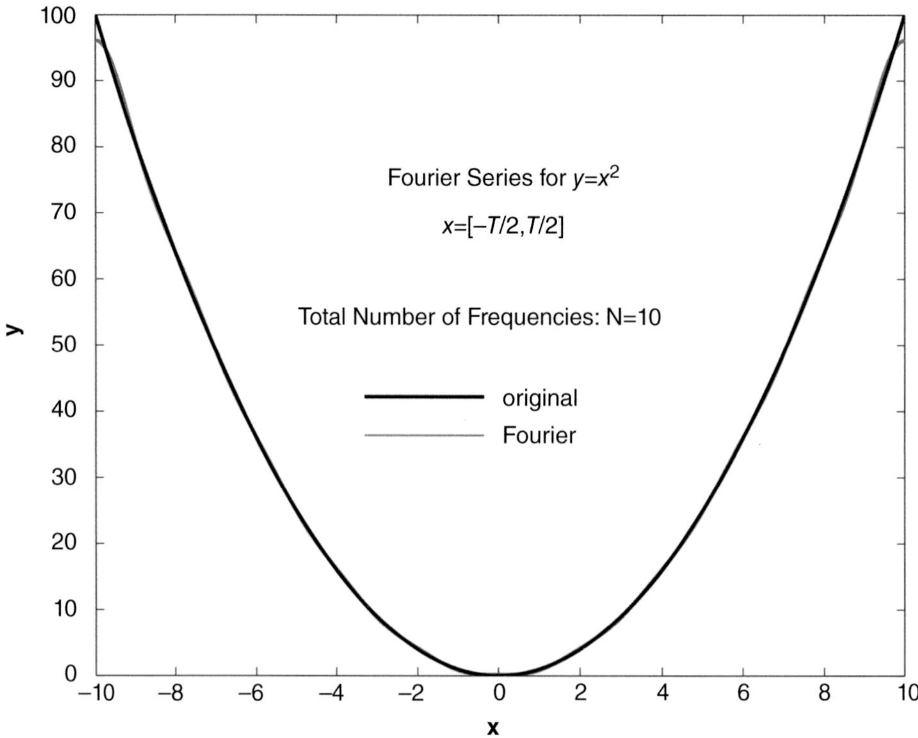

Figure 12.6 The third example for M = 10. Fourier series and the original function.

```
end
figure
subplot(2,1,2)
plot(x,y,x,y1,'LineWidth',1)
subplot(2,1,1)
plot(x,y - y1,'LineWidth',1)
```

The result is shown in Fig. 12.7. The truncated Fourier series and original function are now almost identical, and there is no way to tell the difference when they are plotted together. To examine the error, we subtract the two (`dy = y - y1`). This clearly shows the error is larger at both ends, although there is no Gibbs Effect. The discontinuity of the first derivative of the original function at the ends of each period makes the difference have an increased oscillation (but not quite the Gibbs Effect), and the original Fourier series does converge.

12.6 Additional Properties of Fourier Series

There are some additional, very useful properties for Fourier series expansions. Here is a list of the major ones:

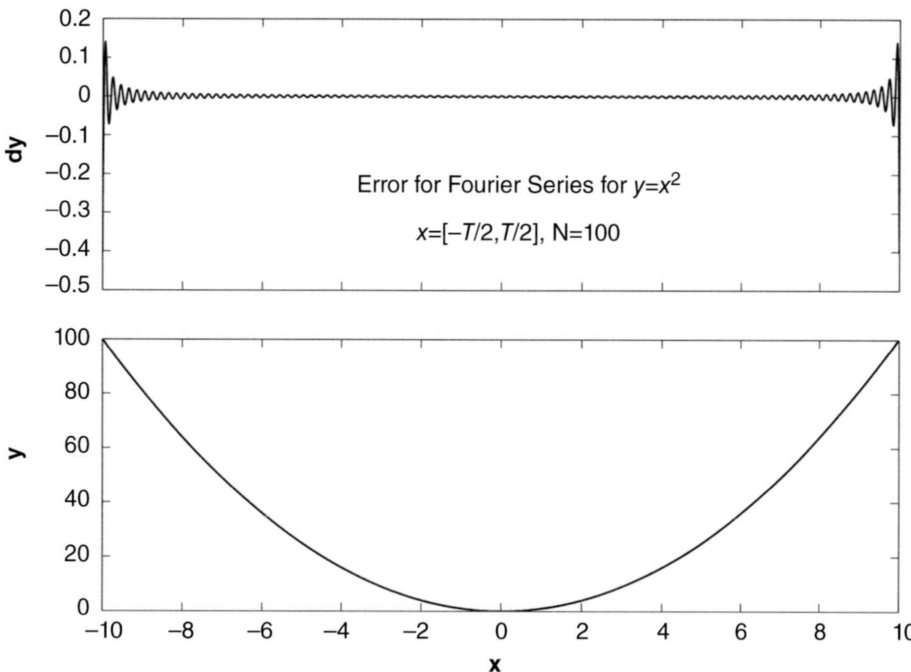

Figure 12.7 The third example for M = 100. The upper panel shows the difference between the Fourier series and the original function. The lower panel shows the original function and the Fourier series (the two are so close that we cannot see any difference visually).

(1) The Fourier series expansion of an even function has only non-zero coefficients for the cosine base functions; coefficients for the sine functions are zeros or all $b_k = 0$.
(2) The Fourier series expansion of an odd function has only non-zero coefficients for the sine functions; coefficients for the cosine functions are zeros, or all $a_k = 0$.
(3) If neither even or odd, both a_k, b_k are non-zeros in general.
(4) If $y(t)$ is discontinuous, the coefficients a_k and b_k all decrease at a rate of $\sim 1/k$ for large k.
(5) If the function is continuous, but its first order derivative is discontinuous, then a_k, b_k decrease at a rate of $\sim 1/k^2$ for large k.
(6) If the function and its first, second, ... , $(m-1)$th order derivatives are continuous, but the mth derivative is discontinuous, then the coefficients a_k, b_k decrease at a rate of $\sim 1/k^{m+1}$.

This means that the smoother the function, the faster its Fourier series expansion converges; the sharper the change of the function, the wider the frequency distribution. A peak in the time series corresponds to more frequencies or a wider range of frequencies.

Review Questions for Chapter 12

(1) Compare the speed of convergence of their Fourier series between two functions: both are continuous, but one is smooth and one has a sharp point. Which one needs fewer terms to reach the same accuracy?

(2) Can Fourier series be used to express an arbitrary function? What is the condition for having a convergent Fourier series expansion for a given function?

Exercises for Chapter 12

(1) Given functions

$$f_1 = \cos(2\pi t),\ f_2 = \sin(2\pi t),\ f_3 = \cos(4\pi t),$$
$$f_4 = \sin(4\pi t),\ f_5 = \cos(6\pi t),\ f_6 = \sin(6\pi t).$$

The fundamental frequency is $f = 1$. What is the corresponding fundamental period T? Use MATLAB to verify that these functions are orthogonal to each other, i.e.

$$\int_0^T f_i(t)f_j(t)dt = 0, \quad (i \neq j, i,j = 1,2,3).$$

To approximate the integration, we use the simplest discrete summation, i.e. we assume that the time interval is a constant *dt* and (from the midpoint rule of numerical integration):

$$\int_0^T f_i(t)f_j(t)dt \approx dt \sum_{k=0}^{N-1} f_i(t_k)f_j(t_k), \quad (i \neq j, i,j = 1,2,3),$$

in which

$$t_k = kdt, \quad N = \frac{T}{dt}, \quad k = 0,1,2,\ldots,N-1.$$

[Hint: Note that the above equation with the summation is really a dot product of the vectors $f_i = [f_i(t_0), f_i(t_1), \ldots, f_i(t_{N-1})]$ and $f_j = [f_j(t_0), f_j(t_1), \ldots, f_j(t_{N-1})]$. If f_1 is defined in MATLAB as `f1`, f_2 is defined as `f2`, both as row-vectors, then the dot product of `f1` and `f2` would simply be `f1 * f2'` in MATLAB.]

(2) Given the same functions as above, prove *analytically* that

$$\int_0^T f_i(t) f_j(t)\,dt = 0, \quad (i \neq j, i, j = 1, 2, 3)$$

and also

$$\int_0^T f_i(t)\,dt = 0, \quad (i = 1, 2, 3).$$

(3) Given $y(x) = x$, $x = [0, T]$, prove that its Fourier series expansion is

$$y(x) = \frac{T}{2} + \sum_{k=1}^{\infty} \left[-\frac{T}{\pi k} \sin\left(\frac{2\pi k}{T} x \right) \right].$$

(4) Given $y(x) = x^2$, $x = [0, T]$, prove that

$$y(x) = \frac{T^2}{3} + \sum_{k=1}^{\infty} \left[-\frac{T^2}{\pi k} \sin\left(\frac{2\pi k}{T} x \right) + \frac{4}{(2\pi k/T)^2} \cos\left(\frac{2\pi k}{T} x \right) \right].$$

(5) Given $y(x) = x^2$, $x = [-T/2, T/2]$, prove that

$$y(x) = \frac{T^2}{12} + \frac{4}{T} \sum_{k=1}^{\infty} \left[\frac{20(-1)^k}{(2\pi k/T)^2} \cos\left(\frac{2\pi k}{T} x \right) \right].$$

13
Fourier Transform

About Chapter 13

This chapter discusses the transition between Fourier series and Fourier Transform, which is the tool for spectrum analysis. Generally speaking, the use of linearly independent base functions allows a wide range of linear regression models that work in a least square sense such that the total error squared is minimized in finding the coefficients of the base functions. A special case of this is sinusoidal functions based on a fundamental frequency and all its harmonics up to infinity. This leads to the Fourier series for periodic functions. In this chapter, we first start from the original Fourier series expression and convert the sinusoidal base functions to exponential functions. We can then conveniently consider the limit when the length of the function and the period of the original function approach infinity (so that the fundamental frequency approaches 0, which includes the case for aperiodic functions), which leads to the so-called Fourier integral and Fourier Transform. We can then define the inverse Fourier Transform. From there we can establish the relationship between the coefficients of Fourier series and the discrete form Fourier Transform. All these are preparations for the fast Fourier Transform (FFT), which is an efficient algorithm of computation of the discrete Fourier Transform in spectrum analysis that is widely used in data analysis for oceanography and other applications. The FFT is discussed in Chapter 16.

13.1 Periodic Function and Discrete Spectrum

For a periodic and continuous function, the Fourier series expansion is convergent. The Fourier series expansion is very useful when we try to decompose a time series signal into the *frequency space* or *frequency domain*, i.e. quantifying the magnitude of variation at different frequencies.

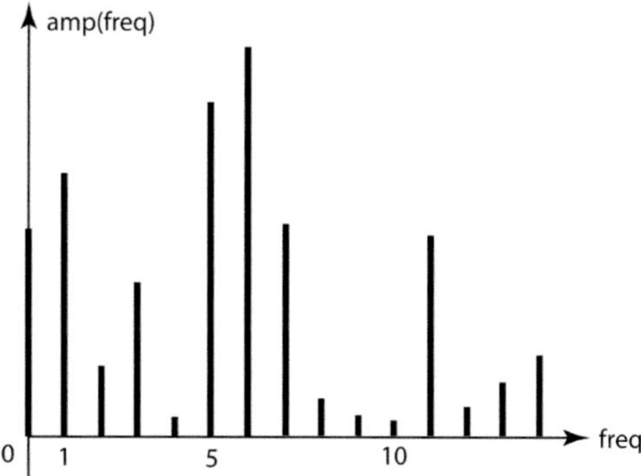

Figure 13.1 Discrete frequency and amplitude diagram. The horizontal axis is frequency normalized by the fundamental frequency $f_0 = 1 / T$. T is the period.

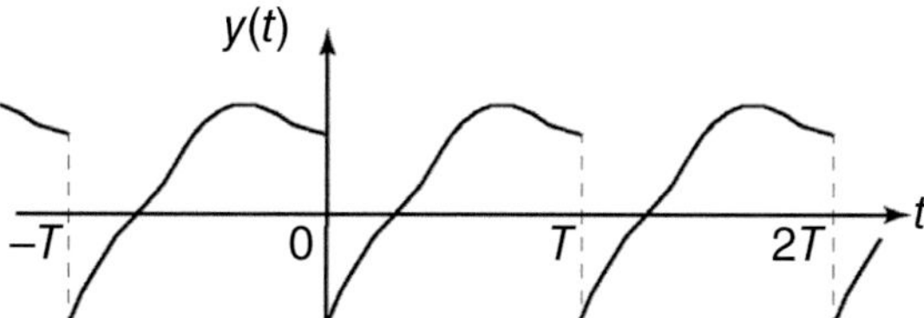

Figure 13.2 A continuous function defined in a segment can be extended periodically by repeating itself.

The base functions for Fourier series expansion are all sine or cosine functions. Note that they are not arbitrary sine or cosine functions: they have well-defined discrete frequencies based on the fundamental frequency $1/T$, or those frequencies are all multiples of $1/T$ (Fig. 13.1), i.e.

$$f_k = \frac{k}{T}, \quad k = 1, 2, 3, \ldots.$$

For a function with a finite segment (the range of t is finite) such as a time series from observations, it is unlikely that it happens to be periodic, and the function at the start point (e.g. at $t = 0$) and end point (at $t = T$) may not have the same value. However, we may still consider it a periodic function, but with a finite jump at $t = 0,\ T,\ 2T,\ \ldots$, i.e. the function repeats itself (Fig. 13.2) between $t = T$ and $t = 2T$, and between $t = nT$ and $t = (n+1)T, \ \ldots \ , (n = 2, 3, \ldots)$. This seems to provide a nice extension of the application of Fourier series expansion. This is especially useful for operational data processing because real data always have a finite segment. We can imagine that we extend the function in both ends to infinity by duplicating the known segment. The extended part of the new (periodic) function may not have anything to do with reality, but it does not matter as long as the Fourier series can represent the time period of the actual observations.

13.2 From Fourier Series to Fourier Transform

Suppose we have an aperiodic function that goes to $t \to +\infty$; the function cannot repeat itself and we cannot copy one segment of the function into another segment, as that will change the function. We can approach this problem with the Fourier series expansion of a given fundamental frequency $1/T$, and then let $T \to +\infty$. This will effectively make the fundamental frequency infinitesimally small, and within a finite frequency range, there are infinite frequencies. If we still use Fourier series expansion, we will effectively change the summation over the discrete values of k into an integral over continuously changing frequency. That leads to what we call a Fourier integral and Fourier Transform.

Before we get to the subject of Fourier Transform, let us first prepare by examining the exponential form of Fourier series.

13.2.1 Exponential Form of Fourier Series

The sine and cosine functions used in the Fourier series can be replaced by exponential functions, which turn out to be more convenient mathematically. It is a different way of expressing the original Fourier series. The base functions and their coefficients are all complex (with real and imaginary parts). This unifies the sine and cosine functions into one category – the exponential form, which makes further development in theory more convenient. It all starts from Euler's formula:

$$e^{ix} = \cos x + i \sin x, \quad e^{-ix} = \cos x - i \sin x, \tag{13.1}$$

in which i is the imaginary unit, or $i = \sqrt{-1}$. Using (13.1), the sine and cosine functions can be expressed as

$$\cos x = \frac{1}{2}(e^{ix} + e^{-ix}), \quad \sin x = \frac{1}{2i}(e^{ix} - e^{-ix}). \tag{13.2}$$

Using the frequency expression:

$$\cos(2\pi kft) = \frac{1}{2}(e^{i2\pi kft} + e^{-i2\pi kft}), \quad \sin(2\pi kft) = \frac{1}{2i}(e^{i2\pi kft} - e^{-i2\pi kft}). \tag{13.3}$$

Recall the Fourier series expansion:

$$y(t) = \frac{a_0}{2} + \sum_{k=1}^{\infty} [a_k \cos(2\pi kft) + b_k \sin(2\pi kft)], \tag{13.4}$$

$$
\begin{aligned}
a_k &= \frac{2}{T}\int_{-T/2}^{T/2} y(t)\cos(2\pi kft)dt,\ k = 0, 1, 2, \ldots \\
b_k &= \frac{2}{T}\int_{-T/2}^{T/2} y(t)\sin(2\pi kft)dt,\ k = 1, 2, \ldots .
\end{aligned}
\tag{13.5}
$$

Here a_k starts from $k = 0$ while b_k starts from $k = 1$. Alternatively, we can say $b_0 = 0$.

Substituting the expression in (13.3) into (13.4), we obtain

$$
\begin{aligned}
y(t) &= \frac{a_0}{2} + \sum_{k=1}^{\infty}\left(\frac{a_k - ib_k}{2}e^{i2\pi kft} + \frac{a_k + ib_k}{2}e^{-i2\pi kft}\right) \\
&= \alpha_0 + \sum_{k=1}^{\infty}(\alpha_k e^{i2\pi kft} + \alpha_{-k}e^{-i2\pi kft}) = \alpha_0 + \sum_{k=1}^{\infty}\alpha_k e^{i2\pi kft} + \sum_{k=-1}^{-\infty}\alpha_k e^{i2\pi kft} \\
&= \sum_{k=-\infty}^{\infty}\alpha_k e^{i2\pi kft}.
\end{aligned}
\tag{13.6}
$$

in which

$$
\alpha_k = \frac{a_k - ib_k}{2}, \alpha_{-k} = \frac{a_k + ib_k}{2}
$$

Now we have

$$
y(t) = \sum_{k=-\infty}^{\infty}\alpha_k e^{i2\pi kft}. \tag{13.7}
$$

This is the *exponential form of the Fourier series*. It is equivalent to the sinusoidal version.

Equation (13.7) deserves a few comments. First, the use of exponential function to replace the sine and cosine functions seems to have simplified the expression. Now it allows negative frequencies! The base functions are also new – they are the exponential functions

$$
f_k(t) = e^{i2\pi kft}. \tag{13.8}
$$

With equation (13.7), we are effectively redistributing the spectrum of cosine and sine onto the exponential base functions in a frequency domain allowing negative and positive values.

It can be verified that

$$\alpha_k = \frac{1}{T}\int_{-T/2}^{T/2} y(t)e^{-i2\pi kft}dt, \quad k = 0, \pm 1, \pm 2, \ldots, \pm\infty, \tag{13.9}$$

in which

$$|\alpha_k| = |\alpha_{-k}| = \frac{\sqrt{a_k^2 + b_k^2}}{2}. \tag{13.10}$$

Note that here $b_0 = 0$, so that $\alpha_0 = a_0/2$.

13.2.2 Cosine Form of Fourier Series

For ocean data and for almost any environmental data, they are real numbers, i.e. $y(t)$ is a real function of time t. Therefore, both a_k and b_k are real, and from (13.6) we see that α_k and α_{-k} are complex conjugates of each other, i.e. we have

$$\alpha_k = \tilde{\alpha}_{-k} = |\alpha_k|e^{j\theta_k}, \quad \alpha_{-k} = |\alpha_k|e^{-j\theta_k}, \theta_k = -\arctan\left(\frac{b_k}{a_k}\right),$$

in which θ_k is the phase for the kth cosine function, and the tilde over α_{-k} is the complex conjugate of α_k. We then have another (expected) way of expressing the Fourier series:

$$y(t) = \alpha_0 + 2\sum_{k=1}^{\infty} |\alpha_k| \cos(2\pi kft + \theta_k).$$

This is the cosine form of the Fourier series. Of course, we can also use sine functions in a similar fashion:

$$y(t) = \alpha_0 + 2\sum_{k=1}^{\infty} |\alpha_k| \sin(2\pi kf\,t + \varphi_k),\ \varphi_k = \arctan\left(\frac{a_k}{b_k}\right)$$

Here the phase φ_k is different from the phase in the cosine function θ_k.

13.2.3 Fourier Transform

Combining (13.7) and (13.9) we have

$$y(t) = \sum_{k=-\infty}^{\infty} \frac{1}{T}\int_{-T/2}^{T/2} y(t')e^{-i2\pi kft'}dt'e^{i2\pi kft}. \tag{13.11}$$

We now define the angular frequency:

$$\omega_k = 2\pi k f, f = \frac{1}{T}, k = 1, 2, 3, \ldots. \tag{13.12}$$

The change in angular frequency is quantum or discrete, or at an interval of

$$\delta\omega = 2\pi f = \frac{2\pi}{T}. \tag{13.13}$$

Equation (13.11) becomes

$$y(t) = \sum_{k=-\infty}^{\infty} \frac{1}{T} \int_{-T/2}^{T/2} y(t')e^{-i\omega t'} dt' e^{i\omega t}. \tag{13.14}$$

If we let $T \to +\infty$ and write

$$d\omega = \frac{2\pi}{T},$$

the summation becomes an integration, and we can rewrite equation (13.14) after letting $T \to +\infty$ as

$$y(t) = \frac{1}{2\pi} \int_{-\infty}^{\infty} e^{i\omega t} \left(\int_{-\infty}^{\infty} y(t')e^{-i\omega t'} dt' \right) d\omega. \tag{13.15}$$

Here we have used ω (continuous) to replace ω_k (discrete). This is the Fourier integral equation for a continuous function $y(t)$ (not necessarily periodic). It has double integrations. The function of angular frequency from the inner integral is

$$Y(\omega) = \int_{-\infty}^{\infty} y(t)e^{-i\omega t} dt. \tag{13.16}$$

which is the *Fourier Transform*, while equation (13.15) or

$$y(t) = \frac{1}{2\pi} \int_{-\infty}^{\infty} Y(\omega)e^{i\omega t} d\omega. \tag{13.17}$$

is the *inverse Fourier Transform* of $y(t)$.

Note that equation (13.16) is from equation (13.9) with $T \to +\infty$, but without the factor $1/T$. The "lost" factor $1/T$ is reclaimed when doing the integration with (13.17), in which the change in angular frequency $d\omega$ is simply a factor of $2\pi/T$. That is why (13.17) has a factor of $1/2\pi$.

Equations (13.16) and (13.17) are one form of the Fourier and inverse Fourier Transform pair. Another way to express this transform pair is to use frequency f rather than the angular frequency ω, i.e.

$$Y(f) = \int_{-\infty}^{\infty} y(t)e^{-i2\pi f t}dt, \tag{13.18}$$

$$y(t) = \int_{-\infty}^{\infty} Y(f)e^{i2\pi f t}d\omega. \tag{13.19}$$

The advantage of using equations (13.18) and (13.19) for the Fourier and inverse Fourier Transforms is that there is a better symmetry of the two equations such that the factor $1/2\pi$ is eliminated. It is easier to remember. Fourier Transform and its inverse transform provide the basis for the Fourier analysis.

13.2.4 Fourier Series and Discrete Fourier Transform

For the Fourier series, the Fourier coefficients expressed in discrete format is

$$\alpha_k = \frac{1}{T}\sum_{j=1}^{M} y(t_j)e^{-i2\pi k f t_j}\ \Delta t_j,\ \ k = 0, 1, 2, \ldots. \tag{13.20}$$

If the time interval is uniform and the total number of elements for the array y is M, then

$$\alpha_k = \frac{1}{M}\sum_{j=1}^{M} y(t_j)e^{-i2\pi k f t_j},\ \ k = 0, 1, 2, \ldots. \tag{13.21}$$

For the Fourier Transform, the discrete form is

$$Y_k = \sum_{j=1}^{M} y(t_j)e^{-i2\pi k f t_j}. \tag{13.22}$$

By comparing the above two equations, we can see that

$$\alpha_k = \frac{1}{M}Y_k. \tag{13.23}$$

This gives the relation between the Fourier series coefficients α_k and the *discrete Fourier Transform* Y_k.

13.3 Power Distribution in Frequency Domain

With Fourier Transform, we can examine the energy distribution in a periodic signal. In physics, the amplitude of a sine wave gives the magnitude of the wave. In fluid dynamics, atmospheric physics, and physical oceanography, wave energy and power are also shown to be proportional to the wave amplitude squared. For a periodic signal, we use the average of the square of the wave within a period as a proxy as the "power," although this quantity does not usually have the unit of power. The mean "power" averaged over a period is

$$P_y = \frac{1}{T}\int_T |y(t)|^2 dt. \tag{13.24}$$

Here the time "period" T could be the real period of a periodic function or, in the case of data analysis, the length of the time series. The integrand $|y(t)|^2$ can be seen as the product of $y(t)$ with its own complex conjugate $\tilde{y}(t)$, even though $y(t)$ is a real function. This is because we would like to take advantage of the Fourier series expression using the complex exponential base functions:

$$P_y = \frac{1}{T}\int_T |y(t)|^2 dt = \frac{1}{T}\int_T y(t)\tilde{y}(t)dt, \tag{13.25}$$

which leads to

$$P_y = \frac{1}{T}\int_T y(t) \sum_{k=-\infty}^{\infty} \tilde{\alpha}_k e^{-i2\pi kft} dt. \tag{13.26}$$

Switch the order of the summation and integration and we have

$$P_y = \sum_{k=-\infty}^{\infty} \tilde{\alpha}_k \left[\frac{1}{T}\int_T y(t) e^{-i2\pi kft} dt\right]. \tag{13.27}$$

What is inside the brackets is simply the Fourier series coefficient again. So, we have

$$P_y = \sum_{k=-\infty}^{\infty} \tilde{\alpha}_k \alpha_k = \sum_{k=-\infty}^{\infty} |\alpha_k|^2. \tag{13.28}$$

The above equation says that the average power P_y of a periodic signal $y(t)$ is a summation of the squares of the magnitude of each of the coefficients of the Fourier series for the function $y(t)$. In a sense, the collection of the squared coefficients provides a *power distribution* within the frequency domain.

The above derivation uses the exponential form of the Fourier series. We can also use the original Fourier series with sinusoidal functions to obtain the same conclusion.

Or, equivalently,

$$P_y = \sum_{k=-\infty}^{\infty} |\alpha_k|^2 = \alpha_0^2 + \frac{1}{2}\sum_{k=1}^{\infty}(a_k^2 + b_k^2) = \left(\frac{a_0}{2}\right)^2 + \frac{1}{2}\sum_{k=1}^{\infty}(a_k^2 + b_k^2). \quad (13.29)$$

To do that, we start from the Fourier series:

$$y(t) = \frac{a_0}{2} + \sum_{k=1}^{\infty}\left[a_k \cos\left(\frac{2\pi kt}{T}\right) + b_k \sin\left(\frac{2\pi kt}{T}\right)\right]. \quad (13.30)$$

Square both sides of the above equation and average over one period. Using an over bar to denote such an average over one period, the first term yields

$$\overline{\left(\frac{a_0}{2}\right)^2} = \left(\frac{a_0}{2}\right)^2. \quad (13.31)$$

The average over a period T of the cosine term squared is

$$\overline{\left(a_k \cos\left(\frac{2\pi kt}{T}\right)\right)^2} = \frac{a_k^2}{2}. \quad (13.32)$$

The average over a period T of the sine term squared is

$$\overline{\left(b_k \sin\left(\frac{2\pi kt}{T}\right)\right)^2} = \frac{b_k^2}{2}. \quad (13.33)$$

The average of the cross-terms are all zeros because of the orthogonality of these sine and cosine functions, and thus we arrive at (13.29). Equations (13.28) and (13.29) are forms of the *Parseval's Theorem*.

Using equation (13.23), we have the relation between the discrete Fourier Transform and the average power of the signal within one period:

$$P_y = \frac{1}{M^2}\sum_{k=-\infty}^{\infty} |Y_k|^2. \quad (13.34)$$

The Fourier series or Fourier Transform sets up a mapping between the original time series function and the coefficients $\left(\alpha_k(f_k)\right)$ of the Fourier series or the Fourier Transform $\left(Y(f)\right)$ as functions of frequency, i.e. (omitting the subscript k)

$$y(t) \leftrightarrow \alpha(f) \text{ or } y(t) \leftrightarrow Y(f). \quad (13.35)$$

Here the mathematical symbol $\leftrightarrow$ means a two-way mapping between time domain functions and frequency domain functions through Fourier Transform. The above relation means that a time function $y(t)$ has a set of Fourier series coefficients α as a function of the frequency f, or $y(t)$ has a Fourier Transform Y as a function of the frequency f. An inverse computation from the Fourier series α or Fourier Transform Y can lead back to the original time function $y(t)$.

Now we can state that the meaning of this mapping is that the "power" of the signal in time can be decomposed in the frequency domain. The power from all frequencies adds up to the total power in time.

13.4 Recap: From Least Squares Method to Fourier Transform

In the last few chapters, we have discussed the base functions and the use of linearly independent base functions to construct linear regression models (Fig. 13.3). The coefficients of the base functions in this regression model are

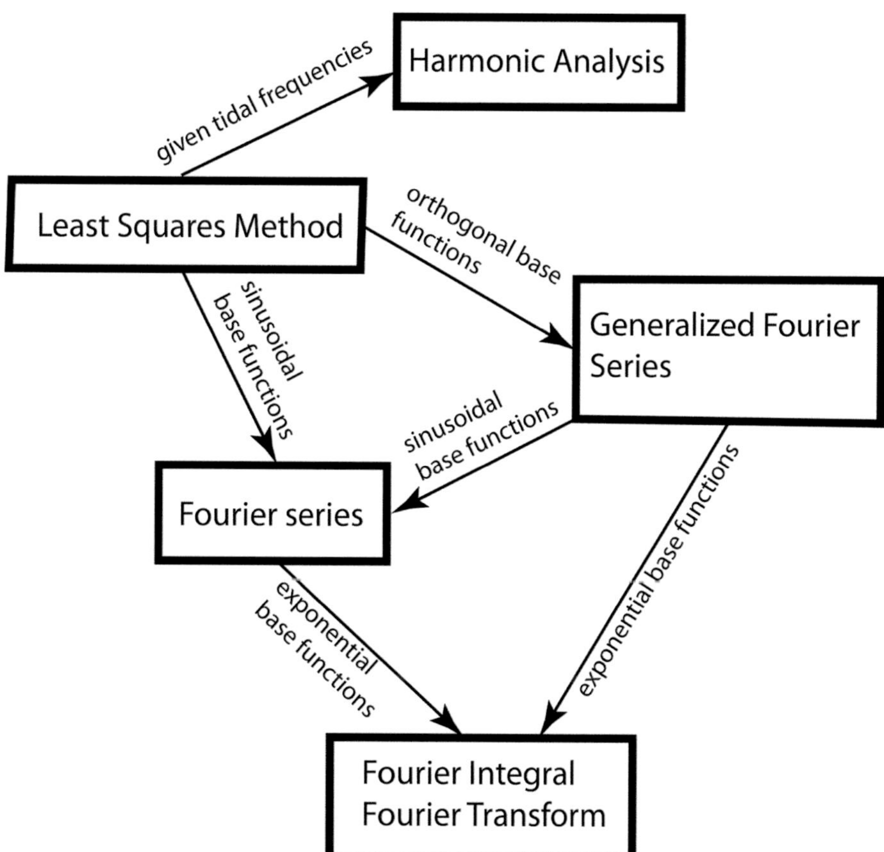

Figure 13.3 Relationships among various methods related to Fourier Transform.

determined by the least squares method. Depending on the selected based functions, we can have the tidal harmonic analysis and Fourier series expansion. In the tidal harmonic analysis, the base functions are sine and cosine functions with given tidal frequencies; while in the Fourier series expansion, the base functions are sine and cosine functions with frequencies determined by the fundamental frequency and all its harmonics. The fundamental frequency is the reciprocal of either the period T of a true periodic function or, in the case of observational data from the ocean, the total length of the record (for most applications) T (Fig. 13.3). The generalized Fourier series expansion is for any orthogonal base functions in which the coefficients are determined explicitly without the matrix inversion, just like in the case of Fourier series expansion.

In this chapter, we started from the Fourier series and used the Euler formula to convert the sinusoidal functions to exponential functions by allowing using complex numbers (with real and imaginary parts). Although the change of base functions may seem to be just another mathematical conversion, it led to the Fourier integral as the total length of record in time (T) approaches infinity. This one step is a giant leap that allows aperiodic functions to be included in a Fourier Transform, which also has an inverse transform. This paves a road to Fourier analysis in frequency domain (vs. time domain for the time series). By Parseval's Theorem, the "power" contained in the time series can be represented by the sum of squares of the amplitudes of the coefficients of base functions in exponential form: each base function at a given frequency contributes part of the total power by the square of the amplitude.

Review Questions for Chapter 13

(1) What is the advantage of using the exponential base functions vs. the sinusoidal functions in Fourier series?

(2) Are the exponential base functions we discussed in this chapter orthogonal functions?

(3) What is a Fourier Transform, and how is it related to Fourier series?

Exercises for Chapter 13

(1) Prove equation (13.30) that the average over time t for a period T of the cosine term squared is

$$\overline{\left(a_k \cos\left(\frac{2\pi kt}{T}\right)\right)^2} = \frac{a_k^2}{2}.$$

Prove equation (13.31) that the average over time t for a period T of the sine term squared is

$$\overline{\left(b_k \sin\left(\frac{2\pi kt}{T}\right)\right)^2} = \frac{b_k^2}{2}$$

for any k and T.

(2) In the cosine form of the Fourier series expansion, we have

$$y(t) = \alpha_0 + 2\sum_{k=1}^{\infty} |\alpha_k| \cos(2\pi k f\, t + \varphi_k).$$

Try to calculate the time average of the square of $y(t)$. Compare your result with (13.28). What can you conclude?

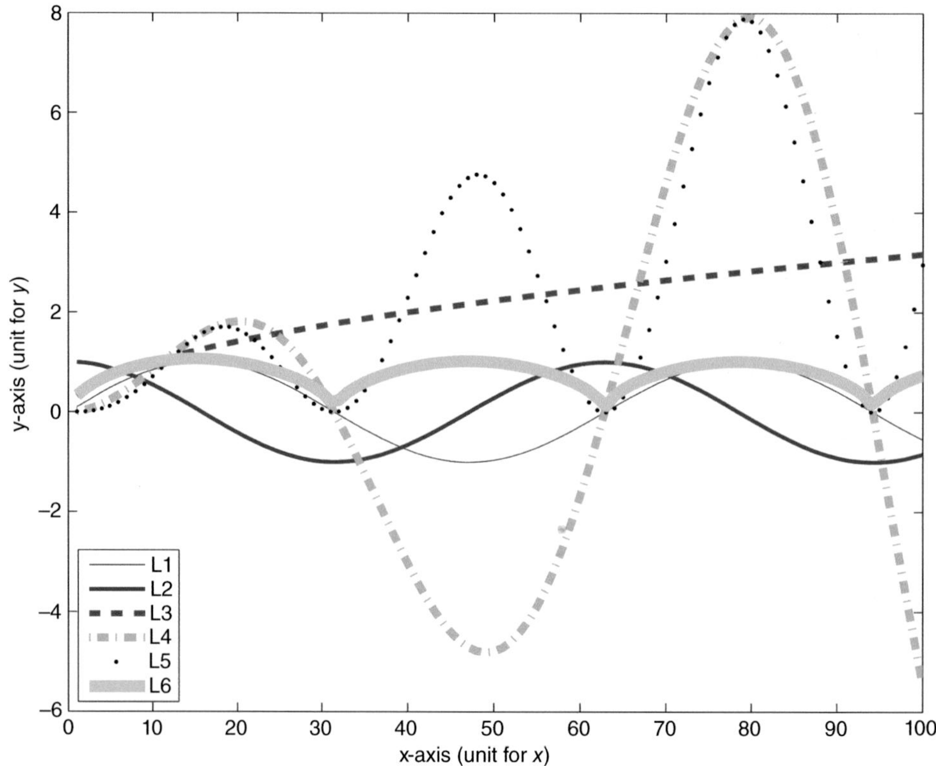

Figure 2.1 Example of figure produced by the MATLAB command `plot`.

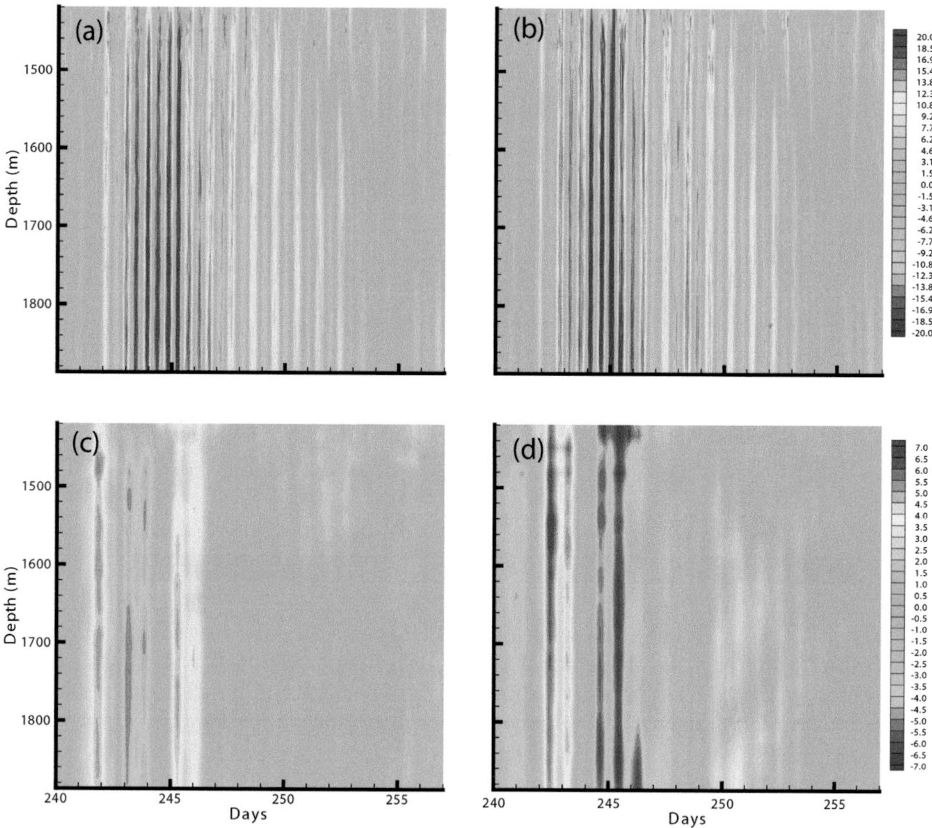

Figure 22.10 High- and low-pass filtered velocity components at 1400–1900 m: (a) high-pass east velocity; (b) high-pass north velocity; (c) low-pass east velocity; and (d) low-pass north velocity.

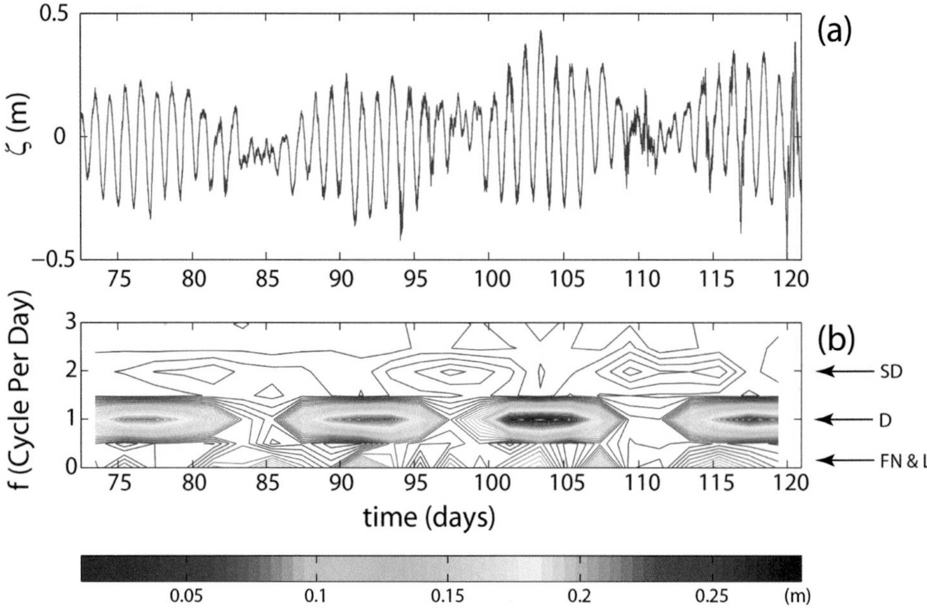

Figure 23.3 An example showing the spectrogram generated by the short-time Fourier Transform. (a) Time series of the water depth variation over the mean value. (b) The STFT. The arrows show the diurnal and semi-diurnal tides, as well as the fortnightly tides and low frequency weather induced oscillations.

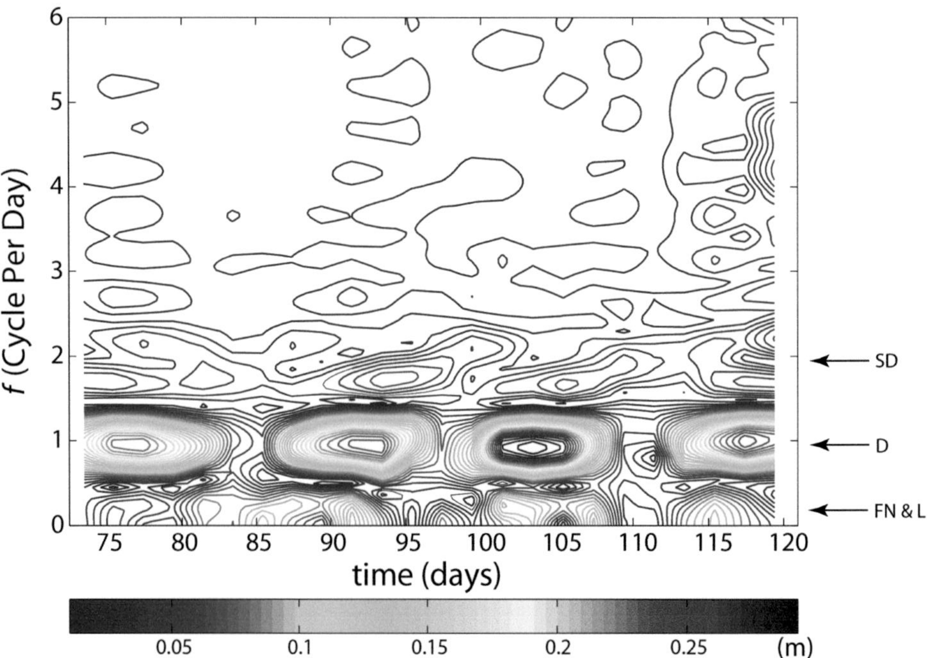

Figure 23.6 The spectrogram from STFT with zero-padding.

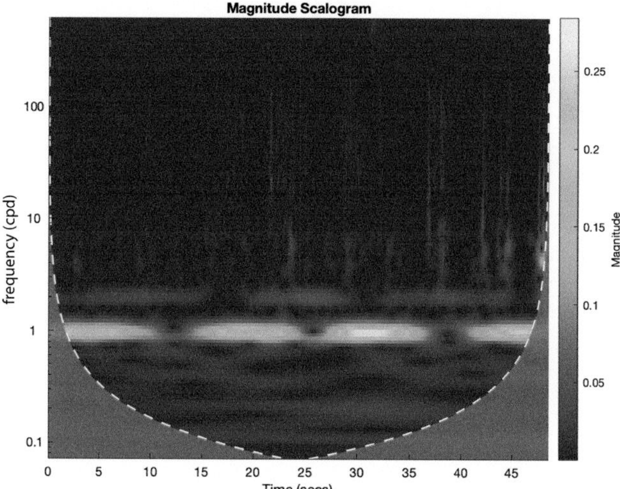

Figure 23.8 Wavelet results: the scalogram for the water depth data of Fig. 23.3a.

14

Discrete Fourier Transform and Fast Fourier Transform

About Chapter 14

This chapter is aimed at introducing the fast Fourier Transform (FFT) for discrete Fourier Transform. It starts from the discretization of the Fourier Transform to its digital expression with constant time intervals. When the integral in Fourier Transform is replaced by a summation, the continuous Fourier Transform is changed to a discrete version. An interesting property is that the discrete Fourier Transform and its inverse are exact relations. An example of the discrete Fourier Transform is discussed for a simple rectangular window function which results in the sinc function, useful for the interpretation of finite sampling effect, which will be further discussed in Chapter 17. A technique of zero-padding is introduced with the discrete Fourier Transform for better visualization of the spectrum. But the computation of discrete Fourier Transform of a long time series can be quite "labor intensive" or costly in computer time with a direct computation. However, since the base functions are periodic, a direct computation can have many duplications in multiplications of terms. Algorithms can be designed to reduce the duplications so that the speed of computation is increased. The reduction of duplicated computations can be repeatedly done through an FFT algorithm. In MATLAB, this is done by a simple command `fft`. The efficiency of FFT is discussed.

14.1 The Discrete Fourier Transform

14.1.1 From Continuous Fourier Transform to Discrete Fourier Transform

We have discussed that the Fourier Transform for a function $y(t)$ is

$$Y(\omega) = \int_{-\infty}^{\infty} y(t)e^{-i\omega t}dt \quad \text{or} \quad Y(f) = \int_{-\infty}^{\infty} y(t)e^{-i2\pi ft}dt. \tag{14.1}$$

Its inverse transform recovers the original function:

$$y(t) = \frac{1}{2\pi} \int_{-\infty}^{\infty} Y(\omega) e^{i\omega t} d\omega \quad \text{or} \quad y(t) = \int_{-\infty}^{\infty} Y(f) e^{i2\pi f t} df. \tag{14.2}$$

If the total length of the record in time is T, and everything outside of the record is assumed zero, or if the function is periodic with a period T, then (14.1) becomes

$$Y(\omega) = \int_{0}^{T} y(t) e^{-i\omega t} dt \quad \text{or} \quad Y(f) = \int_{0}^{T} y(t) e^{-i2\pi f t} dt. \tag{14.3}$$

In actual data analysis, the above integrations are replaced by summations. For real problems, the time period is always finite. Neglecting dt, the discrete form of this equation is

$$Y_m = \sum_{k=1}^{M} y(t_k) e^{-i2\pi m f_0 t_k}. \tag{14.4}$$

in which $m = 0, 1, 2, ..., M - 1$. This is an expression of the *discrete Fourier Transform*. Here both t and the frequency f are discrete real numbers. The frequency is $f = mf_0$, where $f_0 = 1/T$ is the *fundamental frequency*. If the time interval is uniform, then t_k is a multiple of the time interval dt:

$$t_k = (k-1)dt, \quad k = 1, 2, \ldots, M. \tag{14.5}$$

Note that

$$T = Mdt. \tag{14.6}$$

Sometimes we choose m to start from 0 instead of 1, then $t_k = kt$ with m starting from 0, in which case equation (14.4) becomes

$$Y_m = \sum_{k=0}^{M-1} y_k e^{-i\frac{2\pi}{M}km}, \quad (m = 0, 1, 2, \ldots, M-1). \tag{14.7}$$

Note that here the time interval dt has been omitted on purpose. The inverse transform is

$$y_k = \frac{1}{M} \sum_{m'=0}^{M-1} Y_{m'} e^{i\frac{2\pi}{M}km'}, \quad (k = 0, 1, 2, \ldots, M-1). \tag{14.8}$$

Here dt in the denominator has also been omitted so that y_k of (14.8) and Y_m of (14.7) are Fourier Transform pairs. The transform is an exact relationship. This can be verified by substituting (14.8) into (14.7):

$$Y_m = \sum_{k=0}^{M-1} \left(\frac{1}{M} \sum_{m'=0}^{M-1} Y_{m'} e^{i\frac{2\pi}{M}km'} \right) e^{-i\frac{2\pi}{M}km}. \tag{14.9}$$

Switch the order of the sums:

$$Y_m = \frac{1}{M} \sum_{m'=0}^{M-1} Y_{m'} \left(\sum_{k=0}^{M-1} e^{-i\frac{2\pi}{M}k(m-m')} \right). \tag{14.10}$$

It can be proven that on the right-hand side of the above equation, the inner sum is 0 if $m' \neq m$ but equals M if $m' = m$, and so the outer sum has only one non-zero term, which is Y_m. This verifies (14.9) and (14.10) and the conclusion that the discrete Fourier Transform and inverse transform are exact relationships.

14.1.2 The Periodicity and Symmetry of Discrete Fourier Transform

It is important to note that Y_m is a periodic function in m with a period of M. This can be verified by the following:

$$\begin{aligned} Y_{m+nM} &= \sum_{k=0}^{M-1} y_k e^{-i2\pi k(m+nM)/M} = e^{-i2\pi kn} \sum_{k=0}^{M-1} y_k e^{-i2\pi km/M} \\ &= \sum_{k=0}^{M-1} y_k e^{-i2\pi km/M} = Y_m \qquad (m = 0, 1, 2, \ldots, M-1) \end{aligned} \tag{14.11}$$

in which n is any integer (positive or negative). Here we have used the mathematical property that

$$e^{-i2\pi kn} = 1. \tag{14.12}$$

For a function $y(t)$ that is real (not a complex), we also have a symmetry. This is clear with the following derivations:

$$\begin{aligned} Y_{M-m} &= \sum_{k=0}^{M-1} y(t_k) e^{-i2\pi k(M-m)/M} \\ &= \sum_{k=0}^{M-1} y(t_k) e^{-i2\pi k} e^{i2\pi km/M} = \sum_{k=0}^{M-1} y(t_k) e^{i2\pi km/M} = Y_m^* \quad (m = 0, 1, \ldots, M-1). \end{aligned} \tag{14.13}$$

14.1.3 The Nyquist Frequency

The above property is very important – it means that the magnitude of Y_m is symmetric about $M/2$ (Fig. 14.1). In terms of frequency, it is

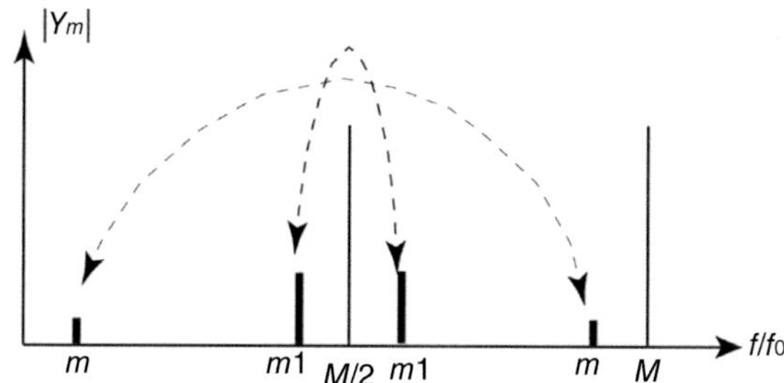

Figure 14.1 Symmetry of the discrete Fourier Transform about the Nyquist frequency $f_N = M/(2T) = 1/(2dt)$.

$$f_N = \frac{M}{2} f_0 = \frac{1}{2dt}. \tag{14.14}$$

In other words, Y_m has only $M/2$ independent complex numbers corresponding to values at frequencies from 0 to f_N. Beyond $m = M/2$ (or $f = f_N$), the discrete function Y_m duplicates itself in magnitude (actual number is a complex conjugate). The frequency f_N is called the *Nyquist frequency*. It is the highest frequency that a Fourier Transform can resolve with given sampling interval dt.

14.1.4 Discrete Fourier Transform: Rectangular Window

Now let us look at an example. Calculate the discrete Fourier Transform of a *rectangular window function* using equation (14.7). The rectangular window is defined such that the function value is 1 within a given width of the window but zero elsewhere. The window function and the magnitude of the transform are shown in Fig. 14.2. The symmetry is about half of M.

The MATLAB script used is as the following:

```
MM = 20;  M = MM * 3; n = (0:M - 1)'; m = n;
y = [ zeros(MM,1); ones(MM,1); zeros(MM,1)] ;
Y = y' * exp(-i * 2 * pi * m * n' / M);
```

Fig. 14.2 shows clearly the symmetric characteristics of the magnitude of the discrete Fourier Transform of the window function. Since $M = 60$ in the above example, the symmetry is about $M/2 = 30$. All $|Y_m|$ values are symmetric about $M/2$, except Y_0, which is the only unpaired single value resulted from a summation of all y values. There are 20 values of 1 in the window function, and thus $Y_0 = 20$.

Now let us modify the way the transform is done. Since the rectangular window function is zero outside of the window, we can extend the number of zeros while not changing anything in the window. We examine how the discrete Fourier

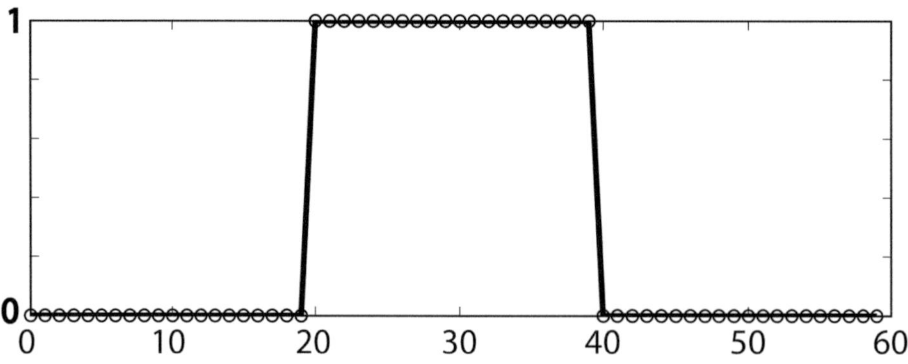

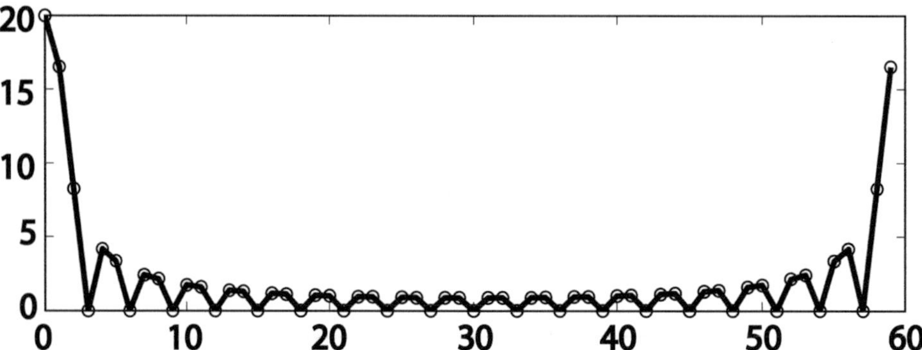

Figure 14.2 Discrete Fourier Transform for a discrete rectangular window function. Upper panel: the discrete time series; lower panel: the magnitude of discrete Fourier Transform of the rectangular window function.

Transform will behave with the longer function. In other words, we add zeros in the time domain before doing the discrete Fourier Transform. This is a technique called *zero-padding*. The MATLAB script is given below:

```
MM = 20;
y = [ zeros(MM,1); ones(MM,1); zeros(10 * MM,1)];
M = length(y); n = (0:M  1)'; m = n;
Y = y' * exp(-i * 2 * pi * m * n' / M);
```

The function and magnitude of the discrete Fourier Transform are shown in Fig. 14.3. The magnitude of this specific discrete Fourier Transform function is much smoother and is described by a function defined by

$$X_T(f) = \frac{\sin(\pi f T)}{\pi f}. \tag{14.15}$$

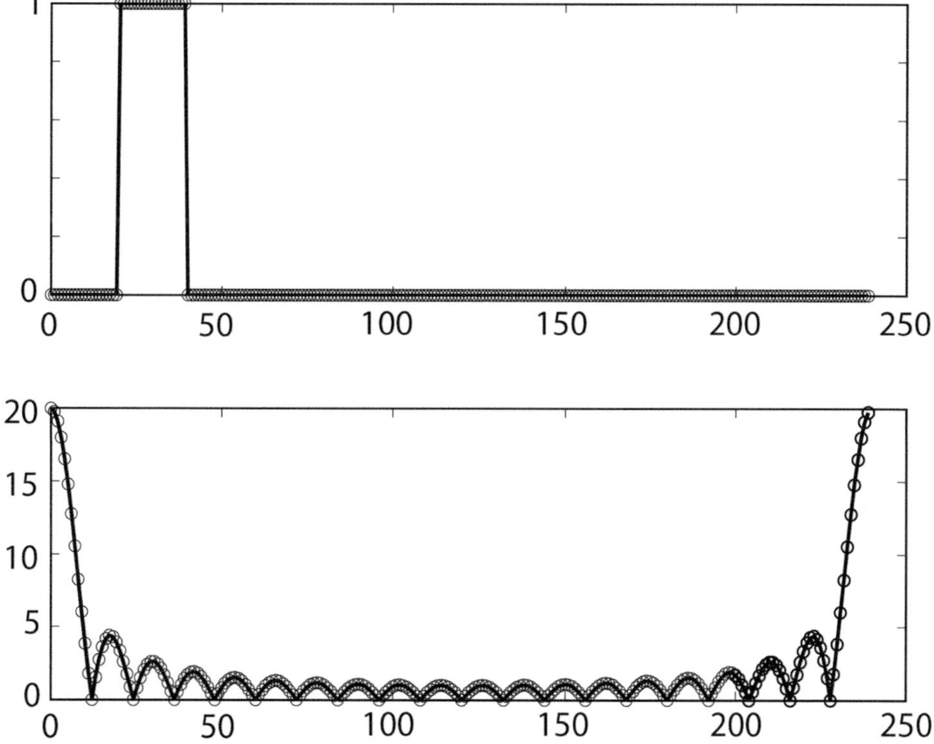

Figure 14.3 Discrete Fourier Transform for a discrete rectangular window function with zero-padding. Upper panel: the discrete time series; lower panel: the magnitude of discrete Fourier Transform of the rectangular window function.

This is called a sinc function in frequency, which is a Fourier Transform of a rectangular window function in time with a window size of T. The sinc function will be further discussed later. Note also that $|Y_0|$ does not change its height – it is still 20.

Zero-padding is a very useful technique used in discrete Fourier Transform. It can increase the resolution of the Fourier Transform visually without changing the original time function.

14.2 The Fast Fourier Transform

14.2.1 The Redundancy of Computations

Because of the periodicity of the discrete Fourier Transform, i.e. equation (14.11), the computations of discrete Fourier Transform can have duplicate terms, which, if eliminated, would reduce the time of computation. To illustrate, let us define

Table 14.1. *Value of* $I = \frac{km}{M}$ for $e^{-i2\pi I}$

$m\ k$	0	1	2	...	$M-2$	$M-1$
0	0	0	0	...	0	0
1	0	$1/M$	$2/M$	...	$(M-2)/M$	$(M-1)/M$
2	0	$2/M$	$4/M$	...	$2(M-2)/M$	$2(M-1)/M$
...	...	...	...	...	...	...
$\boldsymbol{M-2}$	0	$(M-2)/M$	$2(M-2)/M$	...	$(M-2)^2/M$	$(M-2)(M-1)/M$
$\boldsymbol{M-1}$	0	$(M-1)/M$	$2(M-1)/M$	...	$(M-1)(M-2)/M$	$(M-1)^2/M$

Table 14.2. *Value of* $I = \frac{km}{M}, M = 6,$ for $e^{-i2\pi I}$

$m\ k$	0	1	2	3	4	5
0	0	0	0	0	0	0
1	0	1/6	1/3	1/2	2/3	5/6
2	0	1/3	2/3	1	4/3	5/3
3	0	1/2	1	3/2	2	5/2
4	0	2/3	4/3	2	8/3	10/3
5	0	5/6	5/3	5/2	10/3	25/6

Table 14.3. *Equivalent value of* $I = \frac{km}{M}, M = 6,$ for $e^{-i2\pi I}$

$m\ k$	0	1	2	3	4	5
0	0	0	0	0	0	0
1	0	1/6	1/3	1/2	2/3	5/6
2	0	1/3	2/3	0	1/3	2/3
3	0	1/2	0	1/2	0	1/2
4	0	2/3	1/3	0	2/3	1/3
5	0	5/6	2/3	1/2	1/3	1/6

$$I = \frac{km}{M}, \quad (k, m = 0, 1, 2, \ldots, M-1). \tag{14.16}$$

This gives a series of real numbers from 0 to $\frac{(M-1)^2}{M}$. For example, assuming that $M = 6$, the following table gives all the possible values of I (Table 14.1).

The shaded cells in Table 14.1 indicate the I value is greater than 1. Because the function $e^{-i2\pi I}$ is periodic for I, any $e^{-i2\pi I}$ with the I value greater than 1 can be folded into a $e^{-i2\pi I'}$, in which I' is the fraction of the I value. With the example of $M = 6$, Table 14.2 gives all the possible I values with various m and k values. The shaded cells show when the I value is greater than 1. Table 14.3 shows the equivalent I' values that are within $[0, 1]$. This table shows that there are duplicate values for I'.

The meaning of this can be seen from the discrete Fourier Transform. With the example $M = 6$, the transform is

$$Y_m = \sum_{k=0}^{5} y_k e^{-i2\pi km/6}, \quad (m = 0, 1, 2, \ldots, 5). \tag{14.17}$$

There are six equations, which are

$$m = 0 : Y_0 = y_0 + y_1 + y_2 + y_3 + y_4 + y_5, \tag{14.18}$$

$$m = 1 : Y_1 = y_0 + y_1 e^{-i2\pi\frac{1}{6}} + \underline{y_2 e^{-i2\pi\frac{1}{3}}} + \underline{\underline{y_3 e^{-i2\pi\frac{1}{2}}}} + \underline{y_4 e^{-i2\pi\frac{2}{3}}} + y_5 e^{-i2\pi\frac{5}{6}}, \tag{14.19}$$

$$m = 2 : Y_2 = y_0 + y_1 e^{-i2\pi\frac{1}{3}} + \underline{\underline{y_2 e^{-i2\pi\frac{2}{3}}}} + y_3 + \underline{\underline{y_4 e^{-i2\pi\frac{1}{3}}}} + y_5 e^{-i2\pi\frac{2}{3}}, \tag{14.20}$$

$$m = 3 : Y_3 = y_0 + y_1 e^{-i2\pi\frac{1}{2}} + y_2 + \underline{\underline{y_3 e^{-i2\pi\frac{1}{2}}}} + y_4 + y_5 e^{-i2\pi\frac{1}{2}}, \tag{14.21}$$

$$m = 4 : Y_4 = y_0 + y_1 e^{-i2\pi\frac{2}{3}} + \underline{y_2 e^{-i2\pi\frac{1}{3}}} + y_3 + \underline{y_4 e^{-i2\pi\frac{2}{3}}} + y_5 e^{-i2\pi\frac{1}{3}}, \tag{14.22}$$

$$m = 5 : Y_5 = y_0 + y_1 e^{-i2\pi\frac{5}{6}} + \underline{\underline{y_2 e^{-i2\pi\frac{2}{3}}}} + \underline{\underline{y_3 e^{-i2\pi\frac{1}{2}}}} + \underline{\underline{y_4 e^{-i2\pi\frac{1}{3}}}} + y_5 e^{-i2\pi\frac{1}{6}}. \tag{14.23}$$

From these equations, it is clear that there are many duplicate terms with multiplications (the underlined terms). For example, $y_2 e^{-i2\pi\frac{1}{3}}$ is present in equation (14.19) for $m = 1$ and occurs again in equation (14.22) for $m = 4$. For these duplicate terms, the multiplication between y_2 and $e^{-i2\pi\frac{1}{3}}$ needs to be done only once to save computer time. This is useful when M is very large and the duplications plenty.

14.2.2 The Cooley-Tukey Theory

The discussion of this section is based on Cooley and Tukey (1965). For convenience of discussion, let us write the exponential complex function without the variables k and m as B, i.e. define

$$B = e^{-i2\pi/M}, \tag{14.24}$$

and thus

$$Y_m = \sum_{k=0}^{M-1} y_k B^{mk}, \quad (m = 0, 1, 2, \ldots, M-1), \tag{14.25}$$

in which

$$B = e^{-i\frac{2\pi}{M}}, \quad B^{mk} = e^{-i\frac{2\pi}{M}mk}. \tag{14.26}$$

This is a collection of M equations, each having M multiplications plus additions. The multiplications take more time than the additions, and therefore we can use the number of multiplications to measure the intensity of computation. As a result, the above equations contain a total of

$$\beta = M^2 \tag{14.27}$$

multiplications.

Now assume that M has two factors r_1 and r_2, i.e.

$$M = r_1 r_2, \tag{14.28}$$

such that to count from 0 to M using two indices m and k in equation (14.26), we use the following expressions

$$\begin{array}{lll} m = m_1 r_1 + m_0, & m_0 = 0, 1, \ldots, r_1 - 1, & m_1 = 0, 1, \ldots, r_2 - 1 \\ k = k_1 r_2 + k_0, & k_0 = 0, 1, \ldots, r_2 - 1, & k_1 = 0, 1, \ldots, r_1 - 1 \end{array}. \tag{14.29}$$

With these indices m_0, m_1, k_0, and k_1, the discrete Fourier Transform is a function of m_0 and m_1:

$$Y_m \equiv Y(m) \equiv Y(m_0, m_1) = \sum_{k=0}^{M-1} y_k B^{mk}, \quad (m = 0, 1, 2, \ldots, M-1). \tag{14.30}$$

For convenience, we write

$$y(k) \equiv y_k. \tag{14.31}$$

Therefore, we have

$$Y(m) = \sum_{k=0}^{M-1} y(k) B^{mk} = \sum_{k=0}^{M-1} y(k_1 r_2 + k_0) B^{m(k_1 r_2 + k_0)}. \tag{14.32}$$

With any given m, there are still M multiplications plus additions, and so far, there is no change of number of multiplications. Since $M = r_1 r_2$, counting from 0 to $M - 1$ can be done by counting blocks of length r_2 with a total of r_1 blocks, i.e. for $k_1 = 0$, count k from 0 to $r_2 - 1$ by varying k_0 from 0 to $r_2 - 1$; then for $k_1 = 1$, count k for r_2 more numbers from r_2 to $2r_2 - 1$ by varying k_0 from 0 to $r_2 - 1$; after which for $k_1 = 2$, count k for r_2 more numbers from $2r_2$ to $3r_2 - 1$, also by varying k_0 from 0 to $r_2 - 1$, ..., until $k_1 = r_1 - 1$; for the last r_2 numbers from $k = (r_1 - 1)r_2$ to $k = (r_1 - 1)r_2 + r_2 - 1 = r_1 r_2 - 1$ by varying k_0 from 0 to $r_2 - 1$. For all the k_1 blocks, the total number is $M = r_1 r_2$, as shown by (14.33).

$$\left.\begin{array}{lcccc} k_1 = 0: & 0 & 1 & \cdots & r_2 - 1 \\ k_1 = 1: & r_2 & r_2 + 1 & \cdots & 2r_2 - 1 \\ \vdots & \vdots & \vdots & \vdots & \vdots \\ k_1 = r_1 - 1: & \underbrace{(r_1 - 1)r_2 \quad (r_1 - 1)r_2 + 1 \quad \cdots \quad r_1 r_2 - 1}_{r_2} \end{array}\right\} r_1. \tag{14.33}$$

With this method of counting, the sum for the M terms can be rearranged into two sums for the two indices k_0 and k_1:

$$Y(m_0, m_1) = \sum_{k_0=0}^{r_2-1} \sum_{k_1=0}^{r_1-1} y(k_0, k_1) B^{mk_1 r_2} B^{mk_0}. \tag{14.34}$$

At this moment, there are still M terms, each with a multiplication, and there are M equations, so there are still M^2 multiplications.

Next, we will make use of the periodic nature of the function B^{mk}, and so duplicate numbers can be reduced. Because

$$B^{mk_1 r_2} = B^{(m_1 r_1 + m_0)k_1 r_2} = B^{m_1 k_1 r_1 r_2 + m_0 k_1 r_2}. \tag{14.35}$$

Note that the periodicity gives

$$B^{m_1 k_l r_1 r_2} = e^{-i\frac{2\pi}{M} m_1 k_1 M} = e^{-i2\pi m_1 k_1} = 1. \tag{14.36}$$

This simplifies $B^{mk_1 r_2}$:

$$B^{mk_1 r_2} = B^{m_0 k_1 r_2}, \tag{14.37}$$

which leads to

$$Y(m_0, m_1) = \sum_{k_0=0}^{r_2-1} \sum_{k_1=0}^{r_1-1} y(k_0, k_1) B^{m_0 k_1 r_2} B^{mk_0}, \quad (m, m_0, m_1 = 0, 1, 2, \ldots, M - 1). \tag{14.38}$$

Now we have two sums, and the inner sum is

$$Z(k_0, m_0) = \sum_{k_1=0}^{r_1=1} y(k_0, k_1) B^{m_0 k_1 r_2}. \tag{14.39}$$

With (14.37), the array Z only depends on k_0 and m_0, not m_1, and so it has a total of M elements; each has r_1 multiplications, and thus it has a total of $r_1 M$

multiplications. Likewise, in addition to these multiplications for the inner sum, all the elements for the array $Y(m_0, m_1)$ have a total of r_2M multiplications through the outer sum. Therefore, the total multiplication now amounts to

$$J = M(r_1 + r_2). \tag{14.40}$$

Note that here the periodicity represented by (14.37) plays a key role in reducing the number of multiplications: without (14.37), Z would be dependent on k_0, m_0, and m_1, and the total number of multiplications for $Y(m_0, m_1)$ would still be M^2.

The above argument can be easily extended to a more general case in which M has more than two factors, r_1, r_2, ... , and r_N:

$$M = r_1 r_2 \cdots r_N, \tag{14.41}$$

in which case the number of multiplications is

$$\alpha = M(r_1 + r_2 + \cdots + r_N). \tag{14.42}$$

This is the main conclusion of the work of Cooley and Tukey (1965).

In the special case that M is a power of 2, i.e.

$$r_1 = r_2 = \cdots = r_N = r = 2, \quad M = 2^N, \tag{14.43}$$

we have

$$N = \log_2 M, \tag{14.44}$$

in which case (14.42) becomes

$$\alpha = 2M\log_2 M. \tag{14.45}$$

And so the ratio between α and β is

$$R = \frac{\alpha}{\beta} = \frac{2\log_2 M}{M}. \tag{14.46}$$

The larger the number M is, the smaller the ratio of R.

14.2.3 Discrete Fourier Transform with M = 16

So far, the discussion remains quite abstract; we have not discussed exactly how we can reduce the number of duplicated multiplications in actual computation. It helps if we start from some examples. Let us choose a discrete Fourier Transform with a length of 16:

$$Y_m = \sum_{k=0}^{15} y_k e^{-i2\pi km/M}, \quad (m = 0, 1, 2, \ldots, 15). \tag{14.47}$$

This series has a period of 16 in terms of m. If we write equation (14.47) out, it is

$$
\begin{aligned}
Y_m = {}& \underline{\underline{y_0}} + \underline{y_1 e^{-i\frac{2\pi}{16}m}} + \underline{\underline{y_2 e^{-i\frac{2\pi}{16}2m}}} + \underline{y_3 e^{-i\frac{2\pi}{16}3m}} + \underline{\underline{y_4 e^{-i\frac{2\pi}{16}4m}}} + \underline{y_5 e^{-i\frac{2\pi}{16}5m}} + \underline{\underline{y_6 e^{-i\frac{2\pi}{16}6m}}} + \underline{y_7 e^{-i\frac{2\pi}{16}7m}} \\
& + \underline{\underline{y_8 e^{-i\frac{2\pi}{16}8m}}} + \underline{y_9 e^{-i\frac{2\pi}{16}9m}} + \underline{\underline{y_{10} e^{-i\frac{2\pi}{16}10m}}} + \underline{y_{11} e^{-i\frac{2\pi}{16}11m}} + \underline{\underline{y_{12} e^{-i\frac{2\pi}{16}12m}}} \\
& + \underline{y_{13} e^{-i\frac{2\pi}{16}13m}} + \underline{\underline{y_{14} e^{-i\frac{2\pi}{16}14m}}} + \underline{y_{15} e^{-i\frac{2\pi}{16}15m}}.
\end{aligned}
\tag{14.48}
$$

For convenience of discussion, we have highlighted the odd and even terms with single and double underlines. Here, odd and even are in reference to the integer in front of m, i.e. m, $3m$, $5m$, $\ldots$, $15m$ indicate odd terms and 0, $2m$, $4m$, $\ldots$, $14m$ indicate even terms.

If we group the odd terms together and even terms together, respectively, the above equation can be changed to

$$
\begin{aligned}
Y_m = {}& \left[y_0 + y_2 e^{-i\frac{2\pi}{8}m} + y_4 e^{-i\frac{2\pi}{8}2m} + y_6 e^{-i\frac{2\pi}{8}3m} + y_8 e^{-i\frac{2\pi}{8}4m} + y_{10} e^{-i\frac{2\pi}{8}5m} + y_{12} e^{-i\frac{2\pi}{8}6m} + y_{14} e^{-i\frac{2\pi}{8}7m} \right] \\
& + e^{-i\frac{2\pi}{16}m} \left[y_1 + y_3 e^{-i\frac{2\pi}{8}m} + y_5 e^{-i\frac{2\pi}{8}2m} + y_7 e^{-i\frac{2\pi}{8}3m} + y_9 e^{-i\frac{2\pi}{8}4m} \right. \\
& \left. + y_{11} e^{-i\frac{2\pi}{8}5m} + y_{13} e^{-i\frac{2\pi}{8}6m} + y_{15} e^{-i\frac{2\pi}{8}7m} \right].
\end{aligned}
\tag{14.49}
$$

The above equation is two half-sized Fourier Transforms. Each has a length of $M/2 = 8$:

$$
U_1 = \underline{\underline{y_0}} + \underline{y_2 e^{-i\frac{2\pi}{8}m}} + \underline{\underline{y_4 e^{-i\frac{2\pi}{8}2m}}} + \underline{y_6 e^{-i\frac{2\pi}{8}3m}} + \underline{\underline{y_8 e^{-i\frac{2\pi}{8}4m}}} + \underline{y_{10} e^{-i\frac{2\pi}{8}5m}} + \underline{\underline{y_{12} e^{-i\frac{2\pi}{8}6m}}} + \underline{y_{14} e^{-i\frac{2\pi}{8}7m}},
\tag{14.50}
$$

$$
V_1 = \underline{\underline{y_1}} + \underline{y_3 e^{-i\frac{2\pi}{8}m}} + \underline{\underline{y_5 e^{-i\frac{2\pi}{8}2m}}} + \underline{y_7 e^{-i\frac{2\pi}{8}3m}} + \underline{\underline{y_9 e^{-i\frac{2\pi}{8}4m}}} + \underline{y_{11} e^{-i\frac{2\pi}{8}5m}} + \underline{\underline{y_{13} e^{-i\frac{2\pi}{8}6m}}} + \underline{y_{15} e^{-i\frac{2\pi}{8}7m}}
\tag{14.51}
$$

$$
Y_m = U_1 + e^{-i\frac{2\pi}{16}m} V_1. \tag{14.52}
$$

Note that the original discrete Fourier Transform Y_m has a period of 16 while both U_1 and V_1 have a period of 8, as can be seen from (14.50) and (14.51); therefore, (14.52) has $M = 16$ multiplications, which are generally different (non-duplicating), for each m once V_1 is computed. There are additional multiplications in U_1 and V_1 as well, which should be added to the M multiplications at this level of (14.52). For convenience, we call the multiplications in (14.52) Layer 1 (Fig. 14.4) computations.

Likewise, we can regroup the even and odd terms of U_1 and V_1, which leads, respectively, to

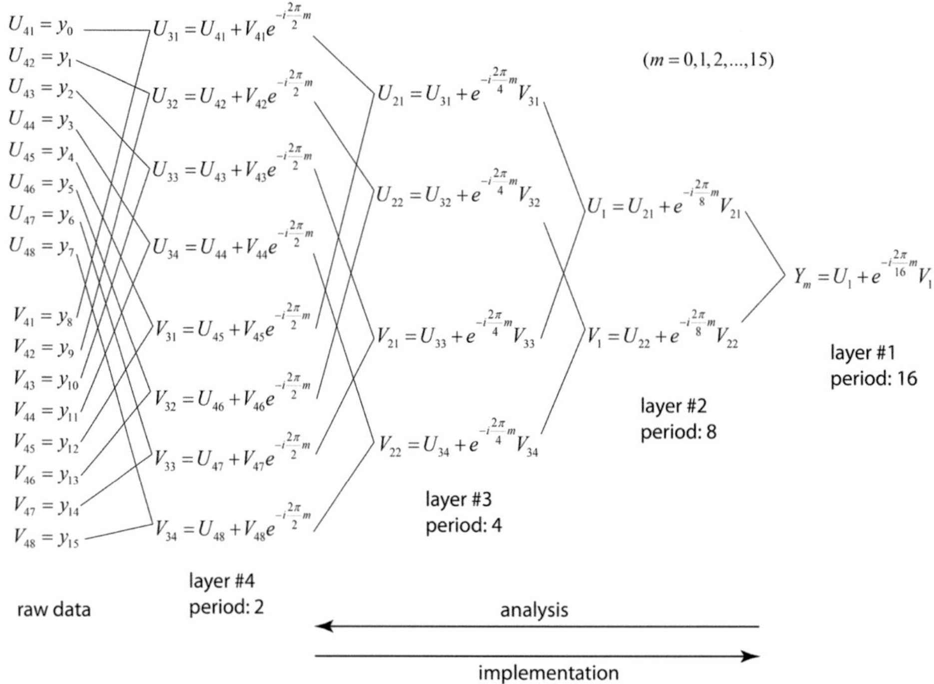

Figure 14.4 A diagram showing the repeated bifurcation of time series with a length of 16 for a fast Fourier Transform, reducing the number of multiplications.

$$\begin{aligned} U_1 = &\left(y_0 + y_4 e^{-i\frac{2\pi}{4}m} + y_8 e^{-i\frac{2\pi}{4}2m} + y_{12} e^{-i\frac{2\pi}{4}3m}\right) \\ &+ e^{-i\frac{2\pi}{8}m}\left(y_2 + y_6 e^{-i\frac{2\pi}{4}m} + y_{10} e^{-i\frac{2\pi}{4}2m} + y_{14} e^{-i\frac{2\pi}{4}3m}\right), \end{aligned} \tag{14.53}$$

$$\begin{aligned} V_1 = &\left(y_1 + y_5 e^{-i\frac{2\pi}{4}m} + y_9 e^{-i\frac{2\pi}{4}2m} + y_{13} e^{-i\frac{2\pi}{4}3m}\right) \\ &+ e^{-i\frac{2\pi}{8}m}\left(y_3 + y_7 e^{-i\frac{2\pi}{4}m} + y_{11} e^{-i\frac{2\pi}{4}2m} + y_{15} e^{-\frac{2\pi}{4}3m}\right). \end{aligned} \tag{14.54}$$

Now we can write

$$U_1 = U_{21} + e^{-i\frac{2\pi}{8}m} V_{21}, \tag{14.55}$$

$$V_1 = U_{22} + e^{-i\frac{2\pi}{8}m} V_{22}, \tag{14.56}$$

in which

$$U_{21} = \underline{\underline{y_0}} + \underline{y_4 e^{-i\frac{2\pi}{4}m}} + \underline{\underline{y_8 e^{-i\frac{2\pi}{4}2m}}} + \underline{y_{12} e^{-i\frac{2\pi}{4}3m}}, \tag{14.57}$$

$$U_{22} = \underline{\underline{y_1}} + \underline{y_5 e^{-i\frac{2\pi}{4}m}} + \underline{\underline{y_9 e^{-i\frac{2\pi}{4}2m}}} + \underline{y_{13} e^{-i\frac{2\pi}{4}3m}}, \tag{14.58}$$

$$V_{21} = \underline{\underline{y_2}} + \underline{y_6 e^{-i\frac{2\pi}{4}m}} + \underline{\underline{y_{10} e^{-i\frac{2\pi}{4}2m}}} + \underline{y_{14} e^{-i\frac{2\pi}{4}3m}}, \tag{14.59}$$

$$V_{22} = \underline{\underline{y_3}} + \underline{y_7 e^{-\frac{2\pi}{4}m}} + \underline{\underline{y_{11} e^{-i\frac{2\pi}{4}2m}}} + \underline{y_{15} e^{-i\frac{2\pi}{4}3m}}. \tag{14.60}$$

It can be seen that U_{21}, V_{21}, U_{22}, and V_{22} all have a period of 4. Obviously, equation (14.55) has eight nonduplicating multiplications with V_{21} for $m = 0$ to $m = 7$. For $m = 8$ to $m = 15$, the multiplications repeat those for $m = 0$ to $m = 7$, because V_{21} has only half of the period of $e^{-i2\pi\frac{m}{8}}$. In other words, the first equation of the Layer 2 computations, equation (14.55), has $M/2 = 8$ nonduplicating multiplications. The second equation of Layer 2 computations, equation (14.56), also has $M/2 = 8$ nonduplicating multiplications. Therefore, Layer 2 has a total of M nonduplicating multiplications.

Now, we further regroup the even and odd terms in the same fashion and have

$$U_{21} = y_0 + y_8 e^{-i\frac{2\pi}{2}m} + e^{-i\frac{2\pi}{4}m}\left(y_4 + y_{12} e^{-i\frac{2\pi}{2}m}\right), \tag{14.61}$$

$$U_{22} = y_1 + y_9 e^{-i\frac{2\pi}{2}m} + e^{-i\frac{2\pi}{4}m}\left(y_5 + y_{13} e^{-i\frac{2\pi}{2}m}\right), \tag{14.62}$$

$$V_{21} = y_2 + y_{10} e^{-i\frac{2\pi}{2}m} + e^{-i\frac{2\pi}{4}m}\left(y_6 + y_{14} e^{-i\frac{2\pi}{2}m}\right), \tag{14.63}$$

$$V_{22} = y_3 + y_{11} e^{-i\frac{2\pi}{2}m} + e^{-i\frac{2\pi}{4}m}\left(y_7 + y_{15} e^{-i\frac{2\pi}{2}m}\right). \tag{14.64}$$

The Layer 3 equations are

$$U_{21} = U_{31} + e^{-i\frac{2\pi}{4}m} V_{31}, \tag{14.65}$$

$$U_{22} = U_{32} + e^{-i\frac{2\pi}{4}m} V_{32}, \tag{14.66}$$

$$V_{21} = U_{33} + e^{-i\frac{2\pi}{4}m} V_{33}, \tag{14.67}$$

$$V_{22} = U_{34} + e^{-i\frac{2\pi}{4}m} V_{34}, \tag{14.68}$$

in which

$$U_{31} = y_0 + y_8 e^{-i\frac{2\pi}{2}m}, \tag{14.69}$$

$$U_{32} = y_1 + y_9 e^{-i\frac{2\pi}{2}m}, \tag{14.70}$$

$$U_{33} = y_2 + y_{10} e^{-i\frac{2\pi}{2}m}, \tag{14.71}$$

$$U_{34} = y_3 + y_{11} e^{-i\frac{2\pi}{2}m}, \tag{14.72}$$

$$V_{31} = y_4 + y_{12} e^{-i\frac{2\pi}{2}m}, \tag{14.73}$$

$$V_{32} = y_5 + y_{13}e^{-i\frac{2\pi}{2}m}, \tag{14.74}$$

$$V_{33} = y_6 + y_{14}e^{-i\frac{2\pi}{2}m}, \tag{14.75}$$

$$V_{34} = y_7 + y_{15}e^{-i\frac{2\pi}{2}m}. \tag{14.76}$$

Obviously, U_{21}, U_{22}, V_{21}, and V_{22} have a period of 4, while the Layer 3 series U_{31}, U_{32}, U_{33}, U_{34}, V_{31}, V_{32}, V_{33}, and V_{34} has a period of 2. Therefore, each of the four Layer 3 equations has $M/4 = 4$ nonduplicating multiplications. The total nonduplicating multiplications for Layer 3 is thus again $M = 16$. Now the Layer 4 equations can be shown:

$$U_{31} = U_{41} + V_{41}e^{-i\frac{2\pi}{2}m}, \tag{14.77}$$

$$U_{32} = U_{42} + V_{42}e^{-i\frac{2\pi}{2}m}, \tag{14.78}$$

$$U_{33} = U_{43} + V_{43}e^{-i\frac{2\pi}{2}m}, \tag{14.79}$$

$$U_{34} = U_{44} + V_{44}e^{-i\frac{2\pi}{2}m}, \tag{14.80}$$

$$V_{31} = U_{45} + V_{45}e^{-i\frac{2\pi}{2}m}, \tag{14.81}$$

$$V_{32} = U_{46} + V_{46}e^{-i\frac{2\pi}{2}m}, \tag{14.82}$$

$$V_{33} = U_{47} + V_{47}e^{-i\frac{2\pi}{2}m}, \tag{14.83}$$

$$V_{34} = U_{48} + V_{48}e^{-i\frac{2\pi}{2}m}, \tag{14.84}$$

in which

$$\begin{array}{ll} U_{41} = y_0 & V_{41} = y_8 \\ U_{42} = y_1 & V_{42} = y_9 \\ U_{43} = y_2 & V_{43} = y_{10} \\ U_{44} = y_3 & V_{44} = y_{11} \\ U_{45} = y_4 & V_{45} = y_{12} \\ U_{46} = y_5 & V_{46} = y_{13} \\ U_{47} = y_6 & V_{47} = y_{14} \\ U_{48} = y_7 & V_{48} = y_{15} \end{array}. \tag{14.85}$$

We have eight equations relating to the Layer 4 series. Layer 4 turns out to be the last layer for this example. The Layer 4 variables are the raw data (Fig. 14.4), as shown above in (14.85). Since the Layer 4 series is now all constants, the Layer 4 equations (14.77)–(14.84) have a total of $(M/8) \times 8 = M = 16$ nonduplicating multiplications. As a result, the four layers of computations have a total of $4M$ nonduplicating multiplications, using this repeated bifurcation method.

The above argument can be extended to arbitrary N bifurcations with $M = 2^N$, for which there would be a total of

$$P = NM = M\log_2 M \tag{14.86}$$

non-duplicating multiplications. This is discussed in more detail below.

14.2.4 The Fast Fourier Transform

Now we examine a general situation in which

$$M = 2^N. \tag{14.87}$$

The base functions are

$$B_M^{km} = e^{-i2\pi km/M}. \tag{14.88}$$

Keep in mind that $k, m = 0, 1, 2, ..., M - 1$. Therefore, the discrete Fourier Transform is expressed as

$$Y_m = \sum_{k=0}^{M-1} y_k B_M^{km}. \tag{14.89}$$

The base functions are functions of k, m, and M. When M is reduced by half, it is equivalent to either k or m being increased by 2:

$$B_{M/2}^{km} = e^{-i2\pi k2m/M} = B_M^{2km}. \tag{14.90}$$

Assume that M is an even number, so we can separate the terms in (14.13) into two groups: one corresponding to the odd terms and one to the even terms:

$$Y_m = \sum_{k=0}^{M/2-1} y_{2k} B_M^{2km} + \sum_{k=0}^{M/2-1} y_{2k+1} B_M^{(2k+1)m} \tag{14.91}$$

or

$$Y_m = \sum_{k=0}^{M/2-1} y_{2k} B_M^{2km} + B_M^m \sum_{k=0}^{M/2-1} y_{2k+1} B_M^{2km}. \tag{14.92}$$

Now, if we apply the property of (14.90), the above equation is

$$Y_m = \sum_{k=0}^{M/2-1} y_{2k} B_{M/2}^{km} + B_M^m \sum_{k=0}^{M/2-1} y_{2k+1} B_{M/2}^{km} \tag{14.93}$$

Here again, $m = 0, 1, 2, \ldots, M - 1$.

The second complex number will multiply the additional factor B_M^m:

$$Y_m = a + B_M^m b, \tag{14.94}$$

in which

$$a = \sum_{k=0}^{M/2-1} y_{2k} B_{M/2}^{km}, \quad b = \sum_{k=0}^{M/2-1} y_{2k+1} B_{M/2}^{km}. \tag{14.95}$$

This says that now the original discrete Fourier Transform has become a combination of two separate discrete Fourier Transforms, both only half the size (i.e. $M/2$). Both of these discrete Fourier Transforms are periodic with a period of $M/2$. The factor outside of the second sum, i.e. B_M^m, however, has a period of M, and thus (14.93) has M multiplications once the two sums are obtained. The total multiplications would be M plus the number of multiplications in the two sums.

Since $M = 2^N$, $M/2$ is still an even number; both a and b can be separated into two discrete Fourier Transform of a size half that of $M/2$, or $M/4$. Each of these new discrete Fourier Transforms has $M/2$ nonduplicating multiplications. So, a total of $M/2 + M/2 = M$ multiplications are needed. It can be seen that this process can be done N times. Each time the discrete Fourier Transform is divided into two half discrete Fourier Transforms, all the smaller discrete Fourier Transforms will have a combined M multiplications. This division is done until it cannot be separated into anything smaller (only one term is left for each "smaller" discrete Fourier Transform, just like in the example with $M = 16$ presented earlier). The total number of multiplications is thus N times M, or

$$P = NM = M\log_2 M. \tag{14.96}$$

Compared to the original M^2 multiplications, this is a significant improvement. The number of multiplications is only

$$\alpha = \frac{\log_2 M}{M} \tag{14.97}$$

of the original effort. This can be significant when M is large. For instance, if $M = 100$, $\alpha = 6.6\%$; if $M = 1000$, $\alpha = 1\%$. MATLAB's implementation of the fast Fourier Transform is a function named `fft`. The following is an example of how to use the function: assume that a function $y(t)$ is defined as a truncated sine with a given frequency (here it has a period of 0.1 hours):

```
clear
t = 0:0.01:10;      % DEFINE TIME (UNIT OF TIME CAN BE DAYS)
f = 24 / 0.1;       % DEFINE A FREQUENCY TO SIMULATE A TIME SERIES.
                    % THE UNIT OF FREQUENCY IS CYCLE PER DAY IF 0.1
                    % IN THE DENOMINATOR MEANS 0.1 HOUR
```

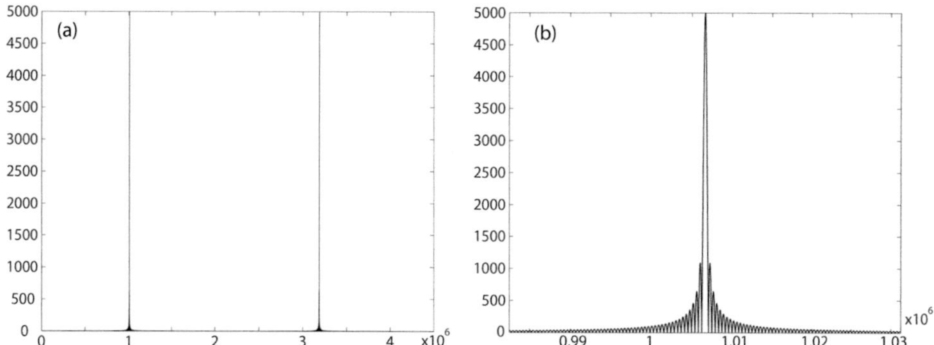

Figure 14.5 Magnitude spectrum of a sine function: (a) the spectrum showing the two peaks; (b) zoomed-in view of the left peak. The ringing (the oscillations away from the main peak) is caused by the truncation of the sine function – an effect further discussed in Chapter 17.

```
y = sin(2 * pi * f * t); % A SINE FUNCTION IN TIME TO BE FFTED
figure
plot(t,y)            % PLOT THE FUNCTION
N = length(t);       % THE LENGTH OF THE TIME SERIES
y1 = [ y zeros(1,N * 10)] ;    % ZERO-PADDING AT THE END OF THE DATA
                               % THE LENGTH OF ZEROS IS 10 TIMES OF
                               % THE ORIGINAL DATA
M = 2^nextpow2(length(y1))     % FIND THE NEAREST NUMBER POWER OF 2
                               % THAT IS ALSO GREATER THAN N
Y = fft(y1,M);                 % DO AN FFT WITH A LENGTH OF POWER OF 2
                               % I.E. THE LENGTH = 2^M
figure
plot(abs(Y))                   % PLOT THE RESULT IN FREQUENCY DOMAIN
```

Fig. 14.5 demonstrates the result of this example. The two peaks represent the single frequency: remember that when the sinusoidal base functions of the Fourier series are converted to the complex exponential function using the Euler formula (Chapter 13), each frequency is redistributed to a positive and a negative frequency and thus two peaks. The two peaks must have the same magnitude (equation 13.13). In discrete Fourier Transform, however, the negative frequency can be consider being flipped to the right of the Nyquist Frequency. To express the result in a negative frequency region, i.e. including $-M/2$ to 0, we can use the MATLAB function `fftshift`.

The MATLAB `fft` function is quite flexible, and the length of the time series doing the FFT does not have to be a power of 2. The drawback is that the speed of computation is slower. With modern computers (even laptops), however, computation speed is usually not of primary concern.

In the rectangular window example in Section 14.1.2, using $MM = 2000$ and $M = 24000$, we test the time it takes to do the discrete Fourier Transform running the following:

```
tic; MM = 2^13;
y = [ zeros(300,1); ones(300,1); zeros(MM-600,1)];
M = length(y); n = (0:M-1)'; m = n;
Y = y' * exp(-i * 2 * pi * m * n' / M); toc
```

It took the computer 2.530307 seconds to finish. If we use `fft` to replace the last line:

```
tic; MM = 2^13;
y = [ zeros(300,1); ones(300,1); zeros(MM-600,1)];
M = length(y); n = (0:M-1)'; m = n;
Y = fft(y); toc
```

the computation only takes 0.008483 seconds, about 0.34% of the first method.

The computation is obviously dependent on the actual computer. In another run using a different computer, the above example gave a time for the discrete Fourier Transform of 2.728188 seconds, while the FFT gave a time of 0.007686 seconds, with a ratio of 0.28%.

There are different implementations of the fast Fourier Transform algorithm. MATLAB's implementation is quite robust. As discussed above, if M is equal to 2^N, the computation of the discrete Fourier Transform can be much faster using the repeated division of the series in two halves, each as a shorter discrete Fourier Transform. If M is not a power of 2, repeated divisions of the series into shorter ones are still possible, but the efficiency may not be as high. MATLAB is robust in the sense that it can allow essentially any length for the time series to do the discrete Fourier Transform using the `fft` function. In fact, the function `fft` allows an arbitrary positive integer, rather than only powers of 2, to specify the length of the series for `fft`:

```
Y = fft(y,N) % N DOES NOT HAVE TO BE A POWER OF 2
```

If the time series $y(t)$ is longer than N, it is truncated to length N; and if $y(t)$ is shorter than N, zeros are padded at the end of the series to make the length equal to N before doing the `fft`. To demonstrate and test the efficiency change and the fact that when N is a power of 2 the computation has the highest efficiency, we can use the following script:

```
y = [ zeros(100,1); ones(100,1); zeros(100,1)];
NN = 33000;
for i = 300: NN
t1 = tic;
fft(y,i);
t(i-299) = toc(t1);
end
```

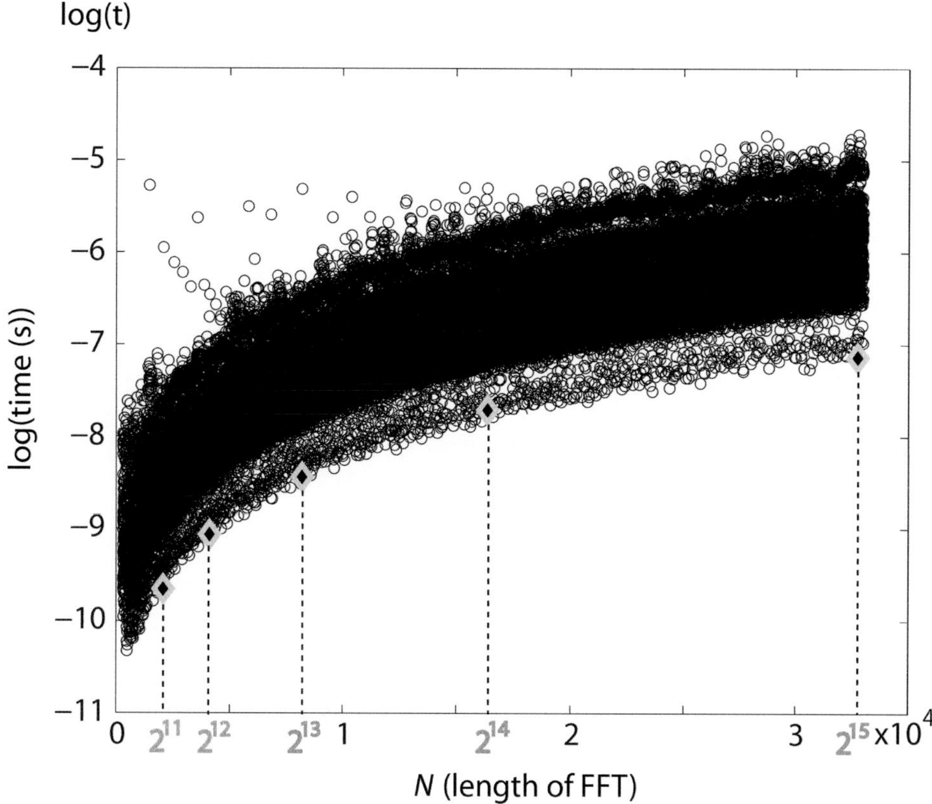

Figure 14.6 Efficiency check of fast Fourier Transform using MATLAB function `fft`.

Fig. 14.6 shows the result from the above calculation. The diamonds show the time of calculation when the length of the series is of power of 2. Note that the computation time by the computer depends on the computer and also the background computation running, which can vary over time. So, there is certain amount of uncertainty here, i.e. Fig. 14.6 may not be exactly the same each time the same MATLAB script is run. However, it clearly demonstrates that the calculation is most efficient when the length of the series is a power of 2.

14.2.5 MATLAB fftshift Function

The MATLAB function `fftshift` is a function that flips the high frequency half (the right half) of the FFT to the negative frequency (or the left side). Let us look at a simple example with a frequency at $f = 12$:

$$y(t) = \cos(2\pi f t), \quad f = 12, \quad t = [0, 5], \quad dt = 0.001.$$

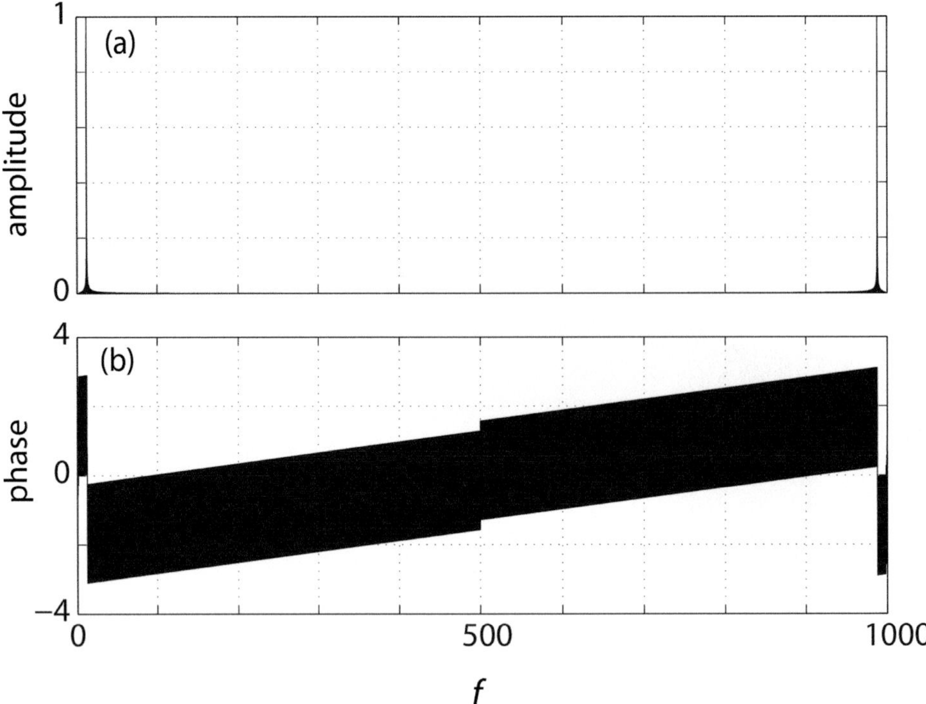

Figure 14.7 Upper panel: magnitude spectrum; lower panel: phase spectrum.

The phase at $f = 12$ should be 0. The phase at other frequencies is meaningless because there is no signal in those frequencies. Fig. 14.7 shows the original FFT result for the amplitude spectrum and phase spectrum, respectively. Fig. 14.8 shows the modified FFT result for the amplitude (or magnitude) spectrum and phase spectrum, respectively, by applying the `fftshift` function. Figs. 14.9 and 14.10 show zoomed-in views of the spectra. The zero phase can be seen at $f = 12$. Phase values at other frequencies do not have any significance. The following script is used to produce the spectrum prior to making Figs. 14.7 and 14.8.

```
dt = 0.001;  % DEFINE TIME INTERVAL
T = 5;       % DEFINE TOTAL LENGTH OF TIME
t = 0:dt:T;  % DEFINE TIME VARIABLE
N = length(t);  % LENGTH OF TIME SERIES
y = cos(2 * pi * 12 * t);  % DEFINE THE SIGNAL
Y = fft([ y zeros(1,length(y) * 10)]);   % ZERO-PADDING % DO FFT
M = length(Y);   % LENGTH OF ZERO-PADDED SERIES
amp = abs(Y);    % AMPLITUDE SPECTRUM
Y1 = fftshift(fft([ y zeros(1,length(y) * 10)])); % USING fftshift
amp1 = abs(Y1);  % AMPLITUDE SPECTRUM OF SHIFTED SPECTRUM
phase = unwrap(angle(Y));  % USING THE MATLAB UNWRAP FUNCTION
phase1 = unwrap(angle(Y1)); % UNWRAP THE PHASE
f = 1 / (M * dt) * (0: M-1); % DEFINE THE FREQUENCY
```

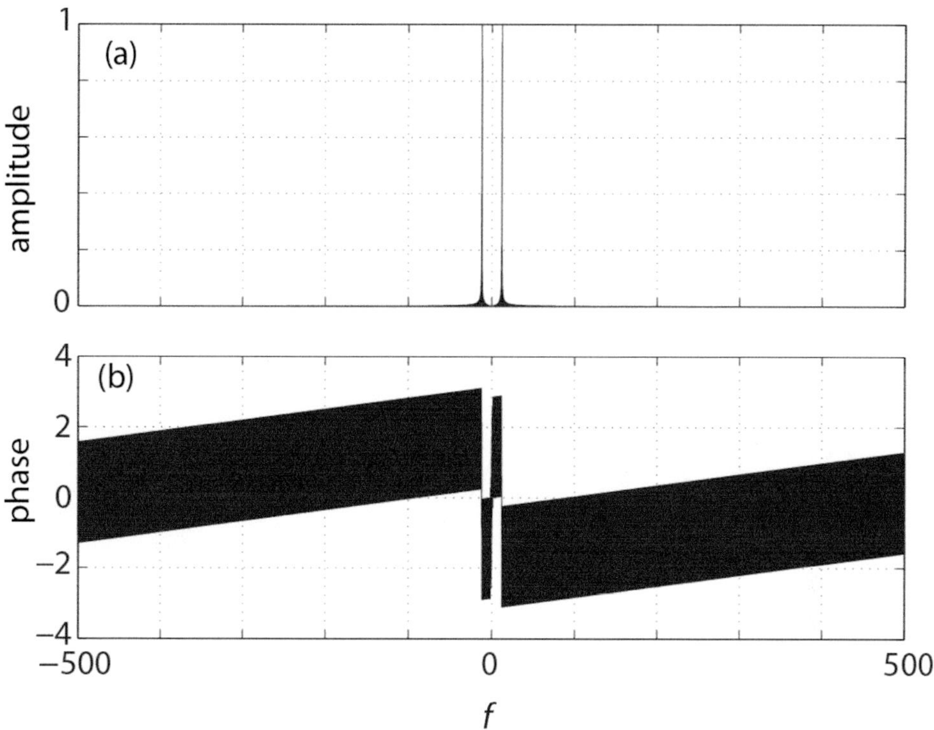

Figure 14.8 Demonstration of MATLAB function `fftshift`. Upper panel: magnitude of spectrum after the shift; lower panel: phase spectrum after the shift.

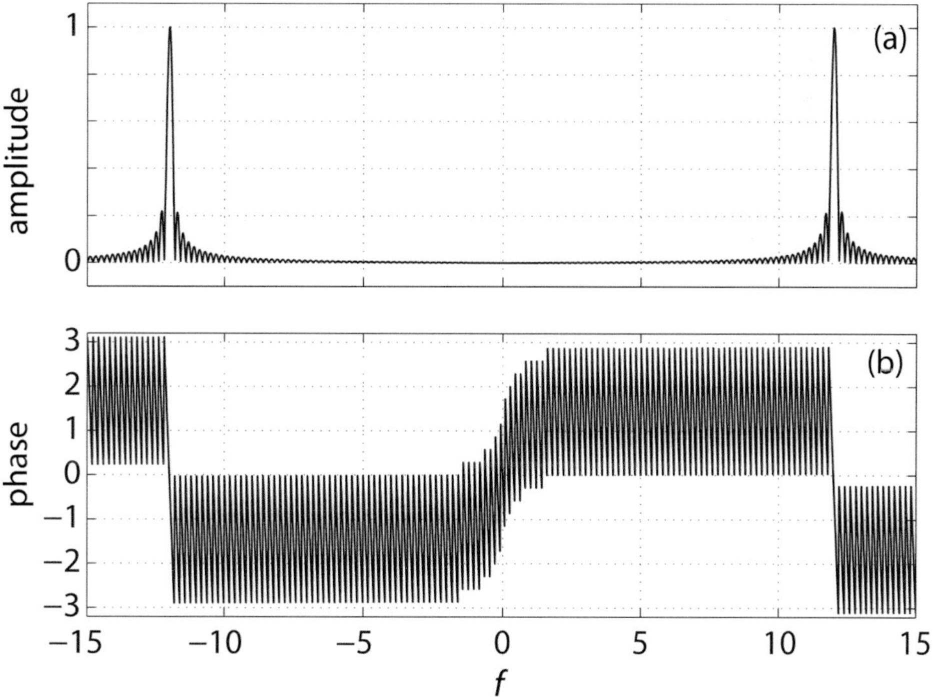

Figure 14.9 Demonstration of MATLAB function `fftshift` with a zoomed-in view. Upper panel: magnitude of `fftshift(fft(y))`; lower panel: phase of `fftshift(fft(y))`.

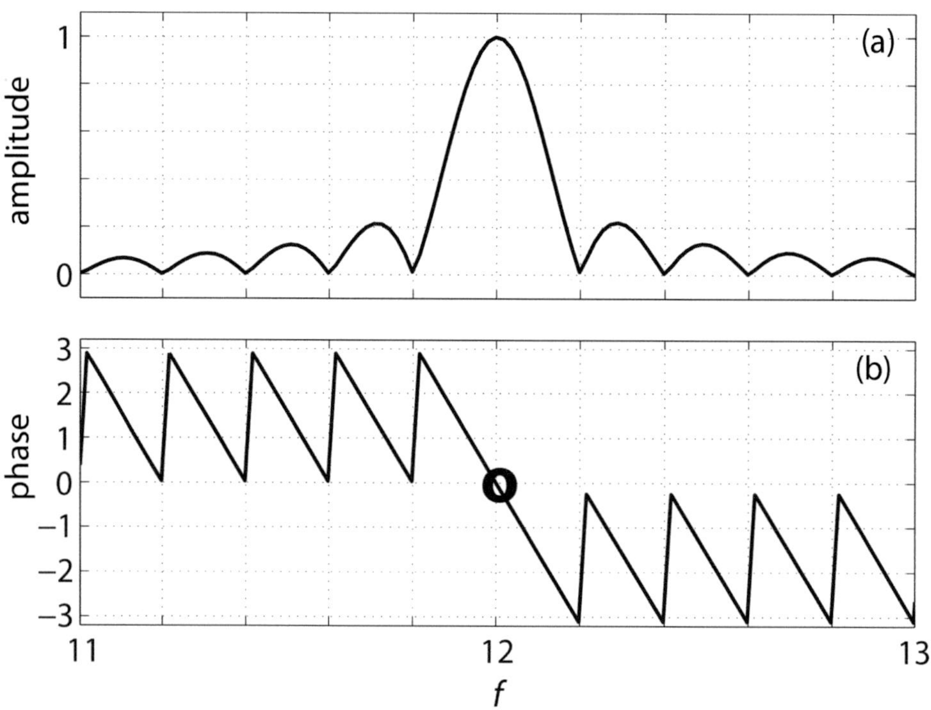

Figure 14.10 Demonstration of MATLAB function `fftshift` – further zoomed-in view. Upper panel: magnitude of `fftshift(fft(y))`; lower panel: phase of `fftshift(fft(y))`.

Review Questions for Chapter 14

(1) Is fast Fourier Transform a new method independent of Fourier Transform?

(2) What is the effect of zero-padding and why it provides better resolution in the frequency domain?

Exercises for Chapter 14

(1) In MATLAB, define a sine function $y(t)$ with a period T and a total length of 10 periods, e.g.

```
t = 0:0.01:100
```

assuming the period is $T = 10$. Define the sine function $y(t)$ in MATLAB. Use MATLAB command

```
Y = fft(y)
```

to do the fast Fourier Transform. Plot the absolute value of Y, i.e. using

```
plot(abs(Y))
```

Discuss what you see [Hint: zoom in on the figure at the left end of the plot].

(2) Following the above work, add 5 times the number of zeros after the function y, i.e. using

```
y = [ y zeros(1,5 * N)] .
```

Now use the MATLAB command

```
Y = fft(y).
```

to do the fast Fourier Transform. Plot the absolute value of Y, i.e. using

```
plot(abs(Y)).
```

Discuss what you see [hint: zoom in on the figure at the left end of the plot].

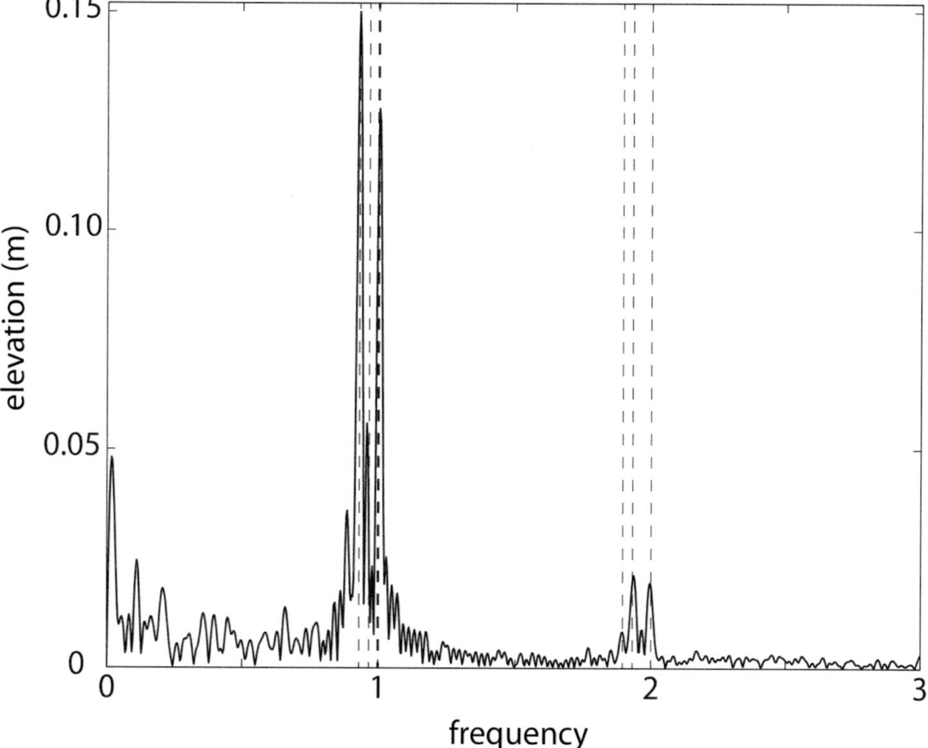

Figure 14.E1 Spectrum plot for Exercise 3.

(3) Work out a spectrum for water elevation.

(a) Load the provided data file named "March-April-2020-depth.mat" into MATLAB. There are two variables: one is water depth anomaly (water elevation in reference to the mean water level) in meters and one is time in days in the year 2020.

(b) Check if the time intervals are constant; if not, an interpolation is needed. Define time interval.

(c) Zero-pad the water depth anomaly by 5 times as many zeros.

(d) Using MATLAB function `fft` to do Fourier Transform for the water depth anomaly.

(e) Determine the frequency array.

(f) Plot the magnitude spectrum of the water depth anomaly as a function of frequency.

(g) Consult Chapter 11 and mark the following tidal frequencies as dashed vertical lines: M2, S2, N2, K1, O1, S1, M1, and P1 tides. The result should look like Fig. 14.E1.

15

Properties of Fourier Transform

About Chapter 15

The objective of this chapter is to discuss the major properties of Fourier Transform for preparation of the subsequent chapters. Fourier Transform has some important properties. These properties are very useful for understanding much of the remainder of this book, such as filtering theory and its applications, the effect of finite sampling, and window functions. We discuss the major properties, but some have more important applications than others, depending on the question at hand. Their usage, if applicable, will become clear in later chapters. This chapter is theoretical and does not have MATLAB implementation, which is deferred to later chapters where relevant.

15.1 Additive Property

Assume that there are two time-functions $y(t)$ and $z(t)$, in which t is time. Their Fourier Transforms are, respectively,

$$Y(\omega) = \int_{-\infty}^{\infty} y(t)e^{-i\omega t}dt, \tag{15.1}$$

$$Z(\omega) = \int_{-\infty}^{\infty} z(t)e^{-i\omega t}dt, \tag{15.2}$$

in which ω is angular frequency ($2\pi f$). The Fourier Transform for the sum of the functions is then the two transforms added together, i.e. if

$$g(t) = y(t) + z(t), \tag{15.3}$$

then

$$G(\omega) = \int_{-\infty}^{\infty} g(t)e^{-i\omega t}dt = Y(\omega) + Z(\omega). \tag{15.4}$$

This is the *additive property of Fourier Transform* and is obvious because integration is a linear operation.

In the above equations, we use the angular frequency ω. We may also use frequency f, in which case equations (15.1), (15.2), and (15.4) can be expressed in terms of f:

$$Y(f) = \int_{-\infty}^{\infty} y(t)e^{-i2\pi ft}dt, \quad Z(f) = \int_{-\infty}^{\infty} z(t)e^{-i2\pi ft}dt, \tag{15.5}$$

$$G(f) = \int_{-\infty}^{\infty} g(t)e^{-i2\pi ft}dt = Y(f) + Z(f). \tag{15.6}$$

Sometimes, we use a symbol $\mathbb{F}$ to represent the mathematical operation of Fourier Transform:

$$\mathbb{F}(y(t)) = Y(f) \quad \text{or} \quad \mathbb{F}(y(t)) = Y(\omega). \tag{15.7}$$

15.2 Symmetric Properties

It is not hard to find that the Fourier Transform and its inverse transform have some kind of *symmetric properties*. This can be seen by switching t with ω. The inverse transform of (15.1) in terms of the angular frequency ω is

$$y(t) = \frac{1}{2\pi}\int_{-\infty}^{\infty} Y(\omega)e^{i\omega t}d\omega. \tag{15.8}$$

Alternatively, in terms of f:

$$y(t) = \frac{1}{2\pi}\int_{-\infty}^{\infty} Y(f)e^{i2\pi ft}df. \tag{15.9}$$

If we use $\mathbb{F}^{-1}$ to denote the mathematical operator of an inverse Fourier Transform, then

$$\mathbb{F}^{-1}(Y(f)) = y(t) \text{ or } \mathbb{F}^{-1}(Y(\omega)) = y(t) \tag{15.10}$$

We sometimes use a two-way arrow to denote this relationship: a Fourier Transform and inverse transform pair:

$$y(t) \Leftrightarrow Y(\omega) \quad \text{or} \quad y(t) \Leftrightarrow Y(f) \tag{15.11}$$

which simply means that $Y(\omega)$ or $Y(f)$ is a Fourier Transform of $y(t)$, which is an inverse Fourier Transform of $Y(\omega)$ or $Y(f)$. We sometimes say that $Y(\omega)$ or $Y(f)$ is an image of $y(t)$. We can also say the above relationship is a *two-way mapping* between $y(t)$ and $Y(\omega)$ or $Y(f)$. This is just one approach to express the *symmetric property of Fourier Transform.*

When ω and t are switched in position, equations (15.1) and (15.8) should still hold, because both ω and t are real numbers ranging from negative infinity to positive infinity. So we have

$$y(\omega) = \frac{1}{2\pi} \int_{-\infty}^{\infty} Y(t) e^{i\omega t} dt, \tag{15.12}$$

$$Y(t) = \int_{-\infty}^{\infty} y(\omega) e^{-i\omega t} d\omega. \tag{15.13}$$

Now let us flip the axis for ω, i.e. using $-\omega$ to replace ω, we have

$$y_1(\omega) = y(-\omega) = \int_{-\infty}^{\infty} \frac{1}{2\pi} Y(t) e^{-i\omega t} dt, \tag{15.14}$$

$$\frac{1}{2\pi} Y(t) = \frac{1}{2\pi} \int_{-\infty}^{\infty} y(-\omega) e^{i\omega t} d\omega = \frac{1}{2\pi} \int_{-\infty}^{\infty} y_1(\omega) e^{i\omega t} d\omega. \tag{15.15}$$

The above two equations mean that the Fourier Transform of function $\frac{1}{2\pi} Y(t)$ is $y(-\omega)$, or

$$Y(t) \Leftrightarrow 2\pi y(-\omega). \tag{15.16}$$

This is another way of expressing the symmetric property of Fourier Transform.

15.3 Similarity Property

If the time function $y(t)$ has its Fourier Transform $Y(\omega)$ or $y(t) \Leftrightarrow Y(\omega)$, and c is a constant, then the Fourier Transform of $y(ct)$ is $\frac{1}{|c|} Y(\frac{\omega}{c})$, or

$$y(ct) \Leftrightarrow \frac{1}{|c|} Y\left(\frac{\omega}{c}\right). \tag{15.17}$$

This is the *similarity property of Fourier Transform.* This means that if the time scale is amplified by a factor of c, the frequency scale as well as the amplitudes at each frequency are reduced by the same factor.

This can be verified by the following:

$$y(ct) = \frac{1}{2\pi} \int_{-\infty}^{\infty} Y(\omega) e^{i\omega ct} d\omega. \tag{15.18}$$

Let

$$\omega c = \omega', \tag{15.19}$$

then

$$\omega = \frac{\omega'}{c}, \tag{15.20}$$

and if $c > 0$, then

$$y(ct) = \frac{1}{2\pi} \int_{-\infty}^{\infty} \frac{1}{c} Y\left(\frac{\omega'}{c}\right) e^{i\omega' t} d\omega' = \frac{1}{2\pi} \int_{-\infty}^{\infty} \frac{1}{|c|} Y\left(\frac{\omega'}{c}\right) e^{i\omega' t} d\omega' \tag{15.21}$$

If $c < 0$,

$$y(ct) = \frac{-1}{2\pi} \int_{-\infty}^{\infty} \frac{1}{c} Y\left(\frac{\omega'}{c}\right) e^{i\omega' t} d\omega' = \frac{1}{2\pi} \int_{-\infty}^{\infty} \frac{1}{|c|} Y\left(\frac{\omega'}{c}\right) e^{i\omega' t} d\omega'. \tag{15.22}$$

On the other hand, we can see from (15.1) that

$$Y\left(\frac{\omega}{c}\right) = \int_{-\infty}^{\infty} y(t) e^{-i\frac{\omega}{c}t} dt. \tag{15.23}$$

Let

$$t' = \frac{t}{c}. \tag{15.24}$$

(15.23) becomes

$$Y\left(\frac{\omega}{c}\right) = |c| \int_{-\infty}^{\infty} y(ct') e^{-i\omega t'} dt'. \tag{15.25}$$

Therefore, there is the Fourier Transform pair denoted by (15.17). This similarity property can help in understanding the relationship between the time domain function and its Fourier Transform counterpart in the frequency domain. For example, when a given signal is stretched in time, the spectrum in the frequency domain will have lower curves and narrower peaks. This can be useful for understanding the Wavelet analysis discussed in a later chapter.

15.4 Time Delay Property

When there is a shift in time, the function's Fourier Transform results in an extra factor:

$$y(t-t') \Leftrightarrow Y(\omega)e^{-i\omega t'}. \tag{15.26}$$

This is called the *time delay property of Fourier Transform.* To prove this, let us look at the Fourier Transform of $y(t-t')$:

$$\mathbb{F}(y(t-t')) = \int_{-\infty}^{\infty} y(t-t')e^{-i\omega t}dt. \tag{15.27}$$

Let $t'' = t - t'$:

$$\mathbb{F}(y(t-t')) = \int_{-\infty}^{\infty} y(t'')e^{-i\omega(t''+t')}dt'' = y(t)e^{-i\omega t'}\int_{-\infty}^{\infty} y(t'')e^{-i\omega t''}dt'' = e^{-i\omega t'}Y(\omega). \tag{15.28}$$

Thus, (15.26) is verified. This property is useful when we use the discrete form of filters later. This theorem is also called a time shifting theorem.

15.5 Frequency Delay Property

When there is a shift in frequency, the inverse Fourier Transform results in an extra factor :

$$y(t)e^{i\omega' t} \Leftrightarrow Y(\omega-\omega'). \tag{15.29}$$

This gives the *frequency delay property for the Fourier Transform.*

This can be verified by the Fourier Transform of $y(t)e^{i\omega' t}$:

$$\mathbb{F}\left(y(t)e^{i\omega' t}\right) = \int_{-\infty}^{\infty} y(t)e^{i\omega' t}e^{-i\omega t}dt = \int_{-\infty}^{\infty} y(t)e^{-i(\omega-\omega')t}dt = Y(\omega-\omega'). \tag{15.30}$$

This is also called a frequency shifting theorem.

15.6 Time Derivative Property

The Fourier Transform of a derivative of a function in time is equal to the Fourier Transform of the original function multiplied by a factor $i\omega$:

$$\frac{dy}{dt} \Leftrightarrow i\omega Y(\omega). \tag{15.31}$$

Therefore,

$$\frac{d^n y}{dt^n} \Leftrightarrow (i\omega)^n Y(\omega), n = 1, 2, ... \tag{15.32}$$

The first equation, (15.31), can be verified by a differentiation of equation (15.5). Assuming that we can switch the order of integration and derivative, we then have:

$$\frac{dy}{dt} = \frac{d}{dt}\left(\frac{1}{2\pi}\int_{-\infty}^{+\infty} Y(\omega)e^{i\omega t}d\omega\right) = \frac{1}{2\pi}\int_{-\infty}^{+\infty} i\omega Y(\omega)e^{i\omega t}d\omega. \tag{15.33}$$

The right-hand side of the above equation is the inverse Fourier Transform of $i\omega Y(\omega)$, which is equivalent to (15.31). If we do the derivative to the above equation $n-1$ more times, we have

$$\frac{d^n y}{dt^n} = \frac{d^n}{dt^n}\left(\frac{1}{2\pi}\int_{-\infty}^{+\infty} Y(\omega)e^{i\omega t}d\omega\right) = \frac{1}{2\pi}\int_{-\infty}^{+\infty} (i\omega)^n Y(\omega)e^{i\omega t}d\omega, \tag{15.34}$$

which is equivalent to (15.32), or the right-hand side of the above equation is the inverse Fourier Transform of $(i\omega)^n Y(\omega)$.

Equation (15.32) is a series of equations and is known as the *time derivative property of Fourier Transform*. This property is most useful for solving problems involving differential equation(s) analytically, but it is included here also for completeness. When solving the frequency distribution in the frequency domain involving a differential equation, this property can simplify the derivative operation in time domain to a simple multiplication in frequency domain, thus significantly simplifies the mathematics. In addition, they can be useful for theoretical development of spectrum methods, either analytically or numerically.

15.7 Frequency Derivative Property

Similarly, we have the *frequency derivative property of Fourier Transform*:

$$-ity \Leftrightarrow \frac{dY(\omega)}{d\omega}, \tag{15.35}$$

$$(-it)^n y \Leftrightarrow \frac{d^n Y(\omega)}{d\omega^n}. \tag{15.36}$$

The first equation, (15.35), can be verified by a differentiation of equation (15.1). Assuming that we can switch the order of integration and derivative, we then have:

$$\frac{dY(\omega)}{d\omega} = \frac{d}{d\omega}\left[\int_{-\infty}^{+\infty} y(t)e^{-i\omega t}dt\right] = \int_{-\infty}^{+\infty} (-it)y(t)e^{-i\omega t}dt. \tag{15.37}$$

The right-hand side of the above equation is the Fourier Transform of $-ity(t)$, which is (15.35). If we further take derivatives with respect to ω of the above equation $n-1$ more times, we have

$$\frac{d^n Y(\omega)}{d\omega^n} = \frac{d^n}{d\omega^n}\left[\int_{-\infty}^{+\infty} y(t)e^{-i\omega t}dt\right] = \int_{-\infty}^{+\infty} (-it)^n y(t)e^{-i\omega t}dt. \tag{15.38}$$

The right-hand side of the above equation is the Fourier Transform of $(-it)^n y(t)$, which is (15.36). The switching of order of integration and differentiation is generally not a problem as far as applications to general oceanographic and environmental time series data are concerned.

15.8 Time Integration Property

Corresponding to the derivative properties, there are also the *time integration properties of the Fourier Transform*. The first is the *time integration property of Fourier Transform*. If a time function $y(t)$ satisfies the following condition:

$$\int_{-\infty}^{+\infty} y(\tau)d\tau = 0, \tag{15.39}$$

then

$$\int_{-\infty}^{t} y(\tau)d\tau \Leftrightarrow \frac{Y(\omega)}{i\omega}. \tag{15.40}$$

When solving an integral-differential equation in a time domain, this property can simplify the operation in the frequency domain to a simple multiplication, thus significantly simplifying the mathematics.

15.9 Frequency Integration Property

In the frequency domain, we have the *frequency integration property of Fourier Transform*, expressed as

$$\frac{y(t)}{-it} \Leftrightarrow \int_{-\infty}^{\omega} Y(\omega)d\omega. \tag{15.41}$$

Similarly, when solving an integral-differential equation in a frequency domain, this property can simplify the operation in the time domain by a multiplication, thereby significantly simplifying the mathematics.

15.10 Time Convolution Property

Convolution is a very useful concept for understanding the theory of filtering. The definition of the convolution of two functions $y(t)$ and $z(t)$ is

$$C(t) \equiv y(t) \otimes z(t) = \int_{-\infty}^{+\infty} y(\tau)z(t-\tau)d\tau. \tag{15.42}$$

Here the mathematical symbol $\otimes$ means convolution.

It can be shown that the convolution of two functions is symmetric, i.e.

$$y(t) \otimes z(t) = z(t) \otimes y(t). \tag{15.43}$$

This can be proven by first letting

$$t - \tau = \tau', \tag{15.44}$$

which gives

$$\tau = t - \tau', \quad d\tau = -d\tau'. \tag{15.45}$$

Substitute these into (15.42):

$$\begin{aligned} y(t) \otimes z(t) &= \int_{-\infty}^{+\infty} y(\tau)z(t-\tau)d\tau = -\int_{+\infty}^{-\infty} y(t-\tau')z(\tau')d\tau' \\ &= \int_{-\infty}^{+\infty} y(t-\tau')z(\tau')d\tau' = z(t) \otimes y(t). \end{aligned} \tag{15.46}$$

This proves (15.43).

A very important theorem is that the Fourier Transform of a convolution of two time-functions is equal to the multiplication of the Fourier Transforms of the individual time-functions, i.e.

$$y(t) \otimes z(t) \Leftrightarrow Y(\omega)Z(\omega). \tag{15.47}$$

This is called the *convolution property of Fourier Transform* or *convolution theorem in the time domain.*

Equation (15.47) says that the Fourier Transform of the convolution of two functions of time is equal to the multiplication of the Fourier Transform of each of the two functions. This theorem can be verified directly from the definitions of Fourier Transform and convolution. More specifically, the Fourier Transform of the convolution, or Fourier Transform of the left-hand side of (15.42), is

$$I = \int_{-\infty}^{+\infty}\int_{-\infty}^{+\infty} y(\tau)z(t-\tau)d\tau e^{-i\omega t}dt = \int_{-\infty}^{+\infty} y(\tau)\left(\int_{-\infty}^{+\infty} z(t-\tau)e^{-i\omega t}dt\right)d\tau. \tag{15.48}$$

Here the right-hand side is obtained by switching the order of integrations of the left-hand side: the integrations over τ and then over t are now changed to integrations over t and then over τ. We then take a factor $e^{-i\omega\tau}$ out of the first integration and add a factor $e^{i\omega\tau}$ in the first integration, which are all independent of t. We then have:

$$I = \int_{-\infty}^{+\infty} y(\tau)\left(\int_{-\infty}^{+\infty} z(t-\tau)e^{-i\omega(t-\tau)}dt\right)e^{-i\omega\tau}d\tau = Z(\omega)\int_{-\infty}^{+\infty} y(\tau)e^{-i\omega\tau}d\tau = Y(\omega)Z(\omega). \tag{15.49}$$

Note that the inner integration of the left-hand side of (15.49) is the Fourier Transform of $z(t)$, i.e. $Z(\omega)$, which can be seen by the fact that:

$$\int_{-\infty}^{+\infty} z(t-\tau)e^{-i\omega(t-\tau)}dt = \int_{-\infty}^{+\infty} z(t-\tau)e^{-i\omega(t-\tau)}d(t-\tau) = Z(\omega). \tag{15.50}$$

15.11 Frequency Convolution Property

With the symmetry of the Fourier Transform and inverse Fourier Transform, there is also the *frequency convolution property of Fourier Transform*: the Fourier Transform of the product of two time-functions is equal to the convolution of the Fourier Transforms of the individual time-functions, i.e.

$$y(t)z(t) \Leftrightarrow Y(f) \otimes Z(f), \tag{15.51}$$

$$y(t)z(t) \Leftrightarrow \frac{1}{2\pi} Y(\omega) \otimes Z(\omega). \tag{15.52}$$

This is also a convolution property of Fourier Transform in the frequency domain, or a *convolution theorem in the frequency domain.*

Equation (15.51) means that the Fourier Transform of the multiplication of two time-functions is equal to the convolution of the Fourier Transform of each of the two functions; or the inverse Fourier Transform of the right-hand side is equal to the left-hand of (15.51). We now verify (15.51). The convolution of $Y(f)$ and $Z(f)$ is

$$Y(f) \otimes Z(f) = \int_{-\infty}^{+\infty} Y(f')Z(f - f')df'. \tag{15.53}$$

The inverse Fourier Transform of (15.53) is

$$\mathbb{F}\Big(Y(f) \otimes Z(f)\Big) = \int_{-\infty}^{+\infty}\int_{-\infty}^{+\infty} Y(f')Z(f - f')df' e^{i2\pi f t} df. \tag{15.54}$$

Switching the order of the integrations, we have

$$\begin{aligned} \mathbb{F}\Big(Y(f) \otimes Z(f)\Big) &= \int_{-\infty}^{+\infty} Y(f')\left(\int_{-\infty}^{+\infty} Z(f - f')e^{i2\pi f t} df\right) df' \\ &= \int_{-\infty}^{+\infty} Y(f')\left(\int_{-\infty}^{+\infty} Z(f - f')e^{i2\pi(f-f')t} df\right) e^{i2\pi f' t} df' \\ &= z(t)\int_{-\infty}^{+\infty} Y(f')e^{i2\pi f' t} df' = y(t)z(t). \end{aligned} \tag{15.55}$$

This theorem is also very useful, e.g. in understanding the effect of finite sampling and the Gibbs Effect.

15.12 Complex Conjugate Property

Taking the complex conjugate on both sides of the Fourier Transform, we get

$$\tilde{y}(t) \Leftrightarrow \tilde{Y}(-\omega), \tag{15.56}$$

$$\tilde{y}(-t) \Leftrightarrow \tilde{Y}(\omega). \tag{15.57}$$

This pair of equations is called the *complex conjugate property of the Fourier Transform*. This can be useful in theoretical development of signal processing, electronics circuits, and other fields in physics.

15.13 Integral Form of Parseval Formula

If the time function $y(t)$ and its squares are integrable in $(-\infty, +\infty)$ for t, there is an integral relation between the integration of the square of the time function $y(t)$ and the integration of the square of the Fourier Transform:

$$\int_{-\infty}^{+\infty} |y(t)|^2 dt = \frac{1}{2\pi} \int_{-\infty}^{+\infty} |Y(\omega)|^2 d\omega. \tag{15.58}$$

If the frequency f is used, an equivalent equation is

$$\int_{-\infty}^{+\infty} |y(t)|^2 dt = \int_{-\infty}^{+\infty} |Y(f)|^2 df. \tag{15.59}$$

Consider that the square of a wave height is proportional to wave energy; the physical meaning of this theorem is that the total energy in a time function equals the total energy partitioned in the frequency domain (energy for each frequency integrated over the entire frequency domain). This theorem is included here, but its verification and further discussion will be given in Chapter 18 when we discuss the Power Spectrum.

Review Questions for Chapter 15

(1) Why are equation (15.8) and (15.9) different by a factor 2π?

(2) Is there any fundamental difference between using the frequency f or angular frequency ω in Fourier Transform?

Exercises for Chapter 15

(1) If a and b are two arbitrary constants, $y(t)$ and $z(t)$ are continuous and integrable time functions. Verify that the Fourier Transform of $ay(t) + bz(t)$ is:

$$\mathbb{F}\left(ay(t) + b(z(t)\right) = a\mathbb{F}\left(y(t)\right) + b\mathbb{F}\left(z(t)\right) = aY(f) + bZ(f).$$

(2) Verify another, more general form of the time-shifting theorem: for real $a \neq 0$ and arbitrary real b, the Fourier Transform of $y(at + b)$ is:

$$\mathbb{F}\big(y(at+b)\big) = \frac{1}{a}e^{\frac{ib\omega}{a}}Y\left(\frac{\omega}{a}\right),$$

in which $Y(\omega)$ is the Fourier Transform of $y(t)$, or $\mathbb{F}(y(t)) = Y(\omega)$.

(3) Verify another, more general form of the frequency-shifting theorem: for real $a > 0$ and arbitrary real b, the Fourier Transform of $e^{ibt}y(at)$ is:

$$\mathbb{F}\left(e^{ibt}y(at)\right) = \frac{1}{a}Y\left(\frac{\omega - b}{a}\right),$$

in which $Y(\omega)$ is the Fourier Transform of $y(t)$, or $\mathbb{F}\big(y(t)\big) = Y(\omega)$.

16

More Discussion on the Harmonic Analysis and Fourier Analysis

About Chapter 16

Harmonic analysis and Fourier analysis are fundamental tools for oceanographic time series data analysis. Both can be derived from the least squares method. They are different, however, in one major respect: Fourier analysis is based on a complete set of base functions, such that the convergence of relevant Fourier series is guaranteed for continuous functions. In contrast, harmonic analysis almost always has non-zero total error squared unless for pure deterministic functions with tidal frequencies only. This chapter provides additional discussion and some examples aimed at a better understanding of the concepts and techniques. The discussion will involve tidal harmonic analysis and Fourier analysis by contrasting them in concept and through some examples for harmonic analysis.

16.1 Comparison between Harmonic and Fourier Analysis

16.1.1 The Scope of Applications

Now that we have discussed harmonic analysis, Fourier series, or Fourier Transform, questions may come up as to which one to use. One reason is that they all use sine and cosine functions or a variant with complex exponential functions as their base functions. So, what makes them different, and under what circumstances would one use one of them but not the other?

Harmonic analysis is more specific for oceanography, while the Fourier series and Fourier Transform are much broadly applicable for signal processing in general, including oceanography. The Fourier series is developed for a periodic function but later extended to nonperiodic functions by allowing the function to be duplicated repeatedly into both positive and negative infinities. The beginning and ending of each period may not have the same value for the time function where there is the Gibbs Effect of the Fourier series. At these discontinuities (assuming at each of the discontinuity points there is only a finite difference of function values), the convergence of the Fourier series is to the average of the two unequal values. With this extension to aperiodic functions, the Fourier series coefficients are obtained very much the same way as for periodic functions – this involves integrations over an entire period.

On the other hand, the Fourier Transform does not need to assume that the function is periodic for an extension (at least in theory). Rather, it allows the extension of the function to an infinite time domain, allowing nonperiodic functions. To do that, the base functions are first converted to exponential forms. As a result, the original function is expressed as a limit of the summation, which becomes an integral involving the complex exponentials. The single integration is over the entire time domain (rather than just one period). The formula is actually a double integration – one over time and one over frequency. The one over time is the Fourier Transform, while the one with an integration over frequency of the first integration is the inverse Fourier Transform. This makes the usage of Fourier Transform more general. As such, we often call the method of using Fourier Transform to examine the signal in the frequency domain the *Fourier analysis.*

16.1.2 The Frequencies

Harmonic analysis for tides or tidal currents is a least squares method by design. The base functions of this analysis are sine and cosine functions with given frequencies related to astronomical factors or relative motions of the Earth, Moon, and Sun. *The harmonic analysis does not require the time to be equally spaced.* As long as there are enough data points and tidal signal is the major component of the time series, the harmonic analysis can be used to resolve the coefficients of the sine and cosine functions with the given tidal frequencies. Since there are many tidal frequencies, the actual selection of which frequencies to use in the analysis may be a little tricky. It is also case dependent: it depends on the length of the data, the nature of the data, and the purpose of the analysis.

It should be noted, however, that in theory, the more frequencies you include in the harmonic analysis, the more accurate your result should be, pertinent to tides. In an idealized world, in which there is no random error and no other processes other than tides (i.e. if tide were a deterministic process), what determines the number of frequencies to choose is mainly the number of data points (degree of freedom), not the length of the record, unless the length of the record is too short. The least squares method can be replaced by a direct solution with a set of equations (a total of $2k + 1$ data points are required for resolving k tidal constituents – if it is overdetermined, i.e. the number of data points M is greater than $2k + 1$, one can still use the least squares method, or one can use any $2k + 1$

of the M data points in theory). For example, if there are 100 data points, the maximum number (k) of tidal frequencies to include would satisfy $2k + 1 = 100$, so the maximum $k \sim 49$.

Would this always work for any tidal signal if there were no random error at all? Not really. Even for deterministic functions, the above statement is more accurate for long time series with sparse data than short time series with many data points (given no randomness or, with a relaxation, with some minor random errors as in some tidal elevation data if weather impact is weak in a certain time period). To state it in a different way, if a tidal signal has one or more frequencies, if the time series is too short (e.g. shorter than a major tidal cycle), the harmonic analysis may not be able to recover the tidal constituents reliably, especially if there is some random error. On the other hand, if the random error is weak, even if the time series is shorter than a major tidal cycle, the harmonic constants may still be reliably recovered by the harmonic analysis. If the time series is long without random error, only sparse data ($2k + 1$ data points or more) are needed.

This may not be the case if significant random errors exist, in which case a time series from a sparse sampling may produce *aliasing* (an effect of folding some higher frequency random error into the signal), which will be discussed further in Chapter 17. Here, "higher frequency" means it has passed a threshold, which is discussed later with the concept of aliasing.

The reality is that there are random errors in addition to other processes (other frequencies). Often only a partial list of the tidal frequencies is included in the harmonic analysis. The reason is that we only need to include the frequencies present in the signal. How do we determine which frequencies are in the signal? That, of course, is partially empirical. The Fourier analysis of a time series of sufficient length can nevertheless reveal the spectrum and the tidal frequencies present in the signal.

In contrast, in the Fourier analysis of a continuous function, the frequency is not selected before the analysis. Ideally, the frequency range is from 0 to infinity for a continuous time function. In reality, for discrete time series data, the frequency range is from 0 to a maximum frequency f_{max} determined by the sampling intervals dt, i.e. $f_{max} = 1/(2dt)$, which will be further discussed in Chapter 17. The frequency interval (or resolution) is reciprocal of the total length of the time series (see next section). In all digital signal time series data analysis practices, the time intervals for Fourier Transform are always constant (unlike that for harmonic analysis, in which the time intervals for the data are not required to be constant). In theory, the time interval does not have to be constant. In the implementation of Fourier analysis, however, the time intervals are constant for ease of calculation using the fast Fourier Transform algorithm.

16.1.3 The Frequency Resolution

An important factor for consideration when it comes to which tidal frequencies to include in the harmonic analysis is the length of the record. Because of random fluctuations in the data, the least squares method is used for a best estimate; the resolution of coefficients for different frequencies depends on the length of the record. The time series is sometimes too short to show all possible tidal frequencies, even though there may be many data points. This is especially true if the signal has some long-period components which would need a long time series to show the oscillations at the low frequencies.

In general, the total length of the record determines the resolution in frequency – in order to separate the effects from two different frequencies, one should have a time record long enough such that its *frequency resolution* $(1/T)$ is smaller than the difference of the two frequencies:

$$\frac{1}{T} < |f_2 - f_1|. \tag{16.1}$$

This is basically the rule of thumb for the Fourier analysis, which can be "borrowed" by harmonic analysis as a reference. Here the frequency resolution is also the frequency intervals in the frequency domain of the Fourier analysis. In Fourier analysis, if the signal has two close frequencies, the length of the record will determine if the data can be used to separate the two frequencies in the Fourier analysis. For harmonic analysis, this rule could be relaxed to allow shorter time series resolving frequencies which are not possible in Fourier analysis. This is only limited to resolving them at tidal frequencies, and only if the signal is mainly tidal. To be on the safe side, however, using the above rule of thumb for resolution estimates is good guidance for harmonic analysis. To emphasize: in practice, the length of the data determines how many frequencies are appropriate to include for harmonic analysis.

For instance, if a velocity sensor is used to measure time series data over a period of only 13 hours, since the length of the data is so short, only a semi-diurnal tidal variation is observable. The time series may look like a semi-diurnal tide and it can be any of the semi-diurnal tidal constituents or a combination of them (Table 16.1). For example, the signal itself may only include S2 with a 12-hour period, or M2 with 12.42-hour period, or any other from the group, or a combination of them. In actual application, there is no way we could separate all of them with a short time series of 13 hours. In theory, however, if there were no random error at all in the observations, then as long as we have enough data, we may still be able to resolve the harmonic coefficients correctly for some of the frequencies in Table 16.1: we only need a minimum of $2k + 1$ data points, in which k is the total number of tidal frequencies. If we have more data, it is even better.

In practice, because of the uncertainty (random error included), it would not be appropriate to include both M2 and S2 in analyzing a 13-hour time series as the two frequencies are quite close in period for this 13-hour dataset, no matter how

Table 16.1. *Semi-diurnal tidal frequencies*

i in f_i	Name	Speed (degree/hour)	Frequency (cycle/day)	Description
1	M2	28.9841042	1.9323	Principal lunar semi-diurnal
2	S2	30	2.0000	Principal solar semi-diurnal
3	N2	28.4397295	1.8960	Larger lunar elliptic semi-diurnal
4	NU2	28.5125831	1.9008	Larger lunar evectional
5	MU2	27.9682084	1.8645	Variational
6	2N2	27.8953548	1.8597	Lunar elliptical semi-diurnal second-order
7	LAM2	29.4556253	1.9637	Smaller lunar evectional
8	T2	29.9589333	1.9973	Larger solar elliptic constituent
9	R2	30.0410667	2.0027	Smaller solar elliptic constituent
10	2SM2	31.0158958	2.0677	Shallow water semi-diurnal constituent
11	L2	29.5284789	1.9686	Smaller lunar elliptic semi-diurnal constituent
12	K2	30.0821373	2.0055	Lunisolar semi-diurnal constituent

many data points there are in the time series, because the condition given by equation (16.1) is not satisfied. This short dataset is not capable of resolving these two frequencies reliably using either harmonic or Fourier analysis.

16.2 Examples

16.2.1 Example 1: No Random Error, Short Record

Suppose we have collected only 13 hours of data from a single point. If the data had no random error, the tidal frequencies only include those from the M2 and S2 tides. The hypothetical time series is defined by:

$$y = \cos(2\pi f_1 t) + \sin(2\pi f_2 t), \tag{16.2}$$

in which f_1 and f_2 are the M2 and S2 tidal frequencies, respectively. The coefficient for the cosine function at the frequency f_1 and that for the sine function at the frequency f_2 are 1; and all other coefficients are 0. The time is uniformly spaced with 5-minute intervals. Given this hypothetical short time series data (Fig. 16.1), we would not know which of the semi-diurnal tidal constituents are present. From Table 11.1, we know that we may have a choice of semi-diurnal tidal frequencies from at least the following list of 12 constituents, which are regrouped in Table 16.1. For simplicity, we ignore diurnal and other tidal frequencies completely and that should not affect the conclusions conceptually.

For the harmonic analysis, a decision needs to be made for the frequencies to be included. As an example, let us consider the following 9 different choices (Experiments 1–9):

(1) choose S2 and M2 only

(2) choose S2 only

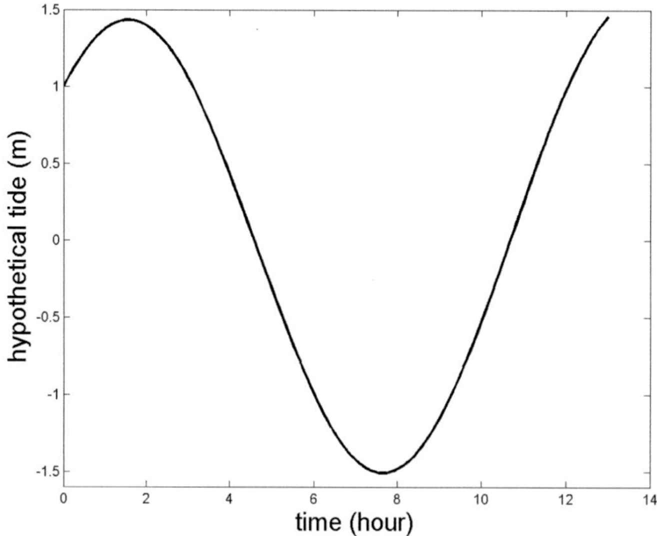

Figure 16.1 Data for Experiments 1–9. Hypothetical tidal time series for 13 hours.

(3) choose M2 only
(4) choose N2 only
(5) choose K2 only
(6) choose L2 only
(7) choose S2, M2, N2, and K2
(8) choose all of the 12 frequencies
(9) choose $f_1, f_2, f_4, f_5, f_{10}, f_{11}$ (or M2, S2, NU2, MU2, 2SM2, and L2)

These are just some examples, and we can have many more choices. The regression is to find the coefficients for the sine and cosine functions of the following equation:

$$\zeta = \zeta_0 + \sum_{i=1}^{K}[A_i \cos(2\pi f_i t) + B_i \sin(2\pi f_i t)]. \quad (16.3)$$

For an idealized situation in which there is no random error at all for a conceptual discussion, equation (16.2) is used to "construct" some hypothetical tidal data. We will then use (16.3) and try to recover the coefficients of selected tidal constituents with each of the nine experiments. With the 5-minute interval and a total of 13-hour long data, the least squares estimates of the coefficients in the above equation are, for each of the nine cases, respectively, tabulated in Tables 16.2–16.4. The following is a major part of the first experiment in MATLAB script, in which M2 and S2 are selected in the harmonic analysis:

Table 16.2. *Experiments 1–7 (no random error)*

Experiment number	Function	a	Total error squared	Comment
1	$a \sin(2\pi f_1)$	0.0000	0	*Choose S2 and M2 only.*
	$a \cos(2\pi f_1)$	1.0000		Accurate result.
	$a \sin(2\pi f_2)$	1.0000		
	$a \cos(2\pi f_2)$	0.0000		
	$a \cdot 1$	0.0000		
2	$a \sin(2\pi f_2)$	1.1106	0.0902	*Choose S2 only.*
	$a \cos(2\pi f_2)$	0.9841		Result seems okay for this short
	$a \cdot 1$	−0.0301		time period.
3	$a \sin(2\pi f_1)$	0.9763	0.4108	*Choose M2 only.*
	$a \cos(2\pi f_1)$	1.1199		Result seems okay for this short
	$a \cdot 1$	0.0080		time period.
4	$a \sin(2\pi f_3)$	0.8953	0.8989	*Choose N2 only.*
	$a \cos(2\pi f_3)$	1.1874		Result seems okay for this short
	$a \cdot 1$	0.0302		time period.
5	$a \sin(2\pi f_{12})$	1.1205	0.1016	*Choose K2 only.*
	$a \cos(2\pi f_{12})$	0.9727		Result seems okay for this short
	$a \cdot 1$	−0.0330		time period.
6	$a \sin(2\pi f_{11})$	1.0511	0.1363	*Choose L2 only.*
	$a \cos(2\pi f_{11})$	1.0486		Result seems okay for this short
	$a \cdot 1$	−0.0130		time period.
7	$a \sin(2\pi f_1)$	0.0000	0	*Choose S2, M2, N2, and K2.*
	$a \cos(2\pi f_1)$	1.0000		Accurate result.
	$a \sin(2\pi f_2)$	1.0000		
	$a \cos(2\pi f_2)$	0.0000		
	$a \sin(2\pi f_3)$	0.0000		
	$a \cos(2\pi f_3)$	0.0000		
	$a \sin(2\pi f_{12})$	0.0000		
	$a \cos(2\pi f_{12})$	0.0000		
	$a \cdot 1$	0.0000		

```
%================================================
% NO ERROR VERSION - Short time series (13 hours)
%================================================
clear
t = (0:5/60/24:13/24)';         %def time (column-vector) at 5-min int.
f = [ 28.9841042                %M2
     30] / 360 * 24             %S2
tide = cos(2 * pi * f(1) * t) + sin(2 * pi * f(2) * t);  %tidal elevation
A = [ sin(2*pi*f(1)*t) cos(2*pi*f(1)*t)  sin(2*pi*f(2)*t)...
     cos(2*pi*f(2)*t) ones(length(t),1)];  %CASE [1] for S2 and M2
                                           %Matrix A
x = A\tide;  %implement left-division (least squares method)
             %to find the harmonic coefficients at
             %chosen frequencies
```

Table 16.3. *Experiment 8*

Experiment number	Function	a	Total error squared	Comment
8	$a \sin(2\pi f_1)$	−0.1983	0	*Choose all 12 frequencies.*
	$a \cos(2\pi f_1)$	0.9752		Rank of A = 13 (< 25).
	$a \sin(2\pi f_2)$	0		Result is not good even though
	$a \cos(2\pi f_2)$	0		there is a perfect fitting with
	$a \sin(2\pi f_3)$	0.0715		0 TES for this short time series.
	$a \cos(2\pi f_3)$	0.0138		
	$a \sin(2\pi f_4)$	0		
	$a \cos(2\pi f_4)$	0		
	$a \sin(2\pi f_5)$	0		
	$a \cos(2\pi f_5)$	0		
	$a \sin(2\pi f_6)$	−0.0114		
	$a \cos(2\pi f_6)$	−0.0030		
	$a \sin(2\pi f_7)$	0		
	$a \cos(2\pi f_7)$	0		
	$a \sin(2\pi f_8)$	0		
	$a \cos(2\pi f_8)$	0		
	$a \sin(2\pi f_9)$	0		
	$a \cos(2\pi f_9)$	0		
	$a \sin(2\pi f_{10})$	−0.0051		
	$a \cos(2\pi f_{10})$	0.0006		
	$a \sin(2\pi f_{11})$	0.3664		
	$a \cos(2\pi f_{11})$	0.0211		
	$a \sin(2\pi f_{12})$	0.7771		
	$a \cos(2\pi f_{12})$	−0.0078		
	$a \cdot 1$	0.0000		

Table 16.4. *Experiment 9*

Experiment number	Function	a	Total error squared	Comment
9	$a \sin(2\pi f_1)$	0.0000	0	*Choose* $f_1, f_2, f_4, f_5, f_{10}, f_{11}$ *(or*
	$a \cos(2\pi f_1)$	1.0000		*M2, S2, NU2, MU2, 2SM2, and*
	$a \sin(2\pi f_2)$	1.0000		*L2).*
	$a \cos(2\pi f_2)$	0.0000		Accurate result.
	$a \sin(2\pi f_4)$	0.0000		
	$a \cos(2\pi f_4)$	0.0000		
	$a \sin(2\pi f_5)$	0.0000		
	$a \cos(2\pi f_5)$	0.0000		
	$a \sin(2\pi f_{10})$	0.0000		
	$a \cos(2\pi f_{10})$	0.0000		
	$a \sin(2\pi f_{11})$	0.0000		
	$a \cos(2\pi f_{11})$	0.0000		
	$a \cdot 1$	0.0000		

16.2.2 Results of Example 1 (Experiments 1–9)

Experiment 1 For the first experiment, the frequencies are correctly selected and the results are accurate. Tidal constituents are successfully recovered. Note that the length of the record is only 13 hours. According to equation (16.1), it would take about two weeks of data to resolve the two frequencies (M2 and S2) in a Fourier Transform. Here we used only a little longer than 0.5 day and we precisely recovered the tidal constituents. This is because the hypothetical data are precisely given with no random error (which would not be true in real applications) and the harmonic analysis is not restricted by equation (16.1) in resolving two very close frequencies. In a deterministic world, this case only has 5 degrees of freedom ($k = 2,\ 2k + 1 = 5$) and there is a need for only 5 data points in theory to fully determine the 5 parameters. With 5-minute intervals for 13 hours, there are a total of 157 data points, much greater than 5. The total error squared (TES) for this case is zero (Table 16.2), which is not a surprise.

Experiments 2–6 For Experiments 2 through 6, there is one thing in common: they all use just one semi-diurnal frequency (either S2, M2, N2, K2, or L2). We could have selected more cases (e.g. the fourth through tenth frequencies of Table 16.1), but the cases presented here are sufficient for discussion. Since the data is only about a semi-diurnal tidal period in length (13 hours), the selection of one frequency seems to be reasonable. The only problem is, without more information or knowledge from historic data, we would not know which frequency to pick. Strictly speaking, none of the choices is accurate since the actual data include two frequencies. In these cases, the TES is no longer zero. The case with the greatest TES is Experiment 4. These single-frequency harmonic analyses all result in a slight departure from the true function. This can be better viewed by the errors by subtracting the true function from the modeled function using the harmonic coefficients (Fig. 16.2).

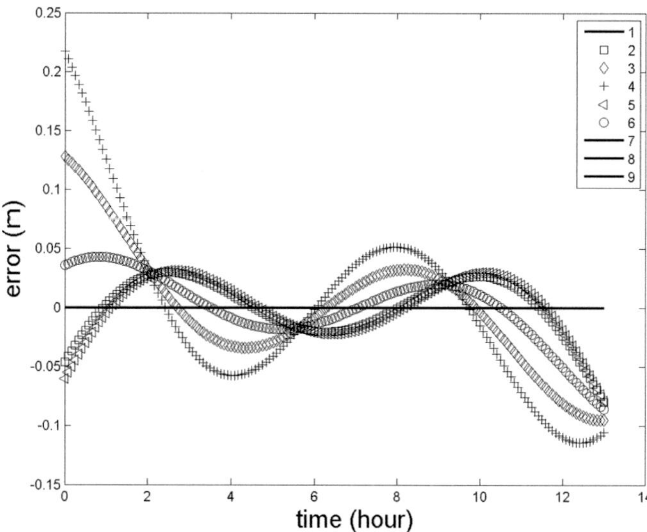

Figure 16.2 Errors for Experiments 1–9.

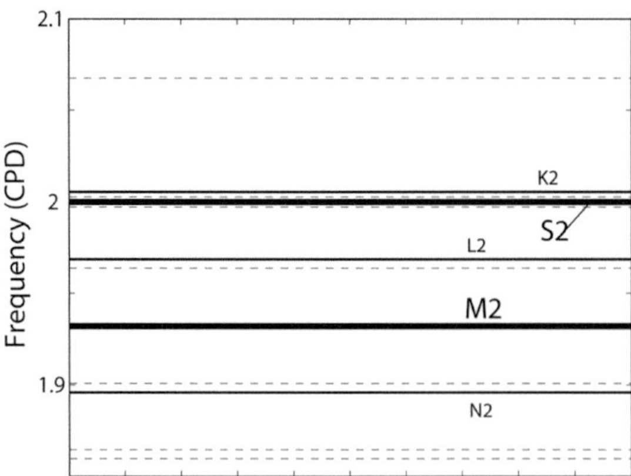

Figure 16.3 Comparison of the semi-diurnal tidal frequencies. The solid lines indicate the frequencies used in the experiments and are marked with the representative letters; the thicker solid lines are those in the signal indicated by larger fonts; while the dashed lines are other semi-diurnal tidal frequencies in Table 16.1 not included in the analysis.

Obviously, just from the look of the error functions in Fig. 16.2, we can see that Experiment 4 (+ symbols) has the largest error, followed by Experiment 3 (diamond symbols), Experiment 6 (circle symbols), Experiment 5 (triangle symbols), and Experiment 2 (square symbols), respectively. This is consistent with the TES values from Table 16.2: the largest TES is from Experiment 4 (TES = 0.8989), followed by Experiment 3 (0.4108), Experiment 6 (0.1363), Experiment 5 (0.1016), and Experiment 2 (0.0902). These are related to the selected frequency relative to the true value (Fig. 16.3). For example, N2 is substantially lower than M2 and the experiment with N2 (Experiment 4) has the largest TES.

Experiment 7 In Experiment 7, both M2 and S2 are selected as well as N2 and K2, which are not present in the signal. The harmonic analysis recovers the coefficients accurately (Table 16.2). Experiment 7 has zero TES, as expected.

Experiment 8 In Experiment 8, all frequencies are included. This is similar to Experiment 7, except that it included all 12 frequencies instead of only 4. However, when the script is run in MATLAB, a warning message appears:

```
Warning: Rank deficient, rank = 13,  tol =   4.3681e-013.
```

This happens when 12 of the 25 columns in the matrix A are "too similar" or essentially the same. Examining Table 16.1, we can see that some are indeed quite close. For example, the fifth and sixth frequencies, or MU2 (f_5 = 1.8645 cpd) and 2N2 (f_6 = 1.8597 cpd); the ninth and twelfth frequencies, or R2 (f_9 = 2.0027 cpd) and K2 (f_{12} = 2.0055 cpd); and the second and eighth frequencies, or S2 (f_2 = 2 cpd) and T2 (f_8 = 1.9973 cpd).

In contrast, there is no warning when Experiment 7 is run. Apparently, 6 of these 12 frequencies are too close for this short time series (13 hours): because of the short time series and the close frequencies, the base functions for different constituents (shown in different columns of matrix A) may be almost in phase during the short time period. As a result, some of the columns are too close. When the time series is longer, the base functions representing different tidal constituents can have greater phase differences as time progresses, and thus the columns of matrix A will be all different. This is discussed in the following sections for long time series experiments. However, it is perhaps unexpected to most that the TES is zero in this case (Table 16.3). This demonstrates that when there is matrix rank deficiency, as in this case, the results are not usable or correct even though the TES is small or even zero (the original signal can be accurately recovered, but the coefficients are "wrong").

Experiment 9 Experiment 9 excluded the close frequencies of Experiment 8. In this experiment, only f_1,f_2,f_4,f_5,f_{10}, and f_{11} are selected. Six other frequencies which are "too close" are excluded. The sixth frequency is not on the list, which eliminates the close pair of fifth and sixth frequencies; the exclusion of f_8 eliminates the close pair of the second and eighth frequencies; and the exclusion of f_9 and f_{12} eliminates the close pair of the ninth and twelfth frequencies of the last experiment. This effectively eliminates the deficiency in the rank of matrix A, and the results are an accurate recovery of the tidal constituents with a zero TES.

16.2.3 Example 2: No Random Error, Longer Time Series (5 Days)

Now we modify the above experiments slightly: all conditions are kept the same except that the time series is extended to 5 days (Fig. 16.4). We call these Experiments 1A–9A. The purpose of these experiments is to examine if there is any change in results if the length of time is longer. Here the length of time, 5 days, is significantly longer than the 13 hours of the first set of nine experiments above but still shorter than the two-week period needed to separate the S2 and M2 tidal constituents according to equation (16.1). Again, let us consider the following nine different choices of frequencies in the harmonic analysis (Experiments 1A–9A) for a hypothetical 5-day time series data at 5 minute intervals defined by (16.2):

(1) choose S2 and M2 only
(2) choose S2 only
(3) choose M2 only
(4) choose N2 only
(5) choose K2 only
(6) choose L2 only
(7) choose S2, M2, N2, and K2
(8) choose all 12 frequencies
(9) choose f_1,f_2,f_4,f_5,f_{10},f_{11} (or M2, S2, NU2, MU2, 2SM2, and L2)

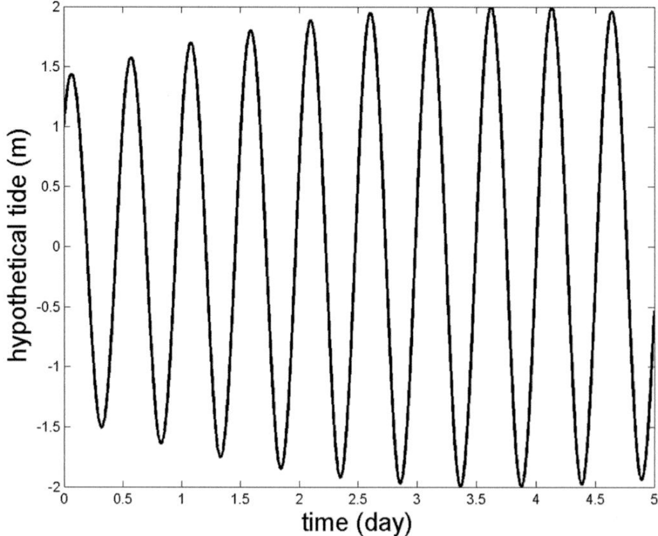

Figure 16.4 Data for Experiments 1A–9A. Hypothetical tidal time series for 5 days.

The results are tabulated in Tables 16.5–16.7 and discussed below.

16.2.4 Results of Example 2 (Experiments 1A–9A)

Experiments 1A, 7A, and 9A For these experiments (Tables 16.5 and 16.7), the frequencies are correctly selected and the results are accurate. The coefficients are correctly recovered and they all have zero TES. The zero TES is because we are using a deterministic function as the hypothetical tide. In reality, this is not the case – some random error is always contained in the signal.

Experiment 8A Again, this experiment is similar to Experiment 8: although the TES is zero, the result is not correct (Table 16.6) for the same reason that the frequencies are too close and the time series is too short.

Experiments 2A–6A For these experiments, there are increased TES values, and the modeled curves seem to be comparable to the hypothetical data (figure omitted), but there are some phase differences. The conclusion is that if some frequency (frequencies) is (are) missing in the selected base functions, the result cannot be reliable, especially for longer time series. This makes sense because the longer the time, the more accumulated effect with a given frequency difference from the true value.

As mentioned, the errors of Experiments 1A, and 7A–9A (Fig. 16.5) are zero. The errors of Experiments 2A–6A are non-zeros, with Experiment 4A having the greatest

Table 16.5. *Experiments 1A–7A*

Experiment number	Function	a	Total error squared	Comment
1A	$a \sin(2\pi f_1)$	0.0000	0	*Choose S2 and M2 only.*
	$a \cos(2\pi f_1)$	1.0000		Accurate result.
	$a \sin(2\pi f_2)$	1.0000		
	$a \cos(2\pi f_2)$	0.0000		
	$a \cdot 1$	0.0000		
2A	$a \sin(2\pi f_2)$	1.7308	230.4	*Choose S2 only.*
	$a \cos(2\pi f_2)$	0.3919		Large error.
	$a \cdot 1$	−0.0141		
3A	$a \sin(2\pi f_1)$	0.3998	224.4	*Choose M2 only.*
	$a \cos(2\pi f_1)$	1.7199		Large error.
	$a \cdot 1$	0.0000		
4A	$a \sin(2\pi f_3)$	−0.5467	808.6	*Choose N2 only.*
	$a \cos(2\pi f_3)$	1.4315		Large error.
	$a \cdot 1$	0.0015		
5A	$a \sin(2\pi f_{12})$	1.7337	301.1	*Choose K2 only.*
	$a \cos(2\pi f_{12})$	0.2297		Large error.
	$a \cdot 1$	−0.0151		
6A	$a \sin(2\pi f_{11})$	1.3551	22.9	*Choose L2 only.*
	$a \cos(2\pi f_{11})$	1.2489		Large error.
	$a \cdot 1$	−0.0071		
7A	$a \sin(2\pi f_1)$	−0.0000	0	*Choose S2, M2, N2, and K2.*
	$a \cos(2\pi f_1)$	1.0000		Accurate result.
	$a \sin(2\pi f_2)$	1.0000		
	$a \cos(2\pi f_2)$	−0.0000		
	$a \sin(2\pi f_3)$	0.0000		
	$a \cos(2\pi f_3)$	0.0000		
	$a \sin(2\pi f_{12})$	−0.0000		
	$a \cos(2\pi f_{12})$	0.0000		
	$a \cdot 1$	−0.0000		

error, followed by 5A, 2A, 3A, and 6A. The errors for Experiment 4A (the greatest among 2A–6A) and Experiment 6A (the smallest among 2A–6A) stand out (Fig. 16.5), while those for Experiments 2A, 3A, and 5A appear to be close at least visually, but the TES gives reliable quantified comparison as shown in Table 16.5.

16.2.5 Example 3: No Random Error, Longer Time Series (15 days)

The next set of experiments is aimed at further extending the time into more than two weeks, which is the minimum to show the biweekly or fortnightly tidal variations. All conditions are the same as the previous two sets of experiments for

Table 16.6. *Experiment 8A*

Experiment number	Function	a	Total error squared	Comment
8A	$a \sin(2\pi f_1)$	−0.0010	0	*Choose all 12 frequencies.* Rank of A = 23 (< 25). Result is not good even though there is an almost perfect fitting for this 5-day time series.
	$a \cos(2\pi f_1)$	0.9982		
	$a \sin(2\pi f_2)$	0		
	$a \cos(2\pi f_2)$	0		
	$a \sin(2\pi f_3)$	0.0001		
	$a \cos(2\pi f_3)$	−0.0020		
	$a \sin(2\pi f_4)$	0.0000		
	$a \cos(2\pi f_4)$	0.0025		
	$a \sin(2\pi f_5)$	−0.0001		
	$a \cos(2\pi f_5)$	0.0001		
	$a \sin(2\pi f_6)$	0.0001		
	$a \cos(2\pi f_6)$	−0.0001		
	$a \sin(2\pi f_7)$	0.0251		
	$a \cos(2\pi f_7)$	0.0161		
	$a \sin(2\pi f_8)$	0.4337		
	$a \cos(2\pi f_8)$	0.0186		
	$a \sin(2\pi f_9)$	0.7819		
	$a \cos(2\pi f_9)$	−0.0336		
	$a \sin(2\pi f_{10})$	0.0000		
	$a \cos(2\pi f_{10})$	0.0000		
	$a \sin(2\pi f_{11})$	−0.0329		
	$a \cos(2\pi f_{11})$	−0.0177		
	$a \sin(2\pi f_{12})$	−0.2069		
	$a \cos(2\pi f_{12})$	0.0178		
	$a \cdot 1$	0.0000		

Table 16.7. *Experiment 9A*

Experiment number	Function	a	Total error squared	Comment
9A	$a \sin(2\pi f_1)$	0.0000	0	*Choose $f_1, f_2, f_4, f_5, f_{10}, f_{11}$ (or M2, S2, NU2, MU2, 2SM2, and L2).* Accurate result.
	$a \cos(2\pi f_1)$	1.0000		
	$a \sin(2\pi f_2)$	1.0000		
	$a \cos(2\pi f_2)$	0.0000		
	$a \sin(2\pi f_4)$	0.0000		
	$a \cos(2\pi f_4)$	0.0000		
	$a \sin(2\pi f_5)$	0.0000		
	$a \cos(2\pi f_5)$	0.0000		
	$a \sin(2\pi f_{10})$	0.0000		
	$a \cos(2\pi f_{10})$	0.0000		
	$a \sin(2\pi f_{11})$	0.0000		
	$a \cos(2\pi f_{11})$	0.0000		
	$a \cdot 1$	0.0000		

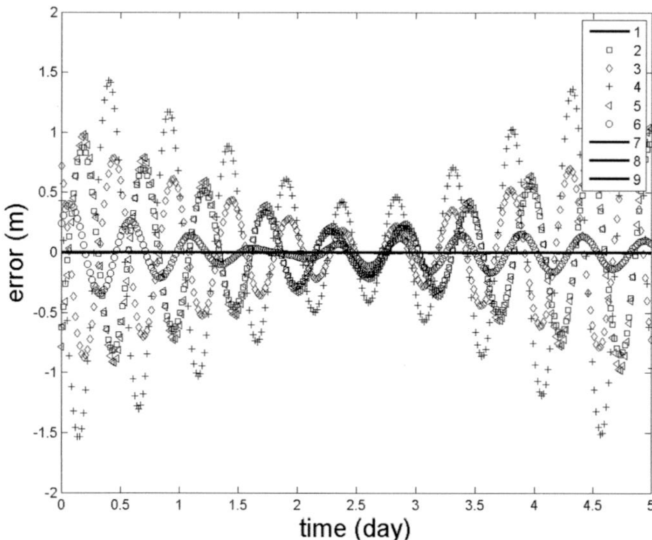

Figure 16.5 Errors of Experiments 1A–9A compared with the original function. Experiments 1A, 7A–9A have zero errors.

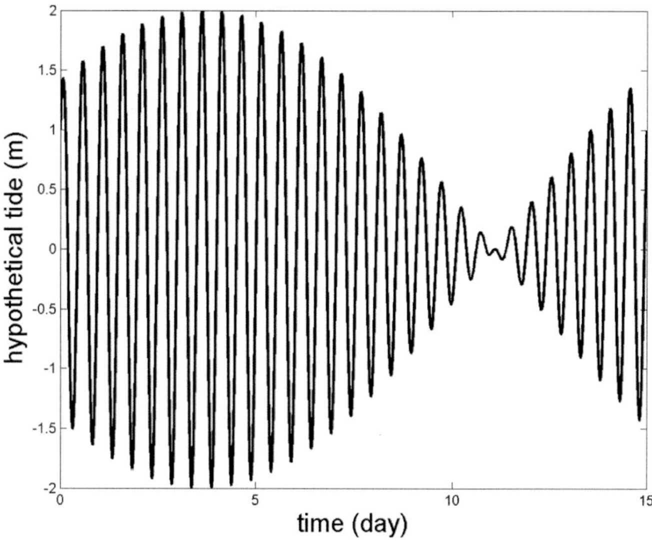

Figure 16.6 Data for Experiments 1B–9B. Hypothetical tidal time series for 15 days.

13 hours and 5 days. We will call these nine experiments 1B–9B. The original function (hypothetical tidal data) is shown in Fig. 16.6. Now we can see a clear modulation of the daily tidal amplitude, which is a typical characteristic of tides (the so-called spring-neap variations) in many regions around the world.

16.2.6 Results of Example 3 (Experiments 1B–9B)

The results are tabulated in Tables 16.8–16.10. Again, Experiments 1B, 7B–9B have zero TES values. This time, however, Experiment 8B gives an accurate result, and there is no deficiency in matrix rank for matrix A. This means that the 15-day time is now sufficient to separate all frequencies in Table 16.1 correctly with harmonic analysis. However, in equation (16.1), the frequency resolution in Fourier analysis requires that to separate e.g. the eighth and ninth frequencies of Table 16.1 we need to have at least a time series of

$$T = \frac{1}{2.0027 - 1.9973} \approx 183 \text{ days.} \tag{16.4}$$

Table 16.8. *Experiments 1B–7B*

Experiment number	Function	a	Total error squared	Comment
1B	$a \sin(2\pi f_1)$	0.0000	0	*Choose S2 and M2 only.*
	$a \cos(2\pi f_1)$	1.0000		Accurate result.
	$a \sin(2\pi f_2)$	1.0000		
	$a \cos(2\pi f_2)$	0.0000		
	$a \cdot 1$	0.0000		
2B	$a \sin(2\pi f_2)$	1.0008	2159	*Choose S2 only.*
	$a \cos(2\pi f_2)$	0.0158		Large error in both TES and
	$a \cdot 1$	−0.0003		coefficients.
3B	$a \sin(2\pi f_1)$	0.0159	2159	*Choose M2 only.*
	$a \cos(2\pi f_1)$	1.0008		Large error in both TES and
	$a \cdot 1$	0.0000		coefficients.
4B	$a \sin(2\pi f_3)$	−0.6059	3496	*Choose N2 only.*
	$a \cos(2\pi f_3)$	0.1234		Large error in both TES and
	$a \cdot 1$	0.0060		coefficients.
5B	$a \sin(2\pi f_{12})$	0.9843	2175	*Choose K2 only.*
	$a \cos(2\pi f_{12})$	−0.1677		Large error in both TES and
	$a \cdot 1$	−0.0006		coefficients.
6B	$a \sin(2\pi f_{11})$	0.6400	2678	*Choose L2 only.*
	$a \cos(2\pi f_{11})$	0.5933		Large error in both TES and
	$a \cdot 1$	−0.0066		coefficients.
7B	$a \sin(2\pi f_1)$	0.0000	0	*Choose S2, M2, N2, and K2.*
	$a \cos(2\pi f_1)$	1.0000		Accurate result.
	$a \sin(2\pi f_2)$	1.0000		
	$a \cos(2\pi f_2)$	0.0000		
	$a \sin(2\pi f_3)$	0.0000		
	$a \cos(2\pi f_3)$	0.0000		
	$a \sin(2\pi f_{12})$	0.0000		
	$a \cos(2\pi f_{12})$	0.0000		
	$a \cdot 1$	0.0000		

Table 16.9. *Experiment 8B*

Experiment number	Function	a	Total error squared	Comment
8B	$a \sin(2\pi f_1)$	0.0000	0	*Choose all 12 frequencies.* Full rank for A. Accurate result.
	$a \cos(2\pi f_1)$	1.0000		
	$a \sin(2\pi f_2)$	1.0000		
	$a \cos(2\pi f_2)$	0.0000		
	$a \sin(2\pi f_3)$	0.0000		
	$a \cos(2\pi f_3)$	0.0000		
	$a \sin(2\pi f_4)$	0.0000		
	$a \cos(2\pi f_4)$	0.0000		
	$a \sin(2\pi f_5)$	0.0000		
	$a \cos(2\pi f_5)$	0.0000		
	$a \sin(2\pi f_6)$	0.0000		
	$a \cos(2\pi f_6)$	0.0000		
	$a \sin(2\pi f_7)$	0.0000		
	$a \cos(2\pi f_7)$	0.0000		
	$a \sin(2\pi f_8)$	0.0000		
	$a \cos(2\pi f_8)$	0.0000		
	$a \sin(2\pi f_9)$	0.0000		
	$a \cos(2\pi f_9)$	0.0000		
	$a \sin(2\pi f_{10})$	0.0000		
	$a \cos(2\pi f_{10})$	0.0000		
	$a \sin(2\pi f_{11})$	0.0000		
	$a \cos(2\pi f_{11})$	0.0000		
	$a \sin(2\pi f_{12})$	0.0000		
	$a \cos(2\pi f_{12})$	0.0000		
	$a \cdot 1$	0.0000		

Table 16.10. *Experiment 9B*

Experiment number	Function	a	Total error squared	Comment
9B	$a \sin(2\pi f_1)$	0.0000	0	*Choose $f_1, f_2, f_4, f_5, f_{10}, f_{11}$ (or M2, S2, NU2, MU2, 2SM2, and L2).* Accurate result.
	$a \cos(2\pi f_1)$	1.0000		
	$a \sin(2\pi f_2)$	1.0000		
	$a \cos(2\pi f_2)$	0.0000		
	$a \sin(2\pi f_4)$	0.0000		
	$a \cos(2\pi f_4)$	0.0000		
	$a \sin(2\pi f_5)$	0.0000		
	$a \cos(2\pi f_5)$	0.0000		
	$a \sin(2\pi f_{10})$	0.0000		
	$a \cos(2\pi f_{10})$	0.0000		
	$a \sin(2\pi f_{11})$	0.0000		
	$a \cos(2\pi f_{11})$	0.0000		
	$a \cdot 1$	0.0000		

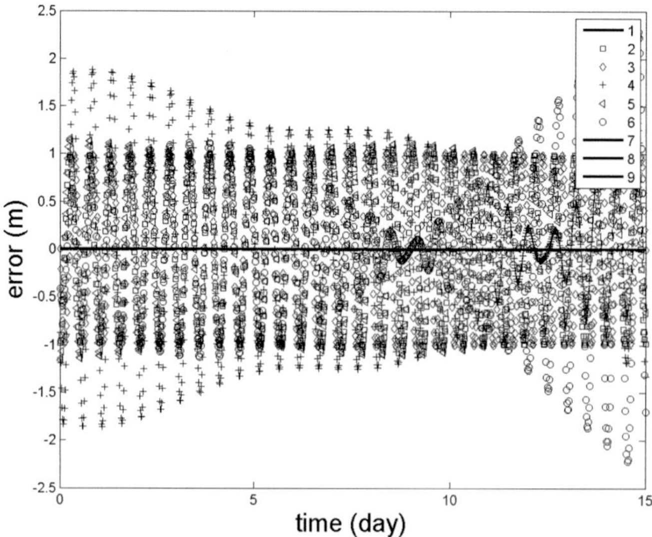

Figure 16.7 Errors of Experiments 1B–9B compared with the original function. Experiments 1B, 7B–9B have zero errors.

The actual data length is only 15 days. This is because the harmonic analysis has more specific information about the frequencies. If the correct frequencies are included, a "sufficiently long time series" can resolve the coefficients associated with these frequencies. Here the 15-day length is apparently "sufficiently long," which is much shorter than that required by equation (16.4).

Experiments 2B–6B, however, all give incorrect results as none of them can resolve the fortnightly variation of the daily tidal amplitude. The errors are shown in Fig. 16.7. For a long time series like this one, with modulations of amplitude due to the interferences (linear superpositions) among different frequencies, the conclusion is that the lack of a certain frequency or frequencies can lead to a significant error in harmonic analysis in reconstructing the time series because some major constituents are missing. Why is this not a problem for short time series? This is because for short time series, the effect of missing frequency has not accumulated enough to be "visible." The phase differences and amplitude modulation are shown with increasing time.

16.2.7 Example 4: With Random Error, a Short Time Series (13 Hours)

Now we add some noise to the original "data" and do harmonic analysis as in the previous examples. In other words, all conditions are the same except that the time series is now given by (Fig. 16.8):

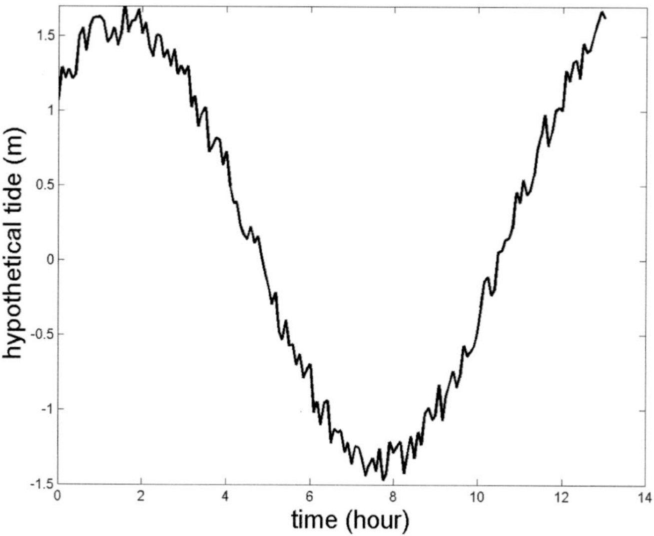

Fig. 16.8 Data for Experiments 1C–9C. Hypothetical tidal time series for 13 hours.

$$y = \cos(2\pi f_1 t) + \sin(2\pi f_2 t) + 0.3r. \tag{16.5}$$

in which r is a uniformly distributed random number between 0 and 1. In generating the hypothetical tide by (16.5), we use the following MATLAB script:

```
t = (0:5/60/24:15)';
tide = cos(2 * pi * f(1) * t) + sin(2 * pi * f(2) * t) + 0.3 * rand(length(t),1);
```

The MATLAB function `rand` is a uniformly distributed pseudorandom number generator. Each time the script is run, it gives a slightly different result for the time series because of the random nature. The error is shown in Fig. 16.9.

16.2.8 Results of Example 4 (Experiments 1C–9C)

With the noise, results for this short time series look similar with comparable TESes: the TES values for the nine experiments are 1.0770, 1.3140, 1.6241, 2.1078, 1.3271, 1.3521, 0.9955, 0. 9201, and 0. 9201, respectively. The first and seventh through ninth have slightly smaller errors. Note that when the eighth experiment is run, it still has a warning with a deficiency in rank for matrix A. The rank of A is 23 (the full rank is 25). The coefficients for Experiments 7C–9C are all unreasonably large (~10^5). Although the reconstructed curve may look good (figure omitted), the results are useless.

```
Warning: Rank deficient, rank = 23,  tol =   1.2146e-011.
```

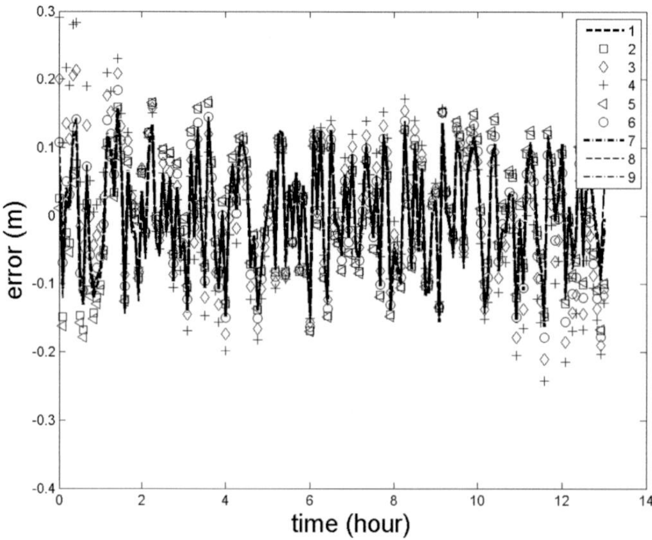

Figure 16.9 Errors of Experiments 1C–9C compared with the original function.

With the noise included in the original signal, the seventh and ninth experiments do not have reasonable results anymore (Tables 16.11, 16.12, and 16.13), even though their errors are relatively small (Fig. 16.9), as shown above with the smaller TES values. This demonstrates that, when doing a harmonic analysis for a short time series with some noise, too many frequencies included in the analysis might render useless results, which we do not see in the earlier examples for deterministic functions.

For Experiments 1C–6C, the results have relatively larger errors but no fundamental problems. The "correct" coefficients are not accurately recovered, but that is expected with the random noise, which essentially changes the signal and new frequencies are introduced. The new frequencies are not tidal but can influence the result.

16.2.9 Example 5: With Random Error, a 5-Day Time Series

Now we extend the time series to 5 days, with all other conditions the same as those in Example 4. Using the same nine different choices, do Experiments 1D–9D. We are still using equation (16.5) for the hypothetical tidal data time series at 5-minute intervals with noise added. The original "data" are given in Fig. 16.10. The errors are shown in Fig. 16.11.

16.2.10 Results of Example 5 (Experiments 1D–9D)

Experiment 1D (Table 16.14) has a reasonable result, as it has included correct frequencies. Experiments 2D–6D (Table 16.14) have large errors because of

Table 16.11. *Experiments 1C–7C*

Experiment number	Function	a	Total error squared	Comment
1C	$a\ \sin(2\pi f_1)$ $a\ \cos(2\pi f_1)$ $a\ \sin(2\pi f_2)$ $a\ \cos(2\pi f_2)$ $a \cdot 1$	−0.0437 1.5929 0.9882 −0.6025 0.1568	1.0770	*Choose S2 and M2 only.* Relatively large error is present.
2C	$a\ \sin(2\pi f_2)$ $a\ \cos(2\pi f_2)$ $a \cdot 1$	1.1205 0.9703 0.1090	1.3140	*Choose S2 only.* Errors increased compared with that without noise.
3C	$a\ \sin(2\pi f_1)$ $a\ \cos(2\pi f_1)$ $a \cdot 1$	0.9868 1.1080 0.1467	1.6214	*Choose M2 only.* Errors increased compared with that without noise.
4C	$a\ \sin(2\pi f_3)$ $a\ \cos(2\pi f_3)$ $a \cdot 1$	0.9059 1.1764 0.1688	2.1078	*Choose N2 only.* Errors increased compared with that without noise.
5C	$a\ \sin(2\pi f_{12})$ $a\ \cos(2\pi f_{12})$ $a \cdot 1$	1.1303 0.9587 0.1061	1.3271	*Choose K2 only.* Errors increased compared with that without noise.
6C	$a\ \sin(2\pi f_{11})$ $a\ \cos(2\pi f_{11})$ $a \cdot 1$	1.0613 1.0357 0.1260	1.3521	*Choose L2 only.* Errors increased compared with that without noise.
7C	$a\ \sin(2\pi f_1)$ $a\ \cos(2\pi f_1)$ . . . and other functions related to f_2, f_3, and f_{12}	−2250 −54550 With unreasonably large values ($\sim 10^5$)	0.9955	*Choose S2, M2, N2, and K2.* Useless result.

Table 16.12. *Experiment 8C*

Experiment number	Function	a	Total error squared	Comment
8C	$a\ \sin(2\pi f_1)$ $a\ \cos(2\pi f_1)$ . . . and other functions	Most coefficients have unreasonably large values ($\sim 10^5$)	0.9201	*Choose all 12 frequencies.* Matrix A has a deficient rank of 23. Small TES but useless result.

Table 16.13. *Experiment 9C*

Experiment number	Function	a	Total error squared	Comment
9C	$a\ \sin(2\pi f_1)$ $a\ \cos(2\pi f_1)$ ... and other functions	Most coefficients have unreasonably large values (~10^{10})	0.9201	*Choose* $f_1, f_2, f_4, f_5, f_{10}, f_{11}$ *(or M2, S2, NU2, MU2, 2SM2, and L2).* Small error but useless results.

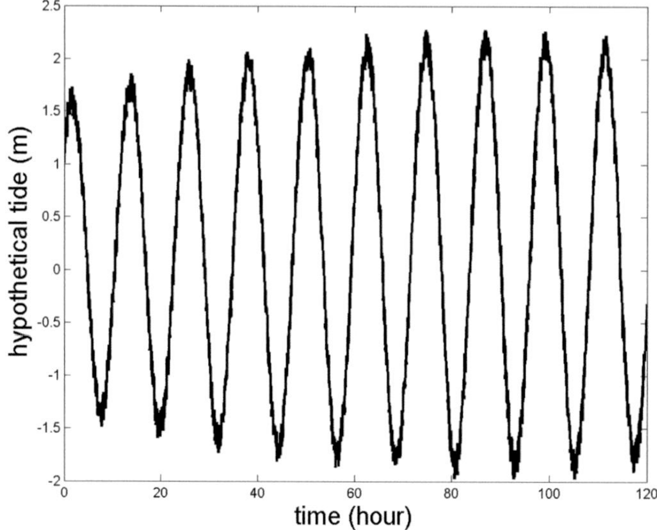

Figure 16.10 Data for Experiments 1D–9D. Hypothetical tidal time series for 5 days.

incomplete frequencies included in the harmonic analysis. Experiment 7D (Table 16.14) has small errors, but the result is incorrect: the inclusion of too many very close frequencies for this short time function (though longer than that of Experiments 7C) has messed up things by redistributing the energy from the true frequencies to other nearby frequencies. Experiment 8D yields useless results as the coefficients are too large (~10^7–10^8) to be reasonable (Table 16.15). When Experiment 8D is run, the following warning is shown by MATLAB:

```
Warning: Rank deficient, rank = 23, tol = 1.2146e-011.
```

Matrix A has a deficiency in rank – the full rank should be 25 for Experiment 8D. Experiment 9E (Table 16.16) has small errors but incorrect results.

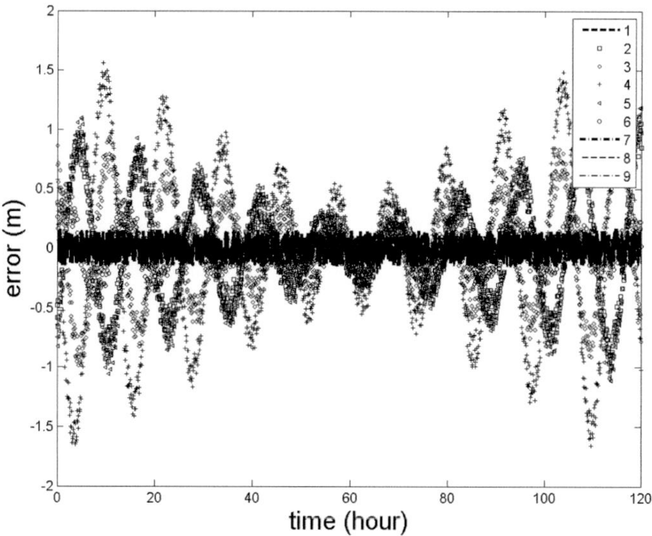

Figure 16.11 Errors of Experiments 1D–9D compared with the original function.

16.2.11 Example 6: With Random Error, a 15-Day Time Series

We now further extend the time series to 15 days. All conditions are otherwise the same, and the hypothetical tidal data are simulated from equation (16.5). With these conditions, we do the same nine experiments – Experiments 1E–9E. Again, because of the random noise, the results can be slightly different each time the MATLAB script is run. The original time series is shown in Fig. 16.12. The error of the experiments is shown in Fig. 16.13. The results of the coefficients of the tidal sinusoidal functions are tabulated in Tables 16.17–16.19.

16.2.12 Results of Example 6 (Experiments 1E–9E)

Experiment 1E recovers the harmonic coefficients quite reliably (Table 16.17) with a relatively small TES value. In contrast, Experiments 2E–6E all produce poor results in recovering the harmonic coefficients. Again, with the long time series having biweekly amplitude variations, the single frequency choices are not appropriate. Experiment 7E now gives a reliable result, unlike Experiment 7D – the extended length of the data allows the resolution of the different frequencies. In theory, had the frequencies been unknown, the length of time of the data to resolve f_2 and f_{12} of Table 16.1 would be

$$T = \frac{1}{2.0055 - 2.0000} \approx 183 \text{ days}. \tag{16.6}$$

Table 16.14. *Experiments 1D–7D*

Experiment number	Function	a	Total error squared	Comment
1D	$a\ \sin(2\pi f_1)$	−0.0051	10.82	*Choose S2 and M2 only.*
	$a\ \cos(2\pi f_1)$	0.9971		Reasonable result.
	$a\ \sin(2\pi f_2)$	1.0089		
	$a\ \cos(2\pi f_2)$	0.0033		
	$a \cdot 1$	0.1514		
2D	$a\ \sin(2\pi f_2)$	1.7356	242.30	*Choose S2 only.*
	$a\ \cos(2\pi f_2)$	0.3977		Large errors.
	$a \cdot 1$	0.1373		
3D	$a\ \sin(2\pi f_1)$	0.3959	233.71	*Choose M2 only.*
	$a\ \cos(2\pi f_1)$	1.7248		Large errors.
	$a \cdot 1$	0.1515		
4D	$a\ \sin(2\pi f_3)$	−0.5513	817.27	*Choose N2 only.*
	$a\ \cos(2\pi f_3)$	1.4319		Large errors.
	$a \cdot 1$	0.1530		
5D	$a\ \sin(2\pi f_{12})$	1.7391	313.23	*Choose K2 only.*
	$a\ \cos(2\pi f_{12})$	0.2349		Large errors.
	$a \cdot 1$	0.1362		
6D	$a\ \sin(2\pi f_{11})$	1.3556	33.53	*Choose L2 only.*
	$a\ \cos(2\pi f_{11})$	1.2562		Large errors.
	$a \cdot 1$	0.1444		
7D	$a\ \sin(2\pi f_1)$	0.1068	10.81	*Choose S2, M2, N2, and K2.*
	$a\ \cos(2\pi f_1)$	1.1310		Smaller error but incorrect result.
	$a\ \sin(2\pi f_2)$	0.2126		
	$a\ \cos(2\pi f_2)$	0.2385		
	$a\ \sin(2\pi f_3)$	−0.0059		
	$a\ \cos(2\pi f_3)$	−0.0826		
	$a\ \sin(2\pi f_{12})$	0.6853		
	$a\ \cos(2\pi f_{12})$	−0.2752		
	$a \cdot 1$	0.1515		

Table 16.15. *Experiment 8D*

Experiment number	Function	a	Total error squared	Comment
8D	$a\ \sin(2\pi f_1)$ $a\ \cos(2\pi f_1)$ ... and other functions	Most coefficients have unreasonably large values (~10^7–10^8)	10.71	*Choose all 12 frequencies.* Useless result.

Table 16.16. *Experiment 9D*

Experiment number	Function	a	Total errors squared	Comment
9D	$a\ \sin(2\pi f_1)$	−14.5084	10.76	*Choose* $f_1, f_2, f_4, f_5, f_{10}, f_{11}$ *(or M2, S2, NU2, MU2, 2SM2, and L2).* Useless result.
	$a\ \cos(2\pi f_1)$	6.0490		
	$a\ \sin(2\pi f_2)$	0.5121		
	$a\ \cos(2\pi f_2)$	5.4452		
	$a\ \sin(2\pi f_4)$	8.8657		
	$a\ \cos(2\pi f_4)$	1.8149		
	$a\ \sin(2\pi f_5)$	−1.2404		
	$a\ \cos(2\pi f_5)$	−1.3538		
	$a\ \sin(2\pi f_{10})$	−0.3333		
	$a\ \cos(2\pi f_{10})$	−0.1957		
	$a\ \sin(2\pi f_{11})$	7.6871		
	$a\ \cos(2\pi f_{11})$	−10.7573		
	$a \cdot 1$	0.1518		

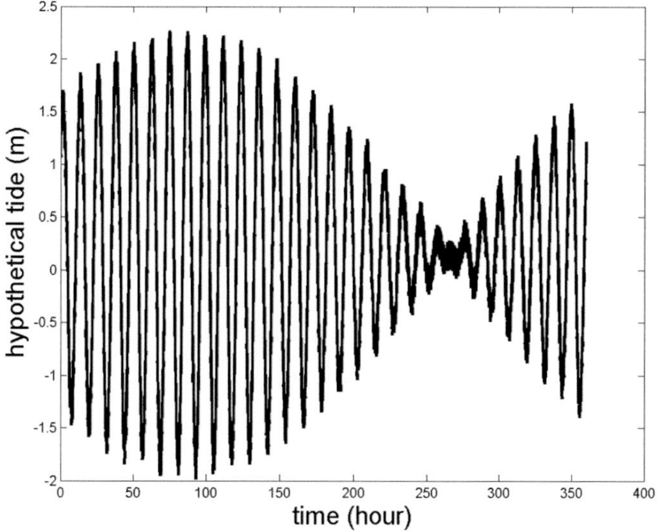

Figure 16.12 Data for Experiments 1E–9E. Hypothetical tidal time series for 15 days.

Again, this length is much longer than our data (15 days). The fact that we know the frequencies has allowed us to use a shorter record to resolve the tidal constituents.

When all the frequencies in Table 16.1 are included in the harmonic analysis (Experiment 8E), the result is unreasonably large (Table 16.18). Apparently, we need an even longer time series to be able to resolve all 12 tidal constituents. In

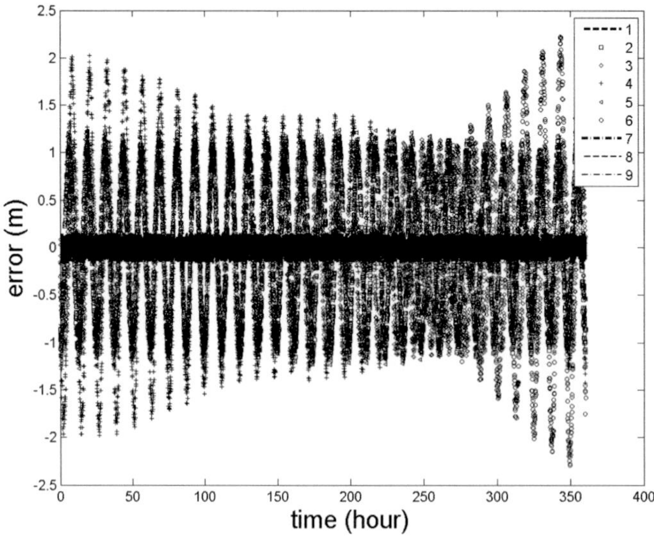

Figure 16.13 Errors of Experiments 1E–9E compared with the original function.

theory, according to equation (16.1), to resolve both the second and eighth frequencies of Table 16.1, we need to have at least 365.26 days of data. When the eighth frequency is eliminated from the list in Experiment 9E, the result is good again (Table 16.19) with reasonable recovery of the coefficients. The small non-zero coefficients are introduced by the random noise.

16.3 Summary

After the discussion on all these experiments (Examples 1 through 6, or Experiments 1–9; 1A–9A, 1B–9B, 1C–9C, 1D–9D, and 1E–9E), we can make the following qualitative conclusions:

(1) When doing harmonic analysis, longer time series is preferred over shorter if resolution of harmonic constants for more frequencies are needed.
(2) We can use equation (16.1), i.e. $\frac{1}{T} < |f_2 - f_1|$, as guidance for the length of data needed to resolve any two frequencies f_1 and f_1, but harmonic analysis can potentially have much better resolution than that because the tidal frequencies are known.
(3) Given specific time series with tidal signals, it is not necessarily true that the more frequencies used in the harmonic analysis, the better. Some frequencies may be too close to allow resolution (the longer the better, as the first conclusion above).

Table 16.17. *Experiments 1E–7E (with random error)*

Experiment number	Function	a	Total error squared	Comment
1E	$a \sin(2\pi f_1)$	−0.0016	32.2	*Choose S2 and M2 only.*
	$a \cos(2\pi f_1)$	0.9999		Good result.
	$a \sin(2\pi f_2)$	1.0009		
	$a \cos(2\pi f_2)$	0.0009		
	$a \cdot 1$	0.1491		
2E	$a \sin(2\pi f_2)$	1.0017	2187.5	*Choose S2 only.*
	$a \cos(2\pi f_2)$	0.0167		Large error.
	$a \cdot 1$	0.1487		
3E	$a \sin(2\pi f_1)$	0.0143	2190.6	*Choose M2 only.*
	$a \cos(2\pi f_1)$	1.0007		Large error.
	$a \cdot 1$	0.1491		
4E	$a \sin(2\pi f_3)$	−0.6055	3526.2	*Choose N2 only.*
	$a \cos(2\pi f_3)$	0.1252		Large error.
	$a \cdot 1$	0.1550		
5E	$a \sin(2\pi f_{12})$	0.9853	2204.8	*Choose K2 only.*
	$a \cos(2\pi f_{12})$	−0.1672		Large error.
	$a \cdot 1$	0.1485		
6E	$a \sin(2\pi f_{11})$	0.6389	2702.1	*Choose L2 only.*
	$a \cos(2\pi f_{11})$	0.5953		Large error.
	$a \cdot 1$	0.1425		
7E	$a \sin(2\pi f_1)$	−0.0084	32.2	*Choose S2, M2, N2, and K2.*
	$a \cos(2\pi f_1)$	1.0048		Good result.
	$a \sin(2\pi f_2)$	0.9675		
	$a \cos(2\pi f_2)$	0.0354		
	$a \sin(2\pi f_3)$	0.0034		
	$a \cos(2\pi f_3)$	0.0080		
	$a \sin(2\pi f_{12})$	0.0227		
	$a \cos(2\pi f_{12})$	−0.0411		
	$a \cdot 1$	0.1491		

Table 16.18. *Experiment 8E*

Experiment number	Function	a	Total error squared	Comment
8E	$a \sin(2\pi f_1)$ $a \cos(2\pi f_1)$ ... and the other functions	Most coefficients have unreasonably large values (~10^4–10^5)	32.1	*Choose all 12 frequencies.* Useless result.

Table 16.19. *Experiment 9E*

Experiment number	Function	a	Total error squared	Comment
9E	$a \sin(2\pi f_1)$	−0.0128	32.2	*Choose* $f_1, f_2, f_4, f_5, f_{10}, f_{11}$ *(or M2, S2, NU2, MU2, 2SM2, and L2).* Good result.
	$a \cos(2\pi f_1)$	1.0287		
	$a \sin(2\pi f_2)$	1.0038		
	$a \cos(2\pi f_2)$	−0.0150		
	$a \sin(2\pi f_4)$	0.0184		
	$a \cos(2\pi f_4)$	0.0080		
	$a \sin(2\pi f_5)$	0.0006		
	$a \cos(2\pi f_5)$	−0.0020		
	$a \sin(2\pi f_{10})$	−0.0027		
	$a \cos(2\pi f_{10})$	−0.0035		
	$a \sin(2\pi f_{11})$	−0.0294		
	$a \cos(2\pi f_{11})$	−0.0041		
	$a \cdot 1$	0.1492		

(4) The results are dependent on the amount of noise.

(5) Sometimes the total error squared may be relatively small, and the reconstructed function might be close to the original data; the tidal constituents, however, may not be correct if some of the frequencies are too close. A rank deficiency warning may be given by MATLAB, but even without the warning, results may still be wrong. This can be avoided if historic data are available to verify the tidal constituents, or data length is increased. It can also be examined by Fourier analysis first, which is then subject to the condition that $\frac{1}{T} < |f_2 - f_1|$.

Review Questions for Chapter 16

(1) What is the resolution of frequency in Fourier analysis with a time series data of 30 days at 30-minute time intervals? What about a time series of 30 days at hourly intervals?

(2) Do you think a Fourier series expansion of Fourier Transform can be done for time series data with uneven time intervals?

Exercises for Chapter 16

(1) A time series data file is obtained with a total length of record of 13 hours at 30-minute intervals. If tide in the region contains both M2 and S2 tidal constituents (i.e. $f_1 = 1.9323$ cpd (cycle per day), $f_2 = 2$ cpd), can you use this dataset to apply a harmonic analysis including both M2 and S2 frequencies and anticipate that you will be able to resolve both?

(2) Create a hypothetical time series data of 13 hours at 30-minute intervals with M2 and S2 constituents; each has an amplitude of 0.5 m:

$$\zeta = 0.5 \cos(2\pi f_1 t) + 0.5 \cos(2\pi f_2 t) + r.$$

Here r is a simulated random error with a maximum of 0.1 m (using the MATLAB command (rand(M,1)−0.5)/5, with M being the total number of data points); then apply the harmonic analysis and show if the results support your conclusion for the above question.

(3) The larger solar elliptic constituent (T2) semi-diurnal tidal frequency is 29.9589333 degree/hour; and the smaller solar elliptic constituent (R2) semi-diurnal tidal frequency is 30.0410667 degree/hour. What are these frequencies in the unit cycle per day? These two frequencies are very close and hard to separate unless we have a sufficiently long time series. What is the minimum length of a time series of tide needed to resolve these two frequencies when doing the fast Fourier Transform (FFT)?

17

Effect of Finite Sampling

About Chapter 17

The objective of this chapter is to discuss a very important issue of the effect of finite sampling with respect to either the finite length of the record or the finite sampling intervals. A few sampling theorems are discussed.

17.1 The Effect of Truncation

Time series data can be viewed as a discrete version of a continuous function of time. The discrete data are finite in two aspects: (1) the length of data is finite (so there is an effect of finite length); and (2) the data have finite time intervals (so there is an effect of discrete sampling). We call the action of obtaining such discrete data *finite sampling*. After finite sampling, the time is no longer continuous. Obviously, the spectral characteristics of the discrete data should be different from the original continuous and infinite function. Here we will examine the effects of finite sampling by looking at these two aspects: the effects of finite length (or truncation of an infinite series) and discrete sampling.

17.1.1 Finite Sampling: Rectangular Window Function

The first effect of sampling, i.e. the truncation of an infinitely long time function into a finite segment of function, is equivalent to multiplying the original function $y(t)$ by a *rectangular window function*, which is a unit step up from 0 and then a unit step down back to 0 after a time period T (Fig. 17.1c). It is defined as the following:

$$x_T(t) = \begin{cases} 1, & |t| \le T/2 \\ 0, & |t| > T/2 \end{cases}, \tag{17.1}$$

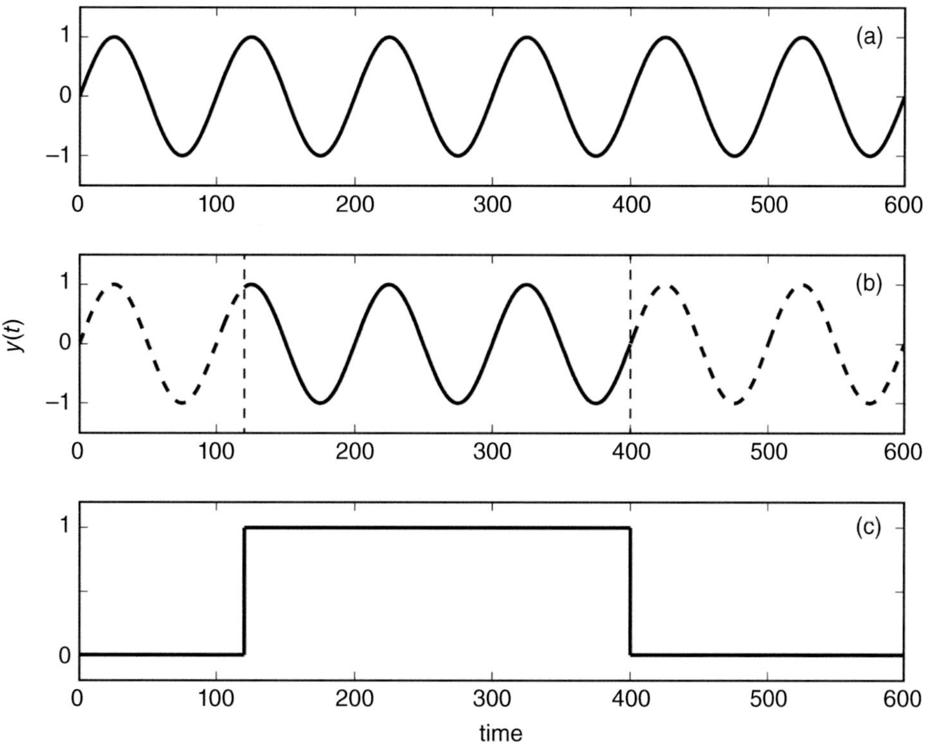

Figure 17.1 Finite sampling (truncation) of an infinitely long function (a), resulting in a finite segment of the original function (b), which is equivalent to multiplying the original function by a rectangular window function (c).

to yield

$$y_s(t) = y(t)x_T(t). \tag{17.2}$$

Here $x_T(t)$ is the rectangular window function, $y_s(t)$ the sampled function, and $y(t)$ the original function. The time for the beginning of the rectangular window function $(t = -T/2)$ is selected for convenience and does not have specific significance. Sometimes, the rectangular window function is also referred to as the *square window function.*

This is depicted by an example of finite sampling of a sine function in Fig. 17.1. In this example,

$$y(t) = \sin(2\pi f_0 t), \tag{17.3}$$

in which f_0 is the frequency of this sine function (here $f_0 = 100$). Fig. 17.1a shows the original function $y(t)$, extending endlessly in the directions of both $-\infty$ and $+\infty$; Fig. 17.1b shows the function after finite sampling $y_s(t)$; and Fig. 17.1c shows the window function $x_T(t)$.

17.1.2 The Sinc Function

When a Fourier Transform is done for the spectrum of some sampled data, what is really done is a Fourier Transform of the function from finite sampling $y_s(t)$, not the original $y(t)$. From what we have learned about the convolution theorem, the Fourier Transform for $y_s(t)$ is $Y_s(f)$, which is the convolution of the Fourier Transform of $y(t)$, $Y(f)$, and the Fourier Transform of $x_T(t)$, $X_T(f)$:

$$Y_s(f) = Y(f) \otimes X_T(f) = \int_{-\infty}^{+\infty} Y(\tilde{f})X_T(f - \tilde{f})d\tilde{f}. \tag{17.4}$$

The Fourier Transform of the rectangular window function can be easily calculated directly:

$$X_T(f) = \int_{-T/2}^{T/2} e^{-i2\pi ft}dt = \frac{e^{-i2\pi ft}\big|_{-T/2}^{T/2}}{-i2\pi f} = \frac{e^{-i\pi fT} - e^{i\pi fT}}{-i2\pi f}. \tag{17.5}$$

Equation (17.5) can be simplified to

$$X_T(f) = \frac{\sin(\pi fT)}{\pi f}. \tag{17.6}$$

This is a continuous function of frequency f with the window size (T) as its parameter. It is called the *sinc function*, as mentioned in Chapter 14. It allows both positive and negative frequency f and is symmetric about $f = 0$. By plotting the absolute value of $X_T(f)$, we see that it ($|X_T(f)|$) has a main peak in the middle at $f = 0$, which is often called the *main lobe*. It has a series of *side lobes* of decreasing magnitude between the zeros at $\pm 1/T, \pm 2/T, \ldots$. It is not hard to see that as T increases, the zeros of the sinc function at $\pm 1/T, \pm 2/T, \ldots$, will get closer to $f = 0$ and thus the main lobe will be narrower.

Conceptually, a constant function running from $-\infty$ to $+\infty$ in time is a zero-frequency function and its spectrum in the frequency domain should be an isolated line at $f = 0$. Because of the truncation, the spectrum becomes the sinc function. The main lobe represents the true peak, while the side lobes are all fake peaks in the spectrum. Unfortunately, we cannot eliminate these side lobes completely – it is the nature of finite sampling.

17.1.3 Finite Sampling Theorem #1

Obviously, even though the magnitude of $X_T(f)$ from (17.5) goes to infinity as $T \to +\infty$, the Fourier series of a constant will be just a constant – there is only a non-zero value at the zero frequency. We have the first finite sampling theorem:

> **Finite Sampling Theorem #1** *The spectrum (Fourier Transform) of a finite segment of a constant function, or a rectangular window function, is a sinc function that has a main lobe representing the correct position of the peak in the true spectrum of the infinite constant function, with a series of spurious side lobes. These side lobes are between $f = n/T$ and $(n + 1)/T$,* or *$-(n + 1)/T$ and $-n/T$, $n = 1, 2, \ldots$ (Fig. 17.2a).*

The finite width of the main lobe and the side lobes is a result of the truncation of the infinite function to a finite segment function. It should be noted that when we use the discrete Fourier Transform with FFT, the output is often expressed as in Fig. 17.2b. One can think of Fig. 17.2b as a different way of visualizing the result by flipping the negative frequency part to that after the Nyquist frequency. Because of the symmetry of the magnitude of Fourier Transform, we sometimes only show half of the sinc function, as in Fig. 17.3.

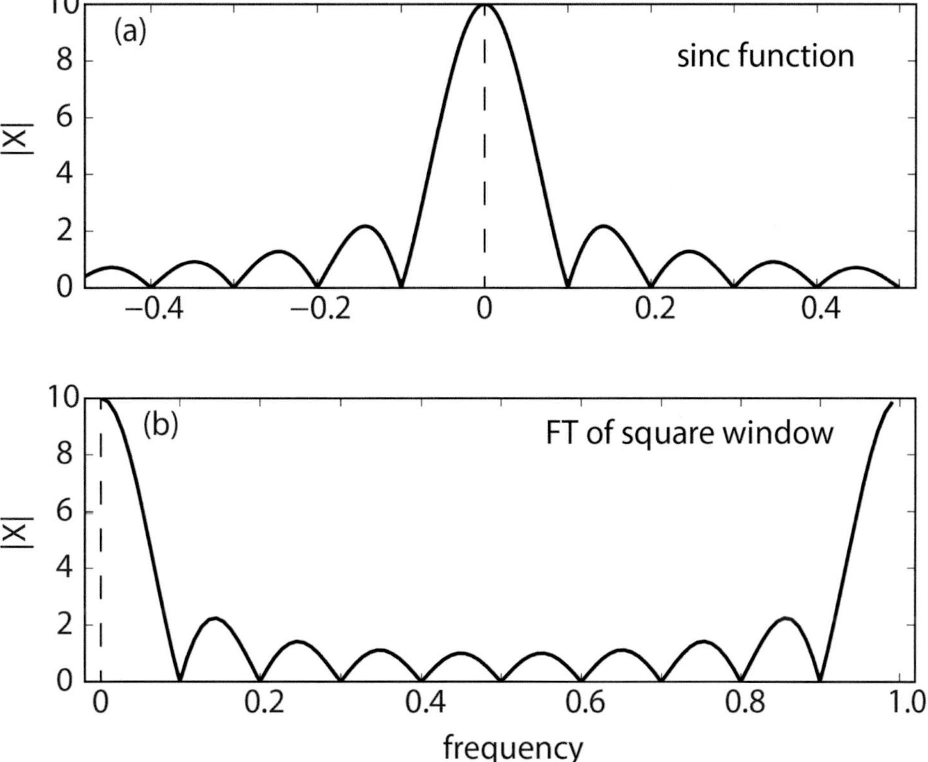

Figure 17.2 (a) The sinc function; (b) Fourier Transform of the rectangular window function.

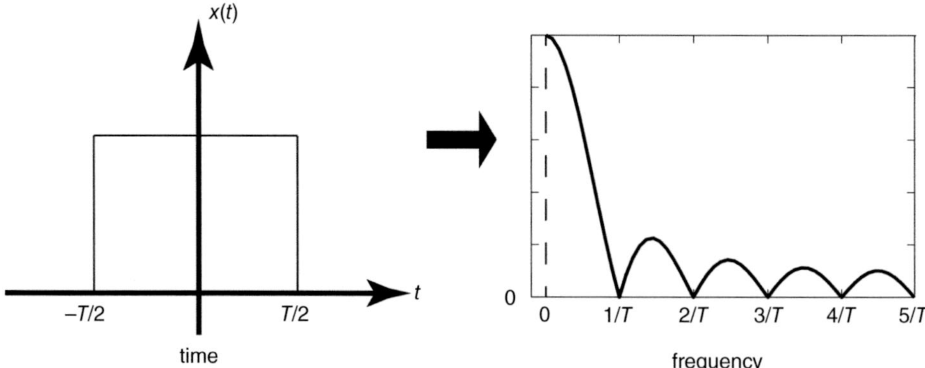

Figure 17.3 From rectangular window function to the sinc function.

17.1.4 The Heights of Side Lobes

The height of the main lobe is T, while the height of each of the side lobes is determined by the solution for

$$|X_T(f)| = \left|\frac{\sin(\pi f T)}{\pi f}\right| = \max, \tag{17.7}$$

for the f values other than $f = 0$, as $f = 0$ is the location of the main lobe. Let $a = \pi f T$; the above equation becomes

$$\left|\frac{\sin(a)}{a}\right| T = \max. \tag{17.8}$$

The necessary condition for this to hold is that the derivative of the left-hand side function with respect to a is zero, or

$$\frac{d}{da}\left(\frac{\sin(a)}{a}\right) = 0.$$

Since

$$\frac{d}{da}\left(\frac{\sin(a)}{a}\right) = \frac{a\cos(a) - \sin(a)}{a^2}, \tag{17.9}$$

we must have

$$g(a) = a\cos(a) - \sin(a) = 0. \tag{17.10}$$

There is no explicit solution for the above equation. We can use Newton's method (Mathews and Fink, 2004) to find the solution. Newton's method uses the following equation to iteratively find the solution of $g(a) = 0$:

$$a = \tilde{a} - \frac{g(\tilde{a})}{g'(\tilde{a})}. \tag{17.11}$$

Here the right-hand side variable $\tilde{a}$ is an initial guess or estimate and is therefore generally different from a on the left. It should be understood that the right-hand side is determined by the given value of $\tilde{a}$ and the left-hand side is an update of the value of a. The updated value is then used as the next estimate $\tilde{a}$ and provided to calculate the left-hand side to update the new value for a, and so on, until the updated value is essentially the same as the estimate, when we say the convergence occurs. Therefore, it is an iterative process. Usually, through a number of iterations, a correct value may be obtained if the initial guess is properly selected.

Since

$$g'(a) = -a \sin(a), \tag{17.12}$$

the iterative equation of (17.11) becomes

$$a = \tilde{a} + \frac{\tilde{a}\cos(\tilde{a}) - \sin(\tilde{a})}{a \sin(\tilde{a})}. \tag{17.13}$$

Using an initial value of $\tilde{a}_1 = 3\pi/2$, we can get an accurate solution for the position of the peak of the first side lobe:

$$a_1 = 4.493409457909064 \approx 4.49, \tag{17.14}$$

after only four iterations with a relative error of

$$er = |(a_1 \cos(a_1) - \sin(a_1))/a_1| = 2.47 \times 10^{-17}.$$

From equation (17.7), the height of the first side lobe is therefore

$$|X(f_1)| = \left|\frac{\sin(a_1)}{a_1}\right| T = 0.217233628211222 \times T \approx 0.22T, \tag{17.15}$$

so the ratio between the first side lobe height and the main lobe height is

$$\beta_1 = \left|\frac{\sin(a_1)}{a_1}\right| \approx 0.22. \tag{17.16}$$

This is a significant value (~22%) and is independent of T. This is true for any side lobe, i.e.

$$\beta_n = \left|\frac{\sin(a_n)}{a_n}\right| \quad \text{is independent of } T. \tag{17.17}$$

Therefore, no matter how long T is, the height of each of the side lobes has a fixed ratio with the height of the main lobe. The width of the main lobe is $2/T$ at the base. In general, the nth side lobe should be between $n\pi$ and $(n+1)\pi$, or near the average of these two values, i.e. the first guess of the maximum value would be near

$$\tilde{a}_n = \frac{2n+1}{2}\pi. \tag{17.18}$$

$\tilde{a}_n$ can be considered as the first guess or initial value of the iteration of Newton's method. For instance, for $\tilde{a}_2 = 5\pi/2$, the second side lobe is at

$$a_2 = 7.725251836937707 \approx 7.73. \tag{17.19}$$

With a relative error of ~3×10^{-15}, the height ratio of the second side lobe with the main lobe is

$$\beta_2 = \left|\frac{\sin(a_2)}{a_2}\right| = 0.128374553525899 \approx 12.83\%. \tag{17.20}$$

Likewise, for $\tilde{a}_3 = 7\pi/2$, the third side lobe is at

$$a_3 = 10.904121659428899 \approx 10.9. \tag{17.21}$$

With a relative error of ~ 8.3×10^{-16}, the height ratio of the third side lobe with the main lobe is

$$\beta_3 = \left|\frac{\sin(a_3)}{a_3}\right| \approx 9.13\%. \tag{17.22}$$

This can be done for any integer n for β_n.

17.1.5 The Limit of Sinc Function: Delta Function

We now take a closer look at the behavior of the sinc function as T approaches infinity. Obviously,

$$\lim_{T\to\infty, f\to 0} |X_T(f)| = \lim_{T\to\infty} \left|\frac{\sin(\pi f T)}{\pi f}\right|_{\max} \to +\infty. \tag{17.23}$$

This maximum value is reached at $f = 0$. At any fixed $f \neq 0$,

$$\lim_{T\to\infty} |X_T(f)|_{f\neq 0} = \lim_{T\to\infty} \left|\frac{\sin(\pi f T)}{\pi f}\right|_{f\neq 0} \to 0. \tag{17.24}$$

This can be seen from Fig. 17.4. As T increases, the first zero after the main lobe at $1/T$ is approaching $f = 0$ (Fig. 17.4a). When the region near the main lobe is zoomed in (Fig. 17.4b), it is shown that for each given value of T, the first side lobe has a value of ~22% of the height of the main lobe (which is equal to T).

It can be shown that

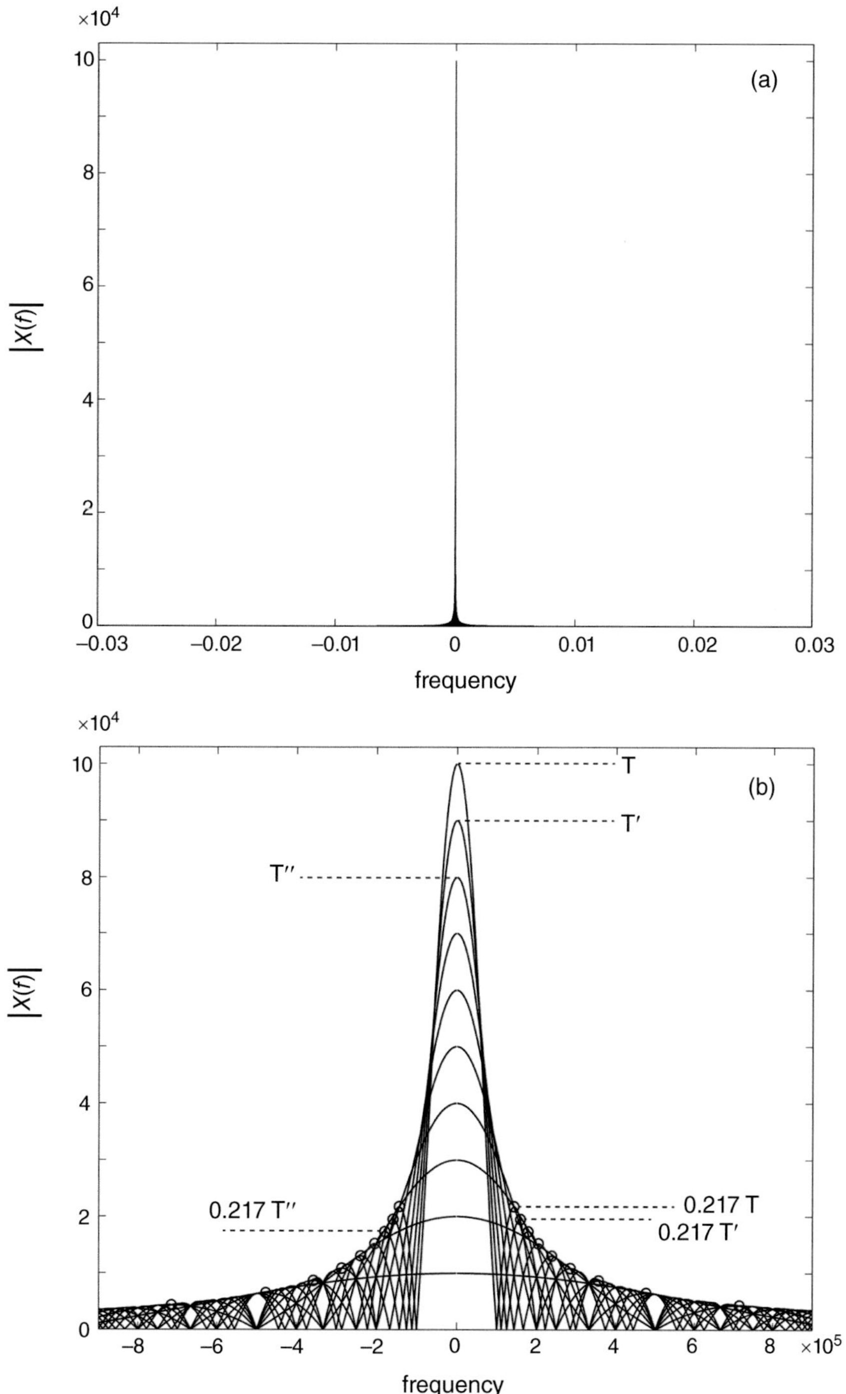

Figure 17.4 (a) The sinc function at high T value: it is a narrow peak with a height of T; (b) zoom-in view of the sinc function, showing the highest points (the circles) for the first side lobes at different T values.

$$\lim_{T \to \infty} X_T(f) = \delta(f) = \begin{cases} \infty, & f = 0 \\ 0, & f \neq 0 \end{cases} \tag{17.25}$$

and

$$\int_{-\infty}^{+\infty} \delta(f) df = 1. \tag{17.26}$$

The function $\delta(f)$ is the so-called *Dirac delta function*. It is a generalized function – a function not in a conventional sense, as it only makes sense when the limit and integration are considered. In other words, the sinc function approaches a Dirac delta function when $T \to \infty$.

The above discussion concludes that the Fourier Transform of 1 is a delta function:

$$\mathbb{F}(1) = \int_{-\infty}^{+\infty} e^{-i2\pi f t} dt = \delta(f). \tag{17.27}$$

From this, we can further extend the conclusion to the Fourier Transform of $e^{i2\pi ft}$:

$$\mathbb{F}(e^{i2\pi f_0 t}) = \int_{-\infty}^{+\infty} e^{i2\pi f_0 t} e^{-i2\pi f t} dt = \delta(f - f_0). \tag{17.28}$$

An important property of the delta function is that for a continuous function $g(x)$, the integration of the product of a continuous function and the delta function $g(x)\delta(x)$ is

$$\int_{-\infty}^{+\infty} g(x)\delta(x) dx = g(0). \tag{17.29}$$

This property will be used in the following.

17.1.6 The Fourier Transform of Truncated Sine Function

Now we come back to equation (17.3) and examine the Fourier Transform of the truncated sine function. In this case, the function $y(t)$ is a sine function $y(t) = \sin(2\pi f_0 t)$. Its Fourier Transform is

$$\mathbb{F}\big(\sin(2\pi f_0 t)\big) = Y(f) = \int_{-\infty}^{+\infty} \sin(2\pi f_0 t) e^{-i2\pi f t} dt. \tag{17.30}$$

Applying the Euler formula, equation (13.1), equation (17.30) can be expressed as

$$Y(f) = \int_{-\infty}^{+\infty} \sin(2\pi f_0 t) e^{-i2\pi f t} dt = \int_{-\infty}^{+\infty} \frac{e^{i2\pi f_0 t} - e^{-i2\pi f_0 t}}{2i} e^{-i2\pi f t} dt. \qquad (17.31)$$

Using (17.28), the above equation is

$$Y(f) = \frac{1}{2i} \int_{-\infty}^{+\infty} \left[e^{-i2\pi(f - f_0)t} - e^{-i2\pi(f + f_0)t} \right] dt = \frac{1}{2i} [\delta(f - f_0) - \delta(f + f_0)]. \quad (17.32)$$

Therefore, equation (17.4) becomes

$$Y_s(f) = Y(f) \otimes X_T(f) = \frac{1}{2i} \int_{-\infty}^{+\infty} [\delta(\tilde{f} - f_0) - \delta(\tilde{f} + f_0)] X_T(f - \tilde{f}) d\tilde{f}. \qquad (17.33)$$

Applying equation (17.29), the above equation becomes

$$Y_s(f) = \frac{1}{2i} \int_{-\infty}^{+\infty} [\delta(\tilde{f} - f_0) - \delta(\tilde{f} + f_0)] X_T(f - \tilde{f}) d\tilde{f} = \frac{1}{2i} [X_T(f - f_0) + X_T(f + f_0)]. \qquad (17.34)$$

This result says that the Fourier Transform of a truncated sine function can be represented by two sinc functions, one centered at f_0 and the other at $-f_0$, each with a half of the magnitude at the center of the sinc function.

17.1.7 Finite Sampling Theorems #2 and #3

The Fourier Transform of a finite length of the sine function (or a truncation from an infinitely long sine function) with a frequency of f_0 is two sinc functions superimposed onto each other, one centered at $f = f_0$ and the other centered at $f = -f_0$. If we only examine the magnitude spectrum for positive frequencies, we can combine the two sinc functions by multiplying by two and write:

$$|Y_s(f)| = |X_T(f - f_0)|. \qquad (17.35)$$

This conclusion can also be reached by using the frequency delay property of the Fourier Transform.

When there is a shift in frequency, the inverse Fourier Transform results in an extra factor: if $Y(\omega)$ is the Fourier Transform of function $y(t)$, then multiplying this function with an exponential function, as in the following, results in a frequency shift of the original Fourier Transform, i.e.

$$y(t)e^{i\omega' t} \Leftrightarrow Y(\omega - \omega').$$

If $y(t)$ here is a window function, we will get a similar result as (17.35).

On the basis of the above discussion, we have our second finite sampling theorem:

> **Finite Sampling Theorem #2** *The magnitude spectrum (magnitude of Fourier Transform) of a finite segment of a single frequency signal, or a sine function with frequency* f_0*, is a sinc function centered at* f_0*.*

This can be extended to a time function with a summation of multiple sinusoidal functions with different frequencies. This leads to our next theorem.

> **Finite Sampling Theorem #3** *The magnitude spectrum (magnitude of Fourier Transform) of a finite segment of a combination of multiple frequency signals or a function with components at frequencies* $f_0, f_1, f_2, \ldots$ *is a series of sinc functions centered at* $f_0, f_1, f_2, \ldots,$ *respectively.*

These theorems can be verified by extending to cosine functions and by including more than one sinusoidal function. We omit the mathematics.

Again, a function running from $-\infty$ to $+\infty$ in time with a series of discrete frequencies $f_0, f_1, f_2, \ldots$ should have a spectrum in the frequency domain represented by isolated lines at $f = f_0, f = f_1, f = f_2, \ldots$. Because of the truncation, the spectrum becomes the sinc functions centered at these true frequencies. The main lobe of each of the sinc functions shows the true peak and the side lobes are all unwanted but resulted from the finite sampling. One can consider that these side lobes are smeared from the true spectrum lines due to the finite sampling.

17.1.8 An Example

As an example, let us assume that a function is a superposition of eight functions:

$$\begin{aligned} s = {} & a_1 \sin(2\pi f_1 t) + a_2 \sin\left(2\pi f_2 t + \frac{\pi}{2}\right) + a_3 \sin\left(2\pi f_3 t + \frac{\pi}{3}\right) + a_4 \sin\left(2\pi f_4 t + \frac{\pi}{4}\right) \\ & + a_5 \sin\left(2\pi f_5 t + \frac{\pi}{5}\right) + a_6 \sin\left(2\pi f_6 t + \frac{\pi}{6}\right) + a_7 \sin\left(2\pi f_7 t + \frac{\pi}{7}\right) + a_8 \sin\left(2\pi f_8 t + \frac{\pi}{8}\right)^{\cdot} \end{aligned} \tag{17.36}$$

in which $f = (f_1, f_2, f_3, f_4, f_5, f_6, f_7, f_8) = (1/20, 1/10, 1/5, 1/3, 1/2, 5, 10, 20)$; and

$$a = (a_1, a_2, a_3, a_4, a_5, a_6, a_7, a_8) = (1, 1.2, 0.6, 2, 0.3, 1, 0.2, 1.1).$$

The time function is shown in Fig. 17.5. The time series (Fig. 17.5a) shows a complicated function, while the zoomed-in view (Fig. 17.5b) shows some of its higher frequency variations. With the following MATLAB script, the Fourier Transform for the given time series is done (Fig. 17.6).

```
t = (0:0.01:100)';
f1 = 1 / 20; f2 = 1 / 10; f3 = 1 / 5; f4 = 1 / 3; f5 = 1 / 2; f6 = 5; f7 = 10; f8 = 20;
s = sin(2*pi*f1*t)+1.2*sin(2*pi*f2*t + pi/2)+0.6*sin(2*pi*f3*t...
+pi/3)+2*sin(2*pi*f4*t+pi/4)+0.3*sin(2*pi*f5*t+pi/5);
s = s + sin(2*pi*f6*t + pi/6)+.2*sin(2*pi*f7*t + pi/7)...
+1.1*sin(2*pi*f8*t+pi/8); Ns = length(s);
s = [ zeros(length(s),1); s; zeros(length(s) * 5,1)];
S = fft(s);
```

Some zero-padding is included for a better visual effect. The spectrum appears to demonstrate some discretely distributed peaks, but a zoomed-in view reveals that they are indeed sinc functions centered at $f = (f_1, f_2, f_3, f_4, f_5, f_6, f_7, f_8)$, respectively (Fig. 17.6).

An alternative way to show the result is to normalize the Fourier Transform by the total number of data points using equation (13.21):

$$\alpha_k = \frac{1}{M} Y_k. \tag{17.37}$$

Fig. 17.7 shows the normalized Fourier Transform. In Fig. 17.7, the magnitude of the Fourier Transform has been multiplied by 2 to include the contribution from the negative frequency. This is called the one-sided spectrum (vs. the two-sided spectrum if the negative frequencies are included). Since the negative frequencies are introduced for convenience when the exponential base functions are used to replace the sine and cosine functions, this one-sided spectrum is more commonly used as it is more concise. The symmetry of the magnitude spectrum about $f = 0$ allows this operation. The advantage of using the one-sided spectrum and normalized values is that they are consistent with the Fourier series coefficients with the same unit as the original time series function.

17.1.9 Finite Sampling Theorem #4

From the above discussion and Theorem 3, we see that when there are multiple frequencies in a signal, sinc functions will replace the true spectrum of individual peaks at those frequencies of the true signal. The truncation of the function into finite length causes each peak to be replaced by a sinc function centered at the

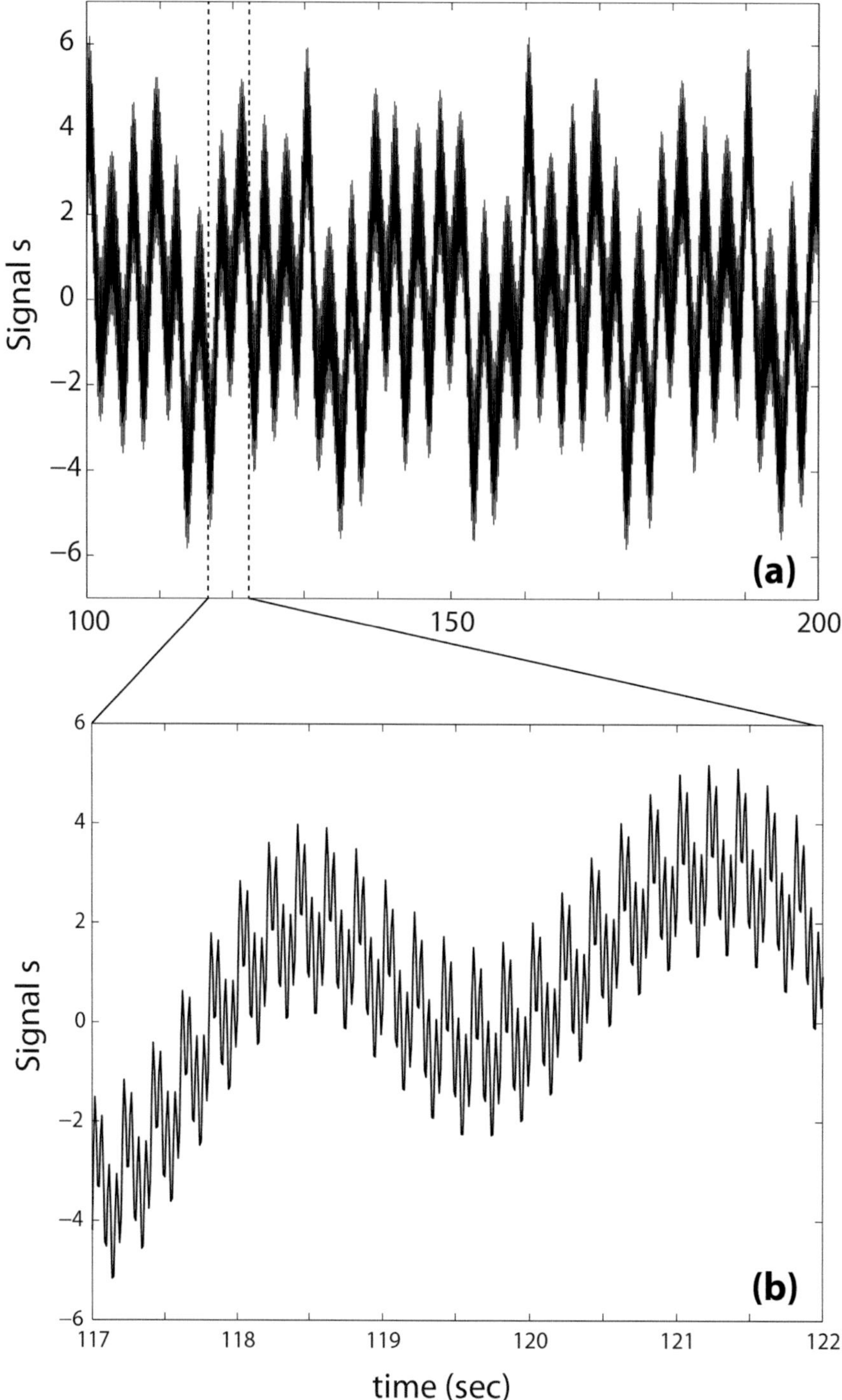

Figure 17.5 An example of a signal with eight different frequencies. (a) The entire series; (b) a zoomed-in segment of the data.

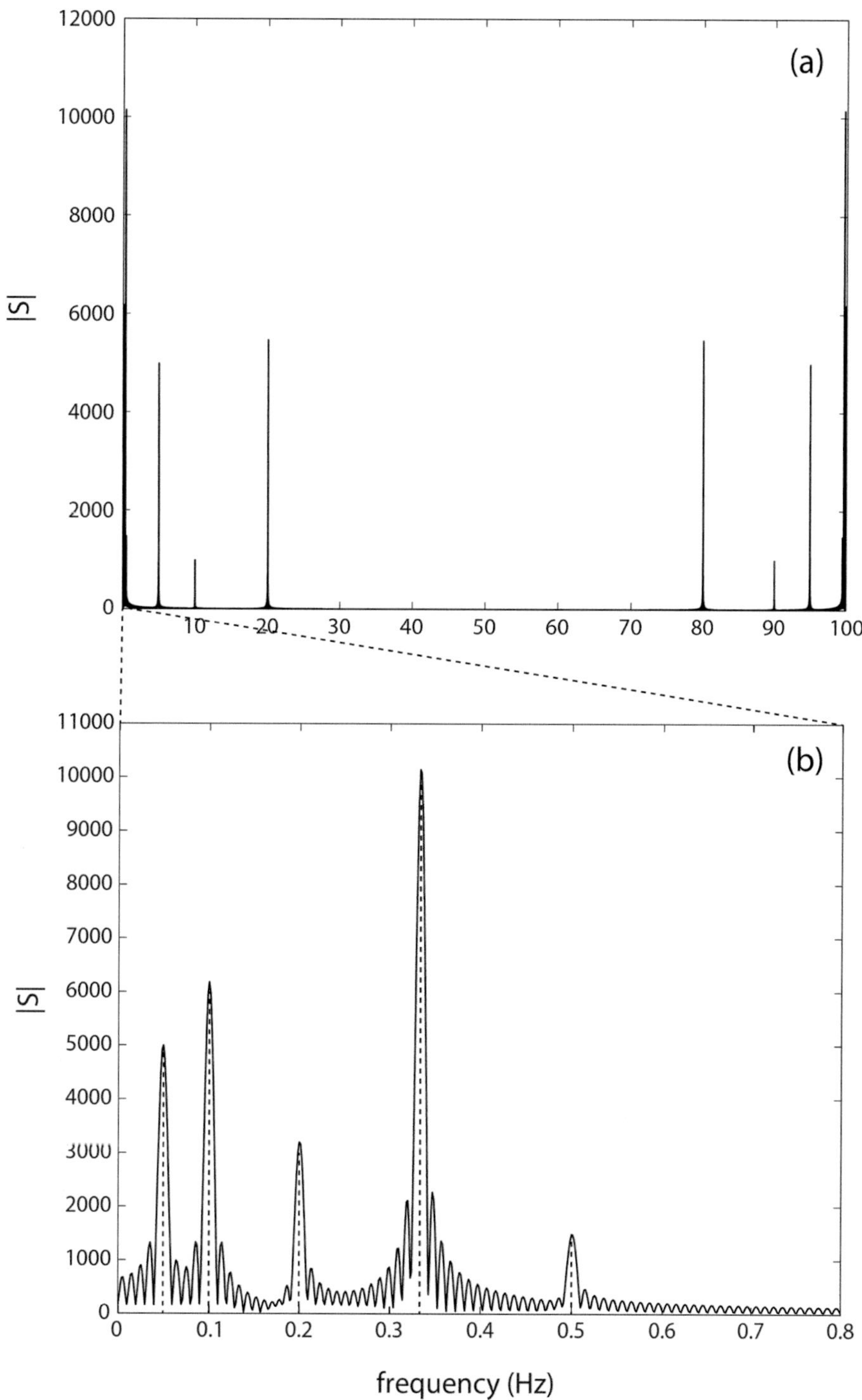

Figure 17.6 An example demonstrating the third theorem. (a) The original spectrum; (b) zoomed-in view at a lower frequency.

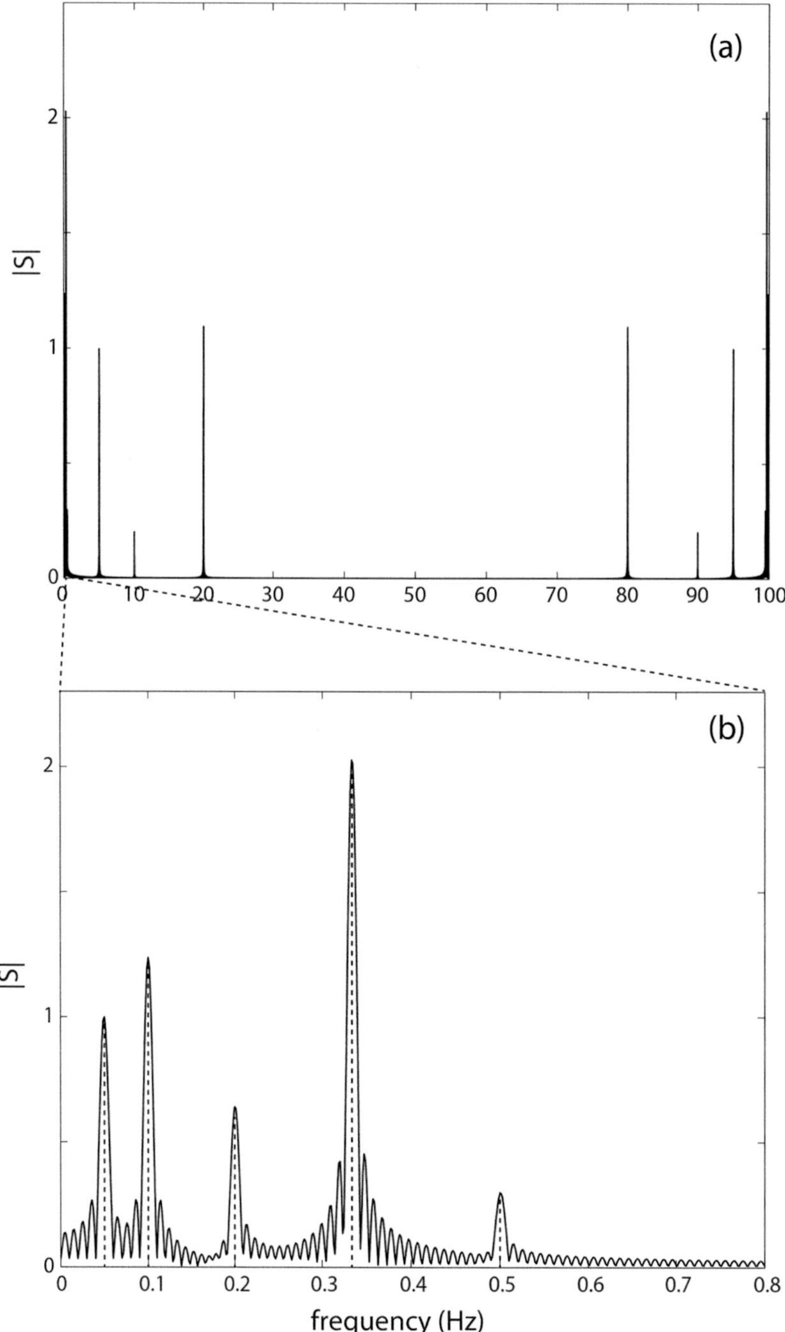

Figure 17.7 Normalized Fourier Transform – the Fourier Transform is divided by the number of data points. They are equivalent to the Fourier series coefficients.

original frequency. The width of one sinc function may overlap with that of another. Since the width of the main lobe is $1/T$, to distinguish between two close peaks, one needs $1/T$ to be less than the difference of the two frequencies. This leads to the next theorem.

Finite Sampling Theorem #4 *The resolution for frequency in Fourier Transform of a function of length T is* $1/T$*. In order to distinguish between two frequencies* f_1 *and* f_2*, we must have*

$$|f_2 - f_1| \geq \frac{1}{T}, \tag{17.38}$$

with a condition that the sampling can also resolve both frequencies, i.e. f_1 and f_2 are resolved individually.

This equation is presented with an ad hoc discussion in Chapter 16, i.e. equation (16.1), and now it is formally presented as a result of the discussion of the sinc function.

How can a given frequency in a signal be resolved by a discrete sampling? This is a subject of the next section, on one of the most important issues of signal data analysis.

17.2 The Effect of Discrete Sampling

Any data analysis using computers will have to use digital data with discrete values. We can consider the real signal as being continuous, and it is sampled at certain intervals. Indeed, natural processes are often considered "continuous," while observations are usually "discrete" with finite sampling intervals. When a continuous function is sampled discretely, we may lose useful information, and as a result the sampled data may have different characteristics (Fig. 17.8). A common-sense example is that the variations between two adjacent sampling points cannot be shown in the sampled data. A "spike" in a process may be missed by observations if the observations are too sparse.

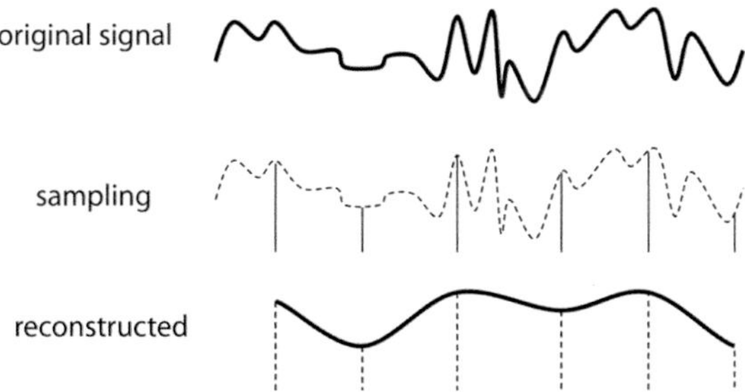

Figure 17.8 Finite sampling of continuous data with complex variability.

For convenience of discussion, we use constant sampling intervals (Δt is constant). In real applications of time series analysis, the sampling intervals may not be a constant or there may be missing data in an otherwise constant interval series. Some treatment may be needed to convert the series to have constant intervals first, e.g. an interpolation of some kind may be used to generate a constant interval series before any subsequent analysis.

17.2.1 Aliasing: A Finite Sampling Effect

In general, if the sampling frequency is slower than a certain frequency contained in a process, some features containing fast variations may be *missed or distorted*, such that the data are misrepresented, and aliasing thus occurs (Fig. 17.9).

Let us again consider a simple example of a sinusoidal function. Assuming it is defined as

$$y(t) = A\cos(2\pi f t + \theta). \tag{17.39}$$

When this time series is sampled with a time interval Δt, we define the sampling frequency f_s to be

$$f_s = 1/\Delta t. \tag{17.40}$$

Thus, the sampling is taken at these discrete times:

$$t_s = n\Delta t = n/f_s, \quad n = 0, 1, 2, \ldots. \tag{17.41}$$

The sampled discrete time series is now determined by the discrete times, i.e.

$$y(t_s) = A\cos(2\pi f t_s + \theta) = A\cos\left(2\pi n\frac{f}{f_s} + \theta\right), \tag{17.42}$$

in which n is an integer $(n = 1, 2, 3, \ldots)$.

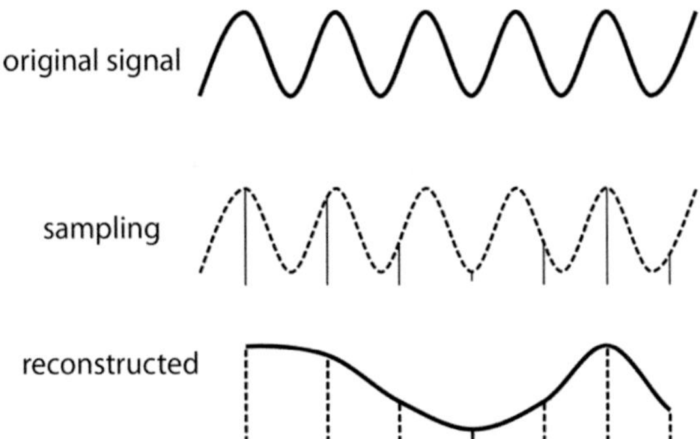

Figure 17.9 Finite sampling of a continuous data with a single frequency. The sampled data appear to have a lower frequency.

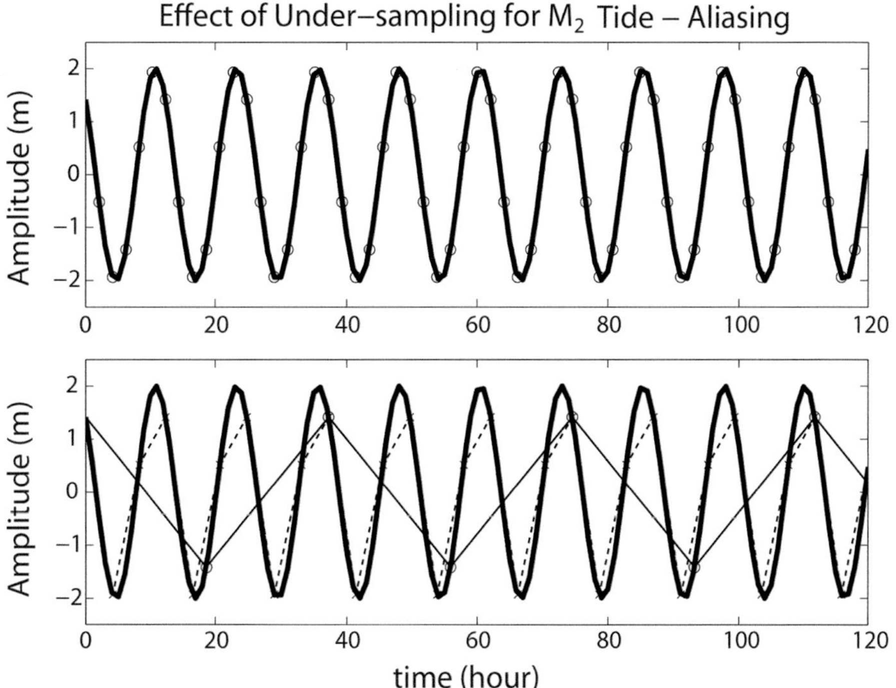

Figure 17.10 Effect of finite sampling: upper panel – sample 6 times a period; lower panel – sample at 3 times per period (red) and 2/3 times per period (black).

Obviously, there are only a limited number of data points within a period $(1/f)$: $\max(n) = [T/\Delta t] + 1$, in which [ ] means the integer part of the number. This is fundamentally different from a continuous time function. The sampling frequency f_s thus cannot be "too small" – otherwise the number of samples within a period $t = [0, 1/f]$ may be too few to reconstruct the original function (Fig. 17.9). As an extreme example: if $f = f_s$ (sampling 1 point in a cycle), then

$$y(t_s) = A\cos(2\pi n + \theta) = A\cos(\theta), \tag{17.43}$$

which is a constant (not a function of n). Obviously, this sampling makes the result beyond recognition (in view of the original function).

Fig. 17.10 shows the effect of sampling of the simple sine function, with 6, 3, and 2/3 samples per period, respectively. Intuitively, the faster the sampling, the more likely that we can reconstruct the correct function. Or we require that Δt is "small enough" or f_s is "large enough" to avoid missing some variations at higher frequencies.

17.2.2 Sampling Theorem #5

Sampling Theorem #5 (the base band form) *If the highest frequency in x(t) is $f_m = B$ and the signal is sampled faster than twice of the highest frequency, i.e. $f_s > 2B$, then $y(t)$ can be exactly recovered from the samples using the interpolation function (Proakis and Manolakis, 1992)*

$$g(t) = \frac{\sin(2\pi Bt)}{2\pi Bt} \tag{17.44}$$

$$y(t) = \sum_{n=-\infty}^{\infty} y\left(\frac{n}{f_S}\right) g\left(t - \frac{n}{f_S}\right). \tag{17.45}$$

Here half of the sampling frequency is the Nyquist frequency:

$$f_N = \frac{f_S}{2} = \frac{1}{2\Delta t}. \tag{17.46}$$

The Nyquist frequency is also called the *folding frequency* – it is the highest frequency resolvable with the given sampling rate. It is determined by the sampling rate.

Another slightly different (perhaps more intuitive) expression of the sampling theorem is the following equation (Proakis and Manolakis, 1992):

$$z(t) = \Delta t \sum_{n=-\infty}^{\infty} y_n \frac{\sin\left(2\pi B(t - t_n)\right)}{\pi(t - t_n)}. \tag{17.47}$$

Here y_n is the discrete sample (or measurement) of the original function $y(t)$, and $z(t)$ is the interpolated function which should be the same as $y(t)$.

Sampling Theorem Restated *If the Nyquist frequency is greater than the maximum frequency in the signal, the original signal can be reconstructed using (17.47) without distortion.*

A discrete sampling cannot reveal any information about the signal at frequencies higher than the Nyquist Frequency. So the conclusion is: *sample as fast as possible.*

In other words, to avoid distortion, we require that

$$f_s = f_{s\min} \geq 2f_{\max}. \tag{17.48}$$

In practice, we may use more than $2f_{\max}$ for $f_{s\min}$ just to be on the safe side. i.e.

$$f_s \geq f_{s\min} = \begin{cases} 2f_{\max} \text{ (theoretical value)} \\ > 2f_{\max} \text{ (practical value)} \end{cases}. \tag{17.49}$$

Otherwise, if $f_s < f_{s\min}$, it is “under sampled.”

17.2.3 Examples for Checking the Sampling Theorem

Now let us discuss some examples for checking the sampling theorem using equation (17.45) or (17.47). Assume that the original continuous function is:

$$y(t) = \sin(2\pi f_1 t) + \sin(2\pi f_2 t) + \sin(2\pi f_3 t) + \sin(2\pi f_4 t) + \sin(2\pi f_5 t), \tag{17.50}$$

in which

$$f_1 = \frac{B}{2},\ f_2 = \frac{B}{3},\ f_3 = \frac{B}{4},\ f_4 = \frac{B}{5},\ f_5 = \frac{8B}{9},\ B = 10,\ t = (-\infty, +\infty). \tag{17.51}$$

In actual calculation, we cannot have an infinite time period. In this example, we choose $t = [-100, 100]$ with a very small dt (0.001, so we can consider it as "continuous"). We then take discrete samples at dt = 0.01 within the same range of t. The "continuous" function and discrete samples are shown in Fig. 17.11. It should be noted that in equation (17.47), B is an upper limit of the frequency

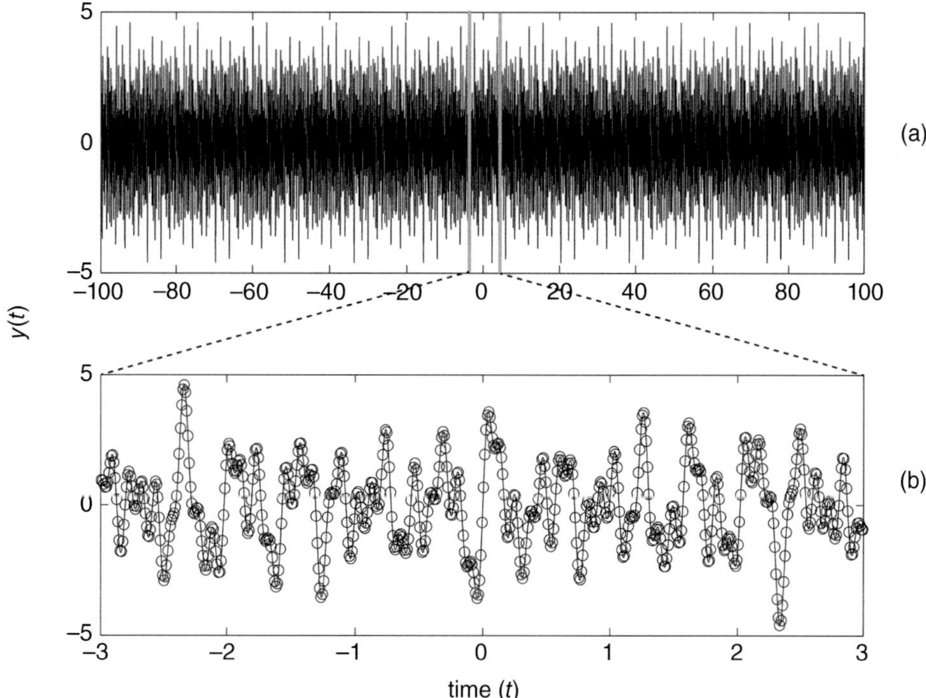

Figure 17.11 A time function from equation (17.50). (a) The original function from (17.50); (b) a zoomed-in view of the original function and the discretely sampled data points in a subset for better visualization.

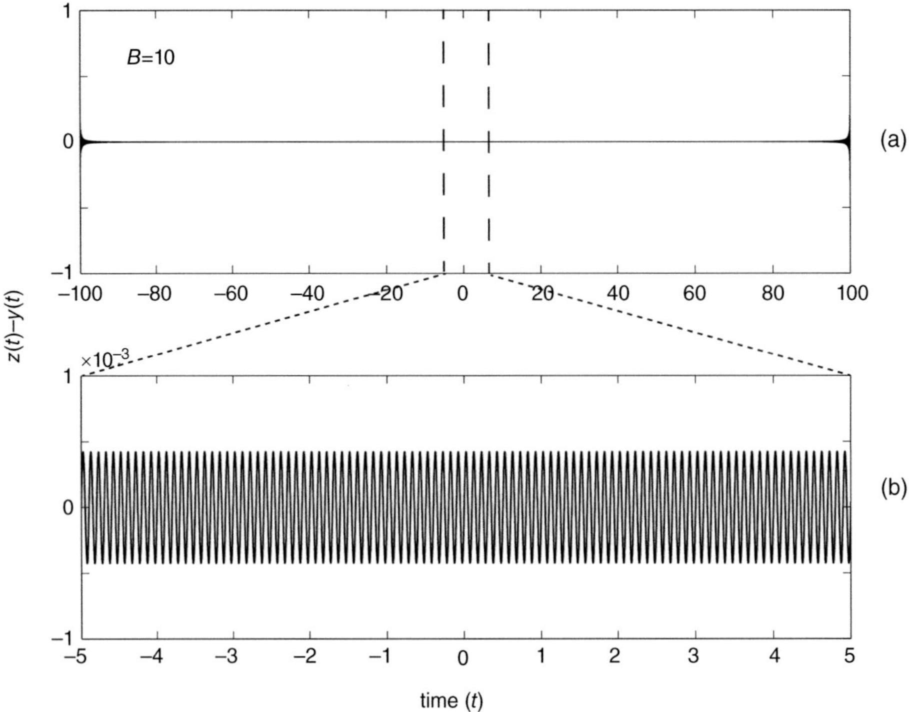

Figure 17.12 The difference between the original function $y(t)$ and the reconstructed function $z(t)$ with the discrete sampling and the sampling theorem using equation (17.47) with $B = 10$. (a) $z(t) - y(t)$; (b) a zoomed-in view of (a).

contained in the signal. As long as B is selected to be greater than the maximum frequency in the signal, equation (17.47) is correct. To show this, we choose different B values (10 and 60). Fig. 17.12 shows the difference between the original function and the reconstructed function after the discrete sampling and the use of equation (17.47) with $B = 10$. Fig. 17.13 is similar but with $B = 60$. The latter appears to have higher accuracy.

17.3 Discussion on Aliasing

Once we choose the sampling frequency (f_s), we can examine the case when the signal has a frequency higher than f_s: if in the original signal, there is a frequency $f > f_s$, then we can always find a $f_1 > 0$, such that

$$f = f_s + f_1 \tag{17.52}$$

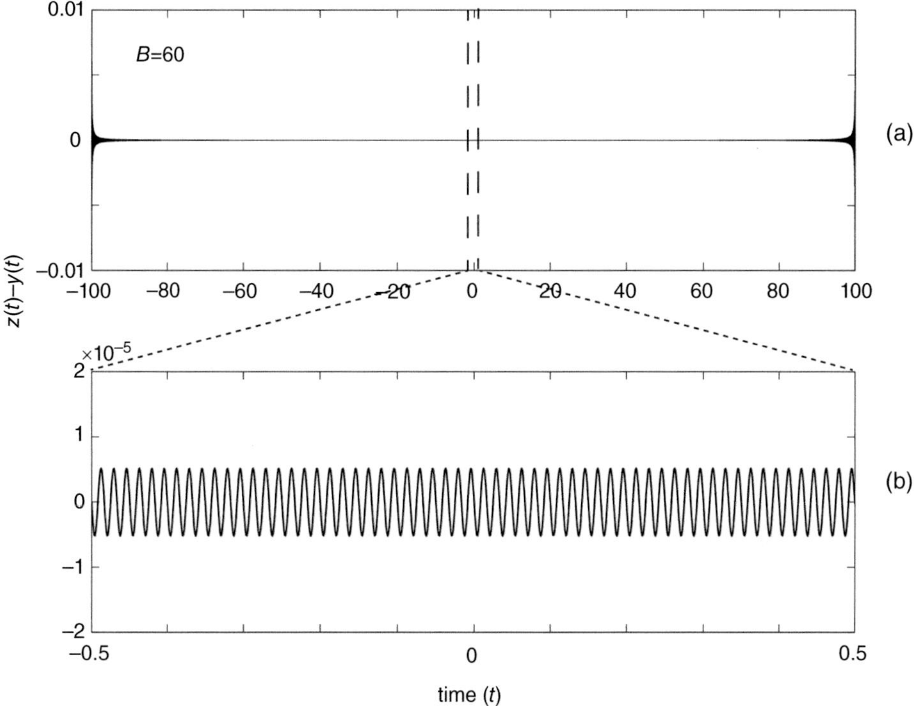

Figure 17.13 The difference between the original function $y(t)$ and the reconstructed function $z(t)$ with the discrete sampling and the sampling theorem using equation (17.47) with $B = 60$. (a) $z(t) - y(t)$; (b) A zoom in view of (a).

and

$$\begin{aligned} y(t_s) &= A\cos\left(2\pi n\frac{f_1 + f_s}{f_s} + \theta\right) \\ &= A\cos\left(2\pi n\frac{f_1}{f_s} + \theta\right) \end{aligned}. \tag{17.53}$$

The above equation tells us that we cannot distinguish between f and f_1. In general, for any $f_k = f + kf_s$, $(k = 1, 2, 3, \ldots)$,

$$\begin{aligned} y(t_s) &= A\cos\left(2\pi n\frac{f + kf_s}{f_s} + \theta\right) \\ &= A\cos\left(2\pi n\frac{f}{f_s} + \theta\right) \end{aligned}. \tag{17.54}$$

Therefore, $f_k = f + kf_s$ is indistinguishable from f. These frequencies are aliases of f. So f should not be greater than f_s. The frequency range

$$0 \le f \le f_s/2 \tag{17.55}$$

is called the *fundamental frequency range*.

Any frequency that is higher than the folding frequency is "folded" into the fundamental frequency range and causes the data to be aliased.

Review Questions for Chapter 17

(1) What do we mean by "finite sampling"?
(2) What parameters are important for finite sampling?
(3) What is the relationship between sinc function and delta function?

Exercises for Chapter 17

(1) Verify in theory that the peak of sinc function at f ¼ 0 is T
(2) Compute the heights of the fourth and fifth side lobes, β_4 and β_5 (equation 17.17), using the method described in this chapter.
(3) To measure water level and resolve an M6 overtide, how fast should one plan to sample?

18

Power Spectrum, Cospectrum, and Coherence

About Chapter 18

The objective of this chapter is to present some important relations between the Fourier Transform and correlation functions. It turns out that the cross-correlation function and autocorrelation function have some useful relationships to Fourier Transform and power spectrum of the individual functions. As a result, cospectrum and coherence (normalized statistical correlation in frequency domain) can be defined.

18.1 Temporal Correlation Functions

18.1.1 Cross-Correlation Function

The Fourier Transform of a function gives the spectrum of the function. The spectrum is often expressed by the *magnitude* of the Fourier Transform. In many applications, such *magnitude spectrum* is sufficient for subsequent analysis and to make proper inferences or conclusions. However, in some applications, it is equally if not more important to examine the so-called Power Spectrum. Sometimes, it is called the *power spectral density function.* Related to this is another important property of Fourier Transform of the correlation function of two functions.

For two functions $y_1(t)$ and $y_2(t)$, their *temporal cross-correlation function* is defined as

$$r_{12}(t) = \int_{-\infty}^{\infty} y_1(\tau) y_2(t+\tau) d\tau. \tag{18.1}$$

This is a function of time t. In general, the temporal cross-correlation function is not symmetric – it depends on the order of the functions inside the integral, i.e.

$$r_{12}(t) \neq r_{21}(t). \tag{18.2}$$

In fact, we can verify that when the order of the functions switched, the cross-correlation function will have to be flipped about the vertical axis, i.e.

$$r_{12}(t) = r_{21}(-t), \tag{18.3}$$

or they are symmetric about $t = 0$.

The verification of the above property is straightforward by a variable change. First, in equation (18.1), use $-t$ to replace t, which leads to

$$r_{12}(-t) = \int_{-\infty}^{\infty} y_1(\tau)y_2(-t+\tau)d\tau. \tag{18.4}$$

On the right-hand side of the above equation, let $\tilde{\tau} = \tau - t$. We get

$$r_{12}(-t) = \int_{-\infty}^{\infty} y_1(\tilde{\tau}+t)y_2(\tilde{\tau})d\tilde{\tau} = r_{21}(t).$$

So, equation (18.3) is verified. This proof is actually just a mathematical realization of a more intuitive interpretation. To illustrate this, note that the cross-correlation function calculation (18.1) is a mathematical operation by overlapping the first function on a translated function of the second function and then multiplying the two functions at their respective positions after the translation. Now the translation is all relative, i.e. translating the first function forward by t is equivalent to translating the second function backward by $-t$. The correlation values will be the same for both cases but at different positions (one at t, one at $-t$), and therefore the property shown in (18.3) becomes obvious (Fig. 18.1).

In the cross-correlation function, if the independent variable t or the amount of shift between the two functions is zero, i.e. $t = 0$, the two functions are directly multiplied together before the integration in equation (18.1). This gives a value that shows how much the two functions are alike without phase shift: if the two functions have a similar trend of variation, the integral is large. An extreme example is when the two functions are the same – the autocorrelation case. Another extreme example is if they are so different that their product is a randomly distributed function, in which case the two functions are uncorrelated and the integral would be relatively small.

Sometimes, the two functions may seem to be uncorrelated but with a phase shift (or time lag), i.e. at certain non-zero t, the cross correlation can be relatively large. In other words, the two functions are somehow correlated but their phase lag

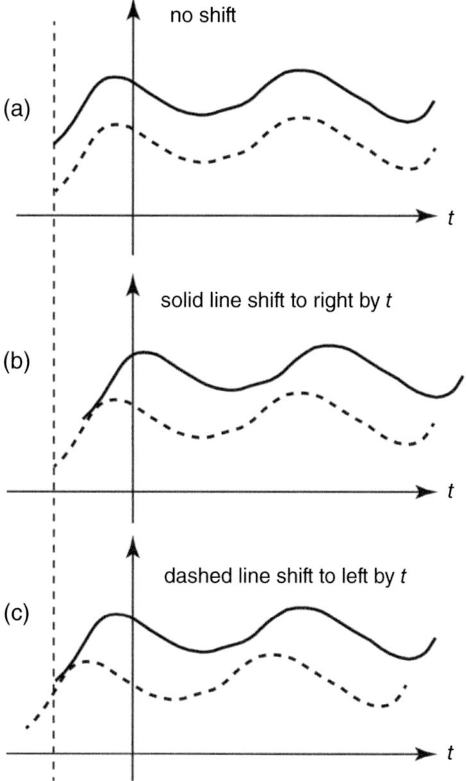

Figure 18.1 Property of cross-correlation function: shifting one function to the right by t is equivalent to shifting the other function to the left by $-t$.

might have obscured that characteristic. That is when the lagged correlation can be examined with the correlation function to reveal the relationship.

18.1.2 Autocorrelation Function

If the two functions are the same, $y_1(t) = y_2(t)$, then we have the *autocorrelation function*:

$$r_{11}(t) = \int_{-\infty}^{\infty} y_1(\tau) y_1(t+\tau) d\tau. \tag{18.5}$$

The autocorrelation function can give information about the self-similarity of a function or the scale over which the function can be correlated to itself. Usually, with a large lag t, the autocorrelation function will have small values compared to the case with no lag or $t = 0$.

18.1.3 Cross-Correlation Fourier Transform Theorem

There is an important theorem on the Fourier Transform of the cross-correlation function:

Assume that we have two Fourier Transform pairs, $y_1(t) \Leftrightarrow Y_1(f)$ and $y_2(t) \Leftrightarrow Y_2(f)$, in which

$$Y_1(f) = \int_{-\infty}^{+\infty} y_1(t)e^{-i2\pi ft}dt, \quad Y_2(f) = \int_{-\infty}^{+\infty} y_2(t)e^{-i2\pi ft}dt \tag{18.6}$$

are the Fourier Transforms of the time functions $y_1(t)$ and $y_2(t)$, respectively. The Fourier Transform of the cross-correlation function of the two time-functions is equal to the complex conjugate of the first Fourier Transform multiplying the Fourier Transform of the second function, i.e.

$$\mathbb{F}\big(r_{12}(t)\big) = Y_1^*(f)Y_2(f). \tag{18.7}$$

here the * means complex conjugate. To verify this theorem, we do Fourier Transform to the cross-correlation function $r_{12}(t)$, which is

$$\int_{-\infty}^{+\infty} r_{12}(t)e^{-i2\pi ft}dt = \int_{-\infty}^{+\infty}\left(\int_{-\infty}^{+\infty} y_1(\tau)y_2(t+\tau)d\tau\right)e^{-i2\pi ft}dt. \tag{18.8}$$

Now we switch the order of the integration:

$$\begin{aligned} \int_{-\infty}^{+\infty} r_{12}(t)e^{-i2\pi ft}dt &= \int_{-\infty}^{+\infty} y_1(\tau)\left(\int_{-\infty}^{+\infty} y_2(t+\tau)e^{-i2\pi ft}dt\right)d\tau = \int_{-\infty}^{+\infty} y_1(\tau)Y_2(f)e^{i2\pi f\tau}d\tau \\ &= Y_2(f)\int_{-\infty}^{+\infty} y_1(\tau)e^{i2\pi f\tau}d\tau = Y_2(f)Y_1(-f) \\ &= Y_2(f)Y_1^*(f) \end{aligned} \tag{18.9}$$

Here we have used the fact that y_1 is a real function (so is y_2), so that

$$Y_1(-f) = Y_1^*(f). \tag{18.10}$$

Therefore, we have

$$r_{12}(t) \Leftrightarrow Y_1^*(f)Y_2(f), \tag{18.11}$$

which is the same as (18.7).

18.1.4 The Wiener-Khinchin Theorem

As a special case of the Cross-Correlation Fourier Transform Theorem (18.7), there is the *Wiener-Khinchin Theorem*, when the two functions are the same, i.e. if

$$y_1(t) = y_2(t) = y(t). \tag{18.12}$$

The above equation gives the Fourier Transform of the autocorrelation function as the magnitude squared of the Fourier Transform.

$$\mathbb{F}(r(t)) = \mathbb{F}\big(r_{11}(t)\big) = |Y(f)|^2, \tag{18.13}$$

in which

$$Y(f) = \int_{-\infty}^{+\infty} y(t)e^{-i2\pi ft}dt, \tag{18.14}$$

The relationship given by equation (18.13) is the Wiener-Khinchin Theorem.

18.1.5 Parseval's Theorem

In the following, we will discuss again the integral form of *Parseval's Theorem*, which is mentioned in earlier chapters in various forms:

$$\int_{-\infty}^{+\infty} |y(t)|^2 dt = \int_{-\infty}^{+\infty} |Y(f)|^2 df. \tag{18.15}$$

This says that the total energy in the time-domain is the same as that in the frequency-domain. We can prove this theorem by using the inverse Fourier Transform of $y(t)$:

$$\begin{aligned}\int_{-\infty}^{+\infty} |y(t)|^2 dt &= \int_{-\infty}^{+\infty} y(t)y^*(t)dt = \int_{-\infty}^{+\infty} y(t)\left(\int_{-\infty}^{+\infty} Y(f)e^{i2\pi ft}df\right)^* dt \\ &= \int_{-\infty}^{+\infty} y(t)\left(\int_{-\infty}^{+\infty} Y^*(f)e^{-i2\pi ft}df\right)dt - \int_{-\infty}^{+\infty} Y^*(f)\left(\int_{-\infty}^{+\infty} y(t)e^{-i2\pi ft}dt\right)df.\end{aligned} \tag{18.16}$$

In the above, we have switched the order of integrations again. It is now obvious that

$$\int_{-\infty}^{+\infty} |y(t)|^2 dt = \int_{-\infty}^{+\infty} Y^*(f)Y(f)df = \int_{-\infty}^{+\infty} |Y(f)|^2 df. \tag{18.17}$$

This verifies (18.15).

18.1.6 Checking the Parseval's Theorem Using FFT

We now check equation (18.15) with a discrete dataset using the discrete Fourier Transform. For discrete Fourier Transform, $Y(f)$ is not calculated directly using (18.14). Instead, the integration for calculating $Y(f)$ is approximated by a summation but without multiplying the factor dt in equation (18.14), i.e. for digital data, the FFT only involves the summation within the parentheses in the following equation:

$$Y(f) = \left(\sum y(t)e^{-i2\pi ft}\right)dt = Y_n(f_n)dt. \tag{18.18}$$

Therefore,

$$|Y(f)|^2 = |Y_n(f_n)|^2(dt)^2. \tag{18.19}$$

So, equation (18.15) becomes (in discrete format)

$$\sum(|y_n|^2)dt = (dt)^2\sum(|Y_n|^2)df = (dt)^2\sum(|Y_n|^2)\frac{1}{T}.$$

Since $T = Ndt$, the above equation leads to

$$\sum(|y_n|^2) = \sum(|Y_n|^2)\frac{1}{N}. \tag{18.20}$$

Now let us use an example to show that the above equation indeed produces the desired results. We check the Fourier Transform of a simple function $x = t^2$ with t varying from 0 to 10.

```
clear
t = 0:0.001:10;      % DEFINE TIME
x = t.^2;            % DEFINE FUNCTION x(t)
X = fft(x);          % X IS FFT OF x
xsum = sum(x.^2)     % SUM IN TIME
Xsum = sum(abs(X).^2) / length(x)   % SUM IN FREQUENCY
```

The output is

```
xsum = 2.0005e+007   (sum in time)
Xsum = 2.0005e+007   (sum in frequency)
```

This verifies equation (18.20) for this example function.

18.1.7 Two-Sided vs. One-Sided Power Spectrum

Since the exponential form of the Fourier series uses negative frequency after the original sine and cosine functions are replaced by the positive and negative exponentials with the Euler formula, the Fourier integral and Fourier Transform all have the naturally inherited negative frequency. Because of that, the function (Power Spectrum) is said to be a *two-sided power spectrum*:

$$P_2 = |Y(f)|^2, \quad f \in (-\infty, +\infty). \tag{18.21}$$

The use of negative frequency, however, does not have any physical meaning for a scalar's spectrum – it is simply a mathematical expression of redistribution of energy into two parts, one associated with the base function $e^{i2\pi ft}$ and the other with $e^{-i2\pi ft}$. For most applications, we can try to combine the terms in the positive f and negative f and construct a *one-sided power spectrum*:

$$P_1 = |Y(f)|^2 + |Y(-f)|^2, \quad f \in (0, +\infty). \tag{18.22}$$

At the zero frequency,

$$P_1(0) = |Y(0)|^2. \tag{18.23}$$

If $y(t)$ is a real function, then

$$|H(f)|^2 = |H(-f)|^2, \quad f \in (0, +\infty). \tag{18.24}$$

Therefore,

$$P_1 = 2|Y(f)|^2, \quad f \in (0, +\infty). \tag{18.25}$$

For the one-sided power spectrum, it is twice the two-sided power spectrum, but the frequency is always positive.

18.2 Usage of Temporal Cross-Correlation Function

18.2.1 Cospectrum and Cross-Correlation Function

The Cross-Correlation Fourier Transform Theorem can be used to calculate (1) the cospectrum of two time series functions; and (2) the cross-correlation function by the Fourier Transform and inverse Fourier Transform. The cospectrum is a function of frequency, while the cross-correlation function is a function of time. The Cross-Correlation Fourier Transform Theorem or equation (18.7) provides an easy way to do both without directly computing the cross-correlation function. More specifically, given two time-functions $y_1(t)$ and $y_2(t)$, the procedure of both applications starts with using equation (18.11):

$$r_{12}(t) \Leftrightarrow Y_1^*(f)Y_2(f).$$

For the right-hand side, the Fourier Transforms for the time functions $y_1(t)$ and $y_2(t)$ are first done using FFT so that $Y_1(f)$ and $Y_2(f)$ are obtained. The right-hand side multiplication of the above equation is then done to give the cospectrum. The implementation of the inverse Fourier Transform on the cospectrum on the right-hand side gives the correlation function on the left-hand side. In MATLAB, the inverse Fourier Transform is the function `ifft`.

18.2.2 An Example of Cospectrum and Correlation Function Computation

Assume that there are two time series functions, $x(t)$ and $y(t)$, defined as the following:

$$x(t) = \sin(2\pi f_1 t) + 0.8\sin(2\pi f_2 t) + 0.5\sin(2\pi f_3 t) + 0.3\sin(2\pi f_4 t), \quad (18.26)$$

$$y(t) = 2\cos(2\pi f_1 t) + \cos(2\pi f_2 t) + 0.8\cos(2\pi f_4 t), \quad (18.27)$$

in which the frequencies f_1, f_2, f_3, and f_4 are

$$f_1 = 1;\ f_2 = 1.9;\ f_3 = 2;\ f_4 = 3. \quad (18.28)$$

If the unit of time t is in days, the frequencies are in cycle per day. The two functions share three of the frequencies: f_1, f_2, and f_4. A sample script for the time series (Fig. 18.2), the spectra of $x(t)$ and $y(t)$ (Fig. 18.3a), cospectrum (Fig. 18.3b), and correlation function (Fig. 18.4) are provided below. In the figures, only the left half of the spectra and cospectrum are shown. The cospectrum (Fig. 18.3b) shows the magnitude of the product of spectra of $x(t)$ and $y(t)$ at all frequencies. It gives information about what frequencies the two functions share (with non-zero amplitude) and the magnitude of the product of amplitudes at these frequencies. Cospectrum is comparable to the correlation function in time. A non-zero value of the cospectrum at a given frequency means that both signals have non-zero amplitude at this frequency.

```
%=====================================================
% THIS IS A SCRIPT SHOWING THE CROSS SPECTRUM AND THE
% INVERSE OF IT, WHICH IS THE CORRELATION FUNCTION.
%=====================================================
t = 0:0.001:30;                    % DEFINE TIME IN DAYS
f1 = 1;f2 = 1.9;f3 = 2;f4 = 3; % DEFINE THE FREQUENCIES FOR SIGNALS
dt = t(2) - t(1);                  % dt
x = sin(2*pi*f1*t)+0.8*sin(2*pi*f2*t)+0.5*cos(2*pi*f3*t)+0.3*sin
(2*pi*f4*t);  % THE
                                          % FIRST SIGNAL (TIME SERIES)
y = 2 * cos(2 * pi * f1 * t) + cos(2 * pi * f2 * t) + 0.8 * cos(2 * pi * f4 * t);
                                         % THE SECOND SIGNAL (TIME SERIES)
figure
plot(t,x,'k','LineWidth',2)     % PLOT THE TIME SERIES x
hold on
plot(t,y,'-k','LineWidth',2)    % PLOT THE TIME SERIES y
axis([ 0 10 -4 4])      % ZOOM IN FOR BETTER VIEW
```

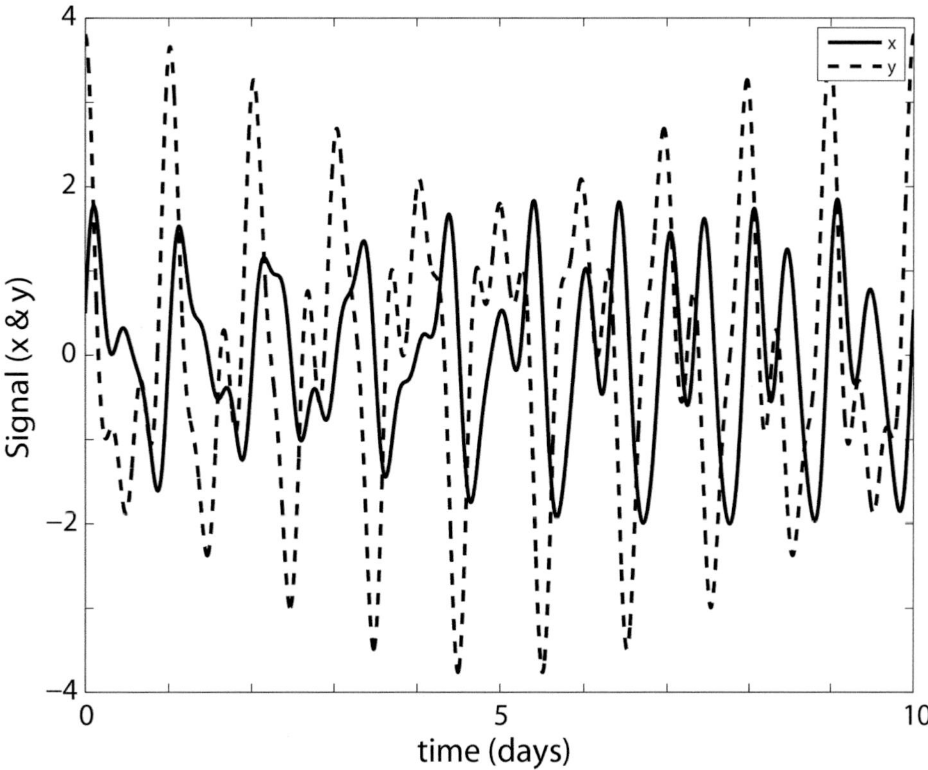

Figure 18.2 Two time series functions.

```
xlabel('time (days)') % LABEL X-AXIS
ylabel('x & y')       % LABEL Y-AXIS
legend('x','y')       % SHOW LEGEND FOR THE LINES

N = length(t);        % NUMBER OF DATA POINTS
X = fft(x);           % DO fft FOR x
Y = fft(y);           % DO fft FOR Y
df = 1 / max(t);      % df
f = df * (0:length(t) - 1);  % DEFINE f
figure
subplot(2,1,1)        % PLOT THE SPECTRA FOR x AND y
plot(f,abs(X) / length(t),'k-','LineWidth',2)
hold on;plot(f,abs(Y) / length(t),'k','LineWidth',2)
legend('|X|','|Y|')
axis([ 0 3.5 0 1.1])
xlabel('frequency (cpd)')
ylabel('Spectrum')
subplot(2,1,2)  % PLOT THE AMPLITUDE OF COSPECTRUM
plot(f,abs(X.*Y) / length(t)^2,'k','LineWidth',2)
axis([ 0 3.5 0 0.6])
```

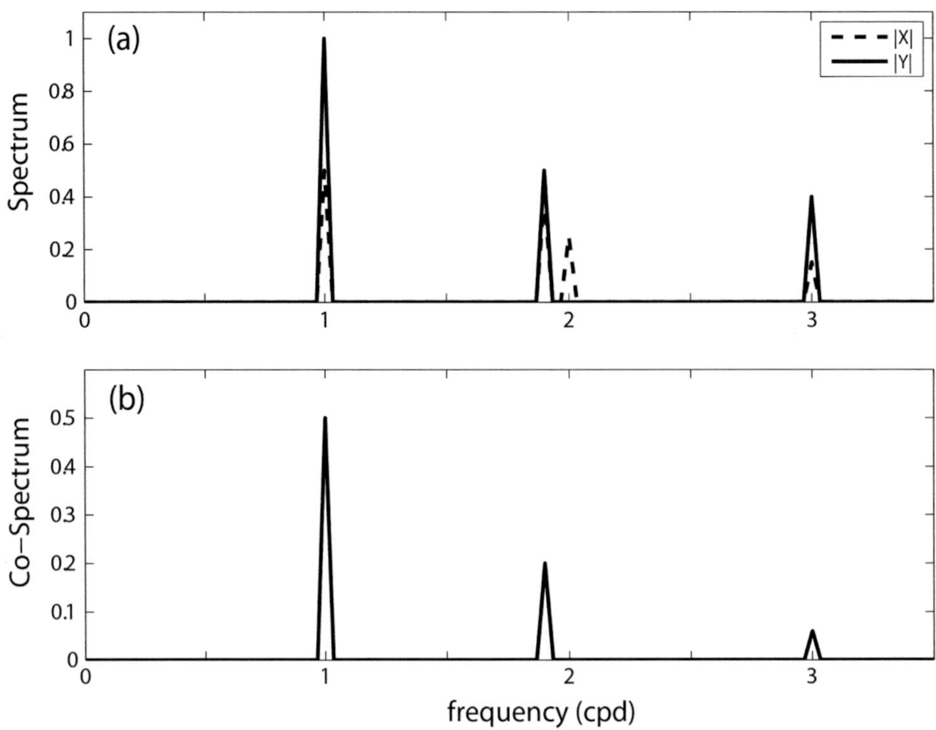

Figure 18.3 Spectra and cospectrum of the two time series functions of Fig. 18.2.

```
xlabel('frequency (cpd)')
ylabel('Cospectrum')

R12 = X'.* Y.';     % COMPUTE THE COSPECTRUM
r = ifft(R12);      % DO INVERSE fft TO GET THE CROSS-CORRELATION
                    % FUNCTION
figure
subplot(2,1,1)  % PLOT MAGNITUDE-INVARIANT CROSS CORRELATION
                % FUNCTION (INDEPENDNET OF N)
plot(t,r / N,'k','LineWidth',2)
xlabel('time (days)')
ylabel('Correlation Func')
subplot(2,1,2)  % RECOVER THE COSPECTRUM AGAIN JUST TO VERIFY
plot(f,abs(fft(r))/ N / N,'k','LineWidth',2)
axis([ 0 3.5 0 0.6] )
xlabel('time (days)')
ylabel('Cospectrum')
%================================
% END OF SCRIPT
%================================
```

18.3 Averaged Periodogram and Coherence

In addition to the power spectrum and cospectrum computed directly, there is often a statistical approach to compute similar quantities. More specifically, the time series can be divided into several segments (Fig. 18.5), and FFT is applied to each segment. These segments can be made without any overlap or with some overlaps, e.g. 30% or 75% overlap. The amount of overlap is adjustable and more or less a

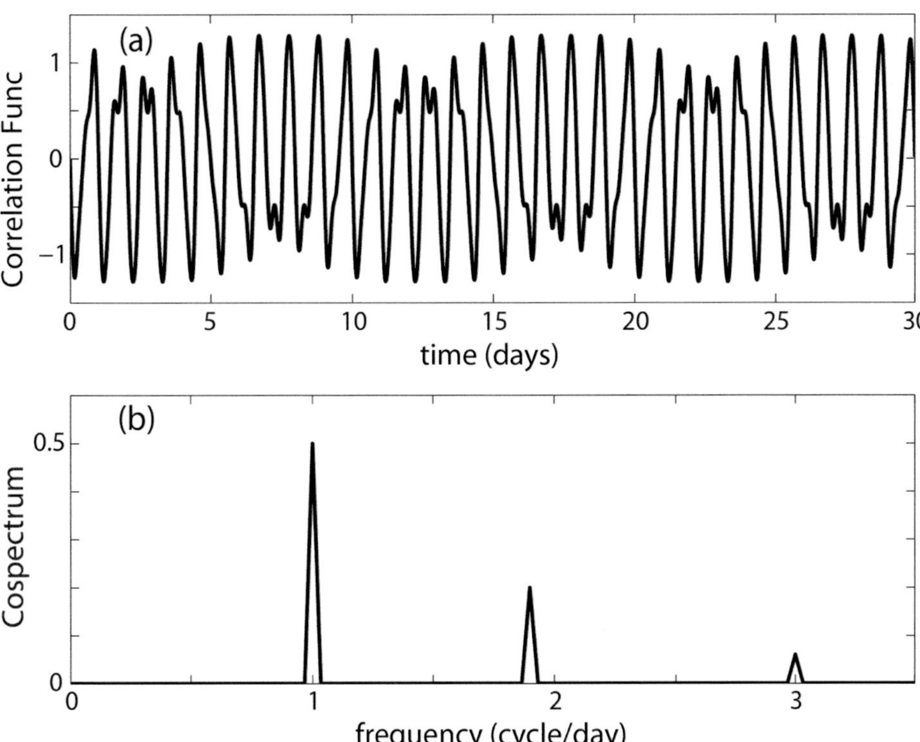

Figure 18.4 Correlation function and recovered cospectrum by doing fft to the cross-correlation function for verification.

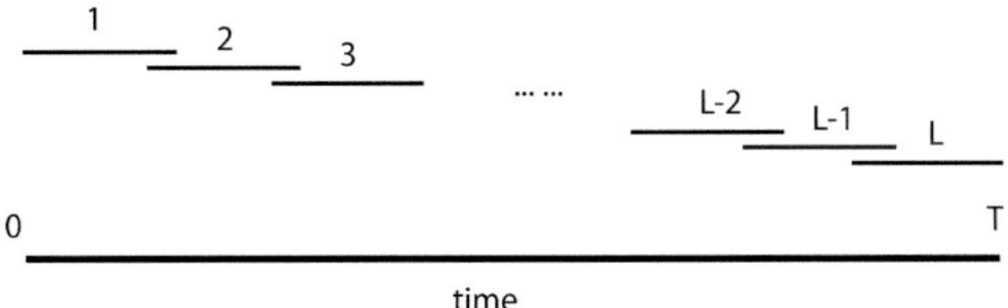

Figure 18.5 Segmented time series for averaged power spectrum (or periodogram) or for computation of coherence.

subjective choice. The advantage of having some overlap is that it allows a greater number of segments. The power spectrum can then be computed by averaging the power spectra of the segments. The result is sometimes called *averaged periodogram* (Welch, 1967). Similarly, the cospectrum can be replaced by a normalized quantity called *coherence function*.

18.3.1 Averaged Periodogram

Assume that a time series $y(t)$ is divided into K equal segments, with or without overlap:

$$y_i(t) = y(t), \quad t \in [t_{i1}, t_{i2}], i = 1, 2, \ldots, K, \tag{18.29}$$

in which t_{i1} and t_{i2} are the start and end times of the ith segment (Fig. 18.5), respectively; $[t_{i1}, t_{i2}]$ is the ith interval in time. Further assume that the Fourier Transform of $y_i(t)$ is $Y_i(f)$; the averaged spectrum is then

$$\bar{Y}(f) = \frac{1}{K}\sum_{i=1}^{K} Y_i(f), \tag{18.30}$$

and the averaged periodogram is

$$P_{yy} = |\bar{Y}(f)|^2 = \frac{1}{K}\sum_{i=1}^{K} |Y_i(f)|^2. \tag{18.31}$$

18.3.2 An Example of Averaged Periodogram

Data file "adcp_data.mat" includes a near surface velocity time series from an estuary. The east and north velocity components are given by the variables u1 and v1; and the variables year, month, day, hour, minute, and second indicate the time strings. Fig. 18.6 shows the time series for both u1 and v1. The time series has an interval of 6 minutes. There are some gaps in the time series. All the gaps are about 12 minutes or less, while the single longest gap is 36 minutes. An interpolation is done first to fill these gaps, and the total length of the time series is 7198 data points for each variable. The times series is then divided into 10 segments, each with a length of 719 data points. The power spectrum for each of the segments is computed using FFT. The average of the power spectrum is then computed by equation (18.31) for both east and north velocity components (Fig. 18.7).

MATLAB has many functions for doing computations related to spectrum, averaged periodogram, and power spectral density functions, e.g. `pspectrum`, `periodogram`, `pburg`, `pcov`, `pmcov`, `pyulear`, `pmtm`, `pwelch`, etc. In older versions of MATLAB, there is a function `spectrum`.

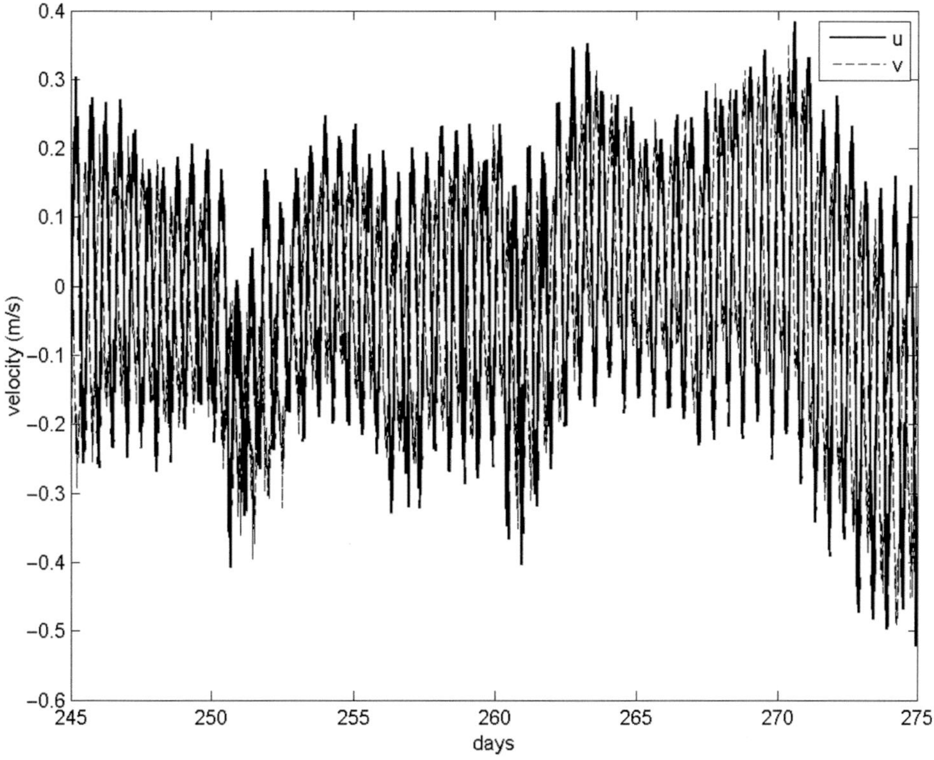

Figure 18.6 Time series of velocity components for the example of averaged periodogram.

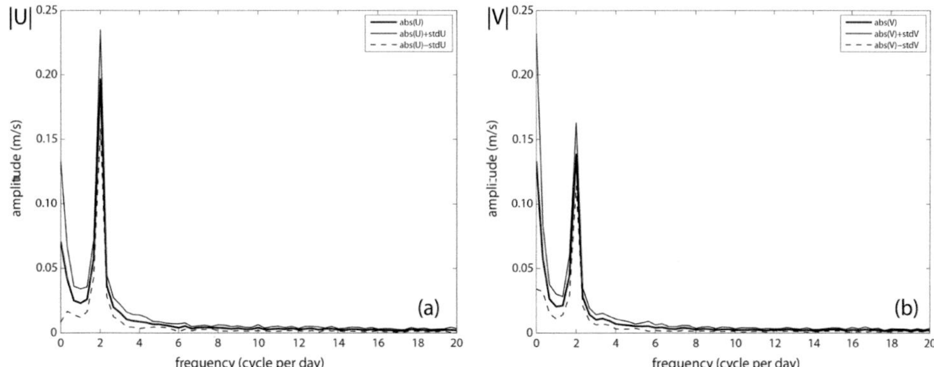

Figure 18.7 The averaged periodogram of the velocity time series of Fig. 18.6. (a) For the east velocity component; (b) for the north velocity component.

18.3.3 Coherence Function

The coherence function or *coherence magnitude squared* of two time series functions $y_1(t)$ and $y_2(t)$ is defined by

$$\gamma_{12}^2(f) = \frac{|P_{12}(f)|^2}{P_{11}(f)P_{22}(f)}, \tag{18.32}$$

in which $P_{11}(f)$, $P_{22}(f)$, and $P_{12}(f)$ are the expected power spectral density functions of $y_1(t)$ and $y_2(t)$, and the cross spectrum of $y_1(t)$ and $y_2(t)$, respectively, i.e.

$$P_{11}(f) = E\big[\mathbb{F}\big(r_{11}(t)\big)\big], \tag{18.33}$$

$$P_{22}(f) = E\big[\mathbb{F}\big(r_{22}(t)\big)\big], \tag{18.34}$$

$$P_{12}(f) = E\big[\mathbb{F}\big(r_{12}(t)\big)\big], \tag{18.35}$$

where

$$\mathbb{F}\big(r_{11}(t)\big) = \int_{-\infty}^{+\infty}\left[\int_{-\infty}^{+\infty} y_1(\tau)y_1(t+\tau)d\tau\right]e^{-i2\pi ft}dt, \tag{18.36}$$

$$\mathbb{F}\big(r_{22}(t)\big) = \int_{-\infty}^{+\infty}\left[\int_{-\infty}^{+\infty} y_2(\tau)y_2(t+\tau)d\tau\right]e^{-i2\pi ft}dt, \tag{18.37}$$

$$\mathbb{F}\big(r_{12}(t)\big) = \int_{-\infty}^{+\infty}\left[\int_{-\infty}^{+\infty} y_1(\tau)y_2(t+\tau)d\tau\right]e^{-i2\pi ft}dt, \tag{18.38}$$

are the Fourier Transforms of r_{11}, r_{22}, and r_{12}, respectively. The expected functions of power spectral density functions and cross-spectrum can be found by dividing the time series into segments as described above (Fig. 18.5). The meaning of (18.32) is similar to the correlation coefficient function, equation (4.34), but here it is a function of frequency. The values are between 0 and 1. Therefore, the coherence function is a normalized correlation, much like the correlation coefficient function, but in frequency domain. If the coherence function has a value of 0 at certain frequency f, the signals $y_1(t)$ and $y_2(t)$ are uncorrelated at that frequency; if, on the other hand, the coherence function has a value of 1 at certain frequency f, the signals $y_1(t)$ and $y_2(t)$ are 100% correlated at that frequency. The closer the value to 1, the more the two functions $y_1(t)$ and $y_2(t)$ are correlated at the relevant frequency.

18.3.4 An Example of Coherence Function

MATLAB has a function, `mscohere`, for the magnitude squared coherence computation using the so-called *Welch method* (Welch, 1967). Here we use a simple example to demonstrate an application. For that purpose, we assume two time-functions $x(t)$ and $y(t)$. The first time series function $x(t)$ includes four frequencies and random noise:

$$x(t) = \sin(2\pi f_1 t) + \sin(2\pi f_2 t) + \sin(2\pi f_3 t) + \sin(2\pi f_4 t) + ns.$$

Here $f_1 = 1/100$, $f_2 = 3/100$, $f_3 = 8/100$, and $f_4 = 20/100$, and ns is random noise.

The second time series function $y(t)$ also includes four frequencies and random noise, but only three are common frequencies shared between $x(t)$ and $y(t)$:

$$y(t) = \sin(2\pi f_2 t) + \sin(2\pi f_3 t) + \sin(2\pi f_4 t) + \sin(2\pi f_5 t) + ns,$$

in which $f_5 = 11/100$, and ns is random noise. The common frequencies shared by both $x(t)$ and $y(t)$ are f_2, f_3, and f_4. Fig. 18.8 shows the time series functions $x(t)$

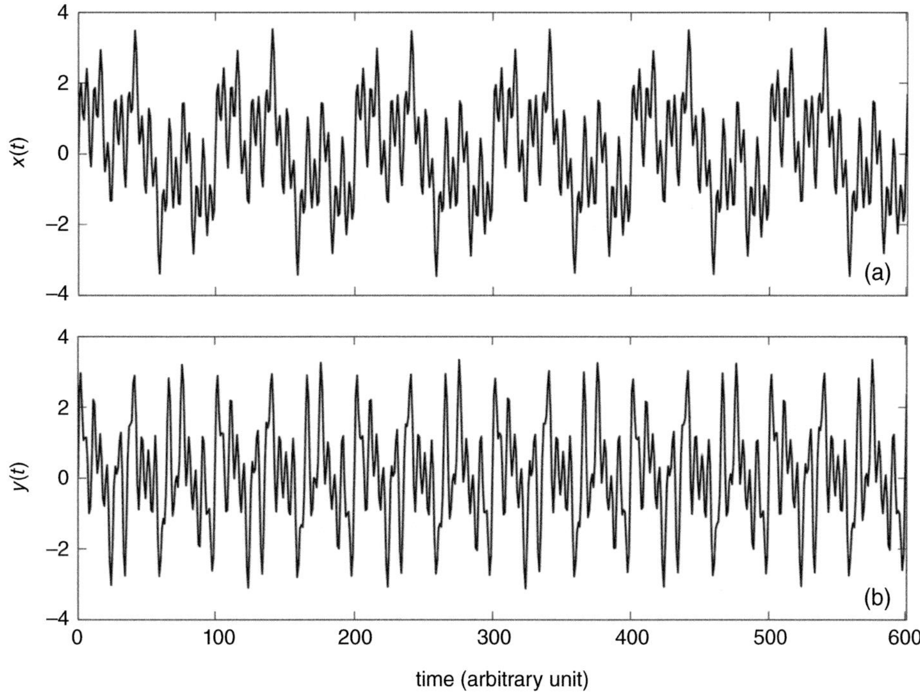

Figure 18.8 Example time series functions (a) $x(t)$, and (b) $y(t)$ for coherence computation.

and $y(t)$. The following are some MATLAB commands doing the coherence computation.

```
t = 0:601;
f1 = 1 / 100; f2 = 3 / 100; f3 = 8 / 100; f4 = 20 / 100; f5 = 11 / 100;
x = sin(2 * pi * t * f1) + sin(2 * pi * t * f2) + ...
    sin(2 * pi * t * f3) + sin(2 * pi * t * f4) + 0.1 * rand(1,length(t));
y = sin(2 * pi * f2 * t) + sin(2 * pi * f5 * t) + ...
    sin(2 * pi * f3 * t) + sin(2 * pi * f4 * t) + 0.2 * rand(1,length(t));
dt = t(2) - t(1);
fs = 1 / dt;
[ gammaxy, f] = mscohere(x,y,hanning(128),64,2^10,fs);
```

The coherence function is shown in Fig. 18.9. It can be seen that, indeed, at their common frequencies f_2, f_3, and f_4, the two signals $x(t)$ and $y(t)$ are completely correlated represented by coherence function values of 1 (A, B, and C marked on Fig. 18.9). At other frequencies, the coherence function values are less than 1 but mostly non-zero because of the random noise.

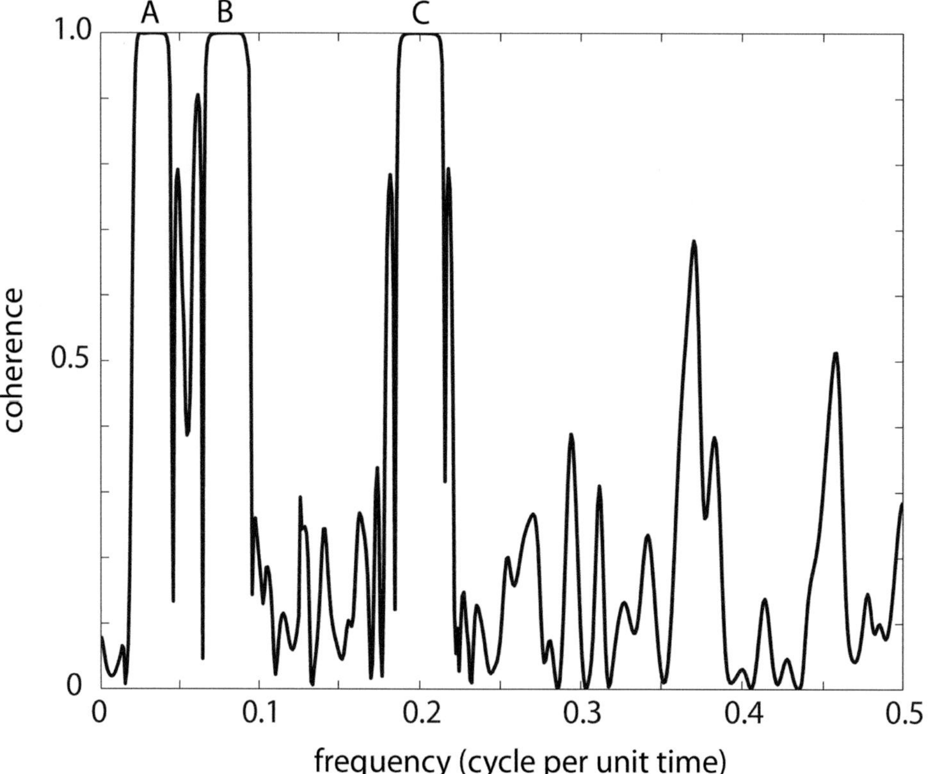

Figure 18.9 Example coherence function of the time series functions $x(t)$ and $y(t)$.

Review Questions for Chapter 18

(1) What is the difference between power spectrum and the averaged periodogram?
(2) Discuss the difference between the cospectrum and coherence.

Exercises for Chapter 18

(1) Verify (18.20) using MATLAB for function $x = t^3$ for $t = [0, 20]$.
(2) Verify (18.20) using MATLAB for function $x = e^{-t}$ for $t = [0, 5]$.
(3) Do averaged periodogram:
  (a) Load data adcp_data.mat.
  (b) Create the time using datenum function.
  (c) Plot the time series.
  (d) Check the time intervals (are they constant?).
  (e) Do interpolations for missing data for u1, v1, w1, using the `interp1` function.
  (f) Plot u1 and v1 after interpolations.
  (g) Divide the time series into 10 equal segments with 50% overlap.
  (h) Do FFT for u1 and v1 for each segment.
  (i) Average the magnitude of FFT in the 10 segments.
  (j) Find the standard deviation of the spectrum for u1 and v1, respectively.
  (k) Plot the mean spectrum for u1 with black color.
  (l) Plot the mean spectrum u1 + standard deviation with red color.
  (m) Plot the mean spectrum u1 – standard deviation with red color.
  (n) Plot the mean spectrum v1 + standard deviation with red color.
  (o) Plot the mean spectrum v1 – standard deviation with red color.
  (p) Repeat the above with 75% overlap and compare the results.

19

Window Functions for Reducing Side Lobes

About Chapter 19

We have learned that, in theory, a single frequency sinusoidal function in a time domain corresponds to an isolated line in the frequency domain. For multiple frequencies with linearly superimposed sinusoidal functions in time, the frequency domain representation is a series of lines (the so-called line spectrum). When finite sampling is performed, the isolated lines will be replaced by sinc functions centered on the locations of the lines in frequency. The lines are "smeared" to become continuous functions; each has a main lobe of finite width ($2T$), and side lobes or a series of decaying ringing toward both smaller and larger frequencies. This chapter discusses some techniques to reduce the ringing (side lobes) away from the main lobes. This is the windowing technique. This usually is done at a price: widening of the main lobe, which is sometimes acceptable in order to substantially reduce the side lobes.

19.1 Drawback of Rectangular Window Function

We have learned that one of the effects of finite sampling is unwanted distortion of the true spectrum. The spectrum is unavoidably modified such that a discrete component in the frequency domain is represented by a sinc function centered at the original frequency. The main problem with the sinc function is that it has "leakage" of energy by the side lobes away from the true frequency (centered at the main lobe). The side lobes are a result of the rectangular window or the abrupt truncation of data into a finite segment. The rectangular window is not a smooth function with gradual change at both ends. The most troublesome problem with these side lobes is their heights: the height of the first and second side lobes are about 22% and 13% of the main lobe, respectively. Even the third side lobe has a height of 9% of the main lobe, as demonstrated in Chapter 17. These side lobes are unwanted and useless. The side lobes can present a false impression of a complicated spectrum: the side lobes may be adding to the main lobe of an adjacent frequency to interfere with the estimate of the magnitudes of closely spaced true peaks in frequency. Fortunately, there are ways to reduce these side lobes. The idea is to change the rectangular window function to one that has much smoother variations and the function tapers off to zero or near zero at both ends. This is an idea to design some artificial *window functions* to multiply the original time series data before doing spectrum analysis.

19.2 The Idea

From the convolution theorem, we have discussed the fact that the Fourier Transform $Y_s(f)$ of a segment of time series $y_s(t)$ truncated from an infinitely long time series $y(t)$ is equal to a convolution between the original Fourier Transform $Y(f)$ and the sinc function $X_T(f)$, which is the Fourier Transform of a rectangular window function with a "width" in time T of the truncated time series, i.e.

$$Y_s(f) = Y(f) \otimes X_T(f) = \int_{-\infty}^{+\infty} Y(\tilde{f}) X_T(f - \tilde{f}) d\tilde{f}.$$

The problem of the resulted spectrum is originated from the function $X_T(f)$. If it were a delta function, that would be perfect. Instead, only the limit of sinc function for infinitely long T is a delta function. The spectrum of the sampled function is smeared by the convolution with the sinc function, which is a Fourier Transform of the rectangular window with sharp changes and discontinuities. As we have learned from the Fourier series expansion, whenever there is a discontinuity in the time function, there is a Gibbs Effect in the reconstructed function from the Fourier series. This is because using smooth sinusoidal functions to exactly represent a function with abrupt changes is logically impossible. The situation here is very similar with the Fourier Transform. The above equation gives us a hint for a possible solution: since the root of the problem is $X_T(f)$ being from a rectangular window, can we somehow smooth the rectangular window and get a "better" behaved function to convolve with the original Fourier Transform so the smearing is less and the heights of side lobes are reduced? The action item would be just to reduce the sharp change of the window function to a smoother version, and the goal is to reduce the side lobes, which is discussed below.

19.3 Window Functions

There are numerous window functions with names. As some examples, the following lists a few commonly used window functions (Fig. 19.1) and their Fourier Transforms (Fig. 19.2) for comparison. Obviously, the Fourier Transforms

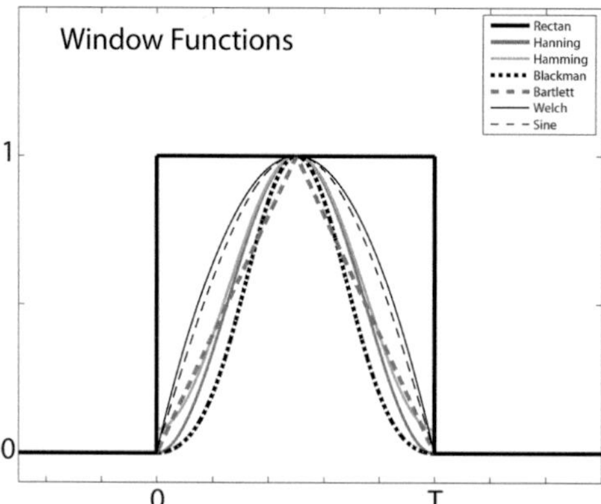

Figure 19.1 Examples of some window functions.

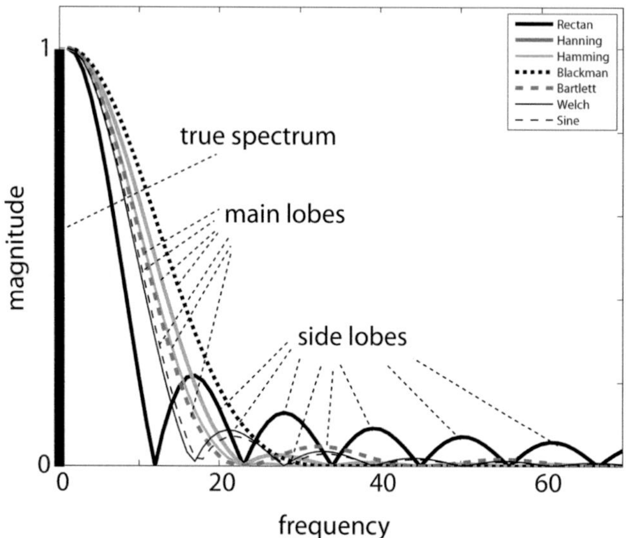

Figure 19.2 Fourier Transforms of the window functions.

of the nonrectangular window functions given here (Fig. 19.1 and equations 19.2–19.7) all show much reduced side lobes (Figs. 19.2 and 19.3). Sometimes we use the logarithm scale (dB, which is $20\log_{10}|Y|$) to better visualize (by exaggerating) the side lobes (Fig. 19.4).

19.3.1 Rectangular Window Function

The rectangular window function is defined as

$$h(t) = \begin{cases} 1, & 0 \le t \le T \\ 0, & t < 0,\ t > T \end{cases}. \tag{19.1}$$

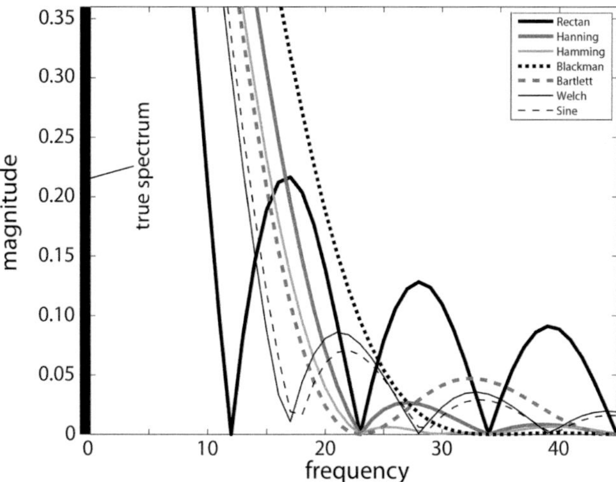

Figure 19.3 Zoomed-in view of Fourier Transform of the window functions.

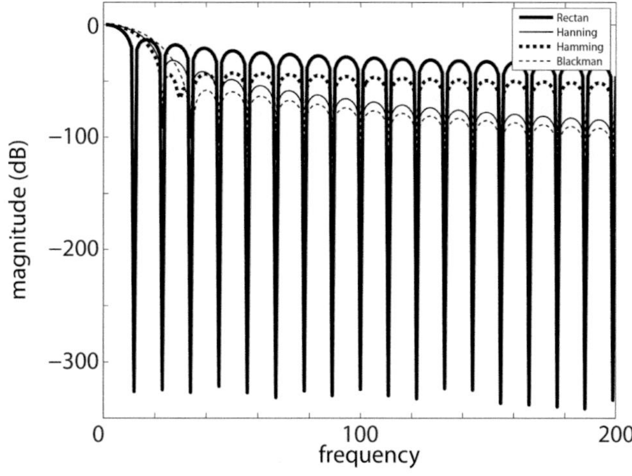

Figure 19.4 Fourier Transforms of the window functions in dB ($20 \log_{10} X$). For clarity, here we only show results for four window functions.

This is the "default" window function, a function of a truncated constant of 1, which is equivalent to doing nothing to the data. FFT is applied directly to the original data. As discussed earlier, because this window has sharp edges (Fig. 19.1), its Fourier Transform (the sinc function) has large side lobes (Fig. 19.2). This is presented here just for comparison purposes.

19.3.2 Hann (Hanning) or Raised Cosine Function

The Hann (Hanning) window function, or *raised cosine window function,* is defined as

$$h(t) = \begin{cases} \frac{1}{2}\left(1 - \cos\left(2\pi \frac{t}{T}\right)\right), & 0 \leq t \leq T \\ 0, & t < 0,\ t > T \end{cases}. \tag{19.2}$$

Its side lobes are much lower than those of the rectangular window function. The main lobe, however, is quite broad.

19.3.3 Hamming Function

The Hamming window function is defined as

$$h(t) = \begin{cases} 0.54 - 0.46 \cos\left(2\pi \frac{t}{T}\right), & 0 \leq t \leq T \\ 0, & t < 0, t > T \end{cases}. \tag{19.3}$$

The Hamming window function is very interesting: it does not go to zero at both ends (Fig. 19.1). However, it has a very low first side lobe ($< 1\%$ of the main lobe), compared to most of the window functions presented here. In fact, the only window function that has a lower first side lobe than this function is the Blackman window function (see below, or Figs. 19.2–19.5). The drawback of this function is that after the first side lobe, other side lobes do not decrease significantly.

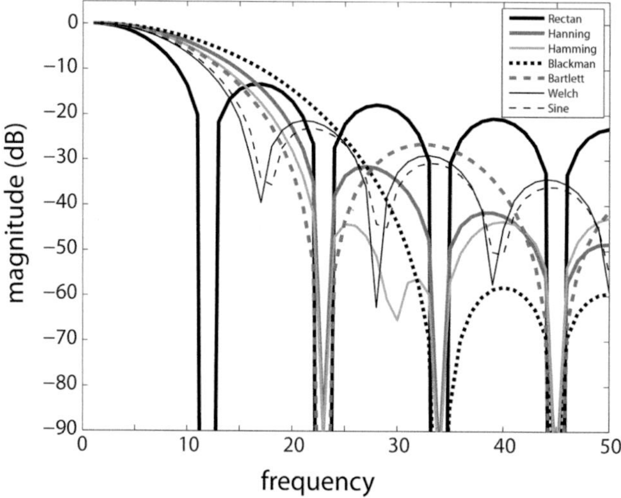

Figure 19.5 Zoomed-in view of Fourier Transforms of the window functions in dB ($20 \log_{10} X$). Here we show results for all seven window functions given above.

19.3.4 Blackman Function

The following function is the *Blackman window function*:

$$h(t) = \begin{cases} 0.42 - 0.5\cos\left(2\pi\frac{t}{T}\right) + 0.08\cos\left(4\pi\frac{t}{T}\right), & 0 \le t \le T \\ 0, & t < 0, t > T \end{cases}. \tag{19.4}$$

The Blackman window function has the lowest side lobes among the seven selected window functions here. At a price: it also has the widest main lobe.

19.3.5 Bartlett Function

The following is the Bartlett window function:

$$h(t) = \begin{cases} 1 - \left|\frac{2t - T}{T}\right|, & 0 \le t \le T \\ 0, & t < 0, t > T \end{cases}. \tag{19.5}$$

The Bartlett window function is a triangular function (Fig. 19.1). It has a relatively narrow main lobe and substantially reduced side lobes compared to the rectangular window. The side lobes, however, are still fairly large: about ~5% of the main lobe. Recall that the first side lobe from the rectangular window is ~22% of the main lobe; 5% is significantly smaller.

19.3.6 Welch Function

Here is the *Welch window function*:

$$h(t) = \begin{cases} 1 - \left|\frac{2t - T}{T}\right|^2, & 0 \le t \le T \\ 0, & t < 0,\ t > T \end{cases}. \tag{19.6}$$

The Welch window function has a relatively narrow main lobe and relatively large side lobes the first side lobe is reduced from the rectangular window function by more than half (Fig. 19.3).

19.3.7 Sine Function

Here is the *sine window function*:

$$h(t) = \begin{cases} \sin\left(\frac{\pi t}{T}\right), & 0 \le t \le T \\ 0, & t < 0, t > T \end{cases}. \tag{19.7}$$

The sine window function has a spectrum similar to the Welch window function but with a slightly wider main lobe. As a result, its side lobes are also slightly smaller.

There are many more window functions, and we will omit more complete coverage. With so many window functions, how do we decide which window function to use? This depends on the applications. The user should ask how important it is to have a low side lobe for a specific application and how important to keep a narrow main lobe. There is a trade-off when these two factors are considered. We generally will not get both. It appears that there is a relationship; given all other factors the same, the product of the width of the main lobe (W_m) and the height of the first side lobe (S_1) maintains a constant, or

$$W_m S_1 \sim \text{constant} \tag{19.8}$$

so that reducing S_1 is balanced by increasing W_m. The more reduction in S_1, the more increase in W_m.

19.4 How to Apply the Window Function

The window function should be used to multiply with the raw data:

$$y_w(t) = y_s(t)x(t), \tag{19.9}$$

in which y_s, $x(t)$, y_w are the original samples or data, the window function, and the resultant array prior to FFT, respectively.

This should be done after making sure that the data time intervals are uniform, i.e. *dt* is a constant. If *dt* is not a constant, an interpolation is often needed. The commonly used MATLAB interpolation functions `interp1` is applicable.

It should also be noted that when a window function is applied to the data before doing FFT, one should also *normalize* the function to get the correct spectrum. For the amplitude spectrum, one should divide the magnitude of the window function $h(t)$ by the mean of the window function:

$$x(t) = \frac{h(t)}{\sum\limits_{n=1}^{N} h(t_n)/N}. \tag{19.10}$$

For example, for the rectangular window function, $h(t)$ is a constant 1 within the window T or for all $n = 1, 2, \ldots, N$. Therefore, $\sum_{n=1}^{N} h(t_n) = N$; therefore, $x(t) = h(t)$, as anticipated.

19.5 Application of the Window Function

In the following, we apply all seven different window functions discussed above to the sine function:

$$y(t) = \sin(2\pi ft),\ f = 2, t \in [0, T],\ T = 10. \tag{19.11}$$

Fig. 19.6 is the magnitude spectrum with linear scale, while Fig. 19.7 is the magnitude spectrum of the function with log scale in the vertical. Fig. 19.8 is a

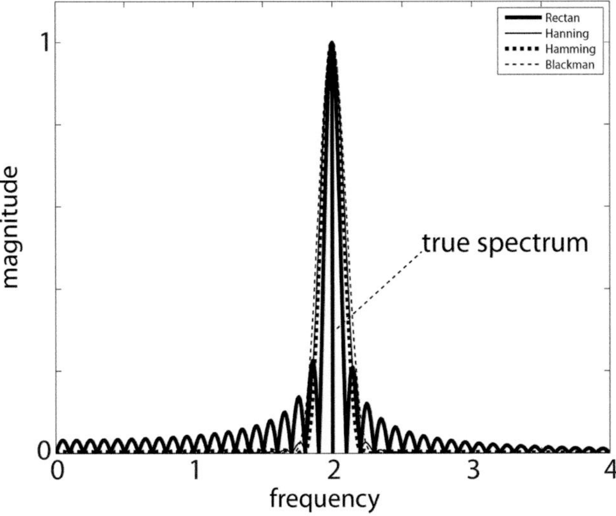

Figure 19.6 The magnitude spectrum of the truncated sine function $y(t) = \sin(2\pi ft),\ f = 2, t \in [0, T],\ T = 10$, with four of the window functions applied, respectively. For clarity, we did not include the results from the other three window functions.

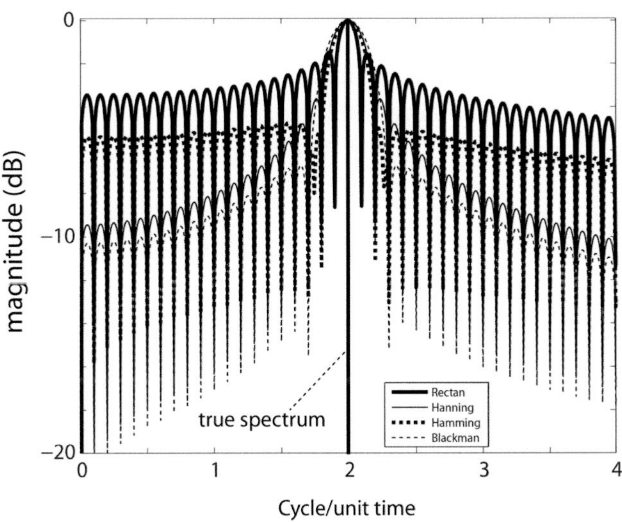

Figure 19.7 Spectrum of a truncated sine function multiplied with a window function. For clarity, we only include four of the seven window functions. Fig. 19.8 provides a zoomed-in view of the spectra of all the seven functions defined earlier.

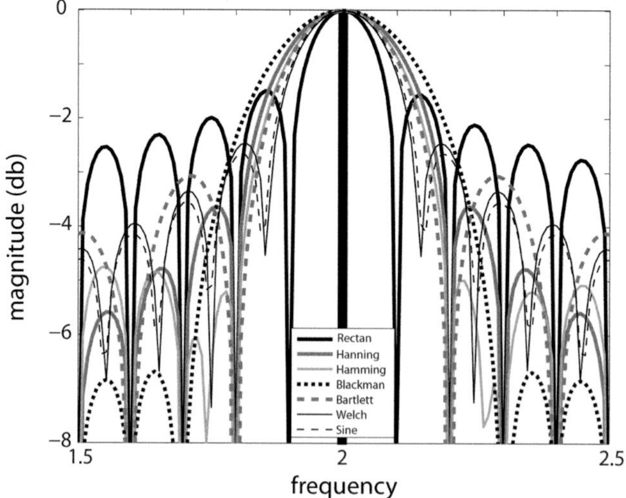

Figure 19.8. Zoom-in view of the spectrum of a truncated sine function multiplied with a window function.

zoomed-in view of Fig. 19.7 but with all results for the seven window functions included. Obviously, the windowed spectra have much lower side lobes. However, they all have wider main lobes, as anticipated.

Review Questions for Chapter 19

(1) Why do we sometimes need a window function for the Fourier Transform?

(2) Can you design a new window function which will reduce both the width of the main lobe and the heights of all the side lobes?

(3) Are the side lobes of any value in data analysis?

Exercises for Chapter 19

(1) Write a MATLAB script and draw the Hanning, Hamming, Blackman, Bartlett, Welch, and sine window functions.

(2) Go to the National Data Buoy Center's webpage (www.ndbc.noaa.gov/) and download some time series of water level at any station with valid data covering a minimum of one month. Do FFT and find the spectrum. Then, apply one of the window functions to the data before doing FFT and find the new spectrum. Compare the two results. What can you conclude?

20

Convolution, Filtering with the Window Method

About Chapter 20

The objective of this chapter is to demonstrate the linkage between convolution and filtering and to discuss preliminary filtering with examples. The simplest low-pass filtering that allows low frequency to pass to the output is a "moving average," which is essentially through a computation involving a rectangular window function (in this case, it is a filter), with a length determined by the cutoff frequency. The filtering action is accomplished by a convolution between the filter and the time series. However, moving average has its drawbacks because of the rectangular window effect or the side lobe effect. It is considered as a "poor man's filter" because of the lack of sophistication in getting rid of the leakage from side lobes. One improvement over the moving average is using a non-rectangular window function in the convolution to reduce the sharp change at the edges. The basic ideas and examples presented here are useful in demonstrating how to do filtering with several MATLAB functions. When a low-pass filter is designed, a high-pass filter can be defined. With two or more low-pass filters, one can also design band-pass and band-stop filters. There are also other filters that can have various controls on the results, which are discussed in the next chapter.

20.1 Convolution

We have learned that a *convolution* is a mathematical operation involving two functions. For filtering, the two functions are quite different: one is a function representing the data, while the other is usually a much shorter function representing the filter. In other words, the first function can be either an infinite function or a truncation of an infinite function, while the second function is non-zero within a much shorter time segment. The description about the second function may be relaxed: it can be infinitely long in theory but decreases rapidly at both ends so the truncation at both ends has relatively small effect. The problem with an infinitely long filter is that it is not practical in implementation and will introduce error when truncated. Here in this chapter, we will discuss a simple method of filtering using window functions.

Convolution is a very useful concept in understanding the theory of *filtering*. The definition of convolution of two functions $y(t)$ and $h(t)$ is

$$C(t) \equiv y(t) \otimes h(t) = \int_{-\infty}^{+\infty} y(\tau)h(t-\tau)d\tau. \tag{20.1}$$

Earlier, we learned through the convolution theorem that the Fourier Transform of the convolution of two functions of time is the product of the Fourier Transforms of each of the two functions respectively, i.e.

$$\mathbb{F}(y(t) \otimes h(t)) = Y(f)H(f). \tag{20.2}$$

This provides a tool to design a filter in the frequency domain, i.e. define a function's Fourier Transform $H(f)$ first. This function can be designed to keep a certain frequency range, by setting $H(f) = 1$ in the wanted-frequency range and eliminating the rest, and by setting $H(f) = 0$ in the unwanted frequency range – an action of filtering, i.e. allowing part of the signal to go through the output and getting rid of the rest. The inverse transform of $H(f)$ is a time function $h(t)$, which is a filter that can be implemented in the time domain by the convolution with the function to be filtered. This gives us a hint that the convolution in time may be considered as a filtering or smoothing operation, if the function $h(t)$ is properly designed.

Since the convolution of two functions is symmetric, i.e. it does not depend on the order of the functions inside the integral, we use the expression given in (20.1) for discussion. The convolution starts with a "flipping" of the function – change the

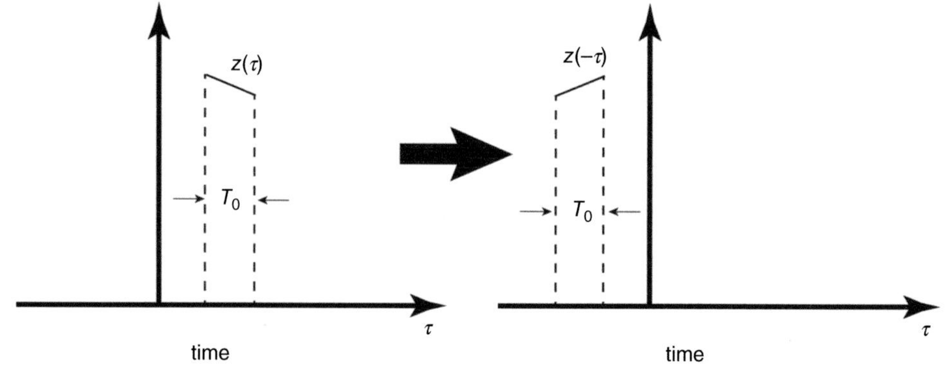

Figure 20.1 Flipping of a function.

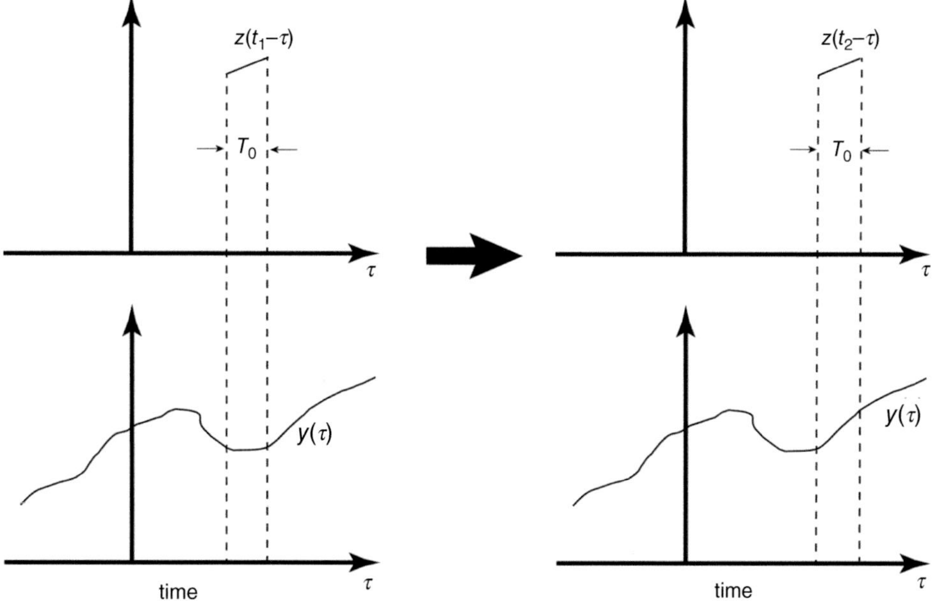

Figure 20.2 Translation of a flipped function (z) by t_1 and multiplication with another function (y). Repeating this operation by a different translation t_2.

function h from $h(\tau)$ to $h(-\tau)$ (Fig. 20.1). It is then translated by t to $h(t-\tau)$. This flipped and translated function $h(t-\tau)$ is then multiplied by the other function $y(\tau)$ to yield the product $y(\tau)h(t-\tau)$. An integration is then performed. These steps are then repeated by using a different translation (or different value of t; Fig. 20.2). After doing the computations for all t values, the convolution function is obtained. How can this be a filter or an operation to smooth a function? Let us look at an example.

20.2 Filtering Using Convolution

20.2.1 Moving Average $h(\tau)$

Assume that the function $h(t)$ is a rectangular window defined by

$$h(t) = \begin{cases} \dfrac{1}{T_0} & 0 < t < T_0 \\ 0 & t < 0, \quad t > T_0 \end{cases}. \tag{20.3}$$

Now substitute the function $h(t)$ defined by (20.3) into (20.1):

$$g(t) = h(t) \otimes y(t) = \int_{-\infty}^{+\infty} y(\tau)h(t-\tau)d\tau = \int_{t-T_0}^{t} y(\tau)h(t-\tau)d\tau = \frac{1}{T_0}\int_{t-T_0}^{t} y(\tau)d\tau. \tag{20.4}$$

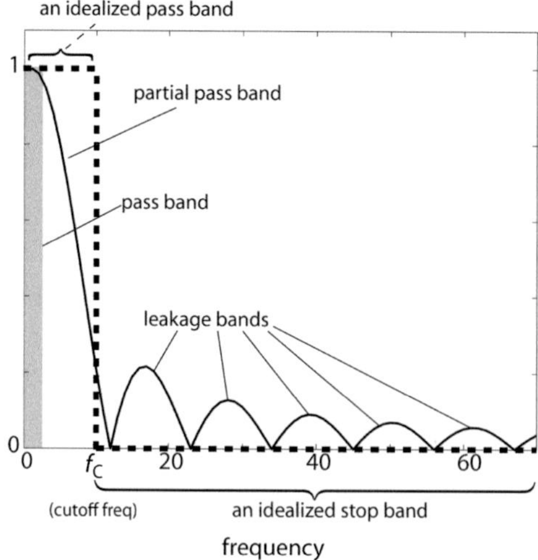

Figure 20.3 Using the rectangular window function as a filter in time domain – its Fourier Transform is a sinc function – is far from the idealized filter because of the side lobes and narrow width of the pass band.

We can see that this new function $g(t)$ after the convolution is a simple *moving average* or *box-car average* of the function $y(t)$ with a window width of T_0 for the averaging. The time interval T_0 is a parameter determining how the filtering or smoothing is done or what the *cutoff frequency* should be. The larger this value, the smoother the resultant function will be. In other words, a long T_0 means a lower cutoff frequency. Recall that the Fourier Transform of a rectangular window function is a sinc function (Fig. 20.3); this moving average is an infinitely long filter in the frequency domain (although in time domain it is a rather convenient function of finite length T_0). In the frequency domain, it is the sinc function $H(f)$ multiplying the Fourier Transform $Y(f)$ of the signal function $y(t)$. The multiplication in the frequency domain suggests that this moving average filter can be problematic: the side lobes will allow some high frequency components to be leaked into the averaged or smoothed function which is supposed to have removed the high frequencies after the averaging. The moving average filter is sometimes referred as the "poor man's filter" because of its lack of quality in performance due to the *side lobe leakage effect.*

From Fig. 20.3, we can see that the sinc function has little resemblance of a window function, particularly considering the side lobes where it is supposed to be zero (the stop band). The pass band of the sinc function is very narrow (the shaded narrow zone on the left edge of Fig. 20.3). The decrease of sinc function from 1 to 0 is gradual, giving a wide partial pass band. In this partial pass band, only part of the energy is kept and part filtered. All of these are signs of a poor filter.

Ideally, in the frequency domain, a *low-pass filter* should keep all frequencies lower than a cutoff frequency f_C, and it should be like a rectangular window in the frequency domain so that the *pass band* in the low frequency end has a constant value of 1 and the *stop band* beyond the cutoff frequency has a constant value of 0 (see the "idealized pass band" and "idealized stop band" in Fig. 20.3):

$$H(f, f_C) = \begin{cases} 1, & f < f_C \\ 0, & f \geq f_C \end{cases}. \tag{20.5}$$

This way, the filtered function has a spectrum of

$$G(f, f_C) = H(f, f_C)Y(f) = \begin{cases} Y(f), & f < f_C \\ 0, & f \geq f_C \end{cases}, \tag{20.6}$$

which is a function truncated to keep a portion of the spectrum left of the cutoff frequency f_C, which is a parameter here for the functions $H(f, f_C)$ and $G(f, f_C)$. The corresponding filter in time domain can be obtained by an inverse Fourier Transform:

$$h(t) = \int_{-\infty}^{+\infty} H(f, f_C)e^{i2\pi ft}df = \int_{-f_c}^{f_c} e^{i2\pi ft}df = \frac{1}{i2\pi t}\left(e^{i2\pi f_c t} - e^{-i2\pi f_c t}\right) = \frac{\sin(2\pi f_c t)}{\pi t}.$$

This filter in time domain is similar to a sinc function, but it is in time domain. This is not a surprise because of the symmetric property between the Fourier Transform and inverse Fourier Transform. The rectangular window in frequency domain has a sinc function like inverse Fourier Transform in time domain, just like a rectangular window function in time has a sinc function Fourier Transform in frequency domain. This sinc function in time is infinitely long and thus not really implementable. Truncations have to be used in practice and thus the idealized effect not achievable.

20.2.2 The Drawback of Moving Average

To analyze the performance of the moving average filter in preparation for more sophisticated and professional filters later, here we present some examples of moving average filtering. Even though it is a simple smoothing technique with significant drawbacks, it is still used frequently. It is a useful reference for improved filter designs using different window functions as well as other, more sophisticated filters. The hypothetical signal is defined as a constant plus several sine functions:

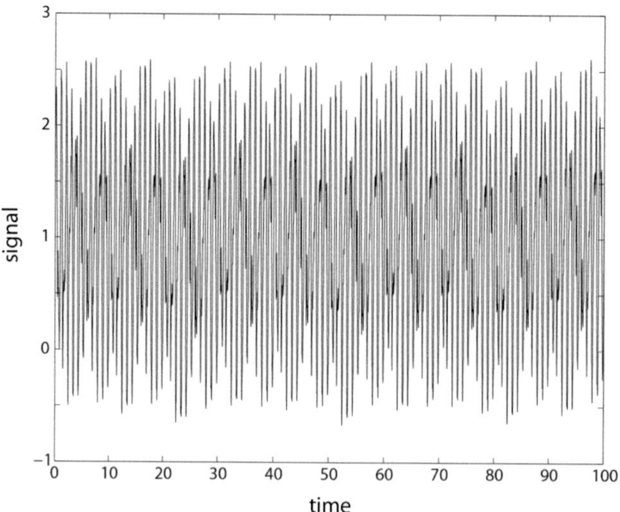

Figure 20.4 Sample time series with a constant plus several sinusoidal functions.

$$y(t) = 1 + \sum_{i=1}^{N} a_i \sin(2\pi f_i t). \tag{20.7}$$

Here we have chosen $N = 8$. These eight frequencies are 1/30, 1/14, 1.1, 2, 3.2, 4, 6, and 8 cycles per unit time (CPUT), respectively. The corresponding amplitudes are arbitrarily selected as 0.05, 0.05, 1, 0.5, 0.1, 0.1, 0.1, and 0.1, respectively, for demonstration purposes. It is basically a superposition of a constant 1 with eight sine functions of specified amplitudes (Fig. 20.4). If one would like to filter out all of the "high frequency" oscillations using the moving average method, one should first consider the cutoff frequency. This, of course, depends on what frequencies are considered high. It is obviously application-dependent and determined by the user. Here we assume that we are interested in keeping the first two low frequencies, 1/30 and 1/14, and getting rid of all other frequencies beyond 1.1, inclusive. To cut all frequencies beyond $f_1 = 1.1$ CPUT, we can choose a cutoff frequency, say, 1.1 CPUT so that all frequencies beyond this frequency would be filtered. This requires that the length of the moving average (or filter) is

$$T_0 = \frac{1}{f_C}, \tag{20.8}$$

so that the sinc function's first zero coincides with the cutoff frequency f_C. Note that we can also choose f_C to be even smaller, but it should be greater than 1/14 in this example.

The number of data points or the length of the moving average filter should be determined by

$$N_0 = \frac{T_0}{dt}, \tag{20.9}$$

in which dt is the sampling interval of the signal $y(t)$, assuming that dt is a constant. In real applications, in order to use filter functions in computer programming such as MATLAB, the time intervals in a given time series should be constant, as the common computer algorithms for filtering of a time series assume constant time intervals. Otherwise, an interpolation may need to be done to have a constant dt before using the MATLAB functions if the time series is not equally sampled in time or if there are data gaps due to instrument malfunction.

In this specific example, the MATLAB scripts we use to define the time series are

```
clear
t = (0:0.01:100)';    %DEFINE TIME
N = length(t);        %LENGTH OF TIME
y = 1;          %DEFINE THE CONSTANT PART
f = [1/30 1/14 1.1 2 3.2 4 6 8] ;  %DEFINE THE
                %FREQUENCIES IN THE HYPOTHETICAL SIGNAL
a = [ 0.05 0.05  1   0.5 0.1 0.1 0.1 0.1] ; %AMPLITUDES
for i = 1:length(f)    %LOOP DEFINING THE TIME SERIES
 y = y + a(i) * sin(2 * pi * f(i) * t);  %ADDING THE SINE FUNCTIONS
end
```

The signal is shown in Fig. 20.4, and the spectrum is shown in Fig. 20.5. Using (20.8) and (20.9), we determine the size of the filter based on our selection of the cutoff frequency.

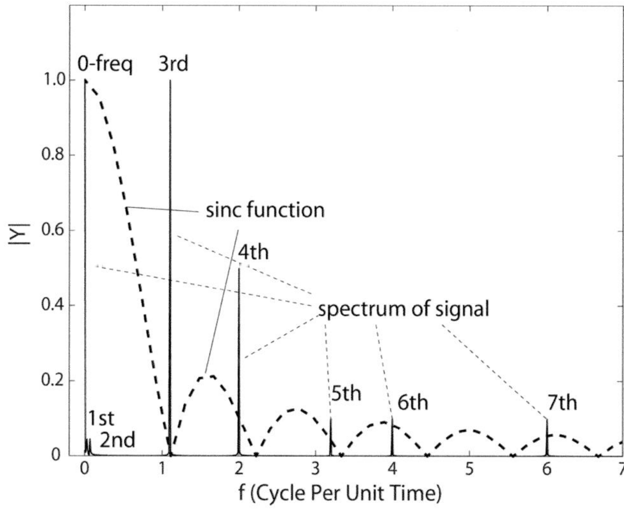

Figure 20.5 Spectrum for the signal defined by (20.7) and shown in Fig. 20.4, compared with the sinc function (the spectrum of a moving average filter).

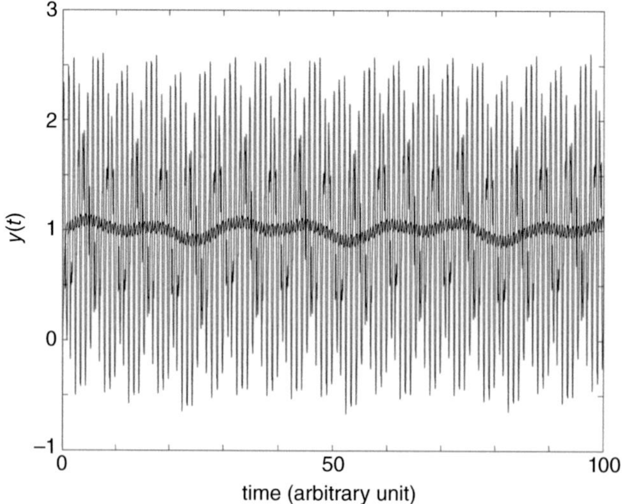

Figure 20.6 Sample time series with a constant plus several sinusoidal functions. The black line is the moving average of the signal implemented by the MATLAB function `filter`.

```
dt = t(2) - t(1);  % TIME INTERVAL
fc = 1.1;          % CUTOFF FREQUENCY
T0 = 1 / fc;       % WIDTH OF AVERAGING OR FILTER LENGTH IN TIME
N0 = fix(T0 / dt); % NUMBER OF DATA POINTS OF SIZE OF FILTER IN #
```

We then define the filter as a time series in the time domain:

```
H = ones(N0,1);       % RECTANGULAR WINDOW
h = h/length(h);      % NORMALIZE THE FILTER
```

We then implement the filter using a MATLAB function `filter` (Figs. 20.6 and 20.7):

```
g1 = filter(h, 1, y);   % USING A MATLAB FUNCTION TO FILTER DATA
```

This can also be done by using another MATLAB function `filtfilt` (Figs. 20.8 and 20.9):

```
g = filtfilt(h, 1, y);   % USING A MATLAB FUNCTION TO FILTER
                         % DATA which RESULTS IN a g with the
                         % SAME LENGTH as y
```

It can be seen that for (Fig. 20.6), after the moving average, there are still variations at high frequencies (Fig. 20.7 for a zoomed-in view) that are meant to be filtered out. This is a result of the leakage of energy due to the unwanted side lobes

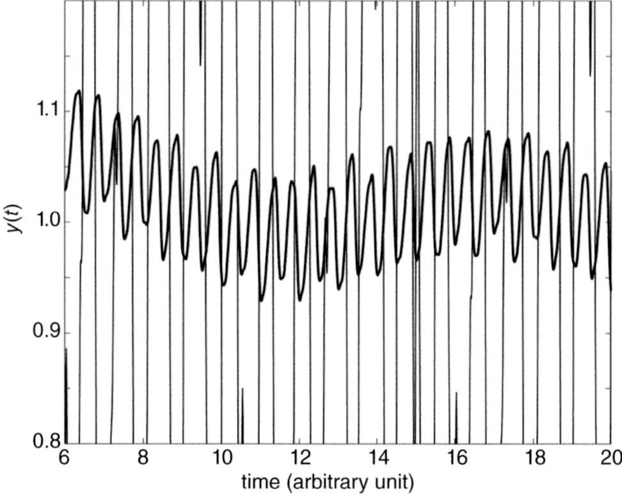

Figure 20.7 A zoom-in view of Fig. 20.6 showing the high frequency oscillations after the low-pass filtering using the moving average.

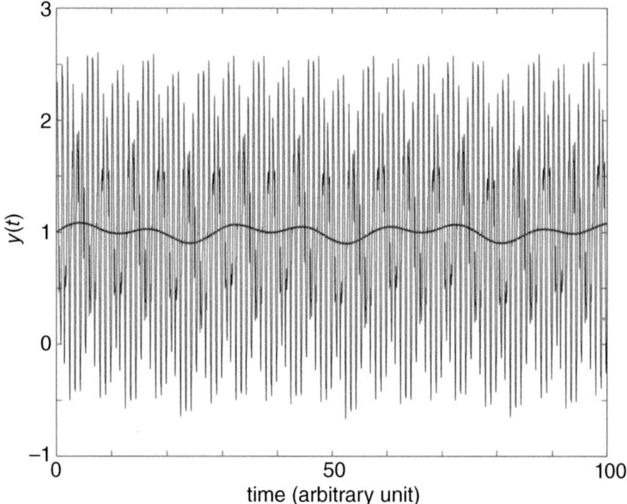

Figure 20.8 Similar to Fig. 20.6 but the filtering implementation is done with the MATLAB function `filtfilt`.

(Fig. 20.5). The MATLAB function `filtfilt` allows the filtering be done twice, one forward and one backward to eliminate phase changes. Fig. 20.8 gives the results using this double filtering function. Fig. 20.9 is a zoomed-in view of Fig. 20.8. It can be seen that the use of the filter twice can eliminate much of the unwanted oscillations, but this is only because of more smoothing, and it still has some residuals due to the side lobe effects (Fig. 20.9). It should be noted that by increasing the length

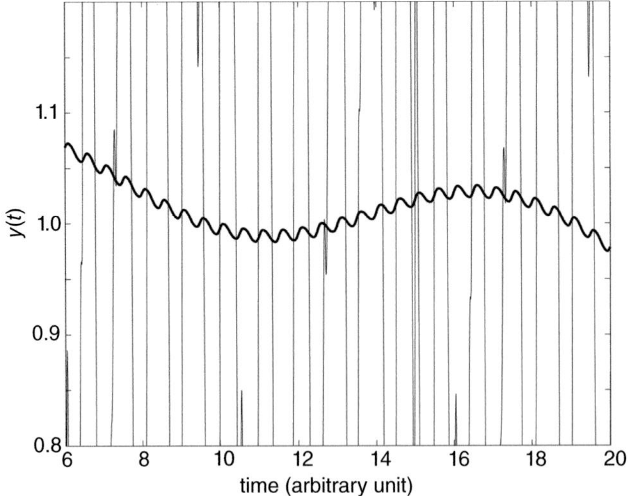

Figure 20.9 A zoomed-in view of Fig. 20.8 showing the high frequency oscillations.

of the moving average or lowering the cutoff frequency, the leakage may be reduced, but the side lobe effect cannot be completely removed.

In addition to the leakage of energy at high frequencies, another drawback of the moving average mentioned earlier is that its pass band is very narrow (Fig. 20.3). This may or may not be a problem, depending on the data and filtering requirement. The gradual transition band is wide, which may reduce the magnitude of spectrum at some frequencies inside the pass band, distorting the frequency components that are meant to be kept.

20.3 Use of Window Functions

20.3.1 Use of Window Functions as Filters

To improve the filter performance, we can use the window functions discussed earlier. For instance, if we choose the Hamming window, instead of the rectangular window, we should have a filter that looks like that in Fig. 20.10. It has very small side lobes, but a wider main lobe. The pass band is still narrow. Now we apply the Hamming window filter. Recall the Hamming window function:

$$h(t) = \begin{cases} 0.54 - 0.46\cos\left(2\pi\dfrac{t}{T}\right), & 0 \leq t \leq T \\ 0, & t < 0, t > T \end{cases}.$$

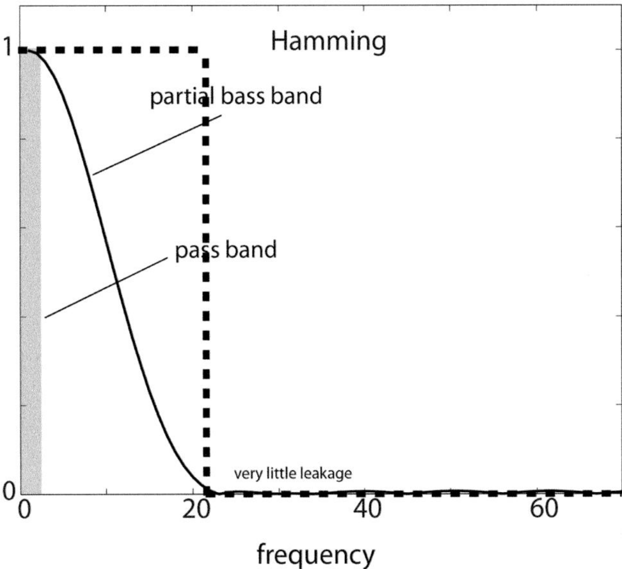

Figure 20.10 Using the Hamming window function as a filter in time domain – its Fourier Transform has much lower side lobes than those of the sinc function. However, it still has only a narrow full pass band.

An example MATLAB code for this filter usage is

```
h = 0.54 - 0.46 * cos(2 * pi * (0:N0 - 1) / (N0 - 1));  % THE HAMMING WINDOW
g = filtfilt(h / sum(h),1,y);                   % USING A MATLAB FUNCTION
                                                % TO FILTER DATA
```

Alternatively, one can use the MATLAB Hamming window function `hamming`, Hanning window function (`hann`), Blackman window function (`blackman`), or the `window` function, a Window design and analysis tool in the Signal Processing Toolbox, to generate any of 17 different window functions (such as the triangular window, Bartlett window, Gaussian window).

The spectra of the signal and filter are shown in Fig. 20.11. Obviously, the new filter has a much nicer stop band, which should take out most of the unwanted frequencies. Now we implement this filter from the Hamming window function to the above problem.

The result is shown in Figs. 20.12 and 20.13. Obviously, the residual high frequency oscillations have disappeared. The filter function is rather smooth in the zoomed-in view (Fig. 20.13). This demonstrates that the window function can indeed improve the low-pass filtering, compared with the simple moving average.

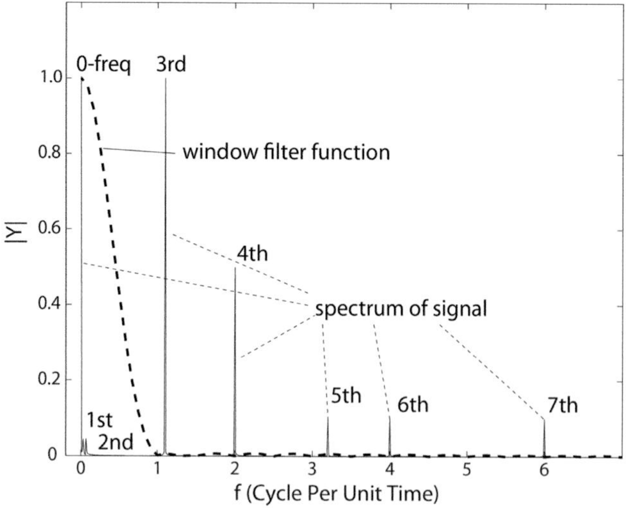

Figure 20.11 Spectrum for the signal in Fig. 20.4 compared with the spectrum of the window function.

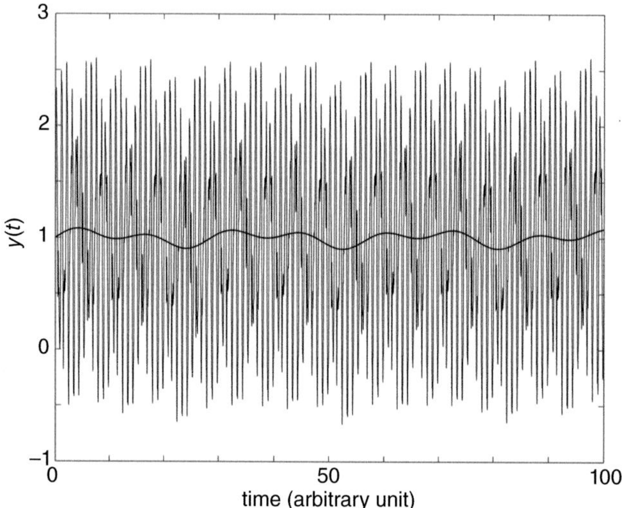

Figure 20.12 Time series (thin line) and low-pass filtered signal (thick line) with the Hamming window function.

The reason is simply because of the significantly reduced side lobes. Other window functions, discussed earlier, can also be used as improved low-pass filters.

With this improved method of using window functions as filters, can we still use the generally poor quality moving average for low-pass filtering? The

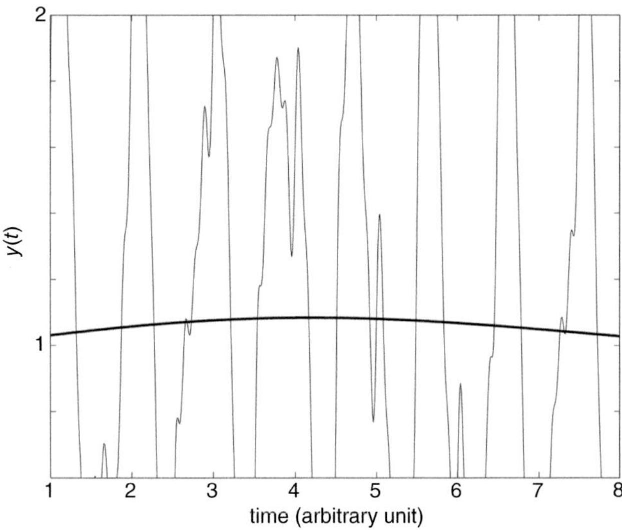

Figure 20.13 A zoom-in view of Fig. 20.12 showing the absence of the high frequency oscillations.

answer is a conditional yes to some problems. For those problems that do not have significant high frequency components coinciding with the side lobes, the moving average may still work well, especially for those problems that do not require a clean removal of high frequencies. However, this is not guaranteed. On the other hand, with so many other choices, such as the use of window function or the more sophisticated filters (see next chapter), there is really no reason to use the moving average as it is not anything significantly easier to implement than other filters and it has such a drawback caused by side lobe leakage.

20.3.2 Steps of Low-Pass Filtering Using Window Functions

In using the window function, including the moving average, for low-pass filtering, the general steps are

(1) Select a type of filter in time domain: $h(t)$.
(2) Determine the cutoff frequency: $f > f_C(\sim 1/T_0)$.
(3) Find the time step dt: same as that of the original data.
(4) Compare the length of filter: $T_0 \sim 1/f_C$.
(5) Choose the size of the filter $N + 1$: $N = [T_0/dt]$ (take the integer part if there is a fraction).

(6) Implement the filter: using a selected MATLAB filtering function (e.g. `filtfilt`, `fftfilt`, `filter`).

20.4 High-Pass, Band-Pass, and Band-Stop Filters

Using a low-pass filter, we can accomplish a high-pass filtering by subtracting the low-pass filtered time series from the original time series, i.e.

$$y_{highpass} = y - y_{lowpass} \tag{20.10}$$

As some examples, if we have two low-pass filters, with different cutoff frequencies

$$f_{C1} < f_{C2} \tag{20.11}$$

We can have a band-pass filtered time series, e.g.

$$y_{bandpass} = y_{lowpass2} - y_{lowpass1} \tag{20.12}$$

which keeps the frequency band between f_{C1} and f_{C2}.

If we need to have a band-stop filter to get rid of the signals between f_{C1} and f_{C2}, we can use:

$$y_{bandstop} = y_{lowpass1} + y_{highpass2} = y_{lowpass1} + (y - y_{lowpass2}) \tag{20.13}$$

One can also try to construct a filter with multiple pass bands, which means that there are multiple stop bands or a mixture of pass bands and stop bands.

Selected MATLAB Filter Implementation Functions

`filter` – a MATLAB filter function, producing an output filtered time series of the same length as the input time series with a phase shift.

`filtfilt` – a MATLAB Signal Toolbox filter function, producing an output filtered time series of the same length as the input time series with no phase shift. It is a double filter (the timeseries is filtered forward and backward, which eliminates the phase shift).

`conv` – a MATLAB Signal Toolbox filter function, producing an output filtered time series that is longer than the input time series with a phase shift.

`fftfilt` – a MATLAB Signal Toolbox filter function, producing an output filtered time series of the same length as the input time series with a phase shift.

Review Questions for Chapter 20

(1) Why is moving average not an ideal filter? What are the pros and cons to using a moving average as a low-pass filter?
(2) How should an ideal low-pass filter behave?
(3) Why can an idealized filter not be exactly realized in applications?

Exercises for Chapter 20

(1) Do low-pass filtering with the following steps:
  (a) Load data "S6-depth-time.mat" (it has two variables useful here: `depth` (m) and `time` (day)).
  (b) Plot the time series of `depth`.
  (c) Select a low-pass cutoff frequency `fc` to be 30 cycles per day for a low-pass filter.
  (d) Find the length in time `T0` for the filter (reciprocal of cutoff frequency).
  (e) What is the value for dt? Use `T0` and `dt` to find the number `N` of data points for the filter.
  (f) Define an array `h`, of length `N`, with a constant value of 1.
  (g) Normalize the array `h` by diving it by its length.
  (h) Use the normalized array `h` as the filter and apply the MATLAB filter function `filtfilt` to get the filtered time series `depthf`.
  (i) Plot the filtered depth `depthf` with the original data `depth`.
  (j) Zoom in on the plot and discuss the effect of filtering.

(2) Do lowpass filtering with the following steps (this is similar to the first question but using a window function in the later steps):
  (a) Construct a time series with randomness and two sine functions at $f_1 = 1$ CPUT, $f_2 = 10$ CPUT, as hypothetical data.

```
t = [0:0.001:20];
y = sin(2*pi*t)+0.5*sin(20*pi*t)+0.2*rand;
```

  (b) Choose a cutoff frequency that allows you to get rid of the second frequency ($f_2 = 10$) in the above signal.
  (c) Design a moving average filter h_1 (similar to the first question).
  (d) Implement the filter using MATLAB filter function or any other filter implementation function (e.g. `filtfilt`).
  (e) Construct a Hamming window h_2.

```
h2 = 0.54 - 0.46 * cos(2 * pi * (0:M) / M)
```

(f) Normalize it by:

```
h2 = h2 / sum(h2)
```

(g) Implement the filter.

(h) Discuss your results and compare the effect of these two filters h_1 and h_2.

(3) Assume there is a signal with frequencies at 0.49, 0.5, 0.6, . . ., and 4.6 cycles per unit time, defined by

$$y(t) = 1 + \sum_{i=1}^{N} a_i \sin(2\pi f_i t) + \text{randomness}$$

and the MATLAB command is:

```
t = (0:0.01:100)';
N = length(t);
y = 1 + 0.5 * sin(2 * pi * 0.49 * t) + ...
0.35 * sin(2 * pi * 0.5 * t) + 0.15 * sin(2 * pi * 0.6 * t) + ...
0.35 * sin(2 * pi * 0.7 * t) + 0.15 * sin(2 * pi * 0.8 * t) + ...
0.35 * sin(2 * pi * 0.9 * t) + 0.15 * sin(2 * pi * 1.0 * t) + ...
0.35 * sin(2 * pi * 1.1 * t) + 0.15 * sin(2 * pi * 1.2 * t) + ...
0.35 * sin(2 * pi * 1.3 * t) + 0.15 * sin(2 * pi * 1.4 * t) + ...
0.35 * sin(2 * pi * 1.5 * t) + 0.15 * sin(2 * pi * 1.6 * t) + ...
0.35 * sin(2 * pi * 1.7 * t) + 0.15 * sin(2 * pi * 1.8 * t) + ...
0.35 * sin(2 * pi * 1.9 * t) + 0.15 * sin(2 * pi * 2.0 * t) + ...
0.35 * sin(2 * pi * 2.1 * t) + 0.15 * sin(2 * pi * 2.2 * t) + ...
0.35 * sin(2 * pi * 2.3 * t) + 0.15 * sin(2 * pi * 2.4 * t) + ...
0.35 * sin(2 * pi * 2.5 * t) + 0.15 * sin(2 * pi * 2.6 * t) + ...
0.35 * sin(2 * pi * 2.7 * t) + 0.15 * sin(2 * pi * 2.8 * t) + ...
0.35 * sin(2 * pi * 2.9 * t) + 0.15 * sin(2 * pi * 3.0 * t) + ...
0.35 * sin(2 * pi * 3.1 * t) + 0.15 * sin(2 * pi * 3.2 * t) + ...
0.35 * sin(2 * pi * 3.3 * t) + 0.15 * sin(2 * pi * 3.4 * t) + ...
0.35 * sin(2 * pi * 3.5 * t) + 0.15 * sin(2 * pi * 3.6 * t) + ...
0.35 * sin(2 * pi * 3.7 * t) + 0.15 * sin(2 * pi * 3.8 * t) + ...
0.35 * sin(2 * pi * 3.9 * t) + 0.15 * sin(2 * pi * 4.0 * t) + ...
0.35 * sin(2 * pi * 4.1 * t) + 0.15 * sin(2 * pi * 4.2 * t) + ...
0.35 * sin(2 * pi * 4.3 * t) + 0.15 * sin(2 * pi * 4.4 * t) + ...
0.35 * sin(2 * pi * 4.5 * t) + 0.15 * sin(2 * pi * 4.6 * t) +0.2* rand(N,1);
```

(a) Choose the cutoff frequency $f_C = 0.4$ cycles per unit time so that all the frequencies in the above signal are filtered.

(b) Design a moving average filter h_1.

(c) Implement the filter using MATLAB filter function or any other filter implementation function (e.g. `filtfilt`).

(d) Construct a Hamming window filter h_2.

```
h2 = 0.54 - 0.46 * cos(2 * pi * (0:M) / M)
```

(e) Normalize it by:

```
h2 = h2 / sum(h2)
```

(f) Implement the filter.

(g) Discuss your results and compare the effects from these two filters.

21

Digital Filters

FIR and IIR Filters

About Chapter 21

The objective of this chapter is to discuss digital filters. We start from a review of theory of Fourier Transform for continuous functions. The continuous Fourier Transform is then discretized. The discretized Fourier Transform and inverse Fourier Transform, however, are not approximate equations – they are exact. Using the shifting theorem, a filter can easily be expressed in the frequency domain. A Finite Impulse Response (FIR) filter is then defined. By adding another implicit convolution to the original convolution for FIR filter, the filtered data depends on not only the input (the original time series) but also the output (the filtered data). This is an iterative relation that forms the Infinite Impulse Response (IIR) filter. These filters are examples of so-called linear systems that have an input and output. The gain is defined by the filter, which is the ratio between the input and output in the frequency domain. Several FIR and IIR filter functions in MATLAB are discussed.

21.1 From Continuous to Discrete Fourier Transforms

21.1.1 The Fourier Transform and Inverse Fourier Transform Pairs

The Fourier Transform of a time function $y(t)$ and its inverse transform as a function of the angular frequency ω are expressed as

$$y(t) = \int_{-\infty}^{+\infty} Y(\omega) e^{i\omega t} d\omega, \tag{21.1}$$

$$Y(\omega) = \frac{1}{2\pi} \int_{-\infty}^{+\infty} y(t) e^{-i\omega t} dt. \tag{21.2}$$

Alternatively, since frequency f and angular frequency ω are related by their definitions,

$$\omega = 2\pi f, \tag{21.3}$$

we can use f instead of ω for the transform pairs as in the following expressions

$$y(t) = \int_{-\infty}^{+\infty} \hat{Y}(f)e^{i2\pi f t}df, \tag{21.4}$$

$$\hat{Y}(f) = \int_{-\infty}^{+\infty} y(t)e^{-i2\pi f t}dt. \tag{21.5}$$

Obviously,

$$\hat{Y}(f) = 2\pi Y(\omega) = 2\pi Y(2\pi f). \tag{21.6}$$

We use the hat to distinguish between $Y(\omega)$ and $\hat{Y}(f)$ only when they appear in the same discussion. In practice, they are usually not used in the same discussion and so there is no need to use the hat, i.e. we can simply write $Y(f)$ instead of $\hat{Y}(f)$.

21.1.2 Discrete Expression

In actual data analysis applications, or in a digital world, the integrations are replaced by a summation, and the continuous functions are replaced by digital arrays. Usually this is considered as an approximation of the integral form. The frequency increment df and time increment dt are replaced by $1/T$ and Δt, respectively, in which T is the time span of the data (the maximum time minus the minimum time) and we have

$$df \sim \frac{1}{T}, \quad dt \sim \Delta t = T/N. \tag{21.7}$$

Here N is the total number of data points, with the assumption that the data are equally spaced. With (21.7), we can write equations (21.4) and (21.5) in discrete format as

$$y(t) = \frac{1}{T}\sum_{m} Y(f_m)e^{i2\pi f_m t}, (f_m = mdf), \tag{21.8}$$

$$Y(f) = \Delta t \sum_{k} y(t_k)e^{-i2\pi f t_k}, (t_k = kdt). \tag{21.9}$$

Note that equation (21.8) is an approximation of equation (21.4) and equation (21.9) is an approximation of (21.5).

If we define

$$\tilde{Y}(f) = \frac{Y(f)}{\Delta t}. \tag{21.10}$$

So that

$$Y(f) = \tilde{Y}(f)\Delta t. \tag{21.11}$$

With (21.7) and (21.11), equation (21.8) becomes

$$y(t) = \frac{1}{N}\sum_{m} \tilde{Y}(f_m)e^{i2\pi f_m t}. \tag{21.12}$$

With equation (21.9), equation (21.10) becomes

$$\tilde{Y}(f) = \sum_{k} y(t_k)e^{-i2\pi f t_k}. \tag{21.13}$$

In the digital world, we use (21.12) and (21.13) for the discrete Fourier and inverse Fourier Transforms, respectively. The advantage of using these equations is that we do not have to worry about multiplying dt or $1/T$, which are all dimensional. This is also why the MATLAB function `fft` does not need to input the time; neither does it need to know the unit for time.

We can put the factor $1/N$ into equation (21.12)

$$y(t) = \sum_{m} \frac{1}{N}\tilde{Y}(f_m)e^{i2\pi f_m t} = \sum_{m} \breve{Y}(f_m)e^{i2\pi f_m t}, \tag{21.14}$$

so that the coefficients of the exponential base functions $e^{i2\pi f_m t}$ are

$$\breve{Y}(f_m) = \frac{1}{N}\tilde{Y}(f_m), m = -(N-1), \ldots, 0, 1, 2, \ldots, N-1. \tag{21.15}$$

Note that, in the continuous world, the Fourier Transform and inverse Fourier Transform have time and frequency intervals all from $-\infty$ to $+\infty$. In the digital world, we often limit them to

$$t \in \left[-\frac{T}{2}, \frac{T}{2}\right], \quad f \in [-f_N, f_N],$$

in which T is the time period or the total length of record of the time series data and f_N the Nyquist frequency. We can also fold half of the time and frequency to the positive side, i.e. the intervals are

$$t \in [0,\ T], \quad f \in [0,\ 2f_N].$$

These ranges make more sense because of the symmetric property of Fourier Transform and because putting time origin at $t = 0$ is a common practice. In time domain, this gives a total of N time values,

$$t = 0,\ dt,\ 2dt,\ \ldots,\ Ndt, \tag{21.16}$$

and N frequency values,

$$f = 0,\ df,\ 2df,\ \ldots,\ Ndf, \tag{21.17}$$

in which $df = 1/T$ and $Ndf = 2f_N$.

21.1.3 The Discrete Frequency and Discrete Base Function Expressions

Now let us look at the discrete form of the exponential base function and note that

$$f_m = m\Delta f, t_n = n\Delta t, T = N\Delta t, \omega = 2\pi f_m \Delta t = 2\pi m \Delta f \Delta t = 2\pi m/N \tag{21.18}$$

and

$$2\pi f_m t = 2\pi m \Delta f n \Delta t = 2\pi mn/N. \tag{21.19}$$

So that

$$e^{-i2\pi f_m t} = e^{-i2\pi mn/N}. \tag{21.20}$$

There are several different quantities that can be used to represent the (discrete) frequency. The integer m is one of them, and it is dimensionless. Another dimensionless frequency is m/N. A dimensional counterpart of it is $2\pi m/N$ or ω which has a dimension of radian. The dimensional quantity of frequency f in digital form can have a few variations: f_m, $m\Delta f$, m/T, $m/(N\Delta t)$ which are all quantitatively the same.

21.1.4 Discrete Fourier Series

The discretization does not have to come from the Fourier integral for continuous functions. The Fourier series can be presented in a discrete (or digital) form from the beginning. It is important that we understand the discrete expression of the Fourier Transform from different angles or with different forms especially for a better understanding of digital filtering. For a periodic function, the discrete Fourier series is

$$y_n = \frac{a_0}{2} + \sum_{m=1}^{N-1} \left[a_m \cos\left(\frac{2\pi nm}{N}\right) + b_m \sin\left(\frac{2\pi nm}{N}\right) \right], \tag{21.21}$$

in which n varies from 1 to $N-1$. It has a finite number of terms because the function $y(t)$ is periodic and thus has a fundamental frequency, $1/T$ and thus $N = T/dt$, with the assumption that dt is a constant. The coefficients a_m and b_m are

$$a_m = \frac{2}{N} \sum_{n=1}^{N-1} y_n \cos\left(\frac{2\pi nm}{N}\right), \quad M = 0, 1, \ldots, N-1, \tag{21.22}$$

$$b_m = \frac{2}{N} \sum_{n=1}^{N-1} y_n \sin\left(\frac{2\pi nm}{N}\right), \quad M = 1, \ldots, N-1. \tag{21.23}$$

The above equations can be verified to be precise, considering the fact that the sinusoidal functions are orthogonal to each other. The proof of the above equations being exact relations is omitted here. In real applications, the periodic function assumption can be relaxed, in which case T is the total length of the time with uniformly spaced time series data.

21.1.5 The Discrete Fourier Series in Exponential Form (Fourier Transform)

As discussed earlier, in Section 21.1.4, even though it is about the Fourier series, we can still view it as the discrete form of Fourier Transform (if we use the exponential form of the Fourier series) because there is little difference in the digital world between them as far as our applications are concerned.

The exponential form of (21.21) is

$$y_n = \frac{1}{N} \sum_m Y_m e^{i\frac{2\pi nm}{N}}, \tag{21.24}$$

in which the coefficients are

$$Y_m = \sum_{n=0}^{N-1} y_n e^{-i\frac{2\pi nm}{N}}, \quad m = 0, 1, 2, \ldots, N-1, \tag{21.25}$$

$$Y_{-m} = Y_m{}', \quad m = 1, 2, \ldots, N-1. \tag{21.26}$$

Here the prime indicates the complex conjugate.

21.1.6 The Periodicity of Discrete Fourier Transform

In the continuous form of Fourier Transform, e.g. equations (21.4) and (21.5), the integrals involve unbounded time and frequency $\left(t, f \in (-\infty, +\infty)\right)$. With the discrete form for digital computation, however, the unbounded domains are not possible, neither necessary. If the total length in time of the record is T, and the data are uniformly spaced in time, i.e. the sampling interval is a constant Δt (sometimes we use dt – they are interchangeable in our context), the highest frequency that is resolvable is the Nyquist frequency $f_N = 1/(2\Delta t)$. Since the frequency resolution in this case is $f_0 = 1/T$, and so the discrete frequency values are $f_0, 2f_0, 3f_0, \ldots$. Therefore, it takes $N/2$ consecutive frequency points to reach f_N, if N is an even number. In addition, recall that given equation (21.25), there is some interesting and useful properties. First, there is the periodicity of the Fourier Transform: for any integer L,

$$Y_{m\pm LN} = \sum_{n=0}^{N-1} y_n e^{-i\frac{2\pi nm\pm LN}{N}} = \sum_{n=0}^{N-1} y_n e^{-i\frac{2\pi nm}{N}} e^{-i2\pi nL} = \sum_{n=0}^{N-1} y_n e^{-i\frac{2\pi nm}{N}} = Y_m. \quad (21.27)$$

So the discrete Fourier Transform expressed in this way is periodic with a repetition period of N numbers (or T in time). The point is, even though the original Fourier Transform is not meant for periodic functions (it is for functions spanning infinitely in both directions of time – past and future), the discrete Fourier Transform has the periodicity in any case and all cases.

In addition, we just concluded that only $N/2$ numbers in frequency are needed for covering all frequencies (i.e. up to the Nyquist frequency f_N), half of the Y_ms must be redundant as well. This is shown by the following equation:

$$Y_{N-m} = \sum_{n=0}^{N-1} y_n e^{-i\frac{2\pi n(N-m)}{N}} = \sum_{n=0}^{N-1} y_n e^{i\frac{2\pi nm}{N}} = Y'_m. \quad (21.28)$$

Indeed, half of the Y_ms are dependent on the values within the first $N/2$ values. This has been shown in Chapter 16 as a property of the Fourier Transform.

21.1.7 The Discrete Inverse Transform

With all these, and the fact that equation (21.25) uses N values of the data (y_n) to obtain N values of the Fourier Transform (Y_m), an inverse transform in the discrete format should also be between these two sets of arrays (NY_ms and Ny_ns). The inverse Fourier Transform can be expressed as

$$y_n = \frac{1}{N}\sum_{m=0}^{N-1} Y_m e^{i\frac{2\pi nm}{N}}, \quad (21.29)$$

which can be verified by substituting (21.25) into the right-hand side of (21.29):

$$\sum_{m=0}^{N-1} Y_m e^{i2\pi mn/N} = \sum_{m=0}^{N-1} \left(\sum_{k=0}^{N-1} y_k e^{-i2\pi mk/N} \right) e^{i2\pi mn/N} = \sum_{m=0}^{N-1} \sum_{k=0}^{N-1} y_k e^{i2\pi m(n-k)/N}. \tag{21.30}$$

We then switch the order of summations:

$$\sum_{m=0}^{N-1} Y_m e^{i2\pi mn/N} = \sum_{k=0}^{N-1} y_k \sum_{m=0}^{N-1} e^{i2\pi m(n-k)/N}. \tag{21.31}$$

Note that the following is true if k and n are not the same:

$$\sum_{m=0}^{N-1} e^{i2\pi m(n-k)/N} = 0, \; k \neq n. \tag{21.32}$$

The above equation can be seen from the following identity:

$$\sum_{m=0}^{N-1} \alpha^m = \frac{1 - \alpha^N}{1 - \alpha} = \frac{1 - e^{i2\pi(n-k)}}{1 - e^{i2\pi(n-k)/N}} = 0. \tag{21.33}$$

Here

$$\alpha = e^{i2\pi(n-k)/N}, \tag{21.34}$$

and we have used the fact that $e^{i2\pi(n-k)} = 1$. However, when $k = n$

$$\sum_{m=0}^{N-1} e^{i2\pi m(n-k)/N} = \sum_{m=0}^{N-1} 1 = N. \tag{21.35}$$

Therefore,

$$\sum_{m=0}^{N-1} Y_m e^{i2\pi mn/N} = \sum_{k=0}^{N-1} y_k \sum_{m=0}^{N-1} e^{i2\pi m(n-k)/N} = N y_n. \tag{21.36}$$

This is the same as (21.29).

The conclusion is: given the discrete Fourier Transform defined exactly by (21.25), the inverse Fourier Transform (21.29) is also an exact equation. In other words, even though the discrete Fourier Transform pairs are approximations of the integrals of the original Fourier Transform and inverse transform pairs, these discrete Fourier Transform pairs are exact relations between themselves. They are exact and there is no error at those discrete data points involved in the transform and inverse transform.

21.2 Filters in Digital World

For the discrete form of Fourier Transform, all the theorems are still valid, but they can be expressed in a slightly different (discrete) form. Here is a review of a couple of the theorems:

The time shift theorem. If $y(t)$ and $Y(f)$ are Fourier Transform and inverse transform pairs, i.e.

$$y(t) \leftrightarrow Y(f), \tag{21.37}$$

then

$$y(t - t_0) \leftrightarrow Y(f)e^{-i2\pi f t_0}. \tag{21.38}$$

In the digital world, if t is the nth time and t_0 the jth time, then

$$y(t_{n-j}) \equiv y_{n-j} \leftrightarrow Y(f)e^{-i2\pi f j \Delta t}. \tag{21.39}$$

21.2.1 The FIR Filters

Now let us look at the filtering using the window method. It is essentially a generalized moving average in which the weighting function is a window function. The calculation can be generally expressed as

$$g_n = \sum_{i=0}^{M-1} h_i y_{n-i}, \quad n = 1, 2, \ldots, N. \tag{21.40}$$

Here g_n is the output, y_k is the input, h_i is weighting function (or filter), and M is the length of the filter. This essentially gives an input–output relationship, or

$$g_n = h_0 y_n + h_1 y_{n-1} + h_2 y_{n-2} + \cdots + h_{M-1} y_{n-(M-1)}. \tag{21.41}$$

The number M is determined by the cutoff frequency. An equivalent form of (21.41) is

$$g(t_n) - h_0 y(t_n) + h_1 y(t_{n\ 1}) + h_2 y(t_{n\ 2}) + \cdots + h_{M-1} y\big(t_{n-(M-1)}\big). \tag{21.42}$$

This equation says that at $t = t_n$; the generalized moving average or filtering is dependent on the previous $M - 1$ values of the signal (or data) $y(t_{n-1}), y(t_{n-2}), \ldots, y\big(t_{n-(M-1)}\big)$. These are values of the signal with a time shift by Δt, $2\Delta t$, ... , and $(M - 1)\Delta t$, respectively. Using the time shift theorem (21.39) repeatedly, we have (after a Fourier Transform)

$$G(f) = h_0 Y(f) + h_1 Y(f)e^{-i2\pi f \Delta t} + h_2 Y(f)e^{-i2\pi f 2\Delta t} + \cdots + h_{M-1} Y(f)e^{-i2\pi f (M-1)\Delta t}. \tag{21.43}$$

Taking out the common factor $Y(f)$, we have

$$G(f) = \left(h_0 + h_1 e^{-i2\pi f\Delta t} + h_2 e^{-i2\pi f 2\Delta t} + \cdots + h_{M-1} e^{-i2\pi f(M-1)\Delta t}\right) Y(f) = H(f)Y(f), \tag{21.44}$$

in which $H(f)$ is the Fourier Transform of the filter $h(t)$ in discrete format:

$$H(f) = h_0 + h_1 e^{-i2\pi f\Delta t} + h_2 e^{-i2\pi f 2\Delta t} + \cdots + h_{M-1} e^{-i2\pi f(M-1)\Delta t}. \tag{21.45}$$

By choosing different weighting functions, i.e. different sets of h_j, we can have different $H(f)$ (different filters). This type of filter is called a *Finite Impulse Response (FIR) filter* or *non-recursive filter*.

21.2.2 The IIR Filter

The name of the FIR filter contrasts with the other kind of filter, which is called an *Infinite Impulse Response filter* or IIR filter, defined by the following equation:

$$g_n = \sum_{i=0}^{M-1} h_i y_{n-i} - \sum_{j=1}^{N-1} b_j g_{n-j}. \tag{21.46}$$

Here again, g_n is the output, y_k is the input, and h_i and b_j form another kind of weighting function (or filter).

This is also an input–output relationship. This equation has two convolutions added together: the first is a convolution between the input (the signal or data) at the present (n) and previous time instances $(n-1, n-2, \ldots, n-(M-1))$ and a filter h; the second is a convolution between the output (g) at the previous instances $(n-1, n-2, \ldots, n-(N-1))$ and another filter b.

Using the Time Shifting Theorem, the Fourier Transform of the above equation is simply

$$\begin{aligned} G(f) = &-\left(b_1 e^{-i2\pi f\Delta t} + b_2 e^{-i2\pi f 2\Delta t} + \cdots + b_{N-1} e^{-i2\pi f(N-1)\Delta t}\right) G(f) \\ &+ \left(h_0 + h_1 e^{-i2\pi f\Delta t} + h_2 e^{-i2\pi f 2\Delta t} + \cdots + h_{M-1} e^{-i2\pi f(M-1)\Delta t}\right) Y(f) \end{aligned}, \tag{21.47}$$

from which we have

$$G(f) = \frac{H(f)}{B(f)} Y(f) = \frac{h_0 + h_1 e^{-i2\pi f\Delta t} + h_2 e^{-i2\pi f 2\Delta t} + \cdots + h_{M-1} e^{-i2\pi f(M-1)\Delta t}}{1 + b_1 e^{-i2\pi f\Delta t} + b_2 e^{-i2\pi f 2\Delta t} + \cdots + b_{N-1} e^{-i2\pi f(N-1)\Delta t}} Y(f), \tag{21.48}$$

in which

$$\begin{aligned} H(f) &= h_0 + h_1 e^{-i2\pi f \Delta t} + h_2 e^{-i2\pi f 2\Delta t} + \cdots + h_{M-1} e^{-i2\pi f (M-1)\Delta t} \\ B(f) &= 1 + b_1 e^{-i2\pi f \Delta t} + b_2 e^{-i2\pi f 2\Delta t} + \cdots + b_{N-1} e^{-i2\pi f (N-1)\Delta t} \end{aligned} . \tag{21.49}$$

The function

$$\mathfrak{T}(f) = \frac{H(f)}{B(f)} \tag{21.50}$$

is called a *transfer function*, which provides a mathematical mapping between the input and output in the frequency domain. For each input Y at a given frequency f, this function gives an output as determined by (21.48). The function (21.50) is a linkage between the input and output and thus has the name of transfer function. The magnitude of the transfer function is also called an amplitude gain, because it gives the ratio between the input and output.

21.2.3 Linear System

The FIR filter equation (21.40) and IIR filter equation (21.46) are examples of linear systems. To discuss this concept, we examine the following equation for input and output:

$$g_k = by_k + ag_{k-1}. \tag{21.51}$$

Here y_k is the digital input, while g_k and g_{k-1} are digital outputs at kth and $(k-1)$th times, respectively.

Because this is a linear relationship, this equation represents a simple "linear system," in which the y_k is the input signal to a "linear system," through which it yields g_k, the output. This equation says that the output depends not only on the input but also the previous output. In signal processing, there are a couple of concepts that should be mentioned here.

The first is a *realizable linear system*: a discrete linear system is realizable if g_k only depends on the previous outputs ($g_{k-1}, g_{k-2}, \ldots$). The second is a *causal linear system*: a discrete linear system is causal if g_k only depends on the present and past inputs ($y_k, y_{k-1}, \ldots$). For a linear realizable system represented by (21.51), it is also causal. This can be seen by repeated use of the definition (21.51):

$$\begin{aligned} g_k &= by_k + ag_{k-1} = by_k + a(by_{k-1} + ag_{k-2}) = by_k + a\big(by_{k-1} + a(by_{k-2} + ag_{k-3})\big) \\ &= by_k + a\Big(by_{k-1} + a\big(by_{k-2} + a(by_{k-3} + \cdots)\big)\Big), \end{aligned}$$

which leads to

$$g_k = by_k + aby_{k-1} + a^2by_{k-2} + a^3by_{k-3} + \cdots a^kby_0 = b\sum_{n=0}^{k} a^n y_{k-n}. \qquad (21.52)$$

Equation (21.52) shows that g_k only depends on the present and previous inputs and thus is causal.

The above simple linear system can be broadened to have variable coefficients:

$$g_k = b_k y_k + a g_{k-1}. \qquad (21.53)$$

Here b is varying with "time" or k, and thus the system is called an *adaptive linear system*. The convolution combination, equation (21.46), is a little more complicated extension of the simplest linear system or an extension of (21.53).

From the above discussion, we can see that the FIR and IIR filters are cases of linear systems with the above properties. Causal and realizable linear systems are useful in real time data processing. For postprocessing, the design of a filter does not need to worry if future data are used in the algorithm. The `filtfilt` function in MATLAB is noncausal and non-realizable – the computation involves both forward and backward operations, and "future" data are used. As an example of application, a navigation algorithm for a drone usually takes in the rapid real-time GPS feed of position data to smooth the track data and obtain a reliable speed and heading so that any adjustment in direction and speed is based on a smooth curve that is an approximate of the true history of positions of previous times. The raw GPS data have random errors which cannot be used directly for navigation control or the drone will fly like crazy as it responds to the random fluctuation of position. This algorithm obviously can only use the current and previous input for the smoothing and the previous output or the previous smoothed positions for the real-time data processing and control. Therefore, the real-time filter for drone navigation control must be causal and realizable. For post-data processing, there is no requirement for a real time output, and the use of data points are not limited to "previous" data because there is no issue about "previous" and "future" data.

21.3 Applications of FIR and IIR Filters in MATLAB

In oceanography, motions can be categorized according to their spectrum. For example, tidal motions refer to the tidal oscillations at tidal frequencies; subtidal motions are those at frequencies lower than the diurnal tidal frequencies; and super-tidal motions are those with frequencies higher than those of semi-diurnal tides. Inertial motions are those caused by Earth's rotational effect with frequencies around that defined by the Coriolis parameter $f = 2\Omega \sin \varphi$, in which Ω and φ are the angular speed of Earth's rotation ($7.29 \times 10^{-5}\ \mathrm{s}^{-1}$) and latitude, respectively.

Sometimes, these inertial motions are called near inertial oscillations (NIO) because advection for background vorticity may modify the frequency to be slightly different from the theoretical local inertial frequency. Sub-inertial motions are those with frequencies lower than the inertial frequency, while the super-inertial motions are higher than the inertial frequency.

As a result, in oceanographic data analysis, filtering is a common technique to separate different parts of the spectrum for further analysis of the dynamics and for interpretations of observed processes. For example, the so-called subtidal variation in water level and current velocity are often obtained by a de-tide process which essentially is a low-pass filtering. In the following, we discuss the applications of several FIR and IIR filter functions in MATLAB.

21.3.1 The Filter Defining Functions in MATLAB

While the filter implementation function was discussed earlier (e.g. `filtfilt`), MATLAB has several FIR and IIR filter defining functions that are easy to use. These include `fir1` and `fir2` for the FIR filters. For the IIR filters, there are the `butter` function for the *Butterworth filter* and `yulewalk` function for the *yulewalk filter*, all are defined in the frequency domain. These are very powerful filters that are excellent choices suitable for most if not all applications in oceanography.

The `fir1` MATLAB function allows the design of low-pass, high-pass, band-pass, band-stop filters, and filters that have alternating pass-bands and stop-bands. All the frequencies used in the function must be normalized by the Nyquist frequency, such that the frequency is always between 0 and 1. The `fir2` MATLAB function allows the design of an arbitrary filter in the frequency domain by specifying the shape of the response function of the filter. The Butterworth filter is very efficient and has nice characteristics of a very smooth pass band and clean stopband, with essentially no leakage. The MATLAB function `yulewalk` is similar to `fir2`, allowing the user to design an almost arbitrary response filter. After the filter is defined/designed, the filter needs to be implemented by using a filter function or convolution function in MATLAB.

21.3.2 Construct High-Pass, Band-Pass, and Band-Stop Filters

Similar to section 20.4, using time series to construct the low-pass, high-pass, stop band, and band-pass filters in time domain, we can construct filters in frequency domain:

(1) To get a high-pass filter, we can use a low-pass filter and do a subtraction:

$$H_{highpass}(W_C) = H_{lowpass}(1) - H_{lowpass}(W_C). \tag{21.54}$$

$H_{lowpass}(1)$ is a "low-pass" filter that passes all signals and therefore can be seen as a low-pass filtering "operation" to the original data with a cutoff frequency at the Nyquist frequency. Alternatively, we can use 1 to replace $H_{lowpass}(1)$, i.e.

$$H_{highpass}(W_C) = 1 - H_{lowpass}(W_C). \tag{21.55}$$

(2) To obtain a *band-pass filter* keeping the signal components within $[W_1, W_2]$, we can subtract the two low-pass filters with W_1 and W_2 as cutoff frequencies, respectively:

$$H_{bandpass}(W_1, W_2) = H_{lowpass}(W_2) - H_{lowpass}(W_1). \tag{21.56}$$

(3) A band-stop filter cutting out the signal components within $[W_1, W_2]$ can be defined by subtracting the above band-pass filter from the low-pass filter that passes all frequencies:

$$H_{bandstop}(W_1, W_2) = H_{lowpass}(1) - H_{bandpass}(W_1, W_2) \tag{21.57}$$

or

$$H_{bandstop}(W_1, W_2) = 1 - H_{bandpass}(W_1, W_2). \tag{21.58}$$

The filter design and implementation procedures in MATLAB are given below.

21.3.3 Low-Pass Filter Design and Implementation

For `fir1` in MATLAB, if a low-pass filter is to be designed, one should first know the cutoff frequency normalized by the Nyquist frequency. In this case, the cutoff frequency is the frequency beyond which the signal should be cut off. So the steps are as follows:

(1) Make sure that the data are equally spaced, i.e. dt is constant. If dt is not constant, one should decide if there is any gap that is too large to interpolate. If the data quality is good and no major gap exists, an interpolation can be done to have constant dt.
(2) The total number of data points should be $N/T/dt$; N is preferably an even number; if not, add 1 or subtract 1.
(3) Find out the Nyquist frequency $f_N = 1/(2dt)$.
(4) Determine (or test and determine) the cutoff frequency f_C.
(5) Convert the cutoff frequency f_C to a nondimensional value by normalizing it with f_C: $W_C = f_C/f_N$, which is a number between 0 and 1.
(6) Choose (or test and choose) the order of the filter.

(7) Obtain the filter by using this MATLAB command: `h = fir1(M, Wc)`.
(8) Implement the filter using one of the filtering functions (`filtfilt`, `fftfilt`, or `filter`), e.g. `filtfilt(h,1,y)`, in which `y` is the data array.

21.3.4 High-Pass Filter Design and Implementation

For a *high-pass filter* design, the steps are the same as above, except that the 7th step should be replaced by a modified MATLAB command: `h = fir1(M,Wc, 'high')` so that the filter h is a high-pass filter with a cutoff frequency of `Wc`. In this context, the cutoff frequency means the frequency below which the signal will be cut off.

21.3.5 Band-Pass Filter Design and Implementation

For a band-pass filter, we can still use `fir1` in MATLAB. Here it is not a single-value cutoff frequency but rather a two-value array `Wn = [W1,W2]` , within which the signal is intended to be preserved and outside of which the signal should be filtered out. The steps are as follows:

(1) Make sure that the data are equally spaced, i.e. *dt* is constant. If dt is not constant, one should decide if there is any gap that is too large to interpolate. If the data quality is good and no major gap exists, an interpolation can be done to have constant *dt*.
(2) The total number of data points should be $N = T/dt$; N is preferably an even number. If not, add 1 or subtract 1.
(3) Find out the Nyquist frequency $f_N = 1/(2dt)$.
(4) Determine (or test and determine) the range of frequency to keep:$[f_1, f_2]$.
(5) Convert the frequency range $f_C = [f_1, f_2]$ to nondimensional values by normalizing it with f_N: $W_n = [f_1, f_2]/f_N = [W_1, W_2]$, which gives two numbers $0 < W_1 < W_2 < 1$.
(6) Choose (or test and choose) the order of the filter M.
(7) Obtain the filter by using this MATLAB command: `h = fir1(M, Wn)`.
(8) Implement the filter using one of the filtering functions (`filtfilt`, `fftfilt`, or `filter`), e.g. `filtfilt(h,1,y)`, in which `y` is the data array.

21.3.6 Band-Stop Filter Design and Implementation

For a band-stop filter design, the steps are the same as above, except that the seventh step should be replaced by a modified MATLAB command: `h = fir1`

(M,Wn, 'stop') so that the filter h is a band-stop filter to eliminate the signal within $[W_1, W_2]$.

As an example, if one decides to use a filter of size 128 to pass frequencies between 0.2 and 0.5 in frequency normalized by the Nyquist, the commands to use in MATLAB are:

```
h = fir1(128, [ 0.2 0.5] );
g = filtfilt(h, 1, y);    % implement the filter,
                          % given y is the signal, g is the filtered array
```

21.3.7 Examples of Performance of MATLAB FIR Filter Functions

The following scripts are used to generate the FFT of a low-pass filter using the MATLAB function fir1. Different orders of the filter (N) are used to examine the behavior of the filter at various orders of the filter.

```
for N = 20:20:100;
Wn = 0.3;
h = fir1(N,Wn);

NN = 10 * N;
NNN = length(h) + NN;
f0 = 2 / (NNN);
f = f0 * (0:NNN - 1);
hpad = [ h zeros(1,NN)] ;
if N == 20
plot(f,abs(fft(hpad)),'k')
end
if N == 40
plot(f,abs(fft(hpad)),'b')
end
if N == 60
plot(f,abs(fft(hpad)),'g')
end
if N == 80
plot(f,abs(fft(hpad)),'m')
end
if N == 100
plot(f,abs(fft(hpad)),'r')
end
hold on
end
legend('N=20','40','60','80','100','Location', 'Best')
plot([ 0.3,0.3] , [0,1] ,'LineWidth',3)
axis([ 0 2 0 1.2] )
xlabel('f / f_N')
```

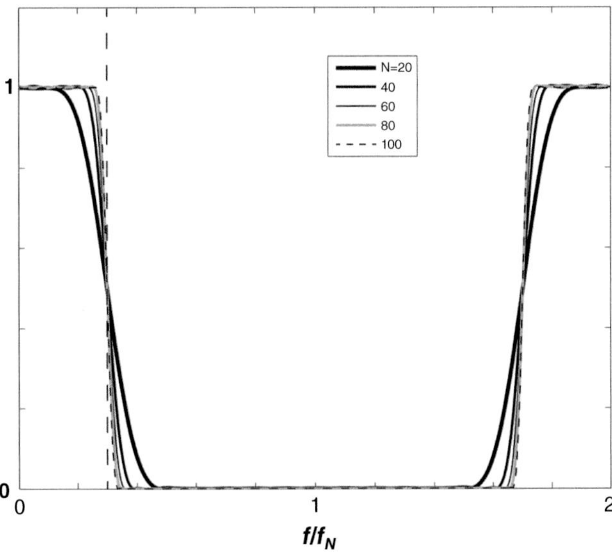

Figure 21.1 A FIR filter shown in frequency domain with a cutoff frequency at 0.3 f_N at the twentieth, fortieth, sixtieth, eightieth, and hundredth, respectively.

In this example (Fig. 21.1), the position of the cutoff frequency (0.3 f_N) is shown by the dashed vertical line. Obviously, the cutoff frequency is within the transition zone. That means that at the cutoff frequency, there is still quite a significant energy going through the filter. The larger the N, the narrower the transition zone, and the closer it is to an idealized rectangular low-pass filter.

When the pass band or stop band is zoomed in (Fig. 21.2), it is obvious that there are some ripples, which is (mostly) a minor problem. These ripples are not easy to get rid of, even with increased order of the filter N. Similarly, we can check the behavior of the band-pass filter (Fig. 21.3), which also shows oscillations at the pass-band and stop-band (as the zoomed-in view in Fig. 21.4).

21.3.8 The Butterworth Filter

The Butterworth filter is a commonly used IIR filter. The MATLAB command for the Butterworth filter is `butter`. The nth order low-pass Butterworth filter with a cutoff frequency of f_C is defined by

$$|H_L(f, f_C)|^2 = \frac{1}{1 + \left(\frac{f}{f_C}\right)^{2N}}. \tag{21.59}$$

This filter has the maximum flatness at pass-band, or both pass-band and stop-band are very smooth (Fig. 21.5), and there are no ripples as the FIR filters demonstrated in Figs. 21.1–21.4. This is an advantage over the FIR filters. Fig. 21.6 is the same

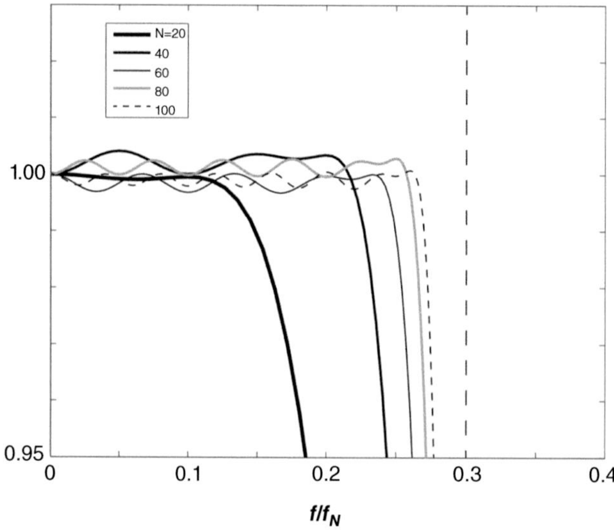

Figure 21.2 A zoomed-in view of Fig. 21.1, showing mainly the pass band ripples.

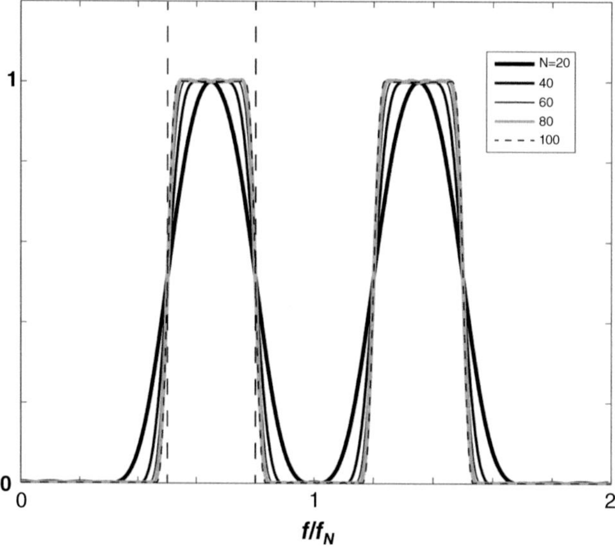

Figure 21.3 A band-pass FIR filter with a pass-band within the frequency range between $0.5\,f_N$ and $0.8\,f_N$ at twentieth, fortieth, sixtieth, eightieth, and hundredth orders, respectively.

as Fig. 21.5 except that here the logarithmic scale with a unit of dB is used. The transition zone between the pass-band and stop-band is narrower when the order is greater: the eighth order filter has much narrower transition zone than the second order one and, as a result, both pass-band and stop-band are wider.

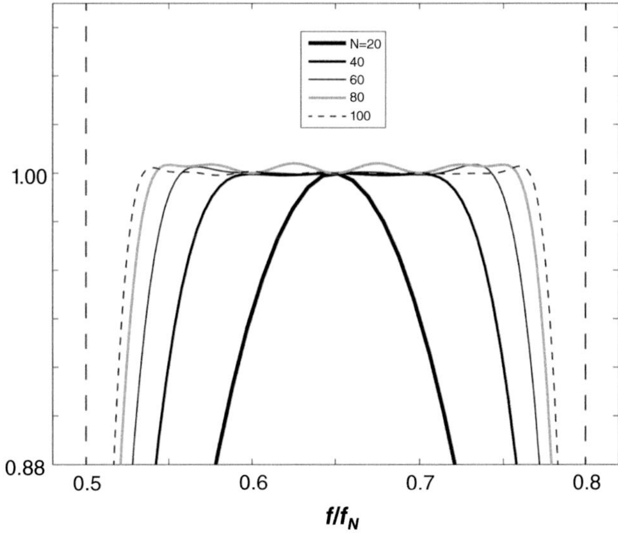

Figure 21.4 A zoomed-in view of Fig. 21.3 with a focus on the pass-band ripples.

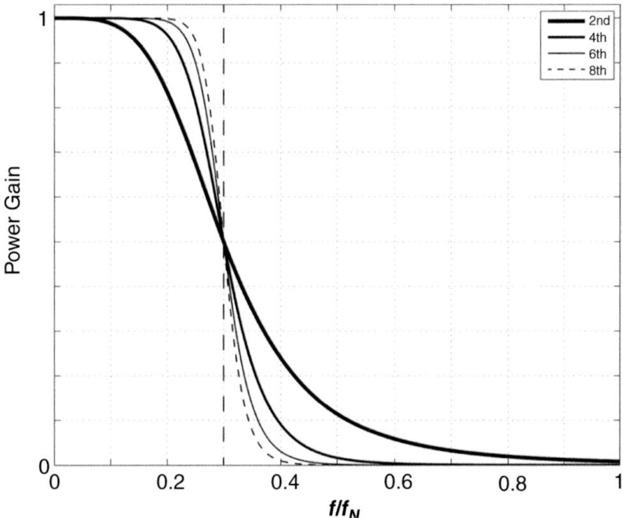

Figure 21.5 A Butterworth low-pass filter: examples of second, fourth, sixth, and eighth order filters with a cutoff frequency of 0.3 f_N as shown by the dashed vertical line.

21.3.9 An Example of FIR Filter

In the following, we examine a hypothetical time series with seven frequencies and apply MATLAB's fir1 filter function to do a band-pass filtering that will eliminate high and low frequencies, a band-stop filter that retains high and low frequencies, as well as

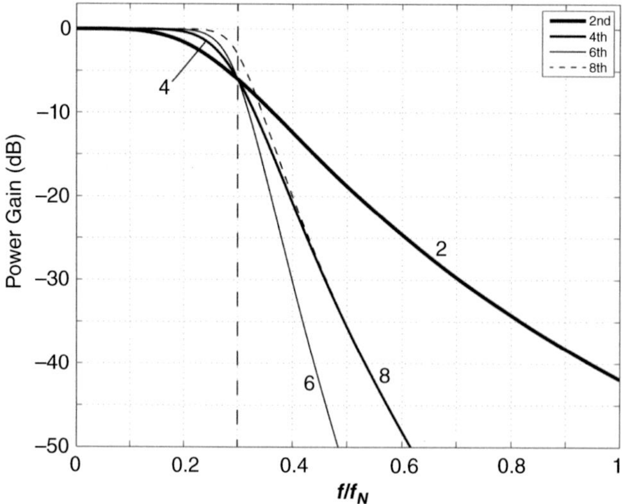

Figure 21.6 A Butterworth low-pass filter – same as Fig. 21.5 but with the impulse response function in dB: examples of second, fourth, sixth, and eighth order filters with a cutoff frequency of 0.3 f_N as shown by the dashed vertical line.

high-pass and low-pass filters. More specifically, the hypothetical signal includes the following frequencies: $f_0 = 1, f_1 = 5, f_2 = 10, f_3 = 50, f_4 = 100, f_5 = 150$, and $f_6 = 200$; the sampling rate should be three times higher than the highest frequency f_6 (i.e. $f_s = 600$), thus $dt = 1/f_s = 0.0017$. The time series is defined by

$$x(t) = 1 + \sin(2\pi f_1 t) + \sin(2\pi f_2 t) + \sin(2\pi f_3 t) + \sin(2\pi f_4 t) + \sin(2\pi f_5 t) + \sin(2\pi f_6 t). \tag{21.60}$$

Suppose that we need to do a band-pass filtering so that the fifth and sixth frequencies are kept, i.e. $f_4 = 100$, $f_5 = 150$ are kept; the rest frequencies need to be filtered out. The Nyquist frequency is $f_N = 1/(2dt) = 300$. The normalized frequencies for the signal are:

$$f_0 = [0, 0.0167,\ 0.0333,\ 0.1667,\ 0.3333,\ 0.5000,\ 0.6667]. \tag{21.61}$$

Therefore, we need to find a two-number array `Wn` to encompass $f_0 = [0.3333, 0.5000]$ in dimensionless frequency or to keep $f_4 = 100$ and $f_5 = 150$ in dimensional frequency. In this example, we choose `Wn =[0.2 0.55],` which will allow the elimination of the first four frequencies and the last frequency (in dimensional frequency, these are $f_0 = 1, f_1 = 5, f_2 = 10, f_3 = 50$, and $f_6 = 200$). The following scripts demonstrate how to do this work with a 11th order filter (Fig. 21.10). Obviously, there are leakages of frequencies into the filtered signal (lower panel of Fig. 21.10, particularly compared with the upper panel). If we increase the order of the filter, e.g. 129th order, the results are significantly improved

(Fig. 21.11). The spectrum at $f_5 = 150$ has a slight decrease in magnitude, which is partly due to the sampling being sparse (four points per cycle).

If the band-pass filter is changed to a band-stop filter, at 129th order, the result is shown in Fig. 21.12. Likewise, we also include an example for a high-pass filtering (Fig. 21.13) retaining the last three frequencies and a low-pass filtering (Fig. 21.14) with a normalized cutoff frequency of $f_C = 0.2$ to get rid of the last three frequencies.

```
%===============================
% EXAMPLE # 1 - bandpass filter
%===============================
% DEFINE THE FREQUENCIES IN THE SIGNAL
f1 = 5;
f2 = 10;
f3 = 50;
f4 = 100;
f5 = 150;
f6 = 200;
dt = 1 / (3 * f6);  % DEFINE THE "SAMPLING RATE"
T = 2;              % DEFINE TOTAL LENGTH OF TIME
t = 0:dt:T;         % DEFINE TIME
fmax = 1 / dt / 2;  % CALCULATE THE NYQUIST FREQUENCY
f0 = [ 0 f1 f2 f3 f4 f5 f6] / fmax    % CALCULATE THE NORMALIZED
                                      % FREQUENCIES (MAXIMUM = 1)
% f0 = [ 0 0.0167 0.0333 0.1667 0.3333 0.5000 0.6667]
% f  = [ 0 5      10     50     100    150    200]

% DEFINE THE TIME SERIES SIGNAL
x = 1 + sin(2 *pi * f1 * t) + sin(2 * pi * f2 * t) + sin(2 * pi * f3 * t) + ...
sin(2 * pi * f4 * t) + sin(2 * pi * f5 * t) + sin(2 * pi * f6 * t);
% DESIGN A BANDPASS FILTER USING "fir1" FUNCTION with N = 10
h = fir1(10, [ 0.2 0.55] );% LET the 5th and 6th frequency pass
                           % and define a filter of size 11
y = filtfilt(h, 1, x);     % implement the filter
%===============================
% EXAMPLE # 2 - bandstop filter
%===============================
% DESIGN A BANDstop FILTER USING "fir1" FUNCTION with N = 10
h = fir1(128, [0.2 0.55], 'stop');    % LET the 5th and 6th
                                      % frequency be stopped
                               % and define a filter of size 129
%===============================
% EXAMPLE # 3 - highpass filter
%===============================
% DESIGN A highPASS FILTER USING "fir1" FUNCTION with N = 10
h = fir1(128, 0.2, 'high');%define a highpass filter of size 129
%===============================
% EXAMPLE # 4 - lowpass filter
%===============================
% DESIGN A lowPASS FILTER USING "fir1" FUNCTION with N = 10
h = fir1(128, 0.2);   % define a lowpass filter of size 129
```

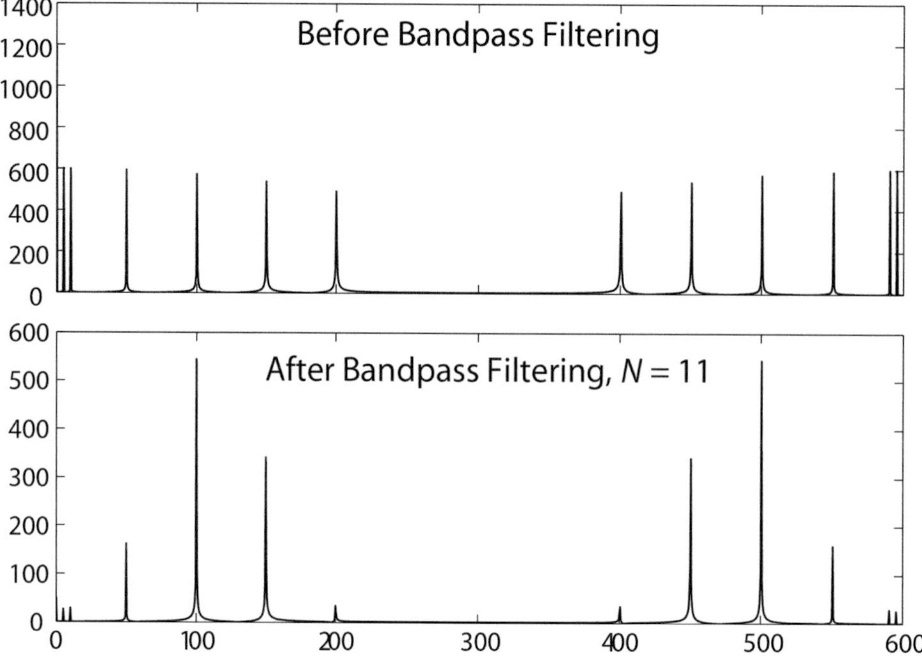

Figure 21.7 Band-pass filtering using the eleventh order MATLAB function `fir1`. Upper panel: the spectrum for original signal. Lower panel: the spectrum for the filtered signal.

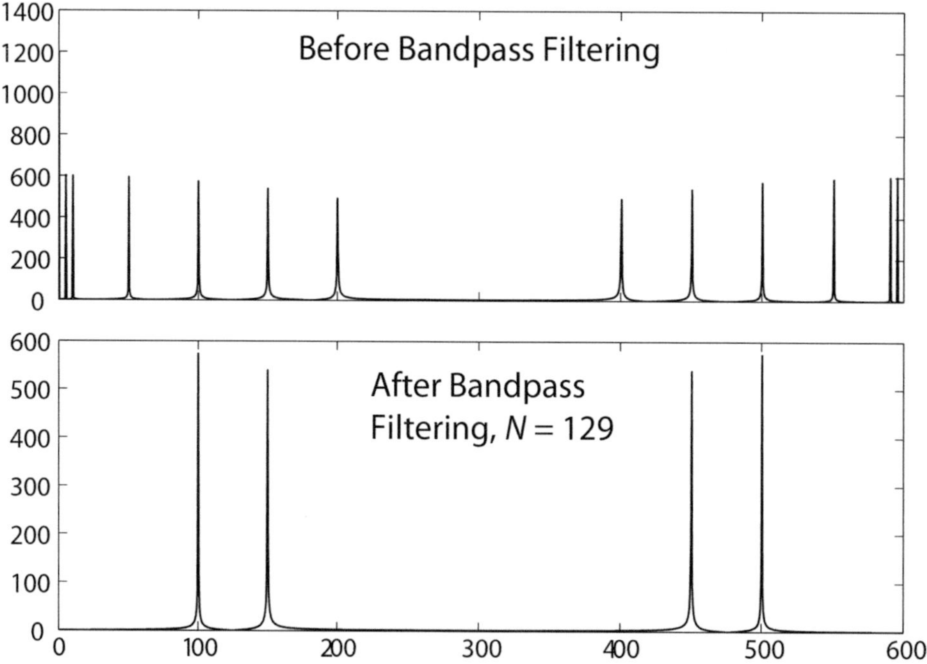

Figure 21.8 Band-pass filtering using the 129th order MATLAB function `fir1`. Upper panel: the spectrum for original signal. Lower panel: the spectrum for the filtered signal.

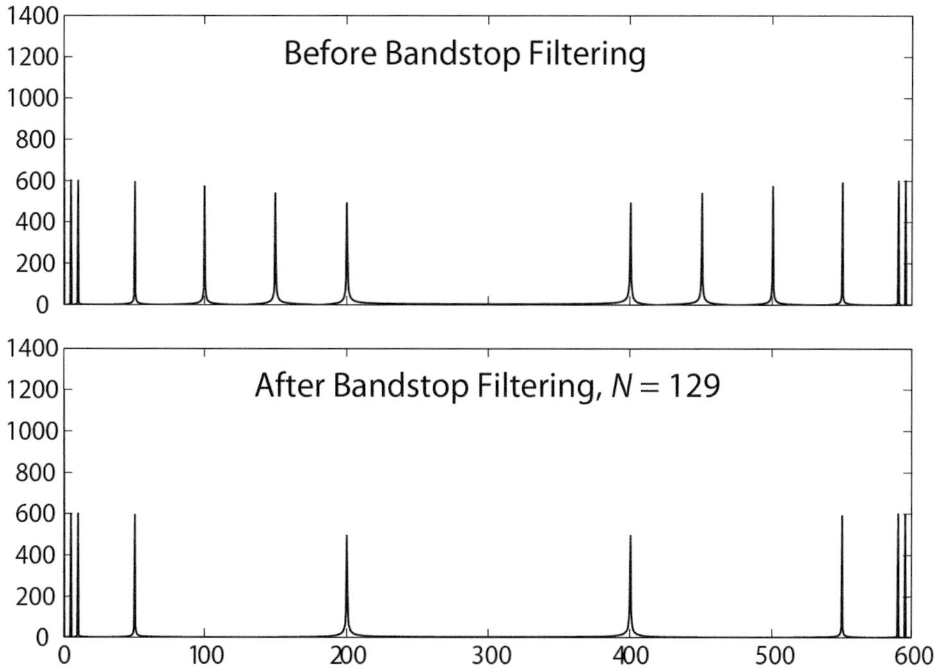

Figure 21.9 Band-stop filtering using the 129th order MATLAB function `fir1`. Upper panel: the spectrum for original signal. Lower panel: the spectrum for the filtered signal.

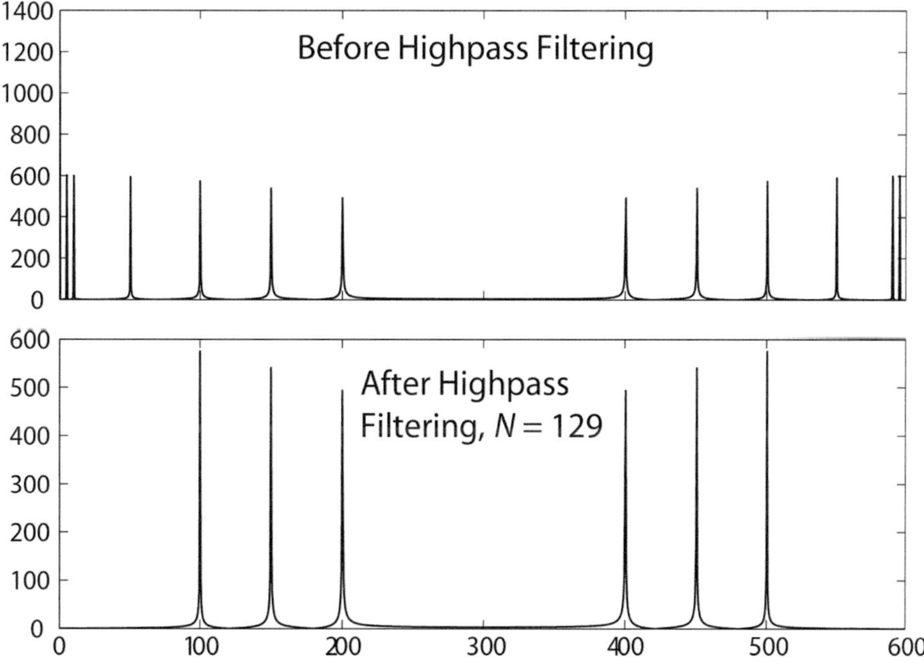

Figure 21.10 High-pass filtering, keeping the last three high frequencies using the 129th order MATLAB function `fir1`. Upper panel: the spectrum for original signal. Lower panel: the spectrum for the filtered signal.

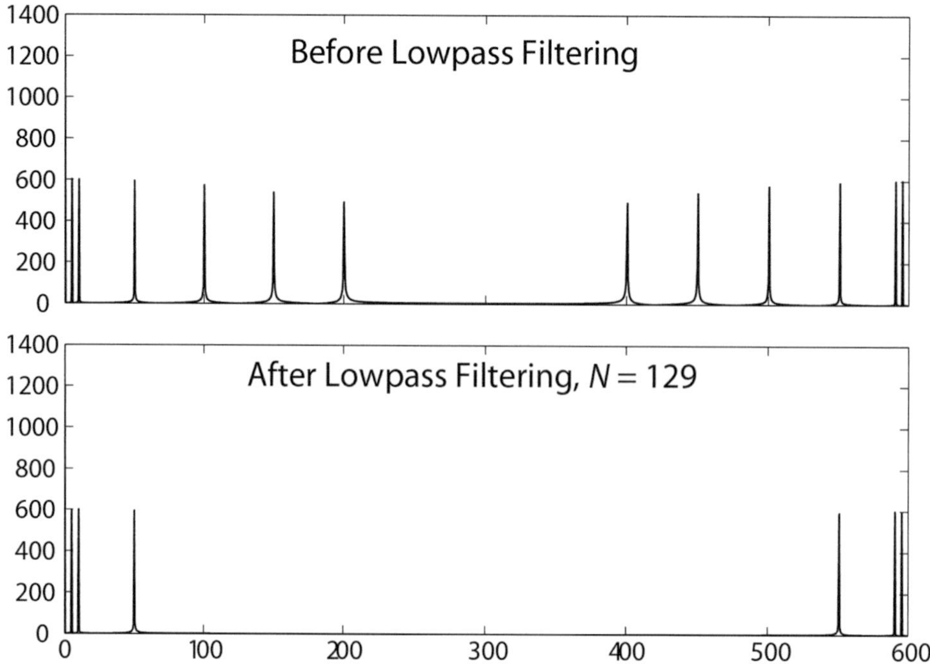

Figure 21.11 Low-pass filtering, eliminating the last three high frequencies using the 129th order MATLAB function `fir1`. Upper panel: the spectrum for original signal. Lower panel: the spectrum for the filtered signal.

Review Questions for Chapter 21

(1) Contrast FIR filters with IIR filters. Which group is better?
(2) Is moving average an FIR filter?
(3) What is a transfer function?

Exercises for Chapter 21

(1) Filtering using MATLAB function butter and yulewalk for hypothetical data:
- **(a)** Construct a 10-year daily time series.
- **(b)** The time series consists of a constant, seven frequencies, and a random error.
- **(c)** The seven frequencies correspond to the following periods: (1) 20 years, (2) 15 years, (3) 3 years, (4) 1 year, (5) 3 months, (6) 21 days, (7) 8 days.
- **(d)** The constant is c0 = 2.6; while the amplitudes for the frequencies are, respectively, c1 = 0.5; c2 = 0.7; c3 = 1.5; c4 = 1.2; c5 = 0.8; c6 = 0.8; c7 = 1.2.

(e) List all your frequencies in an increasing order.
(f) Normalize your frequencies by the Nyquist frequency.
(g) Select a normalized cutoff frequency such that the annual frequency and higher frequencies are filtered.
(h) Construct a Butterworth filter by using the MATLAB function `[b,a] = butter(N,Wc)`, in which N is the order of the filter and Wc is the cutoff frequency.
(i) Plot the frequency response of the filter by `[h1, w] = freqz(b, a, 528);plot(w/pi,abs(h1), 'o')`.
(j) Plot your original data; hold on; then add your low-pass filtered data.
(k) Compare your low-pass filtered data with the hypothetical time series.
(l) Design a low-pass filter to filter out the 21-day oscillations and higher frequency components using the MATLAB `yulewalk` function.

(2) Filtering using MATLAB function butter for observational ocean current data:
(a) Load the data "adcp_data.mat."
(b) Define the time variable by using

```
time = datenum(year, month,day, hour,minute,second)-
datenum(2000,1,1) + 1.
```

(c) Check if dt is constant; if not, do interpolation to the velocity components u1 (east component, m/s), and v1 (north component, m/s).
(d) Do FFT for both components of the velocity and plot the spectra.
(e) Use MATLAB function `fir1` to do a low-pass filtering to get rid of tidal signals.
(f) Use MATLAB function `fir2` to do a low-pass filtering to get rid of tidal signals.
(g) Use MATLAB function `butter` to do a low-pass filtering to get rid of tidal signals.
(h) Use MATLAB function `yulewalk` to do a low-pass filtering to get rid of tidal signals.
(i) Compare with the results from different filters.

(3) Combining filtering, harmonic analysis, and Fourier analysis:
(a) Load the file named "Fourchon-water-level.mat," which has two variables: time in days with a name t1; and water level in meters with a name h1.
(b) Check for time intervals and determine if they are a constant; if not, do interpolation so that the time interval is a constant of 15 minute.
(c) Do a de-trend, or alternatively a high-pass filtering to take out the low-frequency trend of the record and keep all tidal constituents.
(d) Do a harmonic analysis with the following eight tidal frequencies: M2, S2, N2, K1, O1, S1, M1, P1.

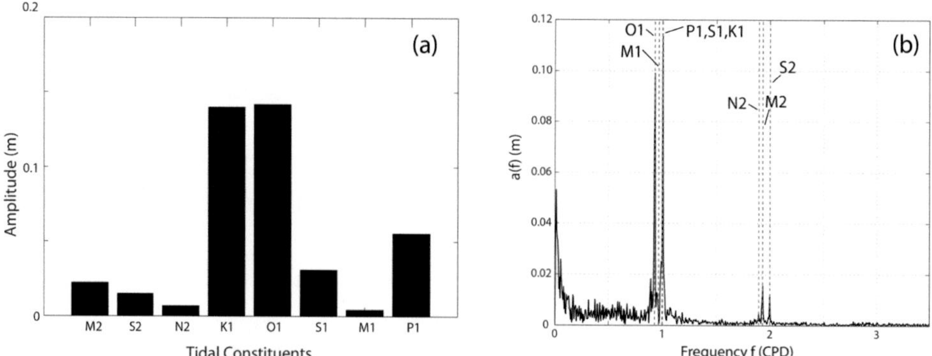

Figure 21.E1 Graphs for Exercise 21.3. The actual graph may be slightly different as shown. (a) Results from the harmonic analysis for tidal amplitude at the given eight frequencies; (b) results from the Fourier analysis.

(e) Do Fourier analysis.

(f) Draw graphs: the tidal amplitude at the eight frequencies (Fig. 21.E1a); and the spectrum with the tidal frequencies marked as shown in Fig. 21.E1b. The graphs should look similar to the figure below.

22

Rotary Spectrum Analysis

About Chapter 22

The objective of this chapter is to introduce rotary spectrum analysis for velocity vector time series. When the two components of a velocity vector have different frequencies, the tip of the displacement vector would draw a figure called a Lissajous Figure. A special case of the Lissajous Figure is when the two components oscillate at the same frequency. Vector time series at a given frequency can only have a few basic patterns or a combination of these patterns: the tip of the vector would draw a line segment back and forth repeatedly, or rotate either clockwise or counterclockwise. This makes it necessary to study the rotary spectra for rotations in both directions. A rectilinear motion is a degenerated version or special case of rotary motion.

22.1 Lissajous Figure

The spectrum analysis we have been discussing is for time series of a scalar. When it comes to spectrum analysis for the time series of a vector, such as wind velocity or ocean current velocity, it is common for the spectrum for each component of the vector to be analyzed and examined separately. This separated approach may not capture the intrinsic characteristics of the vector time series. Often, the components of a vector are somehow related to each other, and they may behave in a coordinated manner.

In physics, we know that if the trajectory of a particle is a result of a combination of motions in two perpendicular directions with two frequencies, a variety of complex figures can be formed. More specifically, the motion in x and y directions is described by the following equation:

$$x = a\sin(\alpha t + \varphi), \tag{22.1}$$

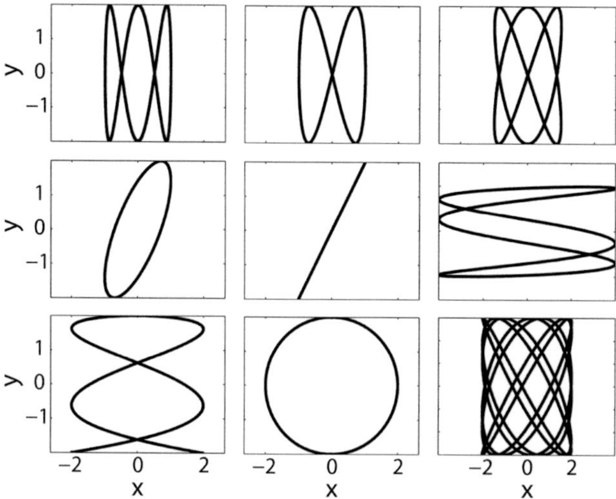

Figure 22.1 Examples for the Lissajous Figure.

$$y = b\sin(\beta t), \tag{22.2}$$

in which a, b, α, β, and φ are constants. The figure drawn by the point (Fig. 22.1) that is defined by the coordinates x and y is called a *Lissajous Figure* (Pain, 2005; Stewart, 2012). When the frequencies $\alpha = \beta$ are the same, the point defined by coordinates x and y draws an ellipse. If $\alpha = \beta$, $a = b$, and $\varphi = \pi/2$, it is a circle. If $\alpha = \beta$, $a = b$, and $\varphi = 0$, it is a line segment.

With a given frequency, the two components of a 2-D vector are both sinusoidal functions. The vector at this single frequency is thus a special case of the Lissajous Figure, and the corresponding time series shows a *rotary motion*. The vector time series at this frequency rotates either clockwise or counterclockwise. Under the influence of the Earth's rotation, rotary vector motion is quite common in both the atmosphere and the ocean.

Because of this, the spectrum analysis can be done to the vector, or, if the spectrum for each component is obtained separately, the analysis is done with all of them considered together. This method is called rotary spectrum analysis. The following gives a few examples of rotary vectors in meteorology and oceanography.

22.2 Rotary Motions

22.2.1 Sea Breeze

As an example of a coordinated variation of different vector components, *sea breeze* (O'Brien and Phillsbury, 1974) is an atmospheric phenomenon in which

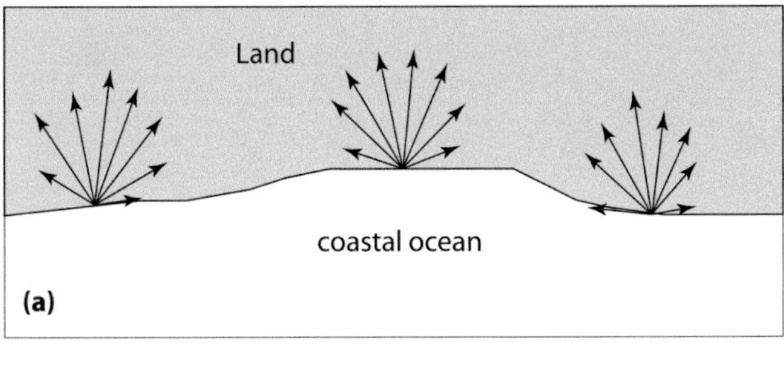

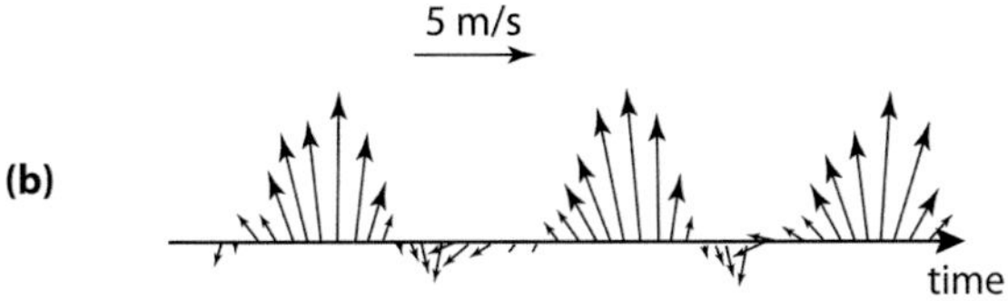

Figure 22.2 A typical sea breeze on the coast. (a) A typical rotating sea-level wind vector at coastal stations under fair weather conditions; (b) time series of the clockwise rotating sea-breeze wind vector in the northern hemisphere.

wind vector varies along the coast under fair weather conditions, with the sea-level wind being landward during the day, maximizing in the afternoon. The wind vector in a sea breeze would rotate (Fig. 22.2) in a clockwise (counterclockwise) sense in the northern (southern) hemisphere due to the Coriolis effect of the Earth's rotation.

22.2.2 Tidal Ellipses

A second example of a coordinated variation of different components in a vector is tidal currents. Tidal currents are oscillations of flow vectors which often appear as rotary currents, i.e. so-called *tidal ellipses* (Fig. 22.3). The term tidal ellipse is defined because the tip of a tidal current vector rotates either clockwise or counterclockwise with the magnitude of the flow varying periodically, drawing an ellipse. Such a tidal ellipse at each given location is usually quite persistent for a given tidal constituent.

22.2.3 Poincare Waves

In the ocean, hydrostatic long waves are affected by the effect of the Earth's rotation. These waves can be either *sub-inertial* or *super-inertial* and propagate in a rotary fashion in the ocean. One of these waves is the so-called *Poincare wave* (Pedlosky, 1987), which is a type of super-inertial shallow water wave modified by

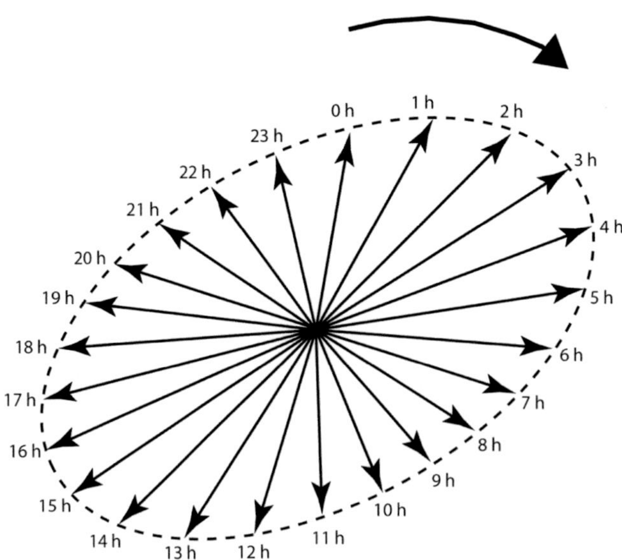

Figure 22.3 A S1 diurnal tidal ellipse. In this example, the tidal current rotates clockwise. 0 h, 1 h, ... , 23 h indicate the tidal current phase in hours within a complete cycle.

the Coriolis effect, assuming we are looking at a problem on an *f-plane* (an approximation only including the Coriolis effect at a fixed latitude but not counting the effect of the variation of Coriolis force with latitude). Poincare waves can be expressed as the following for its components in the direction of the wave propagation and in the direction perpendicular to the wave propagation, respectively:

$$u_{\|} = \frac{\zeta_0}{H} C \cos\left(kx + ly - \omega t + \phi\right), \tag{22.3}$$

$$u_{\perp} = \frac{\zeta_0 f}{H\,\omega} C \sin\left(kx + ly - \omega t + \phi\right), \tag{22.4}$$

in which $u_{\|}$ and $u_{\perp}$ are the velocity components in the directions parallel and perpendicular to the wave propagation, respectively; ζ_0 is the wave height amplitude; H the mean depth; ω the angular speed of the wave; f the Coriolis parameter on a f-plane; x, y are the horizontal coordinates; t is time; k, l are the wave numbers in the x and y directions, respectively; ϕ is the phase difference between the water level and the velocity; and C is the speed of propagation:

$$C = \frac{\omega}{K} = \pm\sqrt{gh + \left(\frac{f}{K}\right)^2}, \qquad K^2 = k^2 + l^2. \tag{22.5}$$

It can be verified that equations (22.3) and (22.4) lead to

$$u_{\|}^2 \left(\frac{f}{\omega} \right)^2 + u_{\perp}^2 = \left(\frac{\zeta_0}{H} \frac{f}{\omega} C \right)^2 \tag{22.6}$$

or

$$\frac{u_{\|}^2}{\left(\frac{\zeta_0}{H} C \right)^2} + \frac{u_{\perp}^2}{\left(\frac{\zeta_0}{H} \frac{f}{\omega} C \right)^2} = 1. \tag{22.7}$$

This is an equation of an ellipse. So, the Poincare wave velocity vector draws an ellipse at a frequency higher than the inertial oscillation frequency f:

$$f = 2\Omega \sin(\theta), \tag{22.8}$$

in which $\Omega = 7.29 \times 10^{-5} \mathrm{s}^{-1}$ is the Earth's angular speed of rotation, and θ the latitude of the wave under consideration.

For a freely traveling Poincare wave, its velocity vector has a clockwise rotation in the northern hemisphere; its main component is in the direction of wave propagation, and the local oscillation of a particle is an ellipse. Its frequency of oscillation is always greater than f (thus super-inertial).

22.2.4 Inertial Oscillation

Theoretically speaking, *inertial oscillations* are free particle motions (under no force) in the ocean or atmosphere. The motion is such that the velocity vector rotates in the clockwise direction at the inertial frequency f defined by equation (22.8). It is a circular motion, which can be described as the following:

$$u = u_0 \cos(ft) \tag{22.9}$$

$$v = u_0 \sin(ft). \tag{22.10}$$

Obviously,

$$u^2 + v^2 = u_0^2, \tag{22.11}$$

which is an equation of a circle.

22.2.5 Near Inertial Oscillation

When there is nonlinearity, e.g. advection due to horizontal velocity shears or vorticity, the inertial oscillation frequency has to be determined dynamically (Mooers, 1975; Kunze, 1985) by

$$\omega \approx \sqrt{f(f+\eta)} \approx f + \frac{\eta}{2}, \tag{22.12}$$

in which η is the component of the relative vorticity due to horizontal velocity shear. The inclusion of the effect of nonlinearity through relative vorticity makes the inertial oscillation frequency change from the fixed value f within a small range. This is a dynamic response and the new frequency depends on the actual flow field. Since the change from f is relatively small, the actual oscillation has a frequency near f. This gives the motion the name *near inertial oscillation* (NIO).

22.3 Rotary Spectrum Analysis

With the above examples, we can see that there are some processes in the atmosphere and ocean with horizontal velocity vectors changing in a rotary fashion. For largescale motions in which the Rossby number is small, the vertical motion is negligible compared to the horizontal velocity components. For these problems, we only need to look at the two-dimensional velocity. For two-dimensional velocity vector time series, a rotary spectrum method (Gonella, 1972; O'Brien and Phillsbury, 1974) is developed to examine the rotation of velocity vectors at different frequencies. A concise way to express a rotary vector in a plane is to use a complex variable. For a two-dimensional velocity in the horizontal plane, e.g. wind or current, we define a complex velocity as

$$w = u + iv, \quad i = \sqrt{-1}. \tag{22.13}$$

This is a *complex velocity*. This complex variable represents the velocity vector on the complex plane. For a given frequency f, if we take out just the component of Fourier Transform at this frequency, we have the sine-cosine format of expression:

$$\begin{aligned} u &= a_1 \cos(2\pi f t) + b_1 \sin(2\pi f t) \\ v &= a_2 \cos(2\pi f t) + b_2 \sin(2\pi f t) \end{aligned}. \tag{22.14}$$

Note that here the frequency f is not necessarily the inertial oscillation frequency. The relationship between the exponential base function and sine-cosine base functions is given by the Euler formula as below.

$$\begin{aligned} e^{i2\pi f t} &= \cos(2\pi f t) + i\sin(2\pi f t) \\ e^{-i2\pi f t} &= \cos(2\pi f t) - i\sin(2\pi f t) \end{aligned}. \tag{22.15}$$

The inverse transform is

$$\begin{aligned} \cos(2\pi f t) &= \frac{e^{i2\pi f t} + e^{-i2\pi f t}}{2} \\ \sin(2\pi f t) &= \frac{e^{i2\pi f t} - e^{-i2\pi f t}}{2i} \end{aligned}. \tag{22.16}$$

These relationships lead to the expression of the complex rotary velocity in the following:

$$\begin{aligned} w &= (a_1 + a_2 i)\cos(2\pi f t) + (b_1 + b_2 i)\sin(2\pi f t) \\ &= (a_1 + a_2 i)\frac{e^{i2\pi f t} + e^{-i2\pi f t}}{2} + (b_1 + b_2 i)\frac{e^{i2\pi f t} - e^{-i2\pi f t}}{2i} \\ &= \left(\frac{a_1 + b_2}{2} + i\frac{a_2 - b_1}{2}\right)e^{i2\pi f t} + \left(\frac{a_1 - b_2}{2} + i\frac{a_2 + b_1}{2}\right)e^{-i2\pi f t} \end{aligned}. \tag{22.17}$$

The above equation shows that the complex velocity has two components as well: one has a time factor of $e^{i2\pi ft}$ while the other a factor of $e^{-i2\pi ft}$. In a complex plane, both of these are unit vectors rotating as time increases. The only difference is that the former rotates counterclockwise (cyclonic) and the latter rotates clockwise (anticyclonic) as time increases. The coefficients in front of the counterclockwise and clockwise rotational components have the following magnitudes, respectively:

$$A = \frac{1}{2}\sqrt{(a_1 + b_2)^2 + (a_2 - b_1)^2} \tag{22.18}$$

and

$$B = \frac{1}{2}\sqrt{(a_1 - b_2)^2 + (a_2 + b_1)^2}. \tag{22.19}$$

The above equations are not directly applicable from the Fourier Transforms of u and v. This is because the Fourier Transform provides only coefficients in front of the exponential-based functions, not those for the sine-cosine functions. Note that

$$\begin{aligned} u &= a_1\frac{e^{i2\pi f t} + e^{-i2\pi f t}}{2} + b_1\frac{e^{i2\pi f t} - e^{-i2\pi f t}}{2i} \\ &= \frac{1}{2}(a_1 - ib_1)e^{i2\pi f t} + \frac{1}{2}(a_1 + ib_1)e^{-i2\pi f t} \end{aligned} \tag{22.20}$$

and

$$\begin{aligned} v &= a_2\frac{e^{i2\pi f t} + e^{-i2\pi f t}}{2} + b_2\frac{e^{i2\pi f t} - e^{-i2\pi f t}}{2i} \\ &= \frac{1}{2}(a_2 - ib_2)e^{i2\pi f t} + \frac{1}{2}(a_2 + ib_2)e^{-i2\pi f t} \end{aligned}. \tag{22.21}$$

We can establish the relations between the Fourier Transform coefficients (for the exponential base functions) and the Fourier series coefficients (for the sine and cosine base functions):

$$\begin{aligned} u &= A_1 e^{i2\pi f t} + B_1 e^{-i2\pi f t} \\ v &= A_2 e^{i2\pi f t} + B_2 e^{-i2\pi f t} \end{aligned}, \tag{22.22}$$

$$\begin{aligned} A_1 &= \frac{1}{2}(a_1 - ib_1),\ A_2 = \frac{1}{2}(a_2 - ib_2) \\ B_1 &= \frac{1}{2}(a_1 + ib_1),\ B_2 = \frac{1}{2}(a_2 + ib_2) \end{aligned}. \tag{22.23}$$

Obviously, $A_1$and B_1 are complex conjugates of each other, and A_2 and B_2 are complex conjugates of each other.

These equations can be used to solve the Fourier series coefficients a_1, a_2, b_1, and b_2:

$$a_1 = A_1 + B_1, \tag{22.24}$$

$$a_2 = A_2 + B_2, \tag{22.25}$$

$$b_1 = i(A_1 - B_1), \tag{22.26}$$

$$b_2 = i(A_2 - B_2). \tag{22.27}$$

So, the procedures of the rotary spectrum calculations can be summarized as the following, given vector data (u, v):

- Step #1: do FFT for u and obtain A_1 and B_1.
- Step #2: do FFT for v and obtain A_2 and B_2.
- Step# 3: calculate a_1, a_2, b_1, and b_2 using (22.24)–(22.27).
- Step #4: calculate A using (22.18), which is a function of f.
- Step #5: calculate B using (22.19), which is a function of f.

Suppose the vector variable is (u, v); then the main computation of rotary spectrum can be done by the following MATLAB script:

```
A1 = fft(u);
B1 = conj(A1);

A2 = fft(v);
B2 = conj(A2);

a1 = A1 + B1;
a2 = A2 + B2;
b1 = (B1 - A1) / sqrt(-1);
b2 = (B2 - A2) / sqrt(-1);

A = sqrt((a1 + b2).^2 + (a2 - b1).^2) / 2;
B = sqrt((a1 - b2).^2 + (a2 + b1).^2) / 2;
```

This can be used to build a MATLAB function for rotary spectrum analysis (this is left as an exercise at the end of the chapter).

22.4 Another Algorithm for Rotary Spectrum Analysis in MATLAB

In addition to the algorithm introduced above for rotary spectrum analysis, we can make use of MATLAB's powerful ability to do complex number calculations to design another algorithm by directly working on the *complex rotary velocity*:

$$w = u + iv. \tag{22.28}$$

Since *the FFT for the complex variable* w consists of terms with negative exponentials and positive exponentials already, we can make use of this to get the clockwise and counterclockwise rotating velocity components directly. As an example, it can be verified that equations (22.9) and (22.10) can be combined to get $w = u + iv = u_0 e^{i2\pi ft}$ which is a cyclonic or counterclockwise rotating vector.

For a signal with a velocity vector time series rotating in the clockwise direction at a frequency of f_1, the time series can be expressed as

$$u_1 = \sin(2\pi f_1 t) \tag{22.29}$$

$$v_1 = \cos(2\pi f_1 t) \tag{22.30}$$

As an example, we choose $f_1 = 2$. Note the difference between equations (22.9) and (22.29) and the difference between equations (22.10) and (22.30). We now have an anticyclonic or clockwise rotating vector:

$$w = u_1 + iv_1 = \frac{e^{i2\pi f_1 t} - e^{-i2\pi f_1 t}}{2i} + i\frac{e^{i2\pi f_1 t} + e^{-i2\pi f_1 t}}{2} = ie^{-i2\pi f_1 t},$$

which is a clockwise rotating vector or $f = -2$. The spectrum of the rotary velocity $w = u + iv$ is shown in Fig. 22.4. In this case, the FFT of w has been shown within negative Nyquist to positive Nyquist frequency. A negative frequency corresponds to a clockwise rotation, while a positive frequency corresponds to a counterclockwise rotation. In this case, it indeed shows a clean peak at $f = -2$. Likewise, an example of a counterclockwise rotating velocity vector at is $f_2 = 3$ described by

$$u_2 = \cos(2\pi f_2 t), \tag{22.31}$$

$$v_2 = \sin(2\pi f_2 t). \tag{22.32}$$

The spectrum for $w = u_2 + iv_2$ is shown in Fig. 22.4. If we add the above two examples together, i.e. superimpose the clockwise rotation at frequency 2 and counterclockwise rotation at frequency 3, we would get

$$u = u_1 + u_2, \tag{22.33}$$

$$v = v_1 + v_2. \tag{22.34}$$

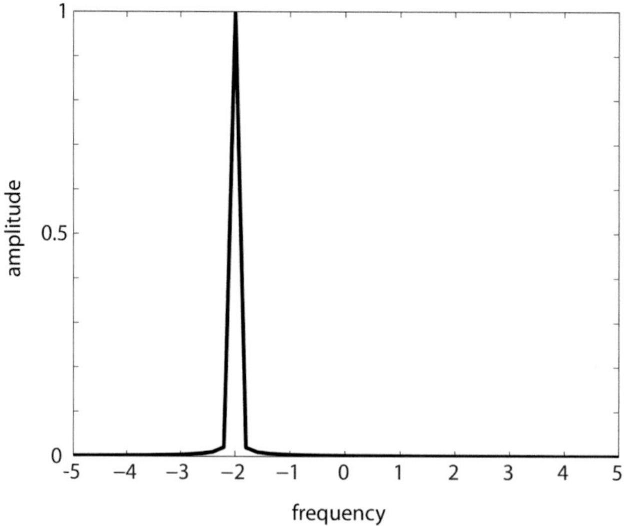

Figure 22.4 Rotary spectrum for a signal with a clockwise rotating flow with a frequency of $f = 2$ by doing FFT to the rotary velocity vector $u + iv$.

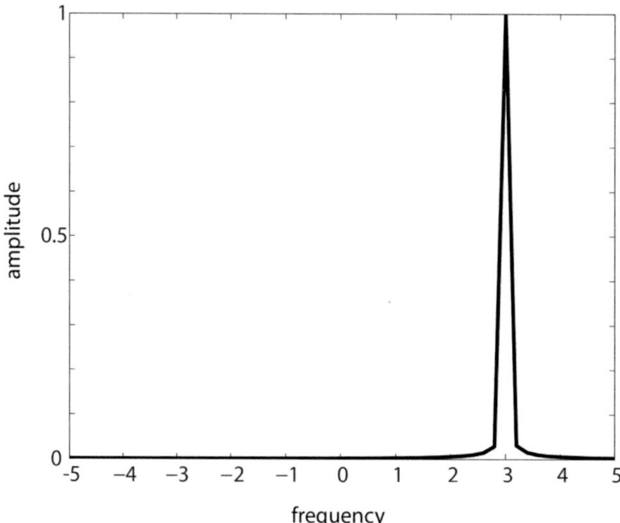

Figure 22.5 Rotary spectrum for a signal with a counterclockwise rotating flow with a frequency of $f = 3$ by doing FFT to the rotary velocity vector $u + iv$.

The spectrum for $w = u + iv$ is shown in Fig. 22.6. Obviously, the two peaks are shown at the negative and positive frequencies as anticipated.

In summary, an even more concise way of doing rotary spectrum analysis is to do FFT to the complex velocity (22.28). A MATLAB function can be easily worked out for this method (this is left as an exercise at the end of chapter).

22.5 A Case Study Example

In 2005, a few strong hurricanes crossed the Gulf of Mexico and initiated some inertial oscillations that were captured by observations in deep water (~ 2000 m). The tracks of these hurricanes are shown in Fig. 22.7. Wind velocity vector time series for that time period is shown in Fig. 22.8. Velocity profile time series data are obtained from a station located at latitude 28.52° N and longitude 88.29° W where the water depth is 1920 m. Strong clockwise oscillations are shown immediately after each of the three hurricanes as shown by the progressive velocity vector plot (Fig. 22.9). These clockwise oscillations were excited as soon as the

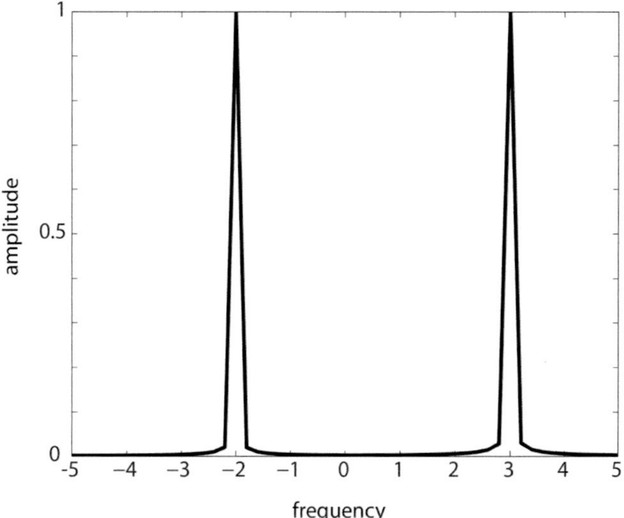

Figure 22.6 Rotary spectrum for a signal with a combination of a clockwise rotating flow with a frequency of $f = 2$ and a counterclockwise flow with a frequency at $f = 3$ by doing FFT to the rotary velocity vector $u + iv$.

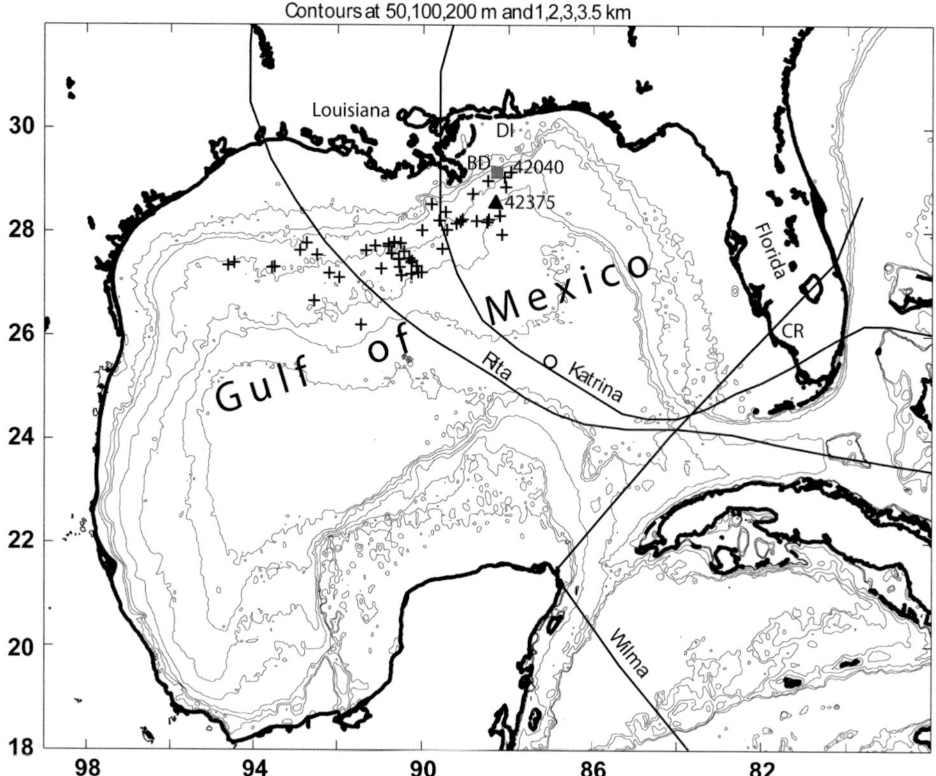

Figure 22.7 Several major hurricane tracks in 2005 in the Gulf of Mexico.

hurricane influenced the area, and during each hurricane event, the oscillation lasted for about 10 days.

Apparently, these are NIOs caused by the weather events. To better visualize the vertical structure of the NIOs caused by the hurricanes, a sixth order Butterworth filter is implemented with a cutoff frequency of 0.6 cycle/day or, equivalently, a 40-hour filter. The high-pass and low-pass filtered vertical profiles of velocity components are shown in Fig. 22.10 for both the east and north components during the time period of Hurricane Katrina. The vertical profiles for Hurricanes Rita and Wilma are similar and are omitted here. Fig. 22.11 shows the magnitude of velocity (m/s) averaged between ~1400 and 1900 m before Hurricane Katrina, during and after Katrina, between Katrina and Rita, during and after Rita, between Rita and Wilma, and during and after Wilma. It is apparent that the weather forcing produced the oscillations that

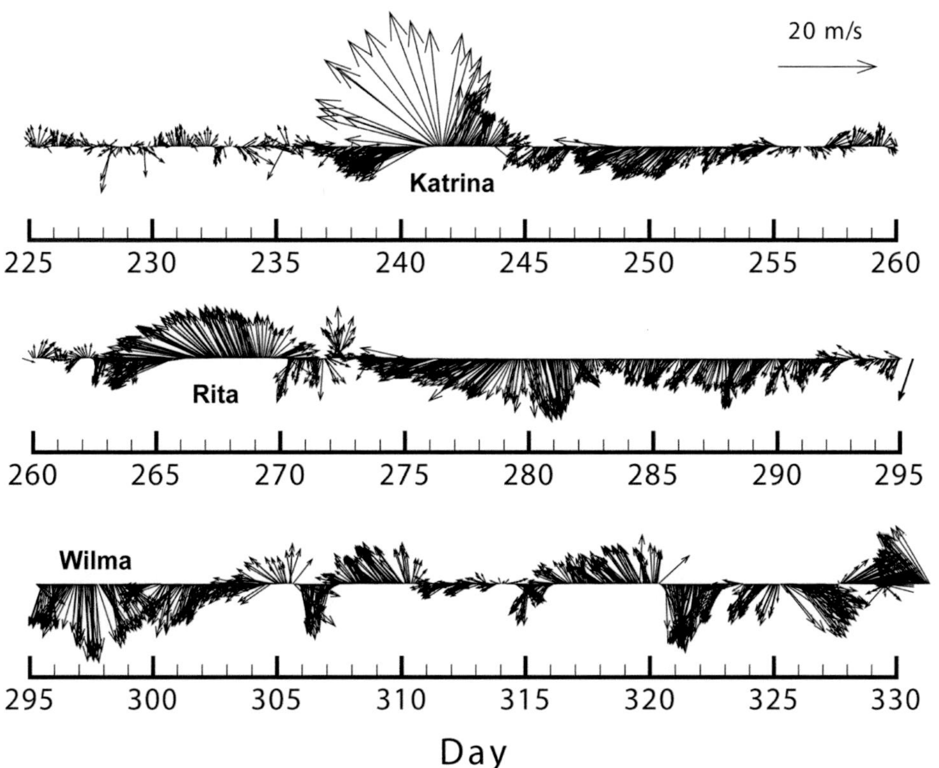

Figure 22.8 Wind vectors time series from Station 42040 (29° 12′ 19″ N 88° 12′ 19″ W), which is ~ 118 km south of Dauphin Island, AL (DI in Fig. 22.3), and ~76 km north of platform 42375 (the triangle on Fig. 22.3).

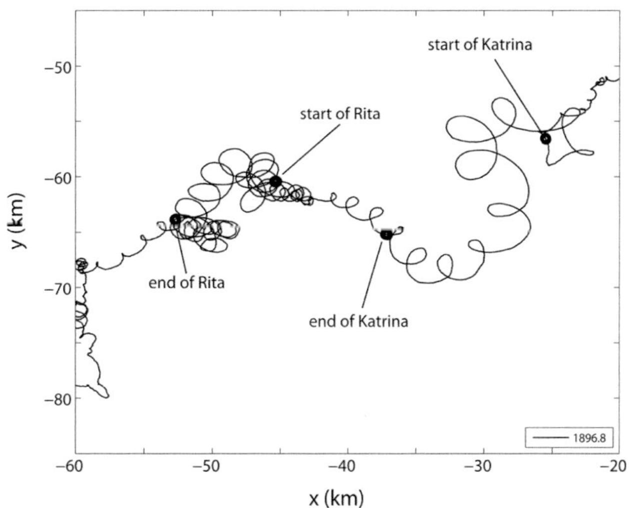

Figure 22.9 Progressive velocity vector diagram showing the NIOs at ~1896.8 m below the sea surface in August and September 2005, covering the period of Hurricanes Katrina and Rita.

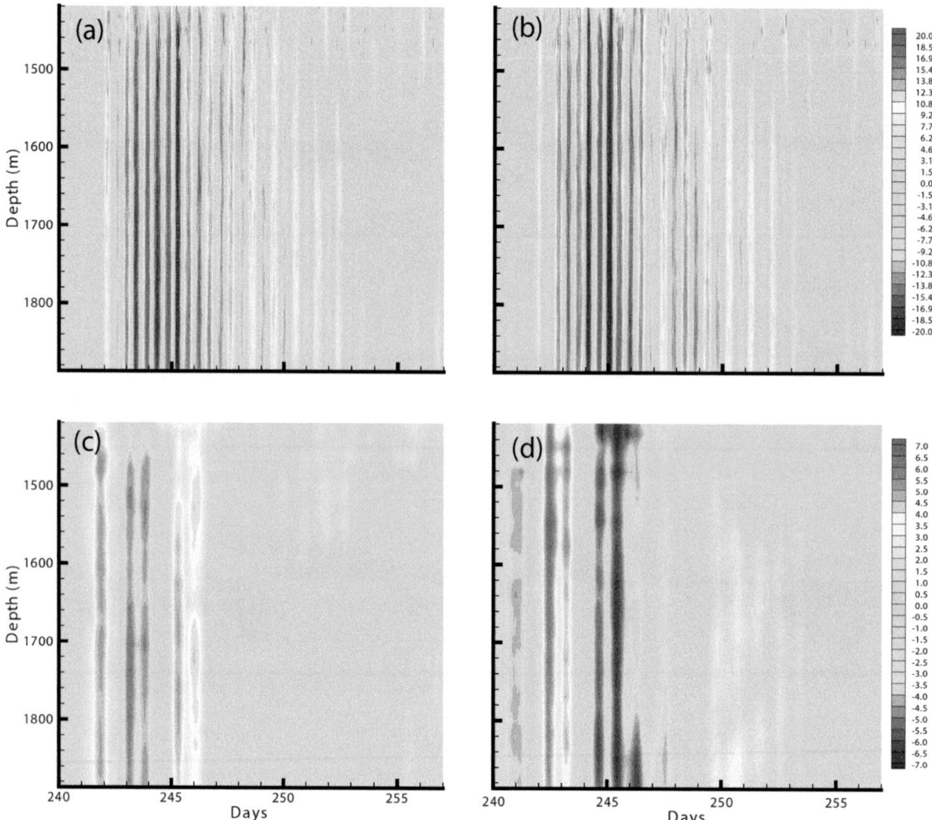

Figure 22.10 High- and low-pass filtered velocity components at 1400–1900 m: (a) high-pass east velocity; (b) high-pass north velocity; (c) low-pass east velocity; and (d) low-pass north velocity. A black-and-white version of this figure will appear in some formats. For the color version, please refer to the plate section.

appear to be the near inertial oscillations. A further confirmation can be made by rotary spectrum analysis.

Fig. 22.12 shows the rotary spectra for the CW rotating component and CC rotating component. It can be seen that the CW component is generally much larger than the CC component; the peak of the CW component is centered around the local inertial period (around 24 hours). There is a spread of NIO periods, mostly between 22 and 26 hours; Hurricane Katrina caused the strongest NIO, followed by Rita, and then Wilma; and the NIO around the local inertial period seemed to have a broader peak during and after the hurricanes than non-hurricane periods.

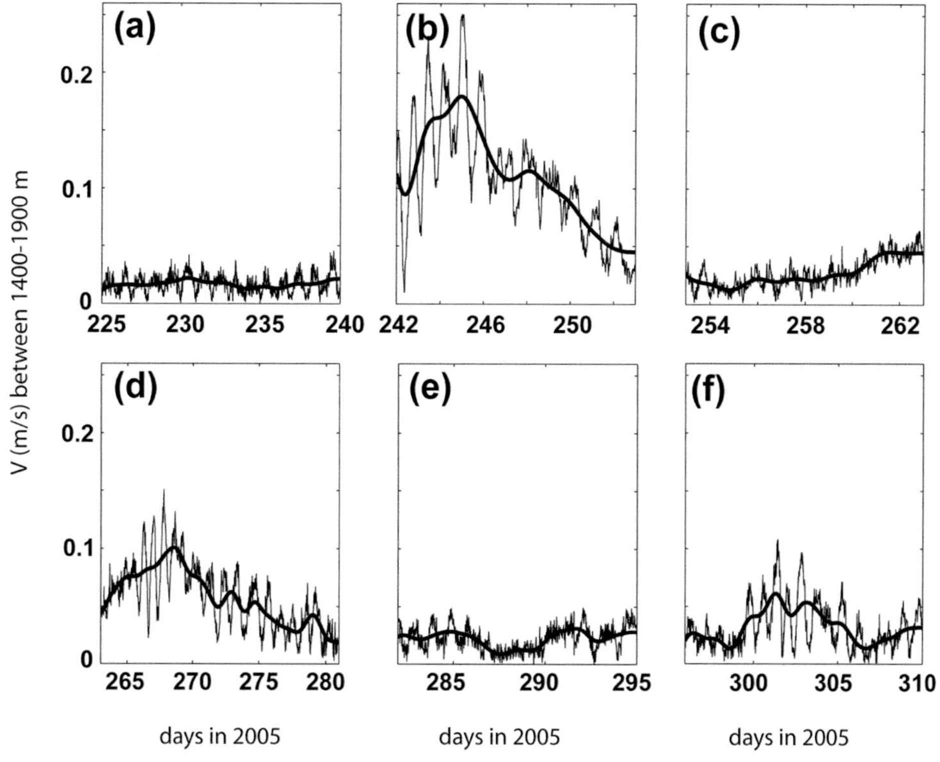

Figure 22.11 Magnitude of velocity (m/s) averaged between ~1400 and 1900 m (a) before Hurricane Katrina; (b) during and after Katrina; (c) between Katrina and Rita; (d) during and after Rita; (e) between Rita and Wilma; and (f) during and after Wilma.

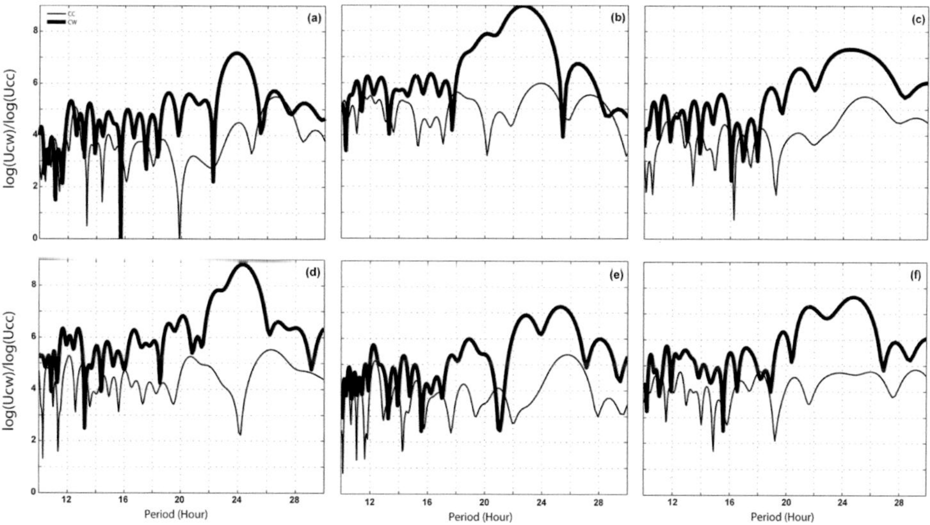

Figure 22.12 Rotary spectra for the depth-averaged velocity (u, v) in log-scale (a) before Hurricane Katrina; (b) during and after Katrina; (c) between Katrina and Rita; (d) during and after Rita; (e) between Rita and Wilma; and (f) during and after Wilma. The thick lines are for the clockwise rotating component and the thin lines are for the counterclockwise rotating component.

Review Questions for Chapter 22

(1) What is a rotary vector?
(2) What kind of rotary motions exist in the atmosphere and ocean?

Exercises for Chapter 22

(1) Write a MATLAB program to draw some Lissajous Figures using equations (22.1) and (22.2) with various parameters so that there are circle, ellipse, line, and more complicated figures (using Fig. 22.1 as an example).
(2) Write a function for rotary spectrum computation following the steps laid out in Section 22.2. You can use the script provided to work out a function or create your own from scratch.
(3) Write a function for rotary spectrum computation using the method of Section 22.3 and test it with an example.
(4) Data file named "ADCP_data_42362_0506.mat" is a MATLAB file containing velocity profile time series data in water of more than 1000 m deep at a location (27.8° N, 90.67° W) in the Gulf of Mexico. The variables include time (t, days), east velocity (ut, cm/s), north velocity (vt, cm/s), and depth of the vertical bins of the data (z, meters). There are 60 vertical bins, ranging from 79.1 m to 1023.1 m below the ocean surface. Do rotary spectrum analysis to the current velocity at each of the 60 levels, and draw line plots and contour plots of the rotary spectra.

23

Short-Time Fourier Transform and Introduction to Wavelet Analysis

About Chapter 23

This chapter discusses the drawback of Fourier analysis and the methods that can overcome its limitations. In general, Fourier analysis does not include information about time, particularly events. A slight modification of Fourier analysis can allow the addition of a dimension in time: by dividing the time series into smaller segments and doing the Fourier Transform for each segment, a method called short-time Fourier Transform (STFT) is introduced. Wavelet analysis is then discussed as a much better alternative to or replacement for STFT. It involves scaled and translated convolution with a short base function (short in the sense that it is essentially non-zero only in a finite interval). Wavelet analysis uses different base functions than the Fourier Transform. They are limited in time (unlike the infinitely long sinusoidal functions) and can be stretched or compressed to represent different scales (equivalent to frequencies). This method will allow the resolution of events at different times and different scales.

23.1 The Limitation of Fourier Analysis and Possible Improvements

23.1.1 The Limitation of Fourier Analysis

With a little bit of exaggeration, Fourier Transform in data analysis is like Newton's Law in physics, in the sense that it is quite generic: given any time series signal, it can almost always be decomposed into simple sinusoidal functions. This is significant particularly considering that the sinusoidal functions are almost perfect, i.e. with enough such functions, the total error squared approaches zero for infinite terms representing a continuous function. The decomposition of a time series results in a spectrum in frequency domain.

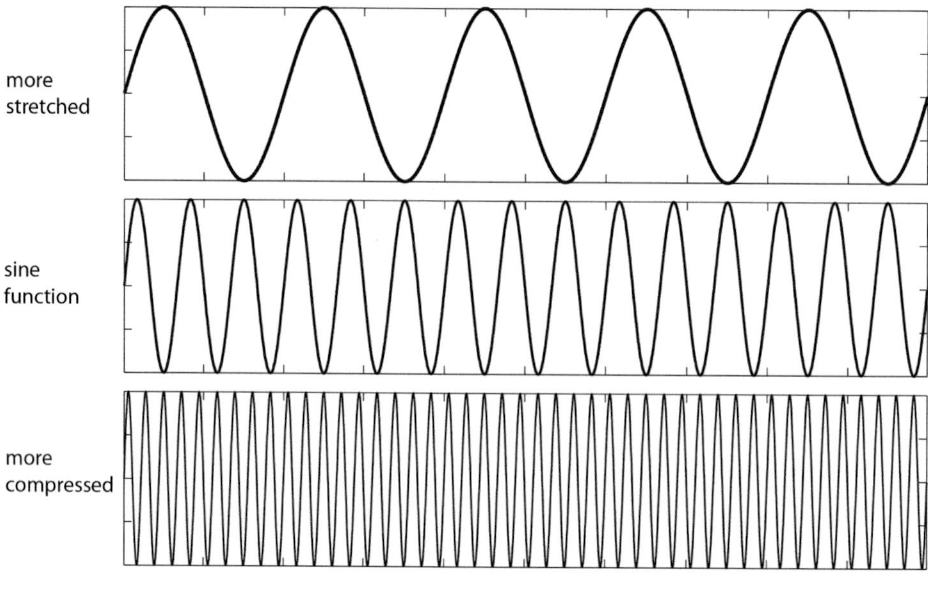

Figure 23.1 Examples for the stretching and compression of the same function.

More specifically, in Fourier Transform, a time series is represented by a linear combination of base functions that are sine and cosine functions, with different frequencies. Sinusoidal functions all have a similar shape, regardless of their frequencies. For higher frequencies, they are more "compressed," while for lower frequencies they are more "stretched." From that point of view, we are using the same kind of function with different "compressions" (Fig. 23.1). This concept of stretching and compression of a given function is similar to *wavelet analysis*, to be discussed later.

In addition, in Fourier Transform we convert a function in time domain to one in frequency domain. The time function includes information about "events" occurring at different times. With Fourier Transform, however, the converted "frequency function" is defined in a range from 0 to the Nyquist frequency, and the information about temporal variation is not explicitly included in the spectrum.

It is through the sine and cosine functions that the definition of frequency is made: $\sin(2\pi ft)$, in which t is time and f is frequency. The main drawback of Fourier Transform is that the information about "events" away from regular oscillations is lost. All we have in the frequency domain is the time independent spectrum. For example, information about a hurricane, a thunderstorm, or a sharp change of water level due to a local transient storm is not represented in the spectrum; although their presence will affect the outcome of the spectrum, there is no way to tell when any events occurred just from the spectrum. Of course, the information contained in the original time series is never lost because we still have the original data. What is missing in the spectrum is a reflection and quantification of the event to the spectrum. Here we are only talking about it in the frequency domain after the Fourier Transform: the spectrum does not provide any information about events. This is a limitation of Fourier analysis.

23.1.2 Short-Time Fourier Transform

One approach to resolve temporal variations of spectrum is to divide a long time series into shorter segments and obtain the spectrum for each segment using Fourier Transform. Each segment therefore has a time stamp, and the spectrum over time should then reflect the temporal variations. This is sometimes called a *short-time Fourier Transform* or STFT. The short time series has its own drawback: coarser frequency resolution and elimination of some of the lower frequency spectrum. In exchange, it allows temporal variation of the spectrum. The graph generated by this method is sometimes referred to as the *spectrogram*.

Assume that the time series is $y(t)$ defined in $0 \le t \le T$, i.e.

$$y = y(t), \quad t \in [0, T]. \tag{23.1}$$

The divided segments of the original time series are denoted as

$$y = y(t_i, \tilde{t}), \quad \tilde{t} \in [t_{i1}, t_{i2}], i = 1, 2, \ldots, K, \tag{23.2}$$

in which the tilde indicates a variable valid for a given time interval; t_{i1} and t_{i2} are the start and end times of the ith interval, respectively; $[t_{i1}, t_{i2}]$ is the ith interval in time; t_i is the average or midpoint of t_{i1} and t_{i2}, i.e.

$$t_i = \frac{t_{i1} + t_{i2}}{2}. \tag{23.3}$$

The Fourier Transform for the ith segment is

$$Y(t_i, f) = \mathbb{F}[y(t_i, \tilde{t})]. \tag{23.4}$$

Here $\mathbb{F}$ is a Fourier Transform operator that only applies to the given segment of time series $y(t_i, \tilde{t})$; the midpoint time t_i is therefore only a parameter and does not participate in the computation. This way, the Fourier Transform is a function of the segment time t_i. These segments are aligned in time but may be allowed to have some overlaps. For nonoverlapping segments, their number of segments is obviously fewer than overlapping ones (Fig. 23.2). The overlap is usually uniform, and the percentage of overlap of STFT is another parameter that the user can adjust. There seems to be no rule of thumb or definitive answer to the "best" value of overlap. Perhaps, for a relatively short time series, more overlap is better because an overlap allows for more segments and higher temporal resolution defined by the difference of t_i:

$$dt_i = t_{i+1} - t_i, \quad i = 1, 2, \ldots, K, \tag{23.5}$$

in which K is the total number of segments. For uniform overlap, dt_i is a constant and does not depend on the index i.

23.1.3 An Example of STFT

The following is an example of STFT for a time series of water depth obtained on the Louisiana continental shelf in March and April of 2020 for about 48.5 days, covering diurnal and semi-diurnal tides, spring-neap variations of tide, and effect of atmospheric frontal passages (Fig. 23.3a). The sampling interval was 60 seconds.

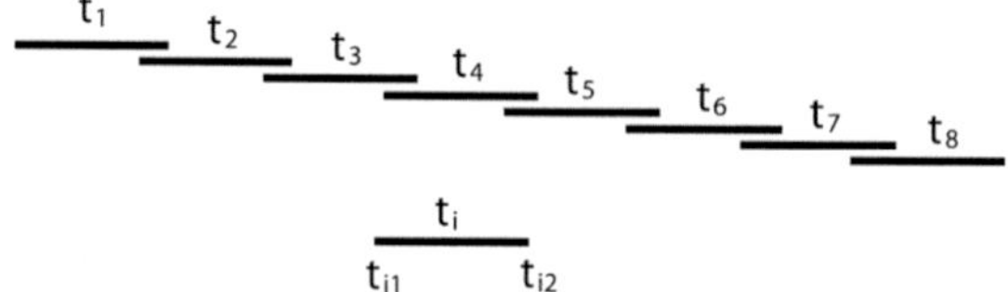

Figure 23.2 Schematic view of an example of segmented time intervals for the STFT. Here there are eight segments ($K = 8$) and a ~20% overlap.

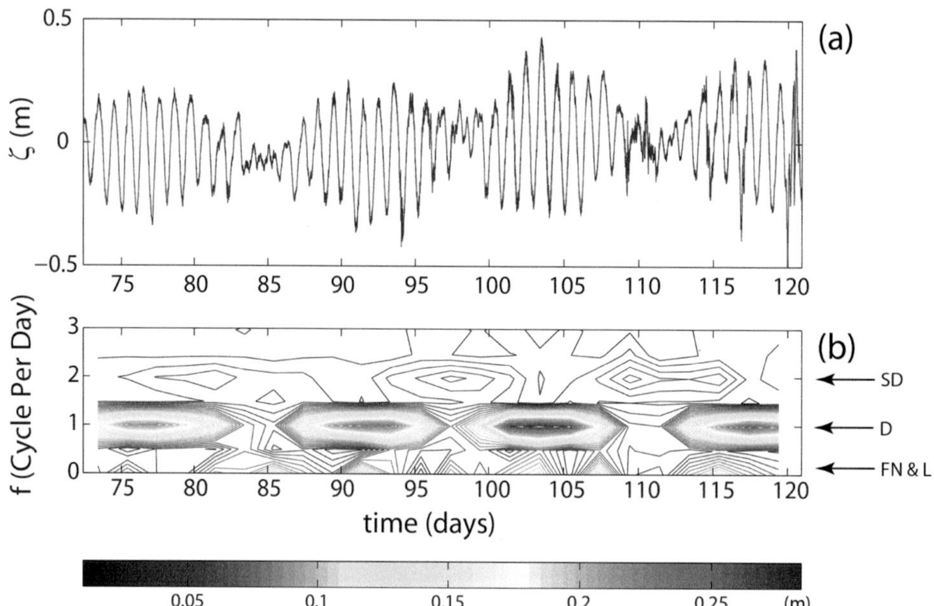

Figure 23.3 An example showing the spectrogram generated by the short-time Fourier Transform. (a) Time series of the water depth variation over the mean value. (b) The STFT. The arrows show the diurnal and semi-diurnal tides, as well as the fortnightly tides and low frequency weather induced oscillations. A black-and-white version of this figure will appear in some formats. For the color version, please refer to the plate section.

A STFT is applied with 24 nonoverlapping segments; each is two days long. The resultant spectrogram shows temporal variations of the diurnal and semi-diurnal tides as well as lower frequency variations (Fig. 23.3b). The segmentation and the application of equation (23.4) are done with the script below.

```
load March-April-2020-depth.mat
%=== divide the data into different segments with no overlap
K = 24; N = fix(length(time) / K);
for i = 1:K
d(1:N,i) = depth((i−1)*N+1:i*N);   % DEFINING THE TIME SERIES
                                   % OF EACH SEGMENT
end
T = time(1)+1:2:time(1)+1+2*(K − 1);
for i = 1:K
D(1:N,i) = abs(fft(d(:,i)));    % THIS IS EQUATION (23.4)
end
```

The added dimension in time allows the spectrum to contain information about different constituents in frequency domain evolving with time, which cannot be done with the original Fourier Transform for the entire time series (Fig. 23.4). The Fourier Transforms for the 24 segments have quite large variabilities. Fig. 23.5 shows all 24 spectra, all of which indicate the tidal peaks and low frequency variations but with different magnitudes. For better visualization, each segment is padded with 10 times long zeros. The figure does not show which line is for which segment as it is aimed at only demonstrating the magnitude of variability. One of them has a relatively large signal, around 4–5 CPD, while the rest have a weaker

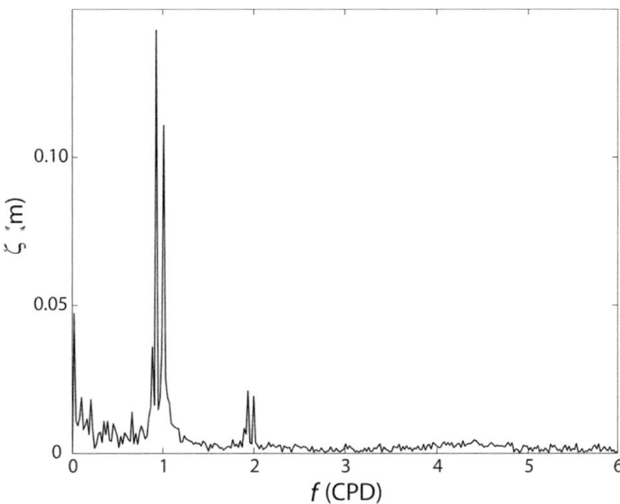

Figure 23.4 Spectrum for the entire time series.

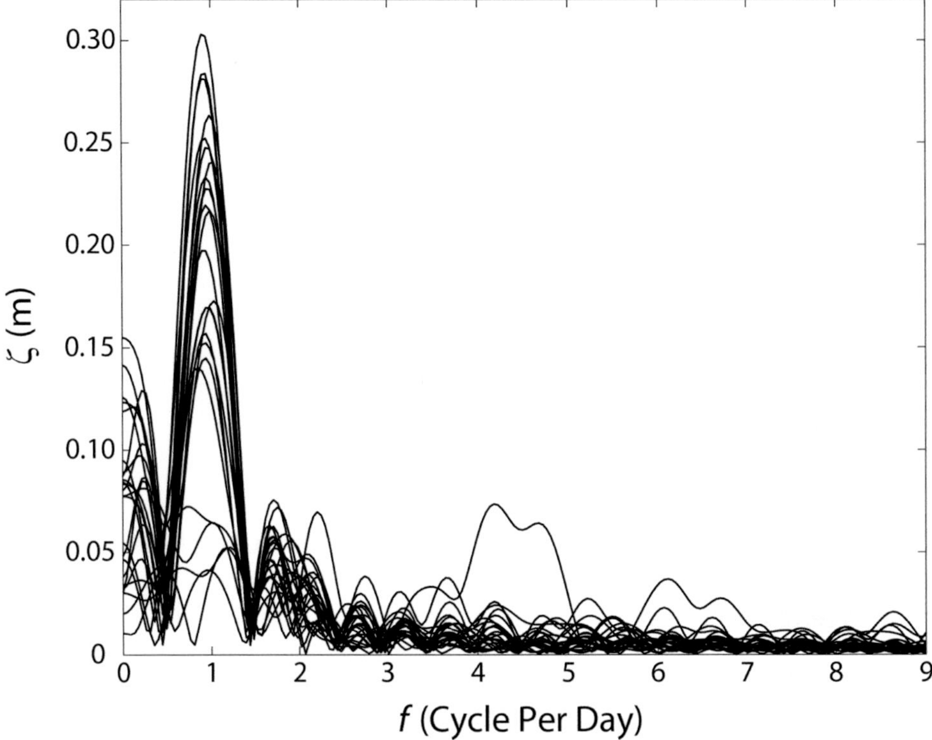

Figure 23.5 Spectra for all segments, showing the large variability.

signal in that frequency band. Obviously, presenting the STFT with Fig. 23.5 is not a good idea as it is hard to visualize the temporal variation of the spectrum. It is better to be presented as a function of the segmentation time t_i and the frequency f as shown by equation (23.4). Fig. 23.6 shows the result of (23.4) with zero-padding. It is essentially the same as Fig. 23.3 but with finer resolution in frequency.

As discussed earlier, zero-padding improves visualization without altering the spectrum. Fig. 23.6 is an improved version of Fig. 23.3b with zero-padding at the end of each segment of the time series. The number of zeros padded is equal to 10 times the length of the segment. Of course, here the two-day segment length is chosen mainly for demonstration purposes, although it is reasonable in showing the major diurnal and semi-diurnal tidal constituents. Users can make their own choices based on the specific purpose of a given application.

While one can program the STFT from scratch, as demonstrated to produce the spectrogram, MATLAB has a function `spectrogram` for that task. The function plots a spectrogram if no output is requested. If output is requested, for one-dimensional time series, the function `spectrogram` returns a two-dimensional

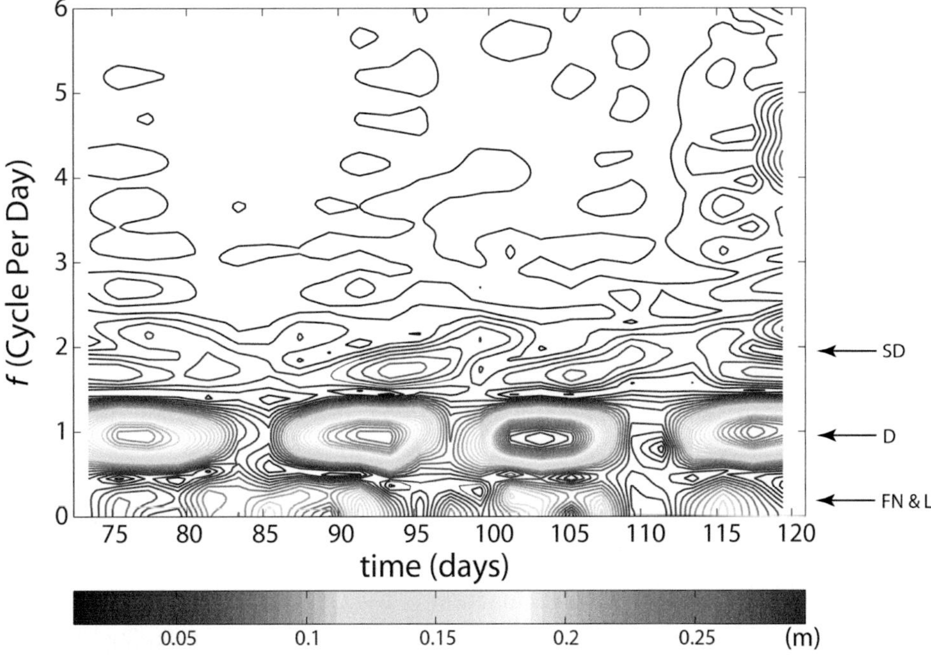

Figure 23.6 The spectrogram from STFT with zero-padding. A black-and-white version of this figure will appear in some formats. For the color version, please refer to the plate section.

array from the STFT (frequency-time). By default, the input time series is divided into eight equal segments. The function also allows the use of Hamming window before the transforms. There is a default overlap of 50% and the overlap can be adjusted by the user.

23.2 Introduction to Wavelet Analysis

23.2.1 Continuous Wavelet Transform

In essence, Fourier Transform involves the computation of correlations of a time series with sinusoidal base functions at various frequencies. Thus, these correlations are functions of frequencies. For those frequencies with large correlation values, the time series must contain significant oscillations at the frequencies. For those frequencies with small or 0 correlation values, the time series does not contain significant oscillations at the frequencies. Since the sinusoidal base functions are infinite in length, the method presents some problems: (1) it cannot resolve "localized" spectrum variations or events, as discussed earlier; and (2) truncation of base functions is unavoidable for practical problems, which will more or less distort the result.

Although the STFT allows the resolution of spectrum variation by events or localized variations, it still uses the infinitely long sinusoidal base functions (with truncation of course). Along that line of thinking, if the infinitely long base functions are replaced by shorter base functions, it may resolve both problems at once. This makes sense because for short time scale variations there is no need to use infinitely long base functions, as truncation will cut them shorter anyway. It is more reasonable to use localized base functions for localized properties. This leads to the *wavelet function* and *wavelet transform*. With wavelet transform, a series of wavelet analysis tools has been developed. Unfortunately, the mathematics of wavelet transform is rather complicated. Here we intend not to cover the method with an in-depth discussion but rather only to provide an introduction. The good news is that with some conceptual understanding, the implementation of wavelet analysis using MATLAB's wavelet toolbox is straightforward.

In wavelet analysis, the base functions are called wavelets. There are many different wavelet functions. Fig. 23.7 shows some example wavelet functions. MATLAB has a function `wavefun` to display these functions. In a given wavelet analysis, only one type of wavelet function is used. Each wavelet has a specific name, e.g. the *Mexican Hat* wavelet, *Morlet wavelet* (Fig. 23.7), *Meyer wavelet*, among many others. The wavelet function is usually expressed as a function of time (or space): $\psi(t)$. For example, the Mexican Hat is expressed as

$$\psi(t) = \frac{2}{\pi^{1/4}\sqrt{3}}(1 - t^2)e^{-t^2/2}, \tag{23.6}$$

and the Morlet wavelet is given by

$$\psi(t) = e^{-t^2/2}\cos(5t). \tag{23.7}$$

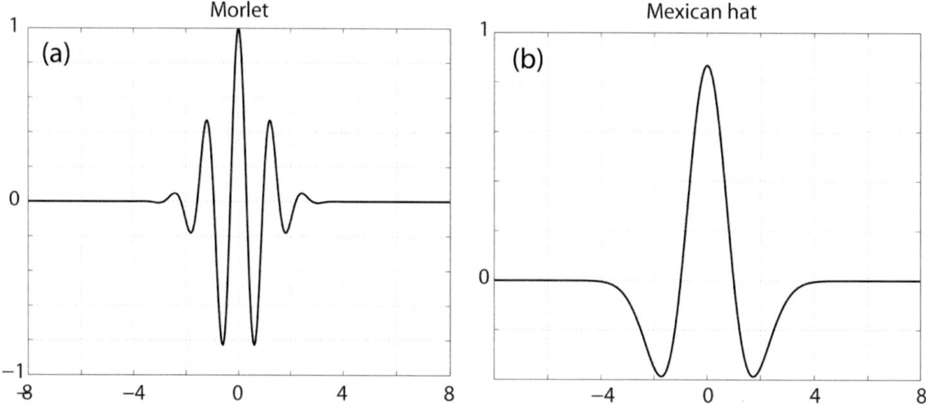

Figure 23.7 Some example wavelet functions: (a) Morlet wavelet; and (b) Mexican Hat wavelet.

Some of the wavelet functions have an associated scale function called father wavelet. The construction of wavelet functions must satisfy some conditions. Typically, a wavelet function must be constrained by the following:

(1) The function must be decaying exponentially at large t, i.e. $|\psi(t)| \leq Me^{-C|t|}$, in which M and C are positive constants.
(2) The integral of the function must be zero, i.e. $\int_{-\infty}^{+\infty} \Psi(t)dt = 0$.

The wavelet function $\psi(t)$ is also called the *mother wavelet*. For wavelet analysis, the mother wavelet function has to be scaled (compressed or stretched or dilated) and put at the "right spot" in time, which is an action called translation, for a correlation computation for obtaining the localized spectrum. When the wavelet function is scaled and translated, the mother wavelet changes its form to a *daughter wavelet*. The daughter wavelet is expressed as

$$\psi(a, t - \tau),$$

in which a is the scale and τ the translation. More specifically, the daughter wavelet is usually more specifically expressed as

$$\psi\left(\frac{t - \tau}{a}\right).$$

It can be seen that both scaling and translations are operations on the time variable t. Sometimes, scale is transformed to an equivalent frequency.

An inner product or correlation of the time function (or time series data in our case) with the daughter wavelet is then performed for the computation. Alternatively, we can view the operation as a convolution between the function being analyzed, i.e. $y(t)$, with the mother wavelet. In mathematics, a *continuous wavelet transform* of a given function $y(t)$ is then given by

$$W_y(a, \tau) = \frac{1}{\sqrt{a}} \int_{-\infty}^{+\infty} y(t)\psi^*\left(\frac{t - \tau}{a}\right)dt. \tag{23.8}$$

Here ψ^* is the complex conjugate of the chosen wavelet function ψ. Equation (23.8) is essentially a convolution between the functions $y(t)$ and a stretched mother wavelet $\psi\left(\frac{t}{a}\right)$. The *inverse wavelet transform* corresponding to the transform of (23.8) is given by

$$y(t) = \frac{1}{C_\psi} \int_{-\infty}^{+\infty} \int_{-\infty}^{+\infty} W_y(a, \tau) \frac{1}{a^2\sqrt{|a|}} \psi\left(\frac{t - \tau}{a}\right) d\tau da, \tag{23.9}$$

in which the coefficient C_ψ is given by

$$C_\psi = 2\pi \int_{-\infty}^{+\infty} \frac{|\mathbb{F}_\psi(\omega)|^2}{|\omega|} d\omega. \tag{23.10}$$

Here $\mathbb{F}_\psi(\omega)$ is the Fourier Transform of the mother wavelet Ψ.

Just like Fourier analysis, wavelet analysis can also have cross-wavelet spectrum and cross-wavelet coherence, which are subjects omitted here. The fundamental ideals are very much the same – the inclusion of a time dimension to the world of spectrum.

It should be noted that in addition to continuous wavelet transform, there is discrete wavelet transform, which involves only specified scales and translations. The operation is dyadic – the scales are not continuous but factors of 2. We will omit the discussion on the discrete wavelet transform, which is perhaps more applicable to subjects like file compression, image processing, etc. New techniques are still being invented with wavelet transform.

23.2.2 An Example for Continuous Wavelet Transform

We now use the same dataset of section 23.1.3 for the STFT and apply the continuous wavelet transform. The following is the code with comments and interpretations. Fig. 23.8 shows the result as pseudo-color plots in the time-frequency domain. It can be seen from this figure that the temporal variation of diurnal tidal amplitude is clearly demonstrated by the continuous wavelet transform. The alternating strong–weak diurnal tidal amplitudes are obvious.

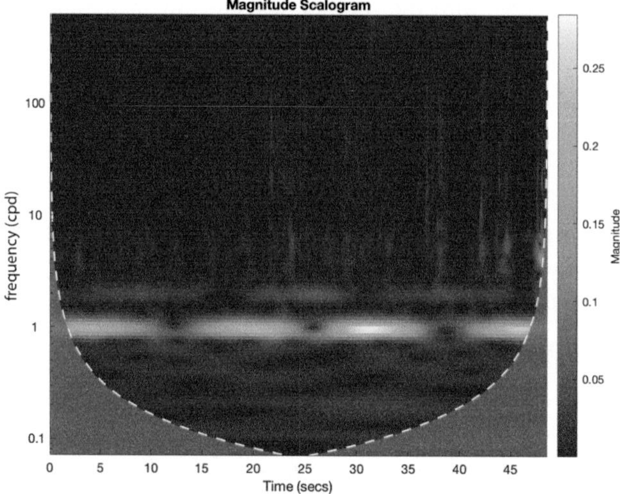

Figure 23.8 Wavelet results: the scalogram for the water depth data of Fig. 23.3a. A black-and-white version of this figure will appear in some formats. For the color version, please refer to the plate section.

The wavelet transform coefficients `cfs` gives the quantitative values of amplitude as a function of time.

```
load March-April-2020-depth.mat % LOAD DATA
Fs = 1 / (time(2) - time(1));   % DEFINE SAMPLING FREQUENCY
[ cfs,f] = cwt(depth,Fs);       % DO CONTINUOUS WAVELET TRANSFORM
                                % & OBTAIN FREQUENCY AND WAVELET
                                % TRANSFORM COEFFICIENTS
figure; cwt(depth,Fs)           % PLOT THE WAVELET TRANSFORM COLOR
                                % CONTOUR
```

Note that these scripts are for demonstration purpose. The default unit for time in the MATLAB function cwt is second and that for frequency is Hz. The user can generate their own scripts for graphs if using different units. The following few lines of MATLAB code plot the wavelet transform coefficient as functions of time to demonstrate the maximization of the coefficient at near the diurnal tidal frequency. The array values `f(94)` and `f(95)` show the frequency at these points being close to those of diurnal tides. Away from these points, at `f(91:93)` and `f(96:98)`, the amplitude is dramatically smaller. The temporal variation is the spring-neap tidal variation.

```
figure
plot(time,abs(cfs(94,:)),'LineWidth',2)
text(time(end), abs(cfs(94,end)),'94')
hold on
plot(time,abs(cfs(93,:)))
text(time(end), abs(cfs(93,end)),'93')
hold on
plot(time,abs(cfs(95,:)),'LineWidth',2)
text(time(end), abs(cfs(95,end)),'95')
```

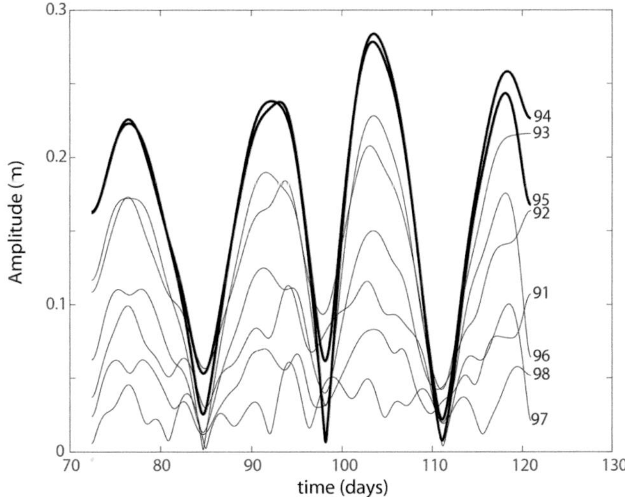

Figure 23.9 Wavelet results: the amplitude as a function of time of the wavelet transform at different frequencies for the water depth data of Fig. 23.3a.

```
plot(time,abs(cfs(96,:)))
text(time(end), abs(cfs(96,end)),'96')
plot(time,abs(cfs(92,:)))
text(time(end), abs(cfs(92,end)),'92')
plot(time,abs(cfs(91,:)))
text(time(end), abs(cfs(91,end)),'91')
plot(time,abs(cfs(97,:)))
text(time(end), abs(cfs(97,end)),'97')
plot(time,abs(cfs(98,:)))
text(time(end), abs(cfs(98,end)),'98')
```

Review Questions for Chapter 23

(1) What are the similarities and differences between Fourier Transform and wavelet transform?

(2) What do you think is the rationale of requiring that the integral of wavelet function must be zero and the wavelet function must decay exponentially?

(3) How do you construct the daughter wavelet from a mother wavelet?

Exercises for Chapter 23

(1) Verify that the Mexican Hat wavelet $\psi(t) = \frac{2}{\pi^{1/4}\sqrt{3}}(1 - t^2)e^{-t^2/2}$ satisfies the condition that $\int_{-\infty}^{+\infty} \Psi(t)dt = 0$.

(2) Apply the continuous wavelet transform to water depth from the time series data in the MATLAB file named "S6-depth-time.mat." Make plots similar to Figs. 23.8 and 23.9.

24

Empirical Orthogonal Function Analysis

About Chapter 24

The analyses we have discussed in previous chapters include the use of base functions, such as sinusoidal functions with specified frequencies, i.e. harmonic analysis; sinusoidal base functions with a frequency range from 0 to the Nyquist frequency with an interval inversely proportional to the total length of time of the data, i.e. Fourier analysis; and wavelet base functions for wavelet analysis. These base functions, however, are chosen regardless of the nature of the variability of the data themselves. In this chapter, we will discuss a different method, in which the base functions are determined empirically, that is dependent on the nature of the data. In other words, this method will find the base functions from the data and these base functions describe the nature of the data. The method is applicable to many types of data, especially to time series data at multiple locations, e.g. a sequence of weather maps or satellite images. There are several variants of the method, but here we will only provide an introduction for the basics.

24.1 The Base Functions

24.1.1 Base Functions for Harmonic, Fourier, and Wavelet Analysis

In Fourier analysis, harmonic analysis, wavelet analysis, and many least squares regression methods, the base functions are chosen before the analysis is done. In the case of harmonic analysis, sinusoidal base functions are chosen with specified tidal frequencies. These tidal frequencies are determined by astronomy, i.e. the relative motions of the Earth, Moon, and Sun. These motions can be decomposed into many different frequencies of astronomical origin. Although the dynamics of the motion of ocean is complicated, the assumption is that *these tidal frequencies will be preserved regardless of the complexity of dynamics*. What the dynamics of ocean will do is act as a natural filter or transfer function so that some of the tidal signals (frequencies) will be amplified and some reduced in magnitude while preserving the frequencies at a given location in the ocean. In addition, there are new frequencies (compound tides such as the overtides and shallow water constituents) generated through nonlinear hydrodynamics processes, especially in shallow waters where nonlinearity is usually larger than in the deep ocean. Regardless of complexity, the *frequencies* of original tidal constituents, overtides, and shallow water constituents are all well known, and the goal of the harmonic analysis is to determine the *amplitude and phase* of each major tidal constituent at a given location. Harmonic analysis does not provide any information for other frequencies, e.g. frequencies induced by storms, nor any information on the interaction between a storm surge and tides. In other words, harmonic analysis gives tidal frequencies special treatment and ignores all other frequencies.

In the case of Fourier analysis, the frequency range is from 0 to the Nyquist frequency $f_N = 1/(2\Delta t)$, in which Δt is the time interval, with an interval inversely proportional to the total length of time of the data T, so that the frequencies include $0, 1/T, 2/T, \ldots$, and $N/2/T = 1/(2\Delta t) = f_N$. No frequency is given special treatment. The analysis can resolve tides as well as nontidal oscillations such as weather-induced variations. Because the frequencies in Fourier analysis are discrete, tidal frequencies are not necessarily exactly resolved. For data with enough length, this is not a problem: tidal signal can be readily picked out from the spectrum of the Fourier analysis. The Fourier analysis method is most suitable for problems with periodic and quasi-periodic motions, but not for events with isolated transient and sometimes abrupt variations.

Wavelet analysis is introduced with localized base functions that are essentially non-zero only in a limited interval. The analysis uses a broader sense parameter called scale, which is a function of time as well. The reciprocal of scale is related to the concept of frequency in Fourier analysis but not specifically meant for sinusoidal motions. The wavelet functions are more complicated, in the sense that there are many more choices. It also has broader applications in e.g. acoustic signal processing, image processing, data compression, etc.

24.1.2 The Natural Base Functions from Data

The above methods have gained a wide range of applications in various fields, and their important roles in science and engineering are obviously unquestionable. The advantages of these methods, however, are also their limitations: the use of specified base functions restricts the analysis of the variabilities in reference only pertinent to these base functions. Decomposing a time series into a series of sinusoidal functions is an analytic way of examining the signal. This analytical way of looking at time series data may obscure some of the holistic picture and the overall data variability, or pattern, which may not be a sinusoidal function or a wavelet function. Although nature has abundant examples of sinusoidal variations (such as the sound waves from music instruments, the electro-magnetic wave from a LASER pointer, pendulum motion, and the motion of a light mass at the end of a spring), many variations are not necessarily naturally sinusoidal. In that case, it may be better to use some intrinsic data structures or patterns, if any, as base functions, rather than force a structure (e.g. sinusoidal functions or wavelet functions) for the data before we even analyze the data.

From a mathematical or geometric point of view, we will try to find the *natural directions* of maximum and minimum variabilities of given data and expand the data around these natural variations. This is the basic idea of *principal component analysis* or *empirical orthogonal function (EOF) analysis*. Here we will start with a commonly used simple example to illustrate (Navarra and Simoncini, 2010; Preisendorfer, 1988).

Assuming a set of scattered data points, with coordinates (x'_k, y'_k), $k = 1, 2, \ldots, n$, which demonstrate certain patterns in space, we would like to describe the pattern or characteristics of the variability. An easy way to describe it is to look at the spatial variability – along which direction does it vary the most, which direction the least?

The averages of the coordinates are

$$\bar{x} = \frac{1}{n}\sum_{k=1}^{n} x'_k, \quad \bar{y} = \frac{1}{n}\sum_{k=1}^{n} y'_k. \tag{24.1}$$

If our interest is the variations away from the mean values, the data need to be de-meaned first, although one can think of cases when the mean should be kept. Let us assume that we are interested in the de-meaned data variability. De-meaning or centering of the data gives

$$x_k = x'_k - \bar{x}, \quad y_k = y'_k - \bar{y}. \tag{24.2}$$

Rotating the coordinate system by an angle θ, the new coordinates are

$$\begin{aligned} \xi_k(\theta) &= x_k \cos\theta + y_k \sin\theta \\ \eta_k(\theta) &= -x_k \sin\theta + y_k \cos\theta \end{aligned} \tag{24.3}$$

in which

$$0 \le \theta \le 2\pi; \quad k = 1, 2, \ldots, n. \tag{24.4}$$

The new horizontal axis is for ξ, vertical axis for η, and the angle θ is between the new horizontal axis (ξ) and the original horizontal axis (x). The projection of a vector $\mathbf{r}_k = (x_k, y_k)$ on the unit vector along the new ξ-axis, $\mathbf{e}$, is $\xi = \mathbf{r} \cdot \mathbf{e}$.

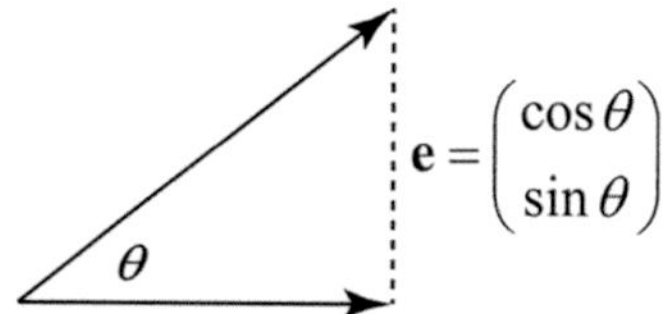

Figure 24.1 Unit vector **e** in the direction of rotation.

The purpose of the rotation is to find a direction θ (Fig. 24.1) so that along the unit vector **e** or along the direction of ξ the variance of coordinates reaches an extreme value:

$$\mathbf{e} = \begin{pmatrix} \cos\theta \\ \sin\theta \end{pmatrix}. \tag{24.5}$$

We will define the variance of coordinates below.

Since the coordinates are de-meaned, we have

$$\sum_{k=1}^{n} x_k = 0, \quad \sum_{k=1}^{n} y_k = 0. \tag{24.6}$$

The new coordinates also have zero means:

$$\sum_{k=1}^{n} \xi_k(\theta) = 0, \quad \sum_{k=1}^{n} \eta_k(\theta) = 0. \tag{24.7}$$

The scatter or variance of coordinates along the hypothetical direction **e** is

$$\begin{aligned} s^2(\theta) &= \frac{1}{n-1}\sum_{k=1}^{n} \xi^2 = \frac{1}{n-1}\sum_{k=1}^{n} (x_k \cos\theta + y_k \sin\theta)^2 \\ &= s_{xx}\cos^2\theta + 2s_{xy}\sin\theta\cos\theta + s_{yy}\sin^2\theta \end{aligned}, \tag{24.8}$$

in which

$$s_{xx} = \frac{1}{n-1}\sum_{k=1}^{n} x_k^2, \quad s_{yy} = \frac{1}{n-1}\sum_{k=1}^{n} y_k^2, \quad s_{xy} = \frac{1}{n-1}\sum_{k=1}^{n} x_k y_k. \tag{24.9}$$

This quantify is an overall measurement of how much spread the data points have. The larger the s^2 value, the greater the average spread of the data points. Imagine the data points are all close together; this value will be closer to zero compared to the case when data points are far apart from each other.

The extreme variance is reached when the derivative of s^2 with respect to θ is 0, i.e.

$$\begin{aligned} \frac{d(s^2)}{d\theta} &= s_{xx} 2\cos\theta(-\sin\theta) + 2s_{xy}(\cos^2\theta - \sin^2\theta) + 2s_{xy}\sin\theta\cos\theta \\ &= (s_{yy} - s_{xx})\sin 2\theta + 2s_{xy}\cos 2\theta = 0 \end{aligned}. \tag{24.10}$$

This leads to

$$\tan 2\theta = \frac{2s_{xy}}{s_{xx} - s_{yy}}. \tag{24.11}$$

There are infinite solutions for this equation:

$$\theta = \frac{1}{2}\arctan\left(\frac{2s_{xy}}{s_{xx} - s_{yy}}\right) + \frac{m\pi}{2}, \quad (m = 0, 1, 2, \ldots). \tag{24.12}$$

However, there are only two independent solutions: e.g. for those with $m = 0$ and $m = 1$, or

$$\theta_1 = \frac{1}{2}\arctan\left(\frac{2s_{xy}}{s_{xx} - s_{yy}}\right), \tag{24.13}$$

$$\theta_2 = \frac{1}{2}\arctan\left(\frac{2s_{xy}}{s_{xx} - s_{yy}}\right) + \frac{\pi}{2}. \tag{24.14}$$

These two directions are perpendicular (orthogonal) to each other. These are the *principal directions*. The new coordinate system formed by the principal directions is the *principal coordinate system*. In addition, the second order derivative of the variance can be shown to be

$$\frac{d^2(s^2)}{d\theta^2} = -4\frac{s_{xy}}{\sin 2\theta}. \tag{24.15}$$

Obviously, because θ_1 and θ_2 have a phase difference of $\pi/2$,

$$\sin 2\theta_2 = \sin 2\left(\theta_1 + \frac{\pi}{2}\right) = -\sin 2\theta_1. \tag{24.16}$$

This means that the second order derivative values at these two angles always have opposite signs. Therefore, if one angle corresponds to the maximum variance, the other must be correspondent to the minimum variance, and vice versa.

In the principal coordinate system, the covariance of the data positions is

$$s_{\xi\eta}(\theta_{1,2}) = \frac{1}{n-1}\sum_{k=1}^{n} \xi_k(\theta_{1,2})\eta_k(\theta_{1,2}), \tag{24.17}$$

in which $\theta_{1,2}$ means either θ_1 or θ_2. This can be shown to be zero:

$$s_{\xi\eta}(\theta_{1,2}) = \frac{1}{2}(s_{yy} - s_{xx})\sin 2\theta_{1,2} + s_{xy}\cos 2\theta_{1,2} = 0. \tag{24.18}$$

The above conclusion is obtained using (24.11). The zero covariance means that the horizontal and vertical coordinates in the principal coordinate system are uncorrelated. The principal directions are denoted by the unit vectors:

$$\mathbf{e}_1 = \begin{pmatrix} \cos\theta_1 \\ \sin\theta_1 \end{pmatrix}, \quad \mathbf{e}_2 = \begin{pmatrix} \cos\theta_2 \\ \sin\theta_2 \end{pmatrix} = \begin{pmatrix} -\sin\theta_1 \\ \cos\theta_1 \end{pmatrix}. \tag{24.19}$$

Again, these two vectors are perpendicular to each other, as can be verified by

$$\mathbf{e}_1 \cdot \mathbf{e}_2 = 0. \tag{24.20}$$

In this example, *the vectors* $\mathbf{e}_1$ *and* $\mathbf{e}_2$ *are the empirical orthogonal functions or EOF*. A data point expressed in the principal coordinate system is

$$\mathbf{z}_k = a_{1k}\mathbf{e}_1 + a_{2k}\mathbf{e}_2, \tag{24.21}$$

in which

$$\mathbf{z}_k = \begin{pmatrix} x_k \\ y_k \end{pmatrix} \tag{24.22}$$

and

$$a_{1k} = \mathbf{z}_k \cdot \mathbf{e}_1, \quad a_{2k} = \mathbf{z}_k \cdot \mathbf{e}_2. \tag{24.23}$$

The above coordinates a_{1k} and a_{2k} in the principal coordinate system are the principal components. Here equation (24.23) is the *analysis equation*, and (24.21) is the *synthesis equation*. Combining (24.19), (24.21), and (24.22) and using (24.3), we can see that

$$\mathbf{z}_k = \xi\mathbf{e}_1 + \eta\mathbf{e}_2. \tag{24.24}$$

This is the relationship between the original coordinates $\mathbf{z}_k = \begin{pmatrix} x_k \\ y_k \end{pmatrix}$ and the new coordinates $\mathbf{z}'_k = \begin{pmatrix} \xi_k \\ \eta_k \end{pmatrix}$, expressed with the unit vectors of the new axes (or eigenvector) $\mathbf{e}_1$ and $\mathbf{e}_2$.

Here we can summarize the procedure of the above work:

(1) Data preparation: define the data scatter or variance.
(2) Finding the EOFs: find the principal directions by minimizing the variance, which gives the EOFs.
(3) Finding the principal components: project the original data onto these EOFs to obtain the principal components (the analysis).
(4) Synthesis: the principal components and EOFs are put together to express the original data (the synthesis).

We can see that step 1 prepares for the data scatter; step 2 resolves the EOFs; step 3 does the analysis; and step 4 does the synthesis. In the Fourier analysis, step 1 is not needed; step 2 is replaced by given sinusoidal functions; step 3 is the analysis or the Fourier Transform; and the last step, i.e. the synthesis, is the inverse Fourier Transform, if needed.

Now let us use an example to show this. Assume that some 2-D data points are defined by

$$y_k = 2.1x_k + x_k^2 + ns, \tag{24.25}$$

in which *ns* is a random noise. The following MATLAB script creates the simulated data points and the computed principal directions or EOFs and principal components.

```
%=== DEFINING RANDOM DATA POINTS SCATTERED ON A 2-D PLANE
x = rand(1,500) * 3 - 2;  % X-AXIS
y = 2.1 * x + x.^2 + 5 * rand(1,length(x));  % Y-AXIS
n = length(x); % LENGTH OF RECORD
figure
plot(x,y,'.')  % PLOT THE SCATTER

xm = mean(x);  % AVERAGE X
ym = mean(y);  % AVERAGE Y

xc = x - xm;   % DE-MEAN X
yc = y - ym;   % DE-MEAN Y
figure
plot(xc,yc,'.')
hold on

sxx = 1 / (n - 1) * xc * xc'; % VARIANCE OF X
syy = 1 / (n - 1) * yc * yc'; % VARIANCE OF Y
sxy = 1 / (n - 1) * xc * yc'; % COVARIANCE OF X AND Y

tan2theta = 2 * sxy / (sxx - syy);
theta = atan(tan2theta) * 180 / pi / 2; % 1ST ANGLE IN DEGREE
theta1 = theta + 90;                    % 2ND ANGLE IN DEGREE

%===FIND COORDINATES FOR THE 1ST DIRECTION
x1 = 2 * cosd(theta);
y1 = 2 * sind(theta);
x10 = 2 * cosd(theta + 180);
y10 = 2 * sind(theta + 180);
%===FIND COORDINATES FOR THE 2ND DIRECTION
x2 = 7 * cosd(theta1);
y2 = 7 * sind(theta1);
x20 = 7 * cosd(theta1 + 180);
y20 = 7 * sind(theta1 + 180);
plot([x10,x1], [y10,y1],'r','LineWidth',2)
plot([x20,x2], [y20,y2],'r','LineWidth',2)
axis([-4 6 -4 6])
set(gca,'DataAspectRatio',[ 1 1 1], 'PlotBoxAspectRatio',[ 1 1 1])

s11 = sxx * cosd(theta)^2 + ...
      2 * sxy * sind(theta) * cos(theta) + syy * sind(theta)^2
```

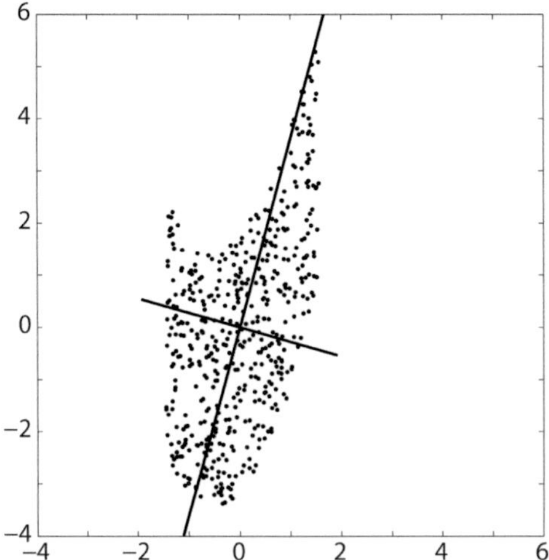

Figure 24.2 Example of scattered data points and principal directions.

```
s22 = sxx * cosd(theta1)^2 + ...
      2 * sxy * sind(theta1) * cos(theta1) + syy * sind(theta1)^2
e1 = [cosd(theta),sind(theta)];
e2 = [cosd(theta + 90),sind(theta + 90)];

[xcs,I] = sort(xc);
ycs = yc(I);
a1 = [xcs' ycs'] * e1';
a2 = [xcs' ycs'] * e2';
```

Fig. 24.2 shows the data points defined by (24.25) in the original coordinate system. The EOFs are unit vectors along the two straight lines, the longer and shorter axes, and are perpendicular to each other. Along the longer axis, the scatter reaches the maximum, while along the shorter axis, the scatter reaches its minimum. Fig. 24.3 shows the time series of the principal components – the coefficients in front of each of the two eigenvectors ($\mathbf{e}_1$ and $\mathbf{e}_2$). The magnitude of each of the coefficients shows the contribution of the corresponding base function (the eigenvector) at any time instance. At any instance, it is a linear superposition of the two orthogonal eigenvectors.

24.2 Principal Component Analysis/Empirical Orthogonal Function (EOF) Analysis: The Theory

24.2.1 Data Matrix

Now, we examine a more general case. We are now looking at scalar time series data at multiple locations. Assume that the original data in space and time can be expressed in a matrix as follows:

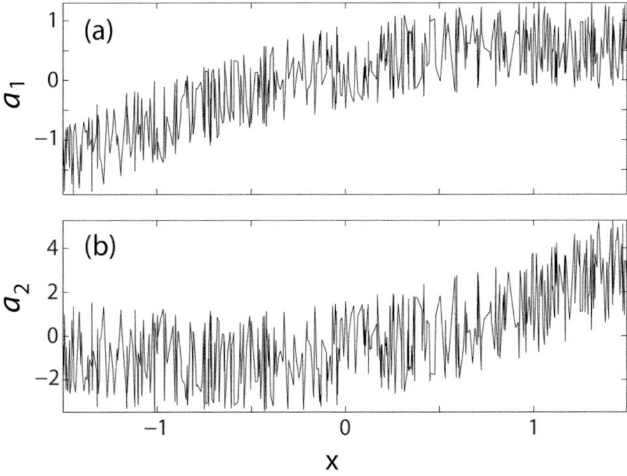

Figure 24.3 The principal components (a) a_1 and (b) a_2.

$$Z' = \begin{pmatrix} z'_{11} & z'_{12} & \cdots & z'_{1p} \\ z'_{21} & z'_{22} & \cdots & z'_{2p} \\ \vdots & \vdots & \cdots & \vdots \\ z'_{n1} & z'_{n2} & \cdots & z'_{np} \end{pmatrix}, \tag{24.26}$$

in which Z' is the original data matrix, which has a mixture of time and space, n is the total number of time instances and p is the total number of positions; the first index denotes time, while the second index denotes location; z'_{ij} is the data value at the ith time and jth location. With this definition, we can view the first row of the matrix as a spatial series (which forms a contour map) at time 1, the second row is a spatial series at time 2, ... , and the last row is a spatial series at time n. In addition, the first column is a time series at location 1, the second column is a time series at location 2, ... , and the last column is a time series at location p. Therefore, the matrix Z' is a way to express a collection of data at a number of locations from different times.

24.2.2 De-mean, De-trend, or Low-Pass Filtering of Data Matrix

On many occasions, we are interested only in the variations of anomaly, and for that purpose, we take out the mean, or we do average for Z' first:

$$\bar{Z} = \frac{1}{n}\left(\sum_{i=1}^{n} z'_{i1} \quad \sum_{i=1}^{n} z'_{i2} \quad \cdots \quad \sum_{i=1}^{n} z'_{ip} \right). \tag{24.27}$$

We now define the anomalies matrix by subtracting the temporal mean from each column, i.e.

$$Z = Z' - \bar{Z} = \begin{pmatrix} z'_{11} & z'_{12} & \cdots & z'_{1p} \\ z'_{21} & z'_{22} & \cdots & z'_{2p} \\ \vdots & \vdots & \cdots & \vdots \\ z'_{n1} & z'_{n2} & \cdots & z'_{np} \end{pmatrix} - \bar{Z}. \tag{24.28}$$

We write the de-meaned data matrix as

$$Z = \begin{pmatrix} z_{11} & z_{12} & \cdots & z_{1p} \\ z_{21} & z_{22} & \cdots & z_{2p} \\ \vdots & \vdots & \cdots & \vdots \\ z_{n1} & z_{n2} & \cdots & z_{np} \end{pmatrix}, \tag{24.29}$$

in which z_{ij} is the de-meaned data value at the ith time and jth location. Sometimes, we can remove the trend or even a low frequency variation of the record first through a de-trend or low-pass filtering and only examine the high frequency variabilities, i.e. if Z_L is the low-pass filtered data, then

$$Z = Z' - Z_L. \tag{24.30}$$

Depending on the actual objective, we may or may not need to do this. The difference is whether the variation is in reference to the mean value, or low-pass filtered baselines, or the origin.

24.2.3 Measure of Data Variability: Scatter Matrix

Now we examine the variance (or scatter) of data. We first assume that the time series in the space has some patterns of variabilities. The patterns or maps are vectors in space. The data variabilities are shown by variations of the data values around these patterns. Imagine we now project a map $\mathbf{z}(t)$ at any time t onto any unit vector $\mathbf{e}$ (representing a map or pattern) in a p- dimensional space. The unit vector $\mathbf{e}$ is expressed as

$$\mathbf{e} = \begin{pmatrix} e_1 \\ e_2 \\ \vdots \\ e_p \end{pmatrix}, \tag{24.31}$$

while the projection is

$$z_e(t) = \mathbf{z}(t) \cdot \mathbf{e}. \tag{24.32}$$

Here $\mathbf{z}(t)$ is any row (a map) in the data matrix Z.

$$\mathbf{z}(t) = \begin{pmatrix} z_{t1} & z_{t2} & \cdots & z_{tp} \end{pmatrix}, \quad t = 1, 2, \ldots, n. \tag{24.33}$$

The projection is an inner product:

$$\mathbf{z}(t) \cdot \mathbf{e} = z_{t1} \cdot e_1 + z_{i2} \cdot e_2 + \cdots + z_{tp} \cdot e_p. \tag{24.34}$$

Or, in matrix format:

$$\mathbf{z}(t)\mathbf{e} = \begin{pmatrix} z_{t1} & z_{t2} & \cdots & z_{tp} \end{pmatrix} \begin{pmatrix} e_1 \\ e_2 \\ \vdots \\ e_p \end{pmatrix} = \begin{pmatrix} e_1 & e_2 & \cdots & e_p \end{pmatrix} \begin{pmatrix} z_{t1} \\ z_{t2} \\ \vdots \\ z_{tp} \end{pmatrix}. \tag{24.35}$$

In such a p-dimensional space, the direction of a (unit) vector $\mathbf{e}$ corresponds to a contour map in the real world. This contour map should present itself with some characteristics – there are some highs and lows in the measured quantity. The magnitude of a p-dimensional vector determines the relative magnitude of those patterns (or highs and lows), but the unit vector in the same direction is enough to represent the patterns. When the direction of such a unit vector is changed, the contour map also changes and the pattern changes. The projection of an arbitrary p-dimensional vector (observations at a given time instance at the p locations) onto this unit vector gives the correlation of the two vectors or the similarity of them. A larger projection means a greater resemblance between the two vectors. Therefore, a measure of similarity of a given dataset at a given time with a given pattern (or unit vector in the p-dimensional space) is the magnitude of the projection onto the unit vector. A measure of the overall similarity of a given dataset at all times with a given pattern (or unit vector in the p-dimensional space) is the total magnitude squared of the projection onto the unit vector, i.e. adding all projections squared. This is quantitatively presented below.

Now we are ready to define the scatter or variance of the data matrix $\mathbf{z}$ along the given direction $\mathbf{e}$ in the p- dimensional space E_p:

$$\psi(\mathbf{e}) = \sum_{i=1}^{n} \left(z_e(t)\right)^2 = \sum_{t=1}^{n} (\mathbf{z}(t) \cdot \mathbf{e})^2. \tag{24.36}$$

For convenience, we call this $\psi(\mathbf{e})$ data scatter. This gives a single value as a function of the unit vector $\mathbf{e}$. It is the sum of projection squared for all time instances – it is a measure of scatter or variance of all data points in the direction of $\mathbf{e}$. For a larger value $\psi(\mathbf{e})$, it means that the given data have a higher frequency of occurrence in the direction of $\mathbf{e}$, or the pattern (map) associated with $\mathbf{e}$ is more frequent than another pattern (a different $\mathbf{e}$) that has a smaller $\psi(\mathbf{e})$.

Since in matrix operation form, at each individual time (24.35) is true, we have

$$(\mathbf{z}(t)\cdot\mathbf{e})^2 = (\mathbf{z}(t)\mathbf{e})^2 = \begin{pmatrix} e_1 & e_2 & \cdots & e_p \end{pmatrix} \begin{pmatrix} z_{t1} \\ z_{t2} \\ \vdots \\ z_{tp} \end{pmatrix} \begin{pmatrix} z_{t1} & z_{t2} & \cdots & z_{tp} \end{pmatrix} \begin{pmatrix} e_1 \\ e_2 \\ \vdots \\ e_p \end{pmatrix}. \tag{24.37}$$

Therefore, by adding the scatters at different times, we have the data scatter expressed in a concise matrix format:

$$\psi(\mathbf{e}) = \mathbf{e}^T S\mathbf{e} = \mathbf{e}^T \sum_{i=1}^{n} S'(t)\mathbf{e}, \tag{24.38}$$

in which $\mathbf{e}^T$ is the transpose of $\mathbf{e}$, and

$$S'(t) = \begin{pmatrix} z_{t1} \\ z_{t2} \\ \vdots \\ z_{tp} \end{pmatrix} \begin{pmatrix} z_{t1} & z_{t2} & \cdots & z_{tp} \end{pmatrix}, \tag{24.39}$$

$$S = \sum_{i=1}^{n} S'(t). \tag{24.40}$$

Equation (24.38) gives a nice matrix expression of the total data projection squared (scatter) associated with the selected unit vector or pattern $\mathbf{e}$. In (24.39), S' is a matrix that is independent of the pattern $\mathbf{e}$ and only dependent on the original data at a given time t:

$$S'(t) = \begin{pmatrix} z_{t1} \\ z_{t2} \\ \vdots \\ z_{tp} \end{pmatrix} \begin{pmatrix} z_{t1} & z_{t2} & \cdots & z_{tp} \end{pmatrix} = \mathbf{z}(t)\mathbf{z}^T(t) = \begin{pmatrix} z_{t1}z_{t1} & z_{t1}z_{t2} & \cdots & z_{t1}z_{tp} \\ z_{t2}z_{t1} & z_{t2}z_{t2} & \cdots & z_{t2}z_{tp} \\ \vdots & \vdots & \cdots & \vdots \\ z_{tp}z_{t1} & z_{tp}z_{t2} & \cdots & z_{tp}z_{tp} \end{pmatrix}. \tag{24.41}$$

By adding the values of (24.41) at all times, we obtain the expression of the matrix S on the right-hand side of equation (24.39):

$$S = \sum_{t=1}^{n} S'(t) = \begin{pmatrix} \sum_{t=1}^{n} z_{t1}^2 & \sum_{t=1}^{n} z_{t1} z_{t2} & \cdots & \sum_{t=1}^{n} z_{t1} z_{tp} \\ \sum_{t=1}^{n} z_{t2} z_{t1} & \sum_{t=1}^{n} z_{t2}^2 & \cdots & \sum_{t=1}^{n} z_{t2} z_{tp} \\ \vdots & \vdots & \cdots & \vdots \\ \sum_{t=1}^{n} z_{tp} z_{t1} & \sum_{t=1}^{n} z_{tp} z_{t2} & \cdots & \sum_{t=1}^{n} z_{tp}^2 \end{pmatrix}. \tag{24.42}$$

This is a matrix that only depends on the original data. For convenience of discussion, we call S the *scatter matrix* or *covariance matrix*. It can be seen that this scatter matrix can be expressed in a simpler way by the original de-meaned data matrix:

$$Z^T Z = \begin{pmatrix} \sum_{t=1}^{n} z_{t1}^2 & \sum_{t=1}^{n} z_{t1} z_{t2} & \cdots & \sum_{t=1}^{n} z_{t1} z_{tp} \\ \sum_{t=1}^{n} z_{t2} z_{t1} & \sum_{t=1}^{n} z_{t2}^2 & \cdots & \sum_{t=1}^{n} z_{t2} z_{tp} \\ \vdots & \vdots & \cdots & \vdots \\ \sum_{t=1}^{n} z_{tp} z_{t1} & \sum_{t=1}^{n} z_{tp} z_{t2} & \cdots & \sum_{t=1}^{n} z_{tp}^2 \end{pmatrix} = \sum_{t=1}^{n} S'(t) = S. \tag{24.43}$$

24.2.4 Properties of Scatter Matrix

To prepare for the calculation of the EOFs, we note that the scatter matrix is symmetric, i.e. the transpose of the matrix is itself:

$$S^T = S. \tag{24.44}$$

This is obvious since

$$S = Z^T Z. \tag{24.45}$$

The transpose gives

$$S^T = (Z^T Z)^T = Z^T Z = S. \tag{24.46}$$

Equation (24.46) proves that S is symmetric.

Therefore, if a and b are two column-vectors, then

$$(a^T S b)^T = b^T S^T a = b^T S a. \tag{24.47}$$

Since the left quantity is a scalar, its transpose is itself, or

$$(a^T Sb)^T = a^T Sb. \tag{24.48}$$

Therefore,

$$a^T Sb = b^T Sa. \tag{24.49}$$

Equation (24.49) says that if the matrix S is symmetric, the order of a and b in computing the value of $a^T Sb$ can be switched.

24.2.5 Finding the EOFs

We now look at the directions $\mathbf{e}$ along which the variance or scatter of the data reach extreme values or maximum or minimum values. To do that, recall that $\psi(\mathbf{e})$ is the scatter along unit vector $\mathbf{e}$. The variation of the scatter $\psi(\mathbf{e})$ due to the change of direction $\mathbf{e}$ is

$$\delta\psi = \psi(\mathbf{e} + \delta\mathbf{e}) - \psi(\mathbf{e}). \tag{24.50}$$

A necessary condition for $\psi(\mathbf{e})$ to reach extreme value is that its derivative with respect to the vector $\mathbf{e}$ is 0, i.e. when $\delta\psi = 0$, $\mathbf{e}$ is the direction of choice. Since the data scatter is

$$\psi(\mathbf{e}) = \mathbf{e}^T S\mathbf{e}, \tag{24.51}$$

a small change in $\mathbf{e}$ results in a change in $\psi(\mathbf{e})$, which can be approximated as

$$\begin{aligned} \psi(\mathbf{e} + \delta\mathbf{e}) &= (\mathbf{e} + \delta\mathbf{e})^T S(\mathbf{e} + \delta\mathbf{e}) = \left(\mathbf{e}^T + (\delta\mathbf{e})^T\right) S(\mathbf{e} + \delta\mathbf{e}) \\ &\approx \mathbf{e}^T S\mathbf{e} + (\delta\mathbf{e})^T S\mathbf{e} + \mathbf{e}^T S\delta\mathbf{e} \end{aligned}, \tag{24.52}$$

which has three terms; the first term is $\psi(\mathbf{e})$, and thus the requirement that

$$\delta\psi = 0 \tag{24.53}$$

leads to

$$(\delta\mathbf{e})^T S\mathbf{e} + \mathbf{e}^T S\delta\mathbf{e} = 0. \tag{24.54}$$

From (24.49), the two terms on the left-hand side of (24.54) are the same, i.e.

$$\mathbf{e}^T S\delta\mathbf{e} = (\delta\mathbf{e})^T S\mathbf{e}. \tag{24.55}$$

Therefore, the symmetry of S leads to

$$(\delta\mathbf{e})^T S\mathbf{e} = 0. \tag{24.56}$$

This conclusion does not give us enough information to find the "direction" $\mathbf{e}$ yet. We need to consider another factor: $\mathbf{e}$ is a unit vector, so its length remains unchanged during the process of direction variation, i.e.

$$\mathbf{e}^T\mathbf{e} = 1, \tag{24.57}$$

$$(\mathbf{e} + \delta\mathbf{e})^T(\mathbf{e} + \delta\mathbf{e}) = 1, \tag{24.58}$$

which yields

$$(\delta\mathbf{e})^T\mathbf{e} + \mathbf{e}^T(\delta\mathbf{e}) + (\delta\mathbf{e})^T\delta\mathbf{e} = 0. \tag{24.59}$$

Since the terms are all scalars, the first and second terms are the same. Neglecting the last term, which has a higher order, we have

$$(\delta\mathbf{e})^T\mathbf{e} = 0, \tag{24.60}$$

i.e. the change of $\mathbf{e}$ must be perpendicular to $\mathbf{e}$. In mathematics, equations (24.56) and (24.60) lead to the conclusion that $\mathbf{e}$ must be the *eigenvector* of the scatter matrix S, i.e. there is a constant λ such that

$$S\mathbf{e} = \lambda\mathbf{e}. \tag{24.61}$$

In other words, the scatter matrix S can be considered as a mapping operation that maps the vector $\mathbf{e}$ to a parallel vector $\lambda\mathbf{e}$. The constant λ is the *eigenvalue* corresponding to the eigenvector $\mathbf{e}$. This can be verified below. Given equation (24.61), multiplying both sides with $(\delta\mathbf{e})^T$, we have

$$(\delta\mathbf{e})^T S\mathbf{e} = (\delta\mathbf{e})^T\lambda\mathbf{e} = \lambda(\delta\mathbf{e})^T\mathbf{e}. \tag{24.62}$$

From (24.60), the above equation leads to (24.56). On the other hand, given equations (24.56) and (24.60), for a constant λ,

$$(\delta\mathbf{e})^T(S\mathbf{e} - \lambda\mathbf{e}) = 0 \tag{24.63}$$

is obvious and thus (24.61) must hold. Therefore, equations (24.56) and (24.60) are equivalent to equation (24.61).

24.2.6 The Orthogonality of EOFs

According to linear algebra, a symmetric matrix S generally has p eigenvectors $\mathbf{e}_j$, corresponding to the p eigenvalues λ_j, $j = 1, 2, \ldots, p$:

$$\begin{pmatrix} S_{11} & S_{12} & \cdots & S_{1p} \\ S_{21} & S_{22} & \cdots & S_{2p} \\ \vdots & \vdots & \cdots & \vdots \\ S_{n1} & S_{n2} & \cdots & S_{pp} \end{pmatrix} \begin{pmatrix} e_{1j} \\ e_{2j} \\ \vdots \\ e_{pj} \end{pmatrix} = \lambda_j \begin{pmatrix} e_{1j} \\ e_{2j} \\ \vdots \\ e_{pj} \end{pmatrix}, \quad j = 1, 2, \ldots p. \tag{24.64}$$

This is a collection of p equations corresponding to p eigenvectors $\mathbf{e}_j$. They can be combined into a single equation. For that purpose, we define the eigenvector matrix as

$$E = (\mathbf{e}_1, \mathbf{e}_2, \ldots \mathbf{e}_p) = \begin{pmatrix} e_{11} & e_{12} & \cdots & e_{1p} \\ e_{21} & e_{22} & \cdots & e_{2p} \\ \vdots & \vdots & \vdots & \vdots \\ e_{p1} & e_{p2} & \cdots & e_{pp} \end{pmatrix}. \quad (24.65)$$

With the eigenvector matrix and equation (24.64), we have a concise expression:

$$SE = EL, \quad L = \begin{pmatrix} \lambda_1 & 0 & \cdots & 0 \\ 0 & \lambda_2 & \cdots & 0 \\ \vdots & \vdots & \ddots & \vdots \\ 0 & 0 & \vdots & \lambda_p \end{pmatrix}. \quad (24.66)$$

Here the eigenvalues $\{\lambda_j, j = 1, 2, \ldots, p\}$ are generally ordered according to their magnitudes: the larger eigenvalues are listed first, and they are ordered in a monotonically decreasing order. The number of non-zero eigenvalues is equal to the rank of the data matrix Z, which is the number of linearly independent rows or columns. In practice, for noisy data, usually the rank of Z is the smaller of $n - 1$ and p, or

$$r = \min\{n - 1, p\}. \quad (24.67)$$

As long as there are a sufficient number of data points in time at each location, or $n - 1 \geq p$, the rank is p and there are p eigenvalues and p eigenvectors.

One of the important properties of the eigenvectors is that they are orthogonal. This can be verified as follows. Assume that $\mathbf{e}_i$ and $\mathbf{e}_j$ are two eigenvectors:

$$\mathbf{e}_i = \begin{pmatrix} e_{1i} \\ e_{2i} \\ \vdots \\ e_{pi} \end{pmatrix}, \quad \mathbf{e}_j = \begin{pmatrix} e_{1j} \\ e_{2j} \\ \vdots \\ e_{pj} \end{pmatrix}. \quad (24.68)$$

We have the following equations:

$$\mathbf{e}_i^T S \mathbf{e}_j = \mathbf{e}_i^T \lambda_j \mathbf{e}_j = \lambda_j (\mathbf{e}_i^T \mathbf{e}_j), \quad (24.69)$$

$$\mathbf{e}_j^T S \mathbf{e}_i = \mathbf{e}_j^T \lambda_i \mathbf{e}_i = \lambda_i (\mathbf{e}_j^T \mathbf{e}_i). \quad (24.70)$$

Since $\mathbf{e}_i^T \mathbf{e}_j$ is a scalar, therefore, $\mathbf{e}_i^T \mathbf{e}_j = \mathbf{e}_j^T \mathbf{e}_i$. Furthermore, since the matrix S is symmetric, recalling (24.49), we have

$$\mathbf{e}_i^T S \mathbf{e}_j = \mathbf{e}_j^T S \mathbf{e}_i. \quad (24.71)$$

Therefore, equations (24.69) and (24.70) combine to yield

$$(\lambda_j - \lambda_i)(\mathbf{e}_i^T \mathbf{e}_j) = 0. \quad (24.72)$$

If the two eigenvalues are different, i.e. $i \neq j, \quad \lambda_j \neq \lambda_i$, we must have

$$\mathbf{e}_i^T \mathbf{e}_j = 0. \tag{24.73}$$

In short, the eigenvectors with different eigenvalues are orthogonal to each other. We can also write the conclusion in the following manner:

$$\mathbf{e}_i^T \mathbf{e}_j = \delta_{ij} = \begin{cases} 0 & (i \neq j) \\ 1 & (i = j) \end{cases}. \tag{24.74}$$

These orthogonal eigenvectors form the natural base functions. If there are duplicate eigenvalues, i.e. $\lambda_i = \lambda_j$, it is still possible to construct proper eigenvectors to satisfy the orthogonality, but the choice is usually not unique and the discussion is omitted here. For noisy data, the chance of having exactly the same eigenvalues is low.

24.2.7 The Analysis and Synthesis of EOF

It can be further verified that the inverse matrix of the eigenvector matrix E defined in (24.65) is its own transpose, i.e.

$$E^T E = \begin{pmatrix} \delta_{11} & \delta_{12} & \cdots & \delta_{1p} \\ \delta_{21} & \delta_{22} & \cdots & \delta_{2p} \\ \vdots & \vdots & \vdots & \vdots \\ \delta_{p1} & \delta_{p2} & \cdots & \delta_{pp} \end{pmatrix} = \begin{pmatrix} 1 & 0 & \cdots & 0 \\ 0 & 1 & \cdots & 0 \\ \vdots & \vdots & \ddots & \vdots \\ 0 & 0 & \vdots & 1 \end{pmatrix} = I_p. \tag{24.75}$$

Using (24.66) and (24.75), we have

$$S = ELE^T. \tag{24.76}$$

The purpose of the above was to find the "direction" in the p-dimensional space E_p along which the scatter or variance of the data reaches its extreme value (or the modal direction which has the information of the spatial structure of the data). After these "directions" are identified by the orthogonal eigenvectors, we can use these orthogonal eigenvectors as base functions to express the original data. The following equation is now evident:

$$Z = Z(EE^T) = (ZE)E^T. \tag{24.77}$$

We define

$$A = ZE \tag{24.78}$$

and call it the *analysis equation*; while (24.77) is called the *synthesis equation*, or in terms of the analysis, it is rewritten as

$$Z = AE^T. \tag{24.79}$$

Equation (24.78) is comparable to the Fourier Transform and (24.79) comparable to the inverse Fourier Transform. If we write out the details of the analysis and synthesis equations, we have

$$A = \begin{pmatrix} z_{11} & z_{12} & \cdots & z_{1p} \\ z_{21} & z_{22} & \cdots & z_{2p} \\ \vdots & \vdots & \cdots & \vdots \\ z_{n1} & z_{n2} & \cdots & z_{np} \end{pmatrix} \begin{pmatrix} e_{11} & e_{12} & \cdots & e_{1p} \\ e_{21} & e_{22} & \cdots & e_{2p} \\ \vdots & \vdots & \vdots & \vdots \\ e_{p1} & e_{p2} & \cdots & e_{pp} \end{pmatrix}, \tag{24.80}$$

$$Z = \begin{pmatrix} a_{11} & a_{12} & \cdots & a_{1p} \\ a_{21} & a_{22} & \cdots & a_{2p} \\ \vdots & \vdots & \cdots & \vdots \\ a_{n1} & a_{n2} & \cdots & a_{np} \end{pmatrix} \begin{pmatrix} e_{11} & e_{21} & \cdots & e_{p1} \\ e_{12} & e_{22} & \cdots & e_{p2} \\ \vdots & \vdots & \vdots & \vdots \\ e_{1p} & e_{2p} & \cdots & e_{pp} \end{pmatrix}. \tag{24.81}$$

These (A and Z) are matrices with dimensions $n \times p$. The analysis equation can also be expressed as a single time series array or the principal component in the following for each location:

$$\mathbf{a}_j = Z\mathbf{e}_j. \tag{24.82}$$

This is sometimes called the time coefficient for the jth EOF.

$$\mathbf{a}_j = \begin{pmatrix} a_{1j} \\ a_{2j} \\ \vdots \\ a_{nj} \end{pmatrix}, \quad \mathbf{e}_j = \begin{pmatrix} e_{1j} \\ e_{2j} \\ \vdots \\ e_{pj} \end{pmatrix}, \quad j = 1, 2, \ldots, p \tag{24.83}$$

or

$$a_{tj} = Z^T(t)\mathbf{e}_j = \sum_{x=1}^{p} Z_{tx}\mathbf{e}_{xj}, \quad t = 1, 2, \ldots, n, j = 1, 2, \ldots p, \tag{24.84}$$

in which

$$Z_{tx} = \sum_{j=1}^{p} a_{tj}e_{xj}, \quad t = 1, 2, \ldots, n; \quad x = 1, 2, \ldots, p. \tag{24.85}$$

The matrix A of (24.80) is also called the "amplitude" matrix, formed by the principal components or amplitude vector in the p-dimensional space E_p.

By definition, the time average of Z is zero:

$$\sum_{i=1}^{n} a_{tj} = 0, \quad j = 1, 2, \ldots, p. \tag{24.86}$$

It is obvious that $\mathbf{a}_j$ has zero mean because Z is de-meaned. We can also see that the transpose of principal component matrix A multiplying itself is

$$A^T A = (ZE)^T ZE = E^T Z^T ZE = E^T SE = E^T EL = L.$$

Therefore,

$$A^T A = L. \tag{24.87}$$

After the eigenvalues and eigenvectors are found, the percentage of variability corresponding to a given eigenvalue can be determined. The amount of variance explained by the pattern associated with the ith eigenvector λ_i is determined by the normalization equation:

$$\tilde{\lambda}_i = \frac{\lambda_i}{\sum_{k=1}^{p} \lambda_k}. \tag{24.88}$$

The uncertainty estimate of the eigenvalues and eigenfunctions (North et al., 1982) are given by

$$\delta\lambda_i = \lambda_i \sqrt{\frac{2}{n}}, \tag{24.89}$$

$$\delta a_i = \frac{\delta\lambda_i}{\lambda_j - \lambda_i} a_j, \tag{24.90}$$

in which λ_j is the closest eigenvalue of λ_i. The computation discussed above is the EOF analysis method.

24.2.8 The Procedures of EOF Analysis

Here are the general steps for doing the EOF analysis.

Step 1. Obtain data in time and space.

Step 2. Order the data such that the positions are in a sequence – each position within that sequence has a time series (ordering scheme).

Step 3. Construct the data matrix Z.

Step 4. De-mean, de-trend, or low-pass filter the data if needed.

Step 5. Find the eigenvalues and eigenvectors of the covariance matrix of the data matrix Z: $S = Z^T Z$ (using the MATLAB command: `eig`).

Step 6. Calculate the time coefficients: $\mathbf{a}_j = Z\mathbf{e}_j$.

- Contour plot the EOF $\mathbf{e}_j$.
- Plot the time coefficient $\mathbf{a}_j$.
- Calculate the percentage of explainable variability (eigenvalues).

24.3 Examples of EOF Analysis

24.3.1 Vertical Structure of Horizontal Velocity

Here we use some time series of more than 200 days of velocity data to demonstrate the EOF analysis. The data were obtained from a bottom mounted ADCP (looking upward) deployed at Rigolets Pass, Lake Pontchartrain, Louisiana in later 2005 until 2006. The velocity profiles were measured at 15-minute intervals continuously for the entire time period. The water depth was around 6 m. The vertical bins size of the ADCP data was 0.5 m. The mid-bin position of the first bin was 1.22 m away from the ADCP transducers. The data file is named "Rigolet-ADCP-east.mat." Since the instrument was deployed in a channel roughly oriented in an east–west direction, the vertical structure of the east velocity component is examined in the analysis. In the data file, the east velocity is named `ueI`. The MATLAB command eig is used to find the eigenvectors and eigenvalues at the same time. The following is the major portion of the script for the analysis.

```
Zp = ueI;   % East velocity in m/s
Z  = detrend(Zp,0);         % Detrend first
S  = Z' * Z;                % Get the covariance matrix
[E, L] = eig(S);   % Get the eigenvalues and eigenvectors of S
lambda = diag(L) / trace(L);% Amount of variance explained by the eigenvector
```

Fig. 24.4 shows the time series of east velocity component at the near bottom bin (the first bin) and the ninth bin (near the surface at about 5.22 m above the ADCP transducers). Although the EOF analysis involves data from all nine bins, Fig. 24.4 only shows the top and bottom bins for clarity. It can be seen that tidal signals are obvious. The first mode explains 99.46% of the variability; while the second mode explains 0.25% of the variability; and the remaining seven modes explain the rest, 0.3%. Obviously, the first two modes are representative of more than 99.7% of the data variability. These two modes are shown in Fig. 24.5: the first mode is a quite uniform structure with a slight decrease of velocity magnitude toward the bottom; while the second mode is a two-layered flow with opposing flows above and below about the mid-depth. Fig. 24.6 shows the time coefficients of these modes. Sometimes the coefficients are positive, sometime negative, indicating tidal oscillations reversing the flow direction in a barotropic way. The two-layered flow is rare, indicating that baroclinic effect did not occur often.

24.3.2 An Example of "Forcing-Response" EOF

Here we discuss an example of a hypothetical time series of vertically averaged ocean current velocity vector and wind vector. Assuming it is in deep ocean and

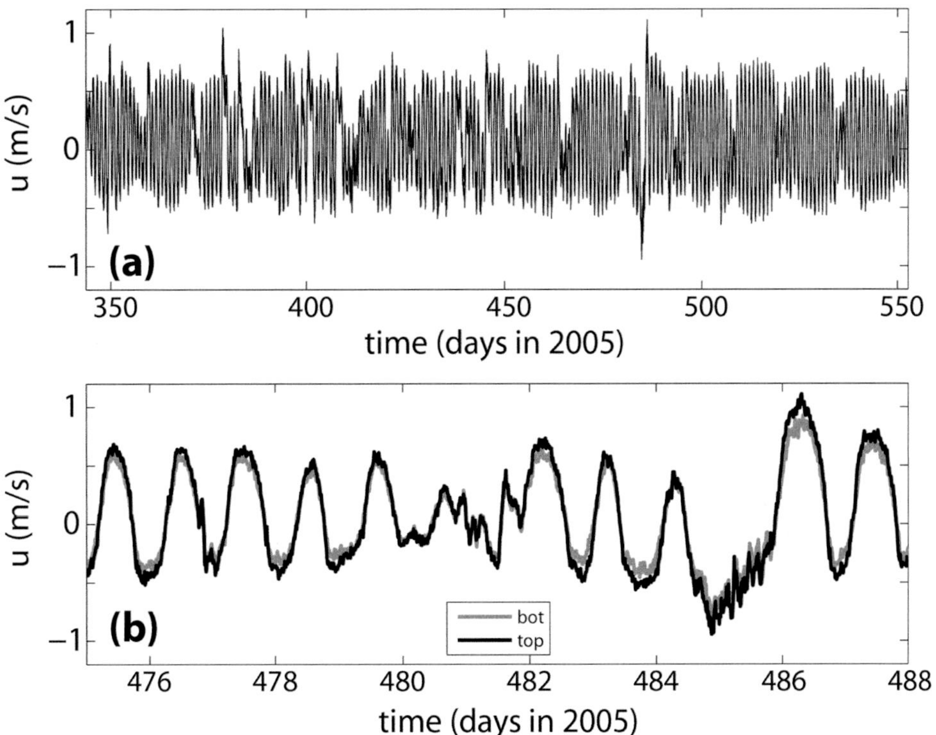

Figure 24.4 Time series of east velocity component obtained at Rigolets Pass, Lake Pontchartrain, Louisiana in 2005. (a) Time series of east velocity at top and bottom; (b) a zoomed-in view.

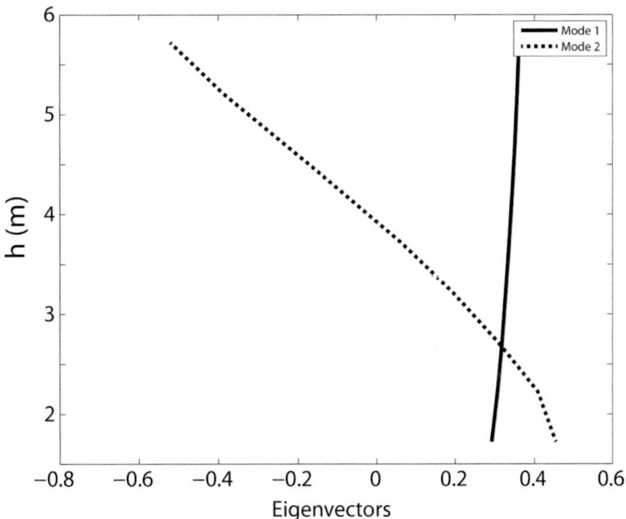

Figure 24.5 The first two modes of the east velocity component obtained at Rigolets Pass, Lake Pontchartrain in 2005.

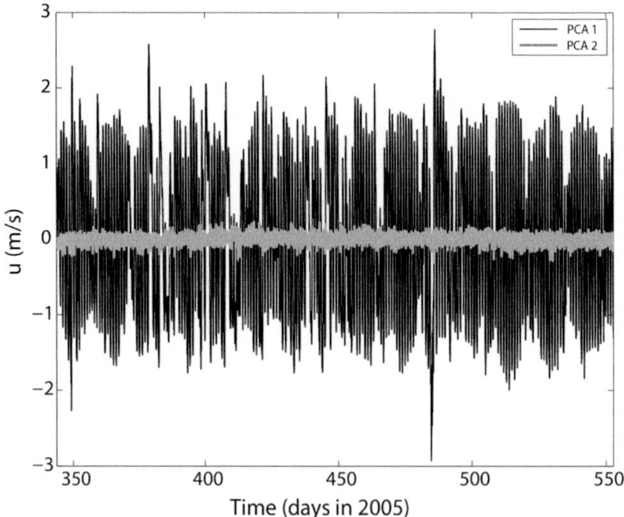

Figure 24.6 The time coefficients of the first two modes for the east velocity component obtained at Rigolets Pass, Lake Pontchartrain in 2005.

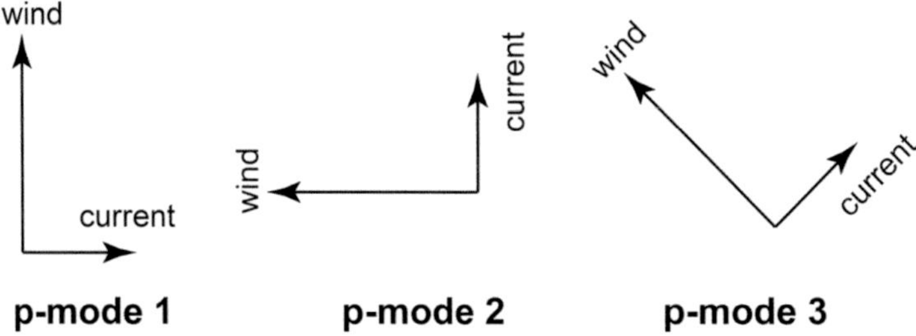

Figure 24.7 Hypothetical physical modes (p-modes) of wind and current. A time series is assumed to have p-mode 1 first, and then p-mode 2, and p-mode 3, in this order.

the depth averaged current velocity is perpendicular to wind under idealized conditions (Ekman theory), the current velocity vector should be pointing 90 degrees to the right (left) of the wind vector in the northern (southern) hemisphere. Assume that current and wind have three different physical modes (p-modes, Fig. 24.7): (1) current is toward the east and wind toward the north (southerly wind); (2) current is toward the north and wind toward the west (easterly wind); and (3) current is toward the northeast and wind toward the northwest (southeasterly wind). We further assume that the data are hourly measured. We assume that for the first 100 hours, data show negative p-mode 1, followed by 100 hours of positive p-mode 1; then 100 hours of negative p-mode 2, 100 hours of positive p-mode 2, 100 hours of negative p-mode 3, and 100 hours of positive

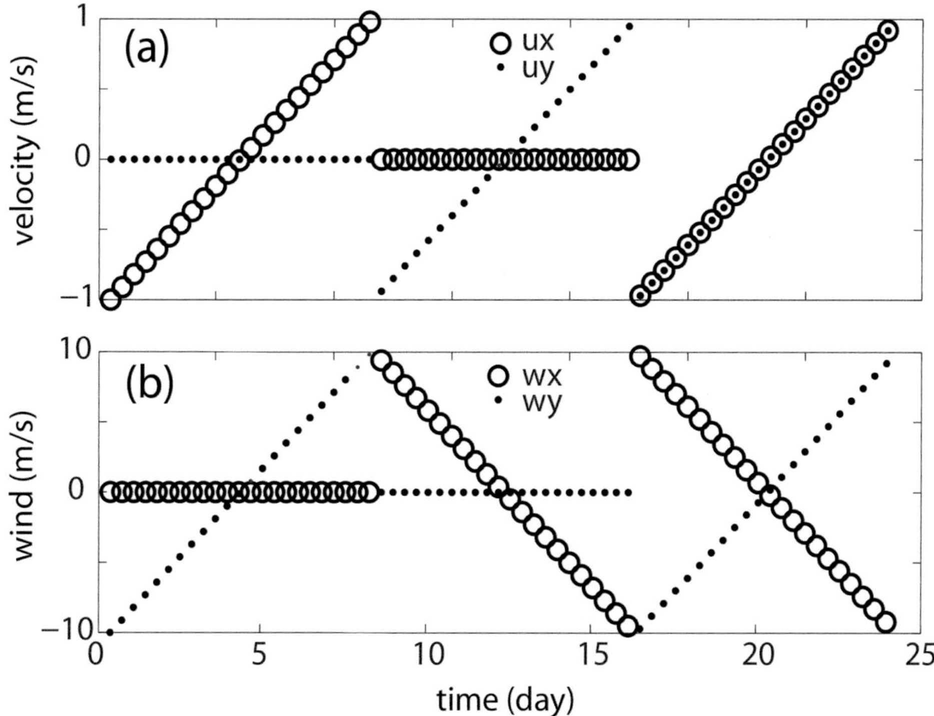

Figure 24.8 Hypothetical time series of current (a) and wind (b). Shown here are the east (ux) and north (uy) components of the current velocity and east (wx) and north (wy) components of the wind velocity.

p-mode 3 (Fig. 24.8). As can be seen, we also assume that the magnitude of the current velocity and wind velocity vary linearly over time though maintaining the same p-mode for every 200 hours. The following MATLAB script defines these hypothetical time series data.

```
%=== DEFINE p-mode 1 WIND AND CURRENT
wind1x = 0 * (-100:100);        % p-MODE 1 WINDX
wind1y = (-100:100) / 10;       % p-MODE 1 WINDY
flow1x = (-100:100) / 100;      % p-MODE 1 FLOWX
flow1y = 0 * (-100:100);        % p-MODE 1 FLOWY
%=== DEFINE p-mode 2 WIND AND CURRENT
wind2x = -(-100:100) / 10;      % p-MODE 2 WINDX
wind2y = 0 * (-100:100);        % p-MODE 2 WINDY
flow2x = 0 * (-100:100);        % p-MODE 2 FLOWX
flow2y = (-100:100) / 100;      % p-MODE 2 FLOWY
%=== DEFINE p-mode 3 WIND AND CURRENT
wind3x = -(-100:100) / 10;      % p-MODE 3 WINDX
wind3y = (-100:100) / 10;       % p-MODE 3 WINDY
flow3x = (-100:100) / 100;      % p-MODE 3 FLOWX
```

```
flow3y = (-100:100) / 100;   % p-MODE 3 FLOWY
%=== DEFINE continuous time series of WIND AND CURRENT
%=== p-mode 1 first, p-mode 2 second, and p-mode 3 last
flowx = [flow1x flow2x flow3x] ';
flowy = [flow1y flow2y flow3y] ';
windx = [wind1x wind2x wind3x] ';
windy = [wind1y wind2y wind3y] ';
%=== CONSTRUCT DATA MATRIX ===
Zp = [ flowx flowy windx windy] ;
```

Using MATLAB script similar to that shown in Section 24.3.1, we obtain the eigenvalues to be

$$\begin{aligned} \lambda_1 &= 2.0504 \times 10^4 \\ \lambda_2 &= 6.8347 \times 10^3 \end{aligned}.$$

The last two eigenvalues are all zeros. The first two EOF modes explain 75% and 25% of the time series data variability, respectively. The time coefficients for the two modes are shown in Fig. 24.9, which shows that for the first 200 hours, there are equal EOF mode 1 and mode 2 coefficients, i.e. the p-mode 1 is a superposition of EOF modes 1 and 2. This is verified in Fig. 24.10, in which it is obvious that EOF mode 1 adding EOF mode 2 gives the p-mode 1 (Fig. 24.7). Indeed, Fig. 24.10 shows that

$$\begin{aligned} \mathbf{V}_1 + \mathbf{V}_2 &= \mathbf{V}_3 \\ \mathbf{W}_1 + \mathbf{W}_2 &= \mathbf{W}_3 \end{aligned}.$$

Fig. 24.9 also shows that for the second 200 hours, there are equal EOF mode 1 and mode 2 coefficients but with opposite signs, i.e. p-mode 2 is a difference of EOF

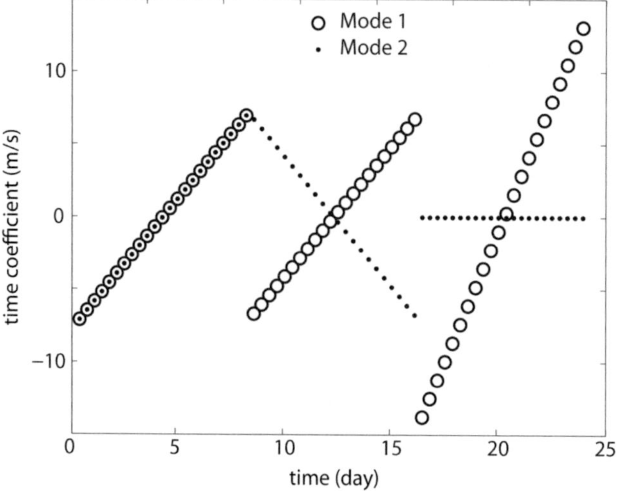

Figure 24.9 Time coefficients for the EOF modes.

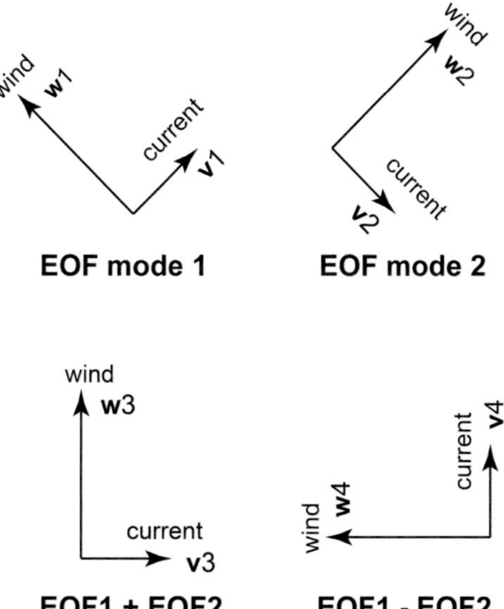

Figure 24.10 EOF modes and how their superposition to generate the first two p-modes by simple addition and subtraction.

modes 1 and 2. This is also verified in Fig. 24.10, in which it can be seen that EOF mode 1 subtracting EOF mode 2 gives the p-mode 2 (Fig. 24.7):

$$\begin{aligned} \mathbf{V}_1 - \mathbf{V}_2 &= \mathbf{V}_4 \\ \mathbf{W}_1 - \mathbf{W}_2 &= \mathbf{W}_4 \end{aligned}.$$

For the last 200 hours, the coefficient for EOF mode 1 is zero and only EOF mode 2 has non-zero coefficient values. Indeed, p-mode 3 (Fig. 24.7) and EOF mode 1 (Fig. 24.10) are identical.

From this example, we can see that the physical modes (p-modes) may or may not be the same as EOF modes. A physical mode may be a combination of EOF modes. This adds a little more challenge in the interpretation of the EOF modes.

24.3.3 An Example of "Forcing-Response" EOF in a Multi-Inlet System

Here another example is demonstrated for the EOF analysis for residual flows through an estuary in the northern Gulf of Mexico – Barataria Bay – with multiple inlets (Li et al., 2019b). The objective of the study was to find the subtidal flows in and out of the bay under the influence of weather, particularly atmospheric fronts.

Barataria Bay (Fig. 24.1) is an irregularly shaped shallow estuary in the central northern Gulf of Mexico. It is located to the south of New Orleans and

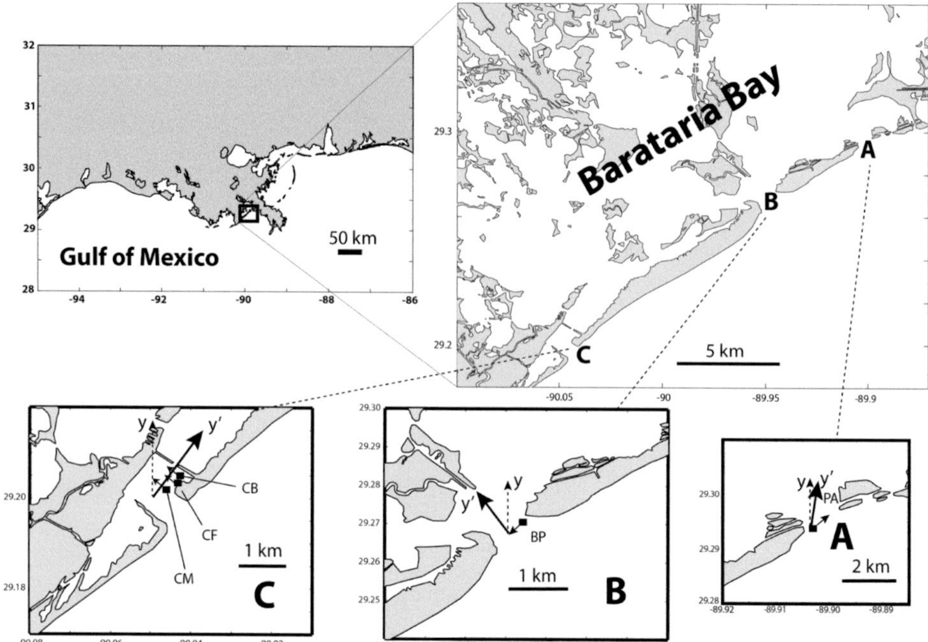

Figure 24.11 Study area and locations of deployment of acoustic Doppler current profilers (ADCPs). The areas marked as A, B, and C are Pass Abel (PA), Barataria Pass (BP), and Caminada Pass (CP), respectively. The three squares in C marked by CM, CF, and CB are the locations at the Caminada Pass in the mid-channel, at the fishing pier, and boat dock, respectively. The arrows indicate the direction in which the transducers of the ADCP are facing.

connected to the coastal ocean through several tidal inlets, including Caminada Pass (800 m wide and 9 m deep), Barataria Pass (800 m wide and 20 m deep), Pass Abel (1.9 km wide and 15 m deep), and Quatre Bayou Pass (2 km wide and 1.3 m deep). Its horizontal scale is approximately ~30–40 km in either north–south or east–west directions, with the major axis oriented in the northwest–southeast directions. Overall, Barataria Bay is very shallow, with an averaged depth of about 2 m.

In this study, five horizontal acoustic Doppler current profilers (ADCP) were deployed at multiple inlets of Barataria Bay (Fig. 24.11). The ADCPs were deployed 2 m below the surface, perpendicular to flow, looking sideways. The deployments were made between December 2013 and February 2014, and between December 2014 and March 2015. They were powered by batteries and solar panels. These ADCPs all have two beams with slant angles of 25 degrees from the axis. The data were saved at 5–15-minute ensemble intervals. The measurements were done acoustically along a horizontal line of about 100 m with five bins at 20 m intervals.

The beginning of the first bin was at ~5 m from the transducer. All instruments recorded high-quality data with no major gaps. The data from all five bins were averaged for the mean velocity along ~100 m length. The velocity components averaged along the 100 m beam length were rotated to the along-channel and cross-channel directions. Only the along-channel velocity component was used in the analysis of subtidal flows. These velocity data were then low-pass filtered for comparison with results from a numerical model.

Used also in this study were wind velocity time series data obtained from the station on Grand Isle at 29.265° N 89.958° W. The weather data had 6-minute ensemble average intervals. Water level data from NOAA at Grand Isle (Station ID 8761724 at 29°15.8″ N and 89°57.4″ W) were also used for model verification.

A 3-D Finite Volume Coastal Ocean Model (FVCOM) was used in this study for numerical simulations of atmospheric cold front-driven circulations in the area with a focus on Barataria Bay. The model covered the entire Gulf of Mexico, roughly from 80.7° W to 97.9° W, and 18.1° N to 30.7° N. The finest resolution was about 20 m in the bay region. There were 40 sigma layers in the vertical (Fig. 24.12). Tidal forcing was provided at the open boundaries with 10 tidal constituents: M2, S2, N2, K2, K1, O1, P1, Q1, MF, and MM. The Climate Forecast System Reanalysis (CFSR) hourly data obtained from the National Centers for Environmental Prediction (NCEP) were used for atmospheric forcing (wind and air pressure).

The model data comparison (Fig. 24.13) showed skill scores of 0.5 to 0.67, demonstrating the model's performance. After the validation of the model, the along-channel velocity component at each of the major inlets was low-pass filtered, after which an EOF analysis was applied.

To determine the modes with which the subtidal along-channel flows responded to the cold fronts, the data matrix Z' for EOF analysis included the low-pass filtered along-channel velocity component at each of the four major inlets and the subtidal wind velocity components:

$$Z' = \begin{pmatrix} v_{r1}(t_1) & v_{r2}(t_1) & v_{r3}(t_1) & v_{r4}(t_1) & W_E(t_1) & W_N(t_1) \\ v_{r1}(t_2) & v_{r2}(t_2) & v_{r3}(t_2) & v_{r4}(t_2) & W_E(t_2) & W_N(t_2) \\ \vdots & \vdots & \vdots & \vdots & \vdots & \vdots \\ v_{r1}(t_n) & v_{r2}(t_n) & v_{r3}(t_n) & v_{r4}(t_n) & W_E(t_n) & W_N(t_n) \end{pmatrix}, \tag{24.91}$$

in which v_{r1}, v_{r2}, v_{r3}, and v_{r4} are the subtidal (rotated) along-channel velocity component for the first (Caminada Pass), second (Barataria Pass), third (Pass Abel), and fourth (Quatre Bayou Pass) inlet, respectively; W_E and W_N are the east and north components of the subtidal wind velocity vector, respectively; $t_1, \ldots, t_n$ are

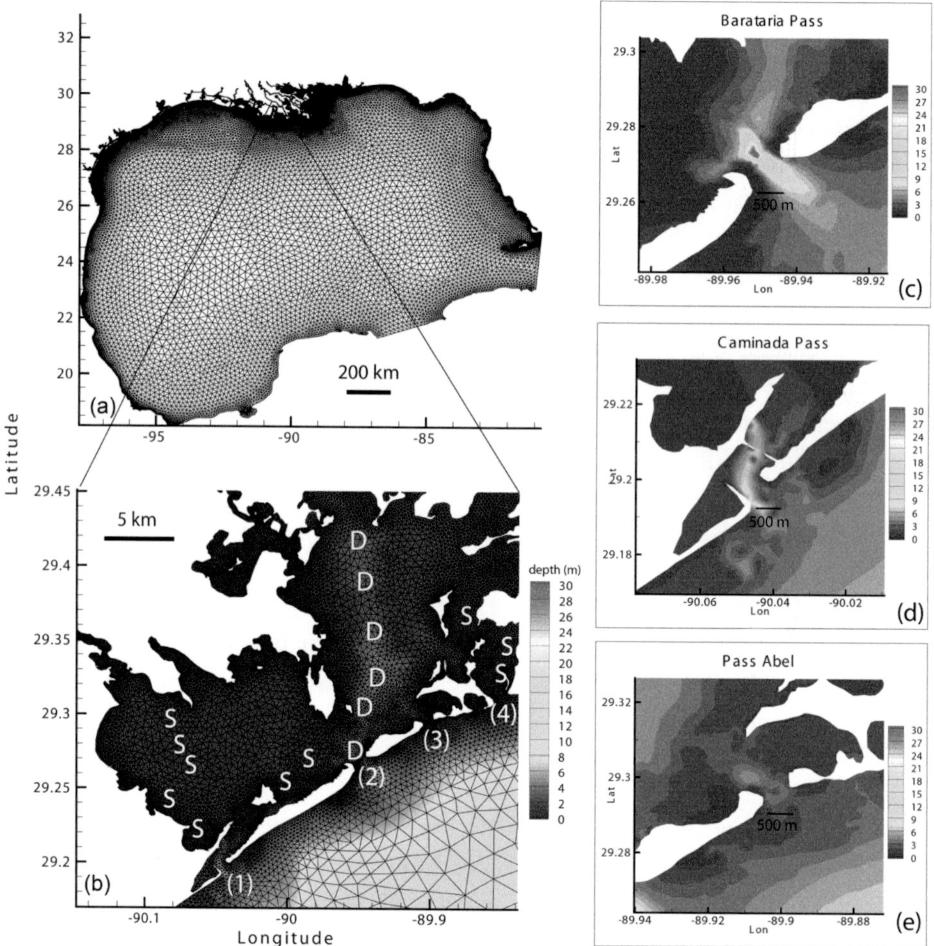

Figure 24.12 Model mesh used for simulation. (a) The entire domain of simulation. (b) Zoomed-in view of Barataria Bay. The inlets (Caminada Pass, Barataria Pass, Pass Abel, Quatre Bayou) are marked by (1) through (4). The bathymetry is shown by the color contours in meters. Some of the deep and shallow waters are marked by S and D, respectively, for convenience of discussion. (c), (d), and (e) are further zoomed-in views showing the bathymetry of the inlets.

times for the first, ... , and nth samples, respectively. This data matrix includes both forcing (W_E and W_N) and response (v_{r1}, v_{r2}, v_{r3}, and v_{r4}). The data matrix is then de-meaned for each column to yield the mean data matrix Z. The EOF analysis is then applied.

The first four EOF modes carry ~45%, 41%, 13%, and 1% variabilities of the data, respectively. Therefore, only the first three modes are discussed here. The three eigenvectors are, respectively (Fig. 24.14),

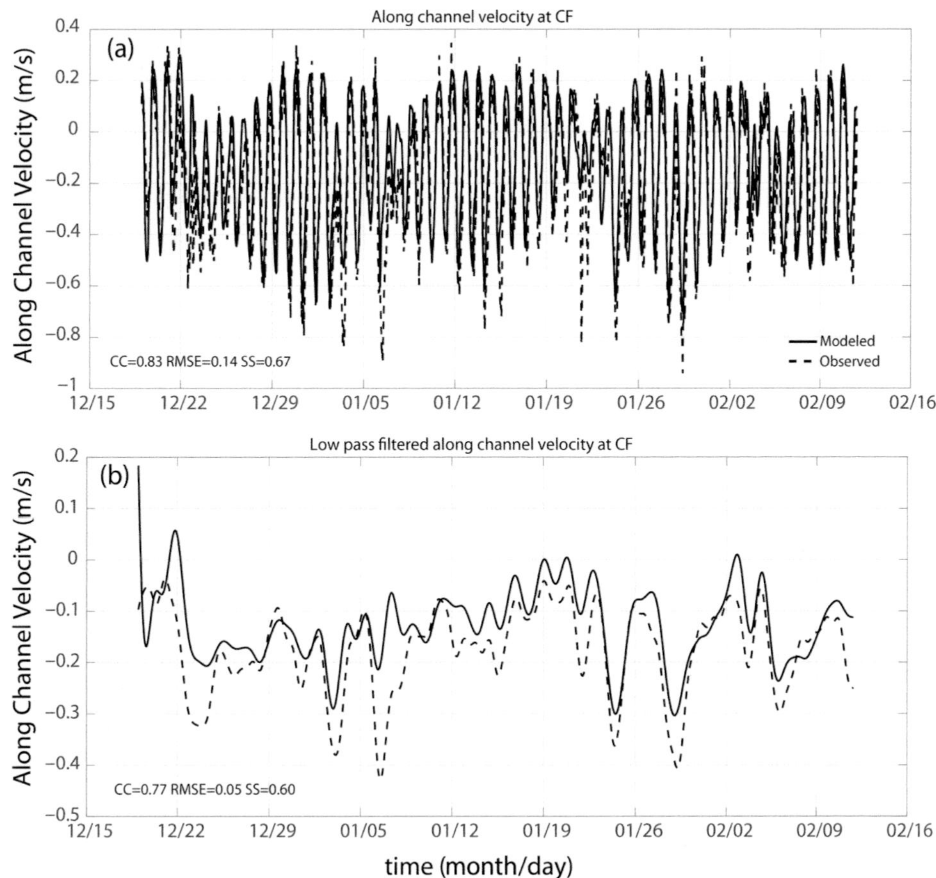

Figure 24.13 Model data comparison for (a) the along-channel velocity and (b) subtidal along-channel velocity.

$$E_1 = \begin{pmatrix} 0.22 \\ 0.59 \\ 0.47 \\ 0.32 \\ -0.35 \\ 0.35 \end{pmatrix}, \quad E_2 = \begin{pmatrix} 0.25 \\ -0.44 \\ -0.10 \\ 0.10 \\ 0.21 \\ 0.82 \end{pmatrix}, \quad E_3 = \begin{pmatrix} 0.23 \\ 0.56 \\ -0.15 \\ -0.19 \\ 0.75 \\ 0.05 \end{pmatrix}.$$

The first mode depicts a pattern with all-in transport at all four inlets under southeasterly wind and all-out transport under northwesterly winds (Fig. 24.14a,b). This mode is dominant when wind is strong and increasing. The second mode (Fig. 24.14c, d) depicts a pattern under south–southwesterly wind that pushes the transport inward at the easternmost and westernmost inlets along the shallow shoals of the bay with counterwind transport through the middle two inlets (Barataria Pass and Pass Abel);

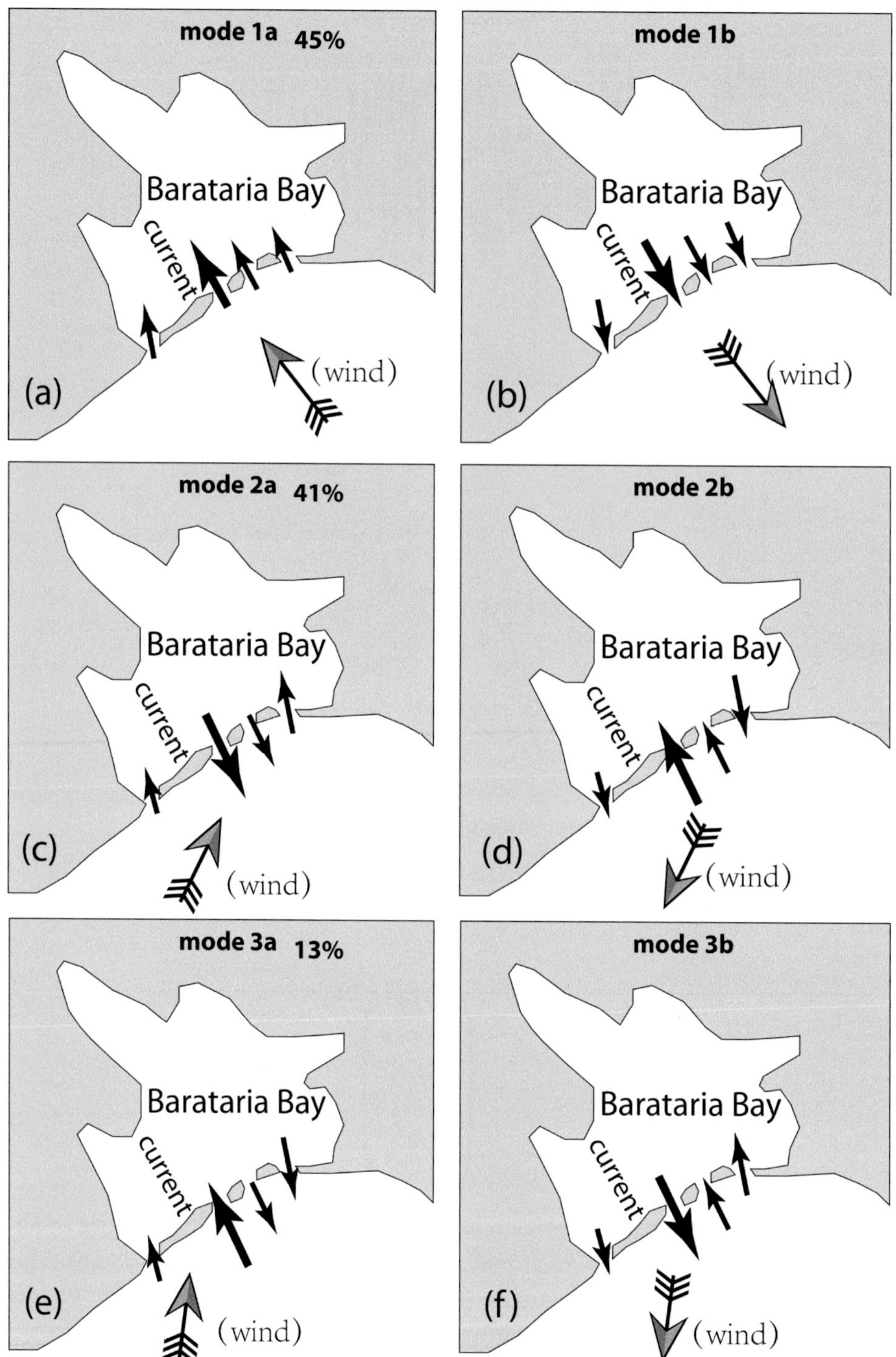

Figure 24.14 The first three modes for the forcing-response joint EOF analysis: (a) and (b) are for the first mode (with positive or negative coefficients); (c) and (d) for the second mode; (e) and (f) are for the third mode.

under north–northeasterly wind, the transport at the easternmost and westernmost inlets along the shallow shoals of the bay is outward and that through the middle two inlets is inward. This mode tends to occur at either equilibrium stages or wind relaxation periods. The third mode (Fig. 24.14e,f) shows inward transport through the Caminada Pass and Barataria Pass with outward transport through Pass Abel and Quatre Bayou Pass under wind from almost due south, and outward transport through the Caminada Pass and Barataria Pass but inward transport through Pass Abel and Quatre Bayou Pass under wind from almost due north. The last mode is consistent with Li (2013) such that wind-driven flows in a multiple inlet system tends to have an inward transport through the upwind inlets and outward transport through the downwind inlets, when the wind vector has an acute angle with the coastline.

Review Questions for Chapter 24

(1) Compare Fourier analysis, wavelet analysis, and EOF analysis. What are the similar aspects and what are the differences?

(2) In the EOF analysis, what does an eigenvector or eigenfunction mean?

(3) Try to explain the meaning of the time coefficients in an EOF analysis.

Exercises for Chapter 24

(1) Data file named "GoM-ADCP.mat" is a MATLAB file containing velocity profile time series obtained from a station at 27.122° N, 91.959° W with a water depth of 1600 m. The data contain several variables: (1) velocity (ut1, vt1) for the (east velocity, north velocity) in cm/s; (2) time (t1) in days; (3) total number of vertical layers (N), and water depth for the N layer (z) in m below the ocean surface. Do the following for the EOF analysis of the vertical structure of the horizontal velocity components ut1 and vt1, respectively:

- **(a)** Define the data matrix (for ut1 and vt1, same below).
- **(b)** Define the scatter matrix.
- **(c)** Find the eigenvalues and eigen vectors for the scatter matrix.
- **(d)** Find how much variability each of the modes explains.
- **(e)** Find the time coefficients for all the EOF modes.
- **(f)** Discuss and contrast the first three modes.

(2) Data file named "ADCP_data_42362_0506.mat" is a MATLAB file containing velocity profile time series data in a water of ~1000 m deep at a location (27.8° N, 90.67° W) in the Gulf of Mexico. The variables include time (t, days), east velocity (ut, cm/s), north velocity (vt, cm/s), and depth of the vertical bins of the data (z, meters). There are 60 vertical bins, ranging from 79.1 to 1023.1 m below the ocean surface. Do the same as above for the EOF modes.

References

Attaway, S. (2013). *Matlab: A Practical Introduction to Programming and Problem Solving*, 3rd ed., Amsterdam: Butterworth-Heinemann.

Bendat, J. S., and Piersol, A. G. (2000). *Random Data – Analysis and Measurement Procedures*, 3rd ed., New York: John Wiley & Sons.

Bevington, P. R., and Robinson, D. K. (2003). *Data Reduction and Error Analysis for the Physical Sciences*, 3rd ed., Boston: McGraw-Hill.

Boon, J. (2004). *Secrets of the Tide: Tide and Tidal Current Analysis and Predictions, Storm Surges and Sea Level*, Oxford: Horwood Publishing.

Bracewell, R. N. (2000). *The Fourier Transform and Its Applications*, 3rd ed., New York: McGraw Hill.

Bronshtein, I. N., Semendyayev, K. A., Musiol, G., and Muehlig, H. (2003). *Handbook of Mathematics*, New York: Springer.

Brummelen, G. V. (2017). *Heavenly Mathematics: The Forgotten Art of Spherical Trigonometry*, Princeton: Princeton University Press.

Butterworth, S. (1930). On the theory of filter amplifiers. *Experimental Wireless – Wireless Engineer*, **7**, 536–541.

Chen, C., Liu, H., and Beardsley, R. C. (2003). An unstructured grid, finite-volume, three dimensional, primitive equation ocean model: application to coastal ocean and estuaries. *Journal of Atmospheric and Oceanic Technology*, **20**, 159–186.

Cooley, J. W., and Tukey, J. W. (1965). An algorithm for the machine calculation of complex Fourier series. *Mathematics of Computation*, 19, 297–301.

Courant, R. (1937). *Differential & Integral Calculus*, Volume 1, 2nd ed. Translated by E. J. McShane, New York: Interscience.

Davis, H. F. (1963). *Fourier Series and Orthogonal Functions*, New York: Dover Publications.

Emery, W. J., and Thomson, R. E. (2004). *Data Analysis Methods in Physical Oceanography*, 2nd ed., Amsterdam: Elsevier.

Glickman, T. S., ed. (2000). *Glossary of Meteorology*, 2nd ed., Boston: American Meteorological Society Boston.

Gonella, J. (1972). A rotary-component method for analysing meteorological and oceanographic vector time series. *Deep-Sea Research*, **19**, 833–846.

Holthuijsen, L. H. (2007). *Waves in Oceanic and Coastal Waters*, New York: Cambridge University Press.

Howse, D. (2003). *Greenwich Time and the Longitude*, London: Philip Wilson Publishers.

Kundu, P. K., and Cohen, I. M. (2004). *Fluid Mechanics*, 3rd ed., New York: Academic Press.

Kunze, E. (1985). Near-inertial wave propagation in geostrophic shear. *Journal of Physical Oceanography*, **15**, 544–565.

Landau, L. D., and Lifshitz, E. M. (1980). *Statistical Physics*, 3rd ed. Translated by J. B. Sykes and M. J. Kearsley, New York: Elsevier.

Landau, L. D., and Lifshitz, E. M. (1987). *Fluid Mechanics*, Volume 6, 2nd ed. Translated by J. B. Sykes and W. H. Reid, Amsterdam: Butterworth-Heinemann.

Li, C. (2003). Can friction coefficient be estimated from cross stream flow structure in tidal channels? *Geophysical Research Letters*, **30**(17). doi:1810.1029/2003GL018087.

Li, C. (2013). Subtidal water flux through a multi-inlet system: observations before and during a cold front event and numerical experiments. *JGR-Oceans*, **118**, 1–16, doi:10.1029/2012JC008109, 2013.

Li, C., Boswell, K. M., Chaichitehrani, N., Huang, W., and Wu, R. (2019a). Weather induced subtidal flows through multiple inlets of an arctic microtidal lagoon. *Acta Oceanologica Sinica*, **38**(3), 1–16.

Li, C., Huang, W., and Milan, B. (2019b). Atmospheric cold front induced exchange flows through a microtidal multi-inlet bay: analysis using multiple horizontal ADCPs and FVCOM simulations. *Journal of Atmospheric and Oceanic Technology*, **36**, 443–472.

Li, C., Valle-Levinson, A., Atkinson, L., and Royer, T. C. (2000). Inference of tidal elevation in shallow water using a vessel-towed ADCP. *Journal of Geophysical Research (Oceans)*, **105**, 26, 225–236.

Li, C., Weeks, E., and Blanchard, B. W. (2010). Storm surge induced flux through multiple tidal passes of Lake Pontchartrain estuary during Hurricanes Gustav and Ike. *Estuarine, Coastal and Shelf Science*, **87**(4), 517–525.

Mathews, J. H., and Fink, K. D. (2004). *Numerical Methods Using MATLAB*, New Delhi: Pearson Prentice Hall.

Milne, R. M. (1921). Note on the equation of time. *The Mathematical Gazette*, **10**(155), 372–375.

Mirsky, L. (1990). *An Introduction to Linear Algebra*, New York: Dover Publications.

Mooers, C. N. K. (1975). Several effects of a baroclinic current on the cross-stream propagation of inertial-internal waves. *Geophysical Fluid Dynamics*, **6**, 245–275.

Navarra, A., and V. Simoncini. (2010). A Guide to Empirical Orthogonal Functions for Climate Data Analysis. *Springer, New York*, pp. 151.

North, G. R., Bell, T. L., Cahalan, R. F., and Moeng, F. J. (1982). Sampling errors in the estimation of empirical orthogonal functions. *Monthly Weather Review*, **110**, 699–706.

O'Brien, J. J., and Phillsbury, R. D. (1974). Rotary wind spectra in a sea breeze regime. *Journal of Applied Meteorology*, **13**(7), 820–825.

Pain, H. J. (2005). *The Physics of Vibrations and Waves*, Chichester: John Wiley & Sons.

Pedlosky, J. (1987). *Geophysical Fluid Dynamics*, 2nd ed., Berlin: Springer-Verlag.

Preisendorfer, R. W. (1988). Principal component analysis in meteorology and oceanography. *Elsevier, New York*, pp. 425.

Proakis, J. G., and Manolakis, D. G. (1992). *Digital Signal Processing: Principles, Algorithms, and Applications*, 2nd ed., New York: Macmillan Publishing.

Press, W. H., Teukolsky, S. A., Vetterling, W. T., and Flannery, B. P. (1992). *Numerical Recipes in C, The Art of Scientific Computing*, 2nd ed., New York: Cambridge University Press.

Proudman, J. (1953). *Dynamical Oceanography*, New York: Wiley.

SiRF Technology. (2005). *NMEA Reference Manual*, SiRF Technology, Inc.

Spiegel, M. R. (1974). *Schaum's Outline of Theory and Problems of Fourier Analysis With Applications to Boundary Value Problems*. New York: McGraw-Hill.

Stearns, S. D., and Hush, D. R. (2011). *Digital Signal Processing with Examples in MATLAB*, 2nd ed., Boca Raton: CRC Press.

Stephenson, F. R., and Morrison, L. V. (1995). Long-term fluctuations in the Earth's rotation: 700 BC to AD 1990. *Philosophical Transactions of the Royal Society of London Series A*, **351**, 165–202.

The MathWorks, Inc. (2020a). MATLAB® Primer. *The MathWorks, Inc.*

The MathWorks, Inc. (2020b). MATLAB® Graphics. *The MathWorks, Inc.*

Stewart, J. (2012). *Multivariable Calculus*, 7th ed., Belmont: Brooks/Cole.

U.S. Government. (1996). NAVSTAR GPS User Equipment Introduction. *Publica release version.*

Vermeille, H. (2002). Direct transformation from geocentric coordinates to geodetic coordinates. *Journal of Geodesy*, **76**(8), 451–454. doi:10.1007/s00190-002-0273-6.

Vincenty, T. (1975). Direct and inverse solutions of geodesics on the ellipsoid with application of nested equations. *Survey Review*, **22**(176), 88–93.

Wahr, J. M. (1988). The Earth's rotation. *Annual Review of Earth and Planetary Science*, **16**, 231–249.

Walker, J. S. (2008). *A Primer On Wavelets and Their Scientific Applications*, 2nd ed., New York: Chapman & Hall/CRC.

Welch, P. D. (1967). The use of fast Fourier transform for the estimation of power spectra: a method based on time averaging over short, modified periodograms. *IEEE Transactions on Audio and Electroacoustic*, **15**, 70–73, https://doi.org/10.1109/TAU.1967.1161901.

Yang, C. T. (2003). *Sediment Transport: Theory and Practice*, Malabar: Krieger Publishing Co.

Index

Printed in the United States
by Baker & Taylor Publisher Services